The Handbook of Global Climate
and Environment Policy

Handbooks of Global Policy Series

Series Editor
David Held
Master of University College and Professor of Politics and International Relations at Durham University

The *Handbooks of Global Policy* series presents a comprehensive collection of the most recent scholarship and knowledge about global policy and governance. Each handbook draws together newly commissioned essays by leading scholars and is presented in a style which is sophisticated but accessible to undergraduate and advanced students, as well as to scholars, practitioners, and others interested in global policy. Available in print and online, these volumes expertly assess the issues, concepts, theories, methodologies, and emerging policy proposals in the field.

Published

The Handbook of Global Climate and Environment Policy
Robert Falkner

The Handbook of Global Energy Policy
Andreas Goldthau

The Handbook of Global Companies
John Mikler

The Handbook of Global Climate and Environment Policy

Edited by

Robert Falkner

A John Wiley & Sons, Ltd., Publication

This edition first published 2013
© 2013 John Wiley & Sons, Ltd

Wiley-Blackwell is an imprint of John Wiley & Sons, formed by the merger of Wiley's global Scientific, Technical and Medical business with Blackwell Publishing.

Registered Office
John Wiley & Sons, Ltd, The Atrium, Southern Gate, Chichester, West Sussex, PO19 8SQ, UK

Editorial Offices
350 Main Street, Malden, MA 02148-5020, USA
9600 Garsington Road, Oxford, OX4 2DQ, UK
The Atrium, Southern Gate, Chichester, West Sussex, PO19 8SQ, UK

For details of our global editorial offices, for customer services, and for information about how to apply for permission to reuse the copyright material in this book please see our web site at www.wiley.com/wiley-blackwell.

The right of Robert Falkner to be identified as the author of the editorial material in this work has been asserted in accordance with the UK Copyright, Designs and Patents Act 1988.

Library of Congress Cataloging-in-Publication Data

The handbook of global climate and environment policy / edited by Robert Falkner.
    p. cm.
  Includes bibliographical references and index.
  ISBN 978-0-470-67324-9 (cloth)
  1. Climatic changes–Government policy.   2. Global warming–Government policy.
3. Environmental policy.   I. Falkner, Robert, 1967-
  QC902.9.H36 2013
  363.7'0561–dc23

                                                                2012045304

A catalogue record for this book is available from the British Library.

Cover image: Oil rig in Beaufort Sea © Ocean/Corbis
Cover design by Design Deluxe

Set in 10/12.5pt Sabon by Aptara Inc., New Delhi, India
Printed in Malaysia by Ho Printing (M) Sdn Bhd

1   2013

Contents

Notes on Contributors

Steinar Andresen is Research Professor at the Fridtjof Nansen Institute in Norway. He has also been Professor of Political Science at the University of Oslo, Norway, and a visiting scholar at Princeton University, University of Washington Seattle, and IISA. He has published extensively on various topics, mostly related to global environmental governance.

Graeme Auld is an Assistant Professor at Carleton University, Ontario, Canada, in the School of Public Policy and Administration. His research examines the emergence, evolution, and impacts of non-state and hybrid forms of global governance. He is co-author (with Ben Cashore and Deanna Newsom) of *Governing through Markets: Forest Certification and the Emergence of Nonstate Authority* (Yale University Press, 2004).

Jessica M. Ayers holds a PhD in climate governance from the London School of Economics and Political Science. At the time of writing, she was a researcher for the Climate Change Group at the International Institute for Environment and Development (IIED), London, UK. She is now a Senior Policy Advisor to the UK Department of Energy and Climate Change (DECC).

Steffen Bauer is a Senior Researcher in the Department for Environmental Policy and Management of Natural Resources at the German Development Institute/Deutsches Institut für Entwicklungspolitik (DIE) in Bonn, Germany. He also serves as Germany's Science and Technology Correspondent to the United Nations Convention to Combat Desertification (UNCCD).

Steven Bernstein is Associate Chair and Graduate Director of the Department of Political Science and Co-Director of the Environmental Governance Lab at the Munk School of Global Affairs, University of Toronto, Canada.

Daniel Bodansky is Lincoln Professor of Law, Ethics, and Sustainability at the Sandra Day O'Connor College of Law, Arizona State University. He is the author of the *Art* *and Craft of International Environmental Law* (Harvard University Press, 2009) and co-editor of *The Oxford Handbook of International Environmental Law* (Oxford University Press, 2007).

Achala Chandani Abeysinghe is a senior researcher and environmental lawyer at the International Institute for Environment and Development (IIED), London. She is the legal advisor to the current Chair of the Least Developed Countries (LDC) group in the United Nations Framework Convention on Climate Change (UNFCCC) negotiations, team leader of the global climate governance program, and head of the European Capacity Building Initiative (ECBI) workshops program at IIED. She is a lead author of the chapter on "Climate Resilient Pathways: Adaptation, Mitigation and Sustainable Development" in the Fifth Assessment Report of the Intergovernmental Panel on Climate Change (IPCC).

Jennifer Clapp is a Professor in the Environment and Resource Studies Department and Associate Dean of Research in the Faculty of Environment at the University of Waterloo, Ontario, Canada. Her recent books include: *Hunger in the Balance: The New Politics of International Food Aid* (Cornell University Press, 2012), *Food* (Polity, 2012), *Paths to a Green World: The Political Economy of the Global Environment*, 2nd edn (with Peter Dauvergne, MIT Press, 2011), and *Corporate Power in Global Agrifood Governance* (co-edited with Doris Fuchs, MIT Press, 2009).

Simon Dalby, formerly at Carleton University, Ontario, Canada, is now CIGI Chair in the Political Economy of Climate Change at the Balsillie School of International Affairs in Waterloo, Ontario, Canada, and is author of *Environmental Security* (University of Minnesota Press, 2002), *Security and Environmental Change* (Polity, 2009) and co-editor of the journal *Geopolitics*.

Radoslav S. Dimitrov is an Associate Professor at Western University in Canada. He is consultant to the World Business Council on Sustainable Development and has served on the European Union delegation at UN climate negotiations. He is the author of *Science and International Environmental Policy* (Rowman and Littlefield, 2006).

Robert Falkner is Reader in International Relations at the London School of Economics and Political Science (LSE). He is an Associate of the Grantham Research Institute on Climate Change and the Environment at LSE and an Associate Fellow of the Energy, Environment and Resources department at Chatham House. He is the author of *Business Power and Conflict in International Environmental Politics* (Palgrave Macmillan, 2008).

Doris Fuchs is Professor of International Relations at the University of Münster, Germany. Her research focuses on corporate structural and discursive power, sustainable development/consumption, and food politics and policy. She is the author of *Business Power in Global Governance* (Lynne Rienner, 2007) and has published numerous articles in peer-reviewed journals such as *Millennium*, *Global Environmental Politics*, *International Interactions*, *Agriculture and Human Values*, *Food Policy*, and *Energy Policy*.

Lars H. Gulbrandsen is Senior Research Fellow and Director of the Global Governance and Sustainable Development research program at the Fridtjof Nansen Institute, Norway. He is the author of *Transnational Environmental Governance: The Emergence and Effects of the Certification of Forests and Fisheries* (Edward Elgar, 2010).

Aarti Gupta is Senior Lecturer (tenured) with the Environmental Policy Group at Wageningen University's Department of Social Sciences, the Netherlands. She is also a Senior Fellow of the Earth System Governance Project and Associate Editor of the journal *Global Environmental Politics*. Her research and publications focus on global risk and environmental governance and the role of science, knowledge, and transparency therein.

Joyeeta Gupta is Professor of Environment and Development in the Global South of the Amsterdam Institute for Social Science Research at the University of Amsterdam and at UNESCO-IHE Institute for Water Education in Delft, the Netherlands.

Stuart Harrop is Professor of Wildlife Management Law and Director of the Durrell Institute of Conservation and Ecology in the School of Anthropology and Conservation at the University of Kent, UK. His research concentrates on the field of international law and policy relating to biodiversity conservation.

Cameron Hepburn is a Senior Research Fellow at the Grantham Research Institute at the LSE and a Fellow of New College, Oxford. He has degrees in law and engineering, a doctorate in economics, and is the author of peer-reviewed publications in economics, biology, philosophy, engineering, and public policy. He is involved in policy formation, including as a member of the Department of Energy and Climate Change (UK) Secretary of State's Economics Advisory Group. He has also had an entrepreneurial career, co-founding two successful businesses and investing in several other start-ups.

Matthew J. Hoffmann is an Associate Professor of International Relations at the University of Toronto, Canada, and Co-Director of the Environmental Governance Lab at the Munk School of Global Affairs there. He is the author of *Climate Governance at the Crossroads: Experimenting with a Global Response after Kyoto* (Oxford University Press, 2012).

David Humphreys is Senior Lecturer in Environmental Policy at The Open University (UK), where he specializes in the global politics of deforestation and climate change. His book *Logjam: Deforestation and the Crisis of Global Governance* (Earthscan, 2006) won the International Studies Association's Harold and Margaret Sprout Award for 2008.

Michael Jacobs is Visiting Professor at the Grantham Research Institute on Climate Change and the Environment at the London School of Economics and Political Science, and in the School of Public Policy at University College London. A former Special Advisor to the UK Chancellor of the Exchequer and Prime Minister (2004–2010), his books include *The Green Economy: Environment, Sustainable Development and the Politics of the Future* (Pluto Press, 1991).

Nico Jaspers is a post-doctoral researcher at the Freie Universität Berlin, Germany. He received a doctorate in international relations from the London School of Economics and Political Science in 2011. He has published widely on nanotechnology policy and in 2009 co-authored an EU-commissioned report on transatlantic cooperation in nanotechnology regulation.

Jonas Meckling is an Associate with the Belfer Center for Science and International Affairs at Harvard University, USA. He is Senior Advisor for Transatlantic Cooperation on Energy and Climate to the German Federal Ministry of the Environment. He is the author of *Carbon Coalitions: Business, Climate Politics, and the Rise of Emissions Trading* (MIT Press, 2011).

Peter Newell is Professor of International Relations and Director of the Centre for Global Political Economy at the University of Sussex, UK. He is author most recently of *Globalization and the Environment: Capitalism, Ecology and Power* (Polity, 2012) and co-author of *Climate Capitalism* (with Matthew Paterson; Cambridge University Press, 2010) and *Governing Climate Change* (with Harriet Bulkeley; Routledge, 2010).

Edward Page is Associate Professor in Political Theory at the University of Warwick, UK. His research interests cover a range of topics in contemporary political theory, environmental politics, applied ethics, and global climate change. He has published articles in journals such as *Environmental Politics, Political Studies, The Monist,* and *International Theory,* and is the author of *Climate Change, Justice and Future Generations* (Edward Elgar, 2006).

Susan Park is a Senior Lecturer in International Relations at the University of Sydney, Australia. She has published in a range of International Relations journals including *International Politics, Global Environmental Politics,* and *Global Governance.* In 2010 she published *World Bank Group Interactions with Environmentalists* (Manchester University Press) and co-edited *Owning Development: Creating Policy Norms in the IMF and the World Bank* (with Antje Vetterlein; Cambridge University Press).

Markus Salomon is a marine biologist working for the German Advisory Council on the Environment, an independent scientific council giving advice to the German federal government. He recently published "Towards a Sustainable Fisheries Policy in Europe," *Fish and Fisheries* (with K. Holm-Müller; 2012).

Miranda Schreurs is director of the Environmental Policy Research Centre (Forschungszentrum für Umweltpolitik) and Professor of Comparative Politics at the Freie Universität Berlin, Germany. She is a member of the German Environment Advisory Council, chair of the European Environment and Sustainable Development Advisory Councils, and was a member of the German Ethics Commission on a Safe Energy Supply.

Henrik Selin is an Associate Professor in the Department of International Relations at Boston University, USA. He is the author of *Global Governance of Hazardous Chemicals* (MIT Press, 2010) as well as a numerous journal articles and book chapters on the politics and management of hazardous substances and wastes.

Benjamin K. Sovacool is an Associate Professor at Vermont Law School, USA, where he also directs the Energy Security and Justice Program at the Institute for Energy and the Environment. He is a contributing author to the forthcoming Fifth Assessment from the Intergovernmental Panel on Climate Change, as well as the author, co-author, or editor of 12 books and almost 200 peer-reviewed articles on energy and climate-change issues.

Hannes Stephan is a Lecturer in Environmental Politics and Policy at the University of Stirling, Scotland. His co-authored article "International Climate Policy after Copenhagen: Towards a 'Building Blocks' Approach" appeared in 2010 in *Global Policy*. Another article on the transatlantic cultural politics of GM foods and crops was published in 2012 in *Global Environmental Politics*.

Johannes Stripple is Associate Professor at the Department of Political Science, Lund University, Sweden. His research interests lie at the intersection of International Relations theory and global environmental politics. His recent research has covered European and international climate policy, carbon markets, renewable energy, sinks, scenarios and governmentalities around climate change, carbon, and the Earth System.

Christopher Wright undertook the research underpinning his contribution to this volume while a Senior Researcher at the Centre for Development and Environment, University of Oslo, Norway. He is currently a Senior Analyst at Norges Bank Investment Management, the division of the Norwegian central bank which manages the Norwegian Government Pension Fund – Global.

Preface

The world faces a growing number of complex global challenges, but global political leadership and international cooperation are in short supply. Climate change and other environmental threats are among the most intractable issues on the global agenda today. Accelerated biodiversity loss, disruptions to food and energy supplies, intensified competition for scarce natural resources, and a warming climate all combine to create global risks that are likely to further destabilize an already unsettled world.

Addressing global environmental threats requires a high degree of international cooperation. As could be witnessed at the recent "Rio+20" UN summit on sustainable development, however, the international community remains divided on how to tackle the most urgent environmental threats. The international institutional architecture for dealing with global environmental problems is fragmented and weak, and global environmental protection efforts are insufficiently funded. It is encouraging that environmental concerns have gained in prominence in international politics, with a large numbers of actors – from concerned scientists to environmental activists and enlightened business and political leaders – now engaged in the search for global solutions. But despite the remarkable growth in global environmental policy-making, the international community appears unable to slow down, let alone reverse, most of the destructive trends of environmental degradation.

Still, some efforts to address specific environmental problems and create innovative institutional solutions are paying off. The Montreal Protocol of 1987, an international agreement to phase out ozone-depleting substances, is one of the outstanding successes of global green diplomacy. Negotiations on a global climate agreement may have proved less successful, but myriad initiatives to halt the growth in greenhouse gas emissions have sprung up at regional, national, and local levels. Environmentalists are mobilizing around the world to limit the destructive side-effects of industrialization and urbanization, while a growing number of corporations are willing to engage with campaigners in efforts to set new rules and standards for environmentally responsible business behavior. And international economic organizations such as the World Trade Organization and the World Bank have begun to integrate environmental concerns more fully into their operations.

As climate change and other environmental issues have moved center stage on the international agenda, global policy practitioners and students are searching for new, innovative solutions and more effective policy approaches. The purpose of this *Handbook* is to help with this search and provide an authoritative guide to recent academic research on global climate and environment policy. The *Handbook* contains 28 chapters that offer in-depth yet accessible surveys of the main global policy issues and approaches emerging from the best research in the field. The *Handbook* is multi-disciplinary in orientation and covers perspectives from international relations and political science as well as economics, environmental studies, geography, and international law. It employs a broad understanding of global climate and environment policy that includes state-centric approaches of international diplomacy, treaty negotiation and law, as well as those transnational political activities that transcend the state-centric system. As such, this volume should appeal to a wide international audience. For policy practitioners of international diplomacy, international organizations and environmental groups, the *Handbook* will provide essential surveys of academic theory and research. For students enrolled in undergraduate or postgraduate degree programs, it will offer the starting point they need for the exploration of particular research fields. For researchers, it will allow easy access to specialized literatures across different topics and disciplines alongside their own.

The contributions to the *Handbook* – all written by world-leading experts in their respective fields – are grouped into four broad parts.

The first part, on *global policy challenges*, consists of seven chapters that review specific environmental issues and the global policy responses and governance systems that have been created to deal with them. The chapters cover climate change, global water governance, biodiversity and conservation, marine environmental protection, deforestation, biotechnology and biosafety, and chemicals safety.

The second part, on *concepts and approaches*, introduces major conceptual and theoretical approaches in the study of global climate and environment policy. The seven chapters discuss the role of global environmental norms, the changing nature of global governance, the concept of global environmental security, developments in international environmental law, discussions surrounding green growth and sustainable consumption, as well as climate change justice.

The third part, on *global actors, institutions, and processes*, introduces the main actors that make up the global policy agenda and examines key processes and institutions through which the international community is addressing global environmental problems. The seven chapters in this part cover the role of the nation-state and international society, NGOs and transnational environmental activism, business actors, international regimes and their effectiveness, international environmental negotiations, regional environmental governance, and the debate surrounding United Nations reform.

The fourth part, on *global economy and policy*, brings together chapters that consider the links between global policy on climate change and environment on the one hand, and major economic trends, institutions, and policy approaches on the other. The seven chapters discuss the concept of economic globalization, the role of private regulation by business actors, the linkages between international trade, environmental protection, and climate change, the environmental dimensions of global finance, linkages between energy policy and climate change, economic instruments for climate change, and the role of international aid in adaptation to climate change.

Robert Falkner

Part I   Global Policy Challenges

Chapter 1

Global Climate Change

Matthew J. Hoffmann

Analysts have struggled to find new and creative ways to describe the scope and complexity of climate change – a problem that finds its sources virtually everywhere, from nearly all kinds of human activity (agriculture, transportation, manufacturing, energy use, land use), and that has effects that are being and will be felt across the globe. Perhaps the most apt characterization has come from Mike Hulme (2009), who eschews the label "problem," preferring to describe climate change as a fundamental part of the modern condition. Yet, no matter how one conceives of climate change, there is little doubt that it is perhaps *the* global challenge of modern times. If climate scientists are correct in their understanding of the dynamics and impact of climate change, then the world needs to essentially decarbonize energy and transportation systems over the course of this century, with the lion's share of progress towards this goal taking place by 2050.

Mitigating climate change,[1] taking the steps necessary to avoid its most dangerous potential impacts, is thus at once elementary (in that we know we need to drastically reduce the emissions of greenhouse gases) and infuriatingly elaborate (in that the pathways to such reductions are fraught with small to enormous technical, economic, social, and political obstacles). This chapter examines the global response to climate change from the perspective of this paradox. I first briefly describe the state of knowledge of climate science and argue that while climate scientists can and do tell us about the nature of the problem, they cannot tell us about what *kind* of a problem it is – i.e. what features are important and what we should do. In fact, deciding what kind of problem climate change presents is an inherently political and fraught process.

These decisions about the nature of the problem are not only difficult, they are also consequential because they shape what kind of a response we can and do formulate. I demonstrate this in the next section by comparing the different foundational

The Handbook of Global Climate and Environment Policy, First Edition. Edited by Robert Falkner.
© 2013 John Wiley & Sons, Ltd. Published 2013 by John Wiley & Sons, Ltd.

understandings of climate change embedded in traditional multilateral and emergent transnational governance responses. These two governance systems differ in how they consider the *global* nature of climate change and in how they focus on proximate (greenhouse gas emissions) or fundamental (carbon dependence) causes of climate change. These differences shape the radically different politics and policy options available in the different processes.

This comparative exercise is not one of whistling past the graveyard or playing a tune as the ship sinks. On the contrary, understanding the foundation of climate mitigation efforts provides context for contemplating and (potentially) hope for developing the paths along which climate governance must (and/or can) proceed in the coming decades. I conclude, therefore, with some brief suggestions for how we can move forward reflexively both in the research and policy-making communities to bring together the two main approaches to climate governance.

Understanding the "Problem(s)" of Climate Change

Just getting one's head around the *problem* of climate change is a stiff challenge precisely because the problem can be conceived in multiple ways. There is the science of climate change – how increasing greenhouse gas concentrations affect global temperatures, ocean chemistry, and vegetation and the associated impacts that emerge from these changes. There are the social-economic-political understandings which focus, among other things, on economic development, the energy system, varied interests of states and other political actors. There is also the ethical dimension that concentrates on who faces the costs of climate change (mitigating it and the effects of it) both now and in the future (Gardiner 2004; Roberts and Parks 2007; Vanderheiden 2008). To further complicate matters, none of these dimensions provide objective understandings of the problem, but are rather wrapped up in the process of *framing* the issue in various ways that legitimate and even necessitate types of policy responses (Kahan *et al.* 2010; Hulme 2011). This brief chapter cannot do justice to all of the dimensions of the problem of climate change, thus this section focuses in on the latest understandings of climate science and how this knowledge can only take us part of the way towards understanding what kind of a problem climate change is because of varied political and economic aspects and framings of the problem.

Climate Science

The scientific logic of the climate change problem is relatively simple to describe (Hoffmann 2011; see also e.g. Maslin 2004; Dessler and Parson 2006; Houghton 2009). The Earth's atmosphere acts as a greenhouse whereby various gases (carbon dioxide, methane, chlorofluorocarbons, water vapor, and others) absorb solar radiation that would otherwise be reflected back into space from the Earth. This greenhouse effect itself is beneficial as it keeps the planet warm and allows life to flourish in the forms with which we are familiar. However, since the industrial revolution humanity has been emitting more and more greenhouse gases (e.g. carbon dioxide, methane, chlorofluorocarbons), mostly through the burning of fossil fuels, increasing their concentrations in the atmosphere and thus increasing the warming effect. Potential effects of increased greenhouse emissions include ocean acidification, along with the global warming that will likely engender sea level rise, increases in the frequency and severity of storms and droughts, changed precipitation patterns, altered disease vectors and trajectories, species migration, reduced agricultural productivity, and more.

The 2007 Intergovernmental Panel on Climate Change (IPCC) report laid out the most comprehensive examination of climate change to date. It found consensus in the scientific community that greenhouse gas emissions have significantly increased due to human activity and further that the modest temperature increases we have already experienced are "very likely due to the observed increase in anthropogenic GHG concentrations" (IPCC 2007). Moving forward, even the relatively conservative IPCC language about the likelihood of further warming in the twenty-first century raises alarms when they note that extant climate models predicted between 2 and 4 °C of warming in the coming century (IPCC 2007). Put simply, in 2007, the scientific community considered that human activity was causing increases in greenhouse gas concentrations and that we could expect significant warming and other effects because of it.

Data and models that have emerged since 2007 have consistently produced more dire predictions about the rate of emissions growth and the warming that we are likely to see. In 2011, the National Research Council (2011) in the USA expanded the range of anticipated warming, noting that now scientists are telling us that:

> Projections of future climate change anticipate an additional warming of 2.0 to 11.5F (1.1 to 6.4C) over the 21st century, on top of the 1.4F already observed over the past 100 years.

The International Energy Agency (2012: 15) concurs and estimates that if current trends of increasing energy use are not altered, the world is headed for at least 6 °C of warming. The current (political) consensus is that constraining global temperature increases to 2 °C is crucial, but that time is rapidly running out to do so. In 2009 a prominent gathering of climate scientists and policy-makers (Copenhagen Diagnosis 2009) declared what has now become a relatively taken-for-granted understanding: "If global warming is to be limited to a maximum of 2 °C above pre-industrial values, global emissions need to peak between 2015 and 2020 and then decline rapidly."

Knowledge about expected warming from current and anticipated concentrations of greenhouse gases is increasingly troubling as the climate science community learns more about the kind of impacts we can expect. Here the news is frankly a bit frightening. The possible impacts of climate change are well known – glaciers melting, sea level rise, altered storm pattern and severity, altered precipitation patterns, and more – but it appears as though at least some impacts are coming sooner than anticipated in earlier models and with greater magnitude. Already in 2009, UNEP (2009) was warning that "The pace and scale of climate change may now be outstripping even the most sobering predictions of the last report of the Intergovernmental Panel on Climate Change (IPCC)." Since 2009, a steady stream of reports have detailed how climate change has already begun and that the impacts like the melting arctic ice cap are coming more quickly than anticipated. The juxtaposition in 2012 of a record-breaking warm winter in North America and bizarre cold snaps in Europe have added an experiential element to the notion that we are already experiencing significant climate change.

However, even with increasingly sophisticated climate science, there are still significant uncertainties that complicate scientific understanding of the problem of climate change. Some of these are inherent uncertainties, in the sense that we simply will not be able to know for sure. These include comprehending and tracing:

- the intervening factors between concentrations of greenhouse gasses, temperature increase, and climatic changes like increased severity and frequency of storms, cycles of droughts and floods, and patterns of precipitation;
- how natural variability in the climate can mask and/or exacerbate the effect of anthropogenic greenhouse gas emissions;
- the uncertain magnitude and geographically variable nature of the effects of climate change;
- the role that feedback effects and tipping points play in offsetting or accelerating the impact of global warming. (Hoffmann 2011: 10–11)

Beyond Climate Science: What Kind of Problem Is Climate Change?

Scientifically, then, we have a pretty good sense of the nature of the problem – its causes and consequences and its uncertainties. But even scientific consensus cannot tell us what kind of a problem climate change is: scientific understanding translates uneasily into policy-making at the global or indeed other levels because it does not make political, economic, technological, and social definitions of the problem obvious (Litfin 1994). In fact, scientific uncertainties, in some ways, pale in comparison to the obstacles and uncertainties that come with understanding what kind of problem climate change is from a social-economic-political perspective. Consider the following:

- *Greenhouse emissions arise from virtually every human activity.* Most current industrial, energy, transportation, and agricultural processes produce greenhouse gases. The world's economy significantly runs on fossil fuel use.
- *Dependence on fossil fuels is uneven.* While the global economy runs on fossil fuels, there is disparity between consumers and producers of fossil fuels – in other words some countries produce a lot of fossil fuels, others consume a lot of fossil fuels, and many that consume less would like to consume more.
- *Per capita greenhouse gas emissions vary significantly.* While absolute emissions from India and China rival those found in the USA and EU, the per capita emissions are wildly divergent. According to the International Energy Agency (2009), in 2007 the average person in the USA produced over 19 t. of carbon dioxide, while the average person in India and China produces 1.2 and 4.6 t. respectively.
- *Historical responsibility for greenhouse gas concentrations is different from future responsibility.* The states that contributed most to the current level of greenhouse gas concentrations (USA, EU) are not going to be the same states that contribute the most to the future level of greenhouse gas concentrations (USA, China, India).
- *Protecting the climate promises diffuse benefits in the future, while engendering concentrated costs now.* Put simply, it is difficult to generate political will,

especially across political jurisdictions, to solve a problem when identifiable groups must pay up-front to generate benefits for the whole world sometime in the future. Scientists agree that the world must take action now to change the nature of our economy and wean itself off fossil fuels so that decades or even a century in the future, our climate remains hospitable for the world's great-grandchildren. This creates an enormous incentive to delay and significantly hampers efforts to generate urgent action in the present.

- *Climate impacts will be felt differentially.* Climate changes will be felt locally, regionally, nationally, and internationally, but with significant variation, and many of the poorest countries are likely to suffer the most dramatic consequences. In addition, the capacity to respond to climate changes also varies significantly. This produces wide disparity in the urgency felt about the problem. (Hoffmann 2011: 10–11)

So what is the problem? Is it a problem of overdevelopment or underdevelopment? Is it a problem of Northern historical responsibility or Southern future responsibility? Is it an economic problem or an environmental problem or an energy problem? Is it a problem of mitigation or adaptation? The very fact that climate change is in many ways objectively undefinable means that the framing of the issue *creates* the kind of issue we are actually dealing with (Hulme 2011). How we understand the problem creates the kind of problem that we try to solve.

Deciding what kind of a problem climate change is means focusing on particular aspects of the problem in formulating responses. This is both difficult and political. It is difficult simply because we cannot know which is the "right" decision. We have no means of ascertaining what aspects of climate change we should focus on and what kind of solutions we should devise to best respond to the problem. It is political because the choice of features and responses to focus on have differential costs and benefits for different groups of people. Actors have very different interests in the climate change problem if it is defined as a problem of mitigation or adaptation, for instance. These decisions are therefore consequential in addition to being difficult because they shape the contours of the global response to climate change. In the next section I demonstrate this by comparing the consequences of two aspects of the foundational understandings of climate change embedded in the multilateral and transnational responses to climate change.

The Global Responses to Climate Change

Traditionally, the multilateral treaty-making process overseen by the UN has been equated with climate governance. Most studies of climate politics are concerned with the negotiation, impact, and effectiveness of this process and center their analyses on the development of major agreements – the UN Framework Convention on Climate Change (UNFCCC, 1992), the Kyoto Protocol (1997), and the more recent attempts to move beyond Kyoto with the Copenhagen Accord (2009) and Durban Agreements (2011). Most public international effort has been directed into this multilateral process as well. Essentially, the UN process has been climate governance, for good or bad, for the last 25 years. More recently, however, a nascent system of transnational governance has emerged to address climate change (Andonova *et al.* 2009; Hoffmann

2011; Abbot 2012; Bulkeley *et al.* 2012). This decentralized approach to climate governance engages multiple actors at multiple levels and is only loosely connected to the multilateral process.

In this section I briefly introduce these two governance mechanisms and compare their understandings of climate change on two dimensions – the definition of the global scope of the problem and whether to focus on proximate or fundamental causes of climate change. This comparison reveals the consequences of choosing what kind of a problem climate change is for politics and policy.

Multilateral Governance

The UN-centered process of multilateral negotiations needs little or no introduction. It has been the key international response to climate change, consisting of annual global conferences and negotiations that produced the UNFCCC, the Kyoto Protocol, and a string of more recent agreements moving towards replacing the Kyoto Protocol. This process has been the subject of intense academic scrutiny, with studies examining, among other areas, the early negotiating phases and regime building (Grubb 1993; Bodansky 1994; Rowlands 1995), the political economy of the negotiations (Grubb 1993; Barrett 2003), the North–South dimensions (Gupta 2000; Roberts and Parks 2007), the rise and inclusion of market mechanisms (Bernstein 2001), and the problems and failures of the process (Victor 2004, 2011; Depledge 2006; Prins and Rayner 2007; Falkner *et al.* 2009). Rather than rehashing a very large literature, this section examines aspects of what kind of a problem climate change is considered to be in the multilateral process and the consequences of that definition.

First, the global scope of the problem has always been emphasized in the multilateral negotiations. From the very beginning of climate change's emergence as an international policy problem, everyone understood that it was a global problem that required a global solution. This seems obvious enough, as climate science tells us that climate change may be the one truly global environmental problem, in that the climate/atmosphere is a global system and that the sources and effects of climate change are found literally everywhere. But this somewhat banal notion of global – of the globe – is an empty signifier that could fit with any number of more specific notions of what kind of a problem climate change actually is. Certainly there may be global effects, but even in the early 1990s, it was fairly clear that at least 75% of the problem could be attributed to fewer than ten states if we consider the EU as a single entity (Hoffmann 2005). Further, even the *global* effects are diversely and differentially distributed regionally and locally. These characteristics of climate change, however, were *not* emphasized when the international community devised a response strategy. Instead, what everyone meant when characterizing climate change as a global problem is that all *states* should participate in the devising of a solution and that all *states* should take responsibility for participating in the solution though the responsibility should be differentiated by development level (Hoffmann 2005).

Second, climate change was clearly defined as an emissions problem. This conception has dominated the global response to climate change in the last 20 years, and the UN process has largely been an effort targeted at negotiating emissions reductions – how far to reduce greenhouse gas emissions, how to distribute reduction commitments, how to achieve reductions, and how to pay the costs of reductions.[2] Clearly climate change does result from the increasing emissions of greenhouse gases. This understanding of the problem is not inaccurate, but it is a specific type of focus that is not the only way to conceive of the problem. A focus on emissions is a focus on proximate causes of the problem. This may appear to be a subtle difference, to focus on the emissions of greenhouse gases rather than the processes that produce them, but it is more than semantics. Defining a problem based on its symptoms (adaptation efforts work from this definition when they look to deal with the consequences of global warming), or its proximate causes, or its fundamental causes makes for very different policy responses.

In fact, both of these foundational conceptions of climate change as a problem (a particular vision of "global" and equating the problem with its proximate causes) are consequential because they constrain the policy tools and politics of the multilateral process. The debates and options that flowed from the underlying definition have remained remarkably stable over the course of the last 20 years, even while progress on an effective global response to climate change has been agonizingly slow (Depledge 2006). The multilateral governance process was constructed as universal interstate negotiations tasked with essentially distributing costs (i.e. emissions reductions), and devising side payments (i.e. development assistance) and flexibility mechanisms (i.e. market measures like cap and trade) to make such costs palatable. Whether the understanding of the problem as one of proximate causes led to the collective action problems or whether the global, multilateral approach made this understanding of the problem inevitable is an open question not fully explored here (see Hoffmann 2005).

From the beginning, all states (even the negotiations in the early 1990s attracted over 100 states) saw themselves as relevant participants in climate governance. "Global" meant universal, interstate governance through negotiation. The lines of contention over emissions reductions in this governance context were clear and had to do with how different states considered the urgency of climate change and costs of emissions reductions. The Europeans, mostly convinced of the urgency of the problem (and beneficiaries of internal diversity of emissions profiles that would make reductions easier to come by within the EU), and small island nations, facing an existential threat, have consistently pushed for significant emissions reductions. The Europeans wanted binding emissions reductions in the UN Framework Convention on Climate Change, took on the deepest emissions reduction commitments in the Kyoto Protocol, and have pledged a 30% reduction of greenhouse gas emissions even in the absence of a legally binding replacement for the Kyoto Protocol.

On the other side of this debate, we find the USA, large developing countries (China, India, Brazil), and oil-producing states. This set of states was concerned about the significant costs of emission reductions to their economies. The USA was the main obstacle to quick action on emissions reductions in these early negotiations, forswearing any moves to include binding greenhouse gas emissions reductions targets in the framework convention. Though the USA changed course in the mid-1990s and agreed to modest emissions reductions in the Kyoto Protocol, it subsequently repudiated those commitments in 2001 and has since rejected binding emissions reductions in international negotiations. China and India, bolstered by the precedent set in the Montreal Protocol for ozone-depleting substances and the accepted principle of common but differentiated responsibilities, urged Northern states to take the lead on significant actions to address climate change and, until very recently, rejected any calls for emissions reductions from the global South.

The result has been stalemate and, from a political economy perspective, not a very surprising one (Hoffmann 2011: 15; see also Barrett 1992, 2003; Sell 1996; Victor 2004). Given its preeminent position as an energy consumer and carbon dioxide producer, the USA does not want to incur what would be significant costs to its economy to deal with the problem, especially in the absence of action by major economic competitors like China. Large developing countries which have rapidly grown in terms of energy consumption and carbon dioxide emissions (in absolute if not per capita terms) prioritize development over action on climate change and also argue that a problem historically caused in the North should be dealt with by Northern states first. The USA is reluctant, at best, to take significant action. The Europeans and major Southern states push for significant actions by Northern states, and the USA and to a lesser extent Japan, Russia, and Canada, work to both reduce and slow the response to climate change and push for concomitant Southern actions. China, India, Brazil, and other developing states are reluctant, at best, to take significant action. The EU, which has taken significant action, has not been able to convince either side to make significant concessions.

The impasse that was already apparent in 2001 is still shaping the climate negotiations of today. The Copenhagen meetings of 2009 were designed to achieve the next step beyond the Kyoto Protocol (ending in 2012) – the next binding emissions reduction treaty. The fact that it failed to do so was not news, given the stalemate that had persisted for the prior decade. The major difference is that the international community has given up, for the time being, on collective emissions targets. After little success in years of trying to take the next binding step beyond Kyoto, the 2009 Copenhagen Accord and subsequent 2010 Cancun Agreement introduced the idea of National Appropriate Mitigation Activities and allowed countries to pledge their own emissions reductions targets and baseline years (UNFCCC 2009, 2010). The focus is still on emissions reductions, but there will be no collective target until at least 2020. The 2011 Durban Agreement pledged only to negotiate a legal instrument by 2015 that would come into force after 2020 (UNFCCC 2011).

The traditional way we go about the international response to climate change – negotiate a treaty among the entire international community to mandate a collective emissions reduction target that is distributed as various national emissions reductions targets (which will include an enforcement mechanism so countries do not cheat) – has led to the impasse. In some ways, focusing on mandated emissions reductions forces the international community into the box of a collective action problem over a joint public good. In other words we define the problem as one where everyone emits greenhouse gases and we have to measurably restrict those in an enforceable way to solve the problem. This fundamental definition of the problem actually creates many of the intractable debates we have seen in the last 20 years – how much to reduce, who is obligated to reduce, what should we do if someone fails to reduce – because it inherently means distributing something costly (emissions reductions).

The fundamental understanding of the problem embedded in multilateral governance also contributed to the boundaries on the policy options available for the global response to climate change. Flexibility became the key term in the negotiation of the Kyoto Protocol as the USA and others sought low-cost mechanisms for achieving the emissions reductions that were under consideration. In this case, and in line with the dominant worldview of liberal environmentalism (Bernstein 2001), flexibility meant the inclusion of market mechanisms into climate governance. Two kinds of carbon markets – credit and allowance – emerged as the main policy tools that would dominate the discussions about achieving emissions reductions (Newell and Paterson 2010; Betsill and Hoffmann 2011).

The USA was the biggest advocate of market mechanisms in the multilateral negotiations that produced the Kyoto Protocol, arguing that they would control the costs of reducing greenhouse gas emissions. The idea of using market mechanisms in service of environmental goals was and remains a familiar motif in USA environmental policy and in the OECD writ large (Raufner and Feldman 1987; Bernstein 2001; Engels 2006; Voß 2007; Newell and Paterson 2010; Paterson 2010).The original vision was to have an integrated global carbon market associated with the Kyoto Protocol consisting of a global cap-and-trade system and a global offset system that engaged both states in the global North (Annex I) that were negotiating to take on emission reduction commitments and those in the global South (non-Annex I) that would not be taking on such commitments (Hoffmann 2011: 125).

The cap-and-trade system was to engage Northern states and facilitate their achievement of the negotiated emission reductions (Hoffmann 2011: chapter 6). Along with a cap-and-trade system, the Kyoto Protocol laid out a complementary credit or offset market. In credit markets actors undertake activities or projects to reduce greenhouse gas emissions from some baseline (plant trees, change land use, invest in energy efficiency or renewable energy, etc.). The reductions are turned into emission credits – tons of greenhouse gases reduced and not emitted – that can be sold to consumers who seek to manage their greenhouse gas emissions (either voluntarily or by mandate). The Kyoto Protocol initiated two credit markets that could be used by Annex I countries to meet their emission reduction commitments. The Joint Implementation initiative was for offsets produced in Annex I countries (especially transitional economies in Central and Eastern Europe). The Clean Development Mechanism (CDM) was negotiated as a way for developing countries to participate in the carbon market – producing credits that could be sold to entities with reduction commitments, simultaneously advancing sustainable development goals. A third type of credit market has recently emerged – the Reduced Emissions through avoided Deforestation and Degradation program (REDD) that produces credits for developing countries that protect their forests (Lederer 2011).

The multilateral process has always been founded on an understanding of climate change as a global (read universal and international) problem of negotiating emissions reductions. Treating climate change as this kind of problem had tangible consequences – namely political dynamics focused on the distribution of costly action and the emergence of particular market-oriented policy options. While the original understanding of the problem is not inaccurate, it is certainly not the only way to apprehend the problem of climate change. A different perspective on what kind of a problem climate change is can be found at the foundation of an alternative global response with significant consequences for the shape of that global response.

Beyond the Multilateral Process

The UN process has thus far failed to produce an effective response to climate change. The future of multilateral negotiations also appears dim given the disappointing outcomes of the last three negotiations. Yet far from lacking a response to climate change as the UN process has floundered, the world is, rather, awash with different approaches (Hoffmann 2011: chapter 1; see also Andonova *et al.* 2009; Bulkeley and Newell 2010; Hoffmann 2011; Bulkeley *et al.* 2012). Global networks of cities are working to alter municipal economies, transportation systems, and energy use. Corporations are forming alliances with environmental NGOs to devise large and small ways to deliver climate-friendly technology and move towards a low-carbon economy. States, provinces, environmental organizations, and corporations are engaged in developing carbon markets that promise low-cost means of reducing emissions. These transnational governance approaches, or what I have called climate governance experiments, are shaping how individuals, communities, cities, counties, provinces, regions, corporations, and nation-states respond to climate change.

These initiatives are more than lobbying efforts looking to shape the multilateral process. On the contrary, they are explicitly engaged in making rules (broadly conceived as including principles, norms, standards, and practices) – and entail a conscious intention to create/shape/alter behavior for a community of implementers (whoever and whatever they may be) to follow. Recent works have explored the emergence and functioning of this new approach to the global response to climate change (Andonova *et al.* 2009; Bernstein *et al.* 2010; Hoffmann 2011; Abbot 2012; Bulkeley *et al.* 2012). Here, I want to explore the foundational understanding of climate change on which this governance approach rests and the implications of this understanding. This is a somewhat more complex task than was the case for multilateral governance because rather than a single, centralized process, transnational governance of climate change is instead a decentralized, networked, self-organized process that does not have a singular focus or direction (Bulkeley 2005). It is a governance approach made up of multiple, often entirely independent, initiatives. That is not to say that the transnational approach is random or chaotic. On the contrary, recent studies have shown that it is fairly structured, with observable patterns in terms of governance functions and activities they engage in (Andonova *et al.* 2009; Bulkeley and Newell 2010; Hoffmann 2011; Bulkeley *et al.* 2012).

This approach to climate governance is founded on a very different understanding of what kind of a problem climate change is. While transnational climate governance also considers climate change to be a global problem, global means something more or different than a universal response by states. Global is understood to mean simultaneously local and global, multilevel. Transnational governance is just as global as multilateral climate governance, it is just global in a very different way and this entails a very different kind of politics. It involves multiple actors and diverse rule-making practices as opposed to set actors (states) and an established, singular means of making rules (multilateral treaty negotiations). It is flexible because there are multiple sites of governance and actors can voluntarily engage in multiple venues where the multilateral process is tied to a formal consensual decision-making. It has areas of questionable political authority instead of the standard authority of international law and sovereignty, but in bringing together like-minded actor around a range of activities, enforcement may be less of a significant issue.

This experimentation entails trying out new configurations and governing political spaces that did not exist before. Transnational climate governance initiatives or experiments function across boundaries whether vertically (local-regional-national-transnational) or horizontally (networks of similar actors across boundaries). Experimentation is thus a process of making rules outside well-established channels. It is cities forming transnational networks (Betsill and Bulkeley 2004; Bulkeley and Kern 2006). It is US states and Canadian provinces cooperating on climate agreements (Rabe 2004, 2007; Selin and Van Deveer 2009). It is NGOs and corporations forming alliances. This kind of governance is making policy, as Hajer (2003) says, without a polity.

It is near impossible, then, to point to a single set of debates that dominates the transnational governance process. The multilevel, decentralized nature of this experimental approach means that there are multiple kinds of politics taking place. Cities are networking with each other and engaged in relationships with higher levels of jurisdiction (up to the global negotiations), trying to get their work and their plights recognized. Corporations and NGOs are forming alliances like The Climate Group's work with Cisco aimed at implementing information and communication technologies in cities to transform transportation, urban design, and energy delivery (http://www.connectedurbandevelopment.org/news). Of course, there is also competition as multiple initiatives work in similar areas like technology deployment in cities, voluntary carbon markets, renewable energy development. The interests (economic and political) are multiplied in the transnational governance system.

Understanding climate change as a simultaneously global and local or inherently multilevel problem is coupled with diverse notions of what counts as addressing climate change. Emissions reductions are not the sole focus. Transnational initiatives are working towards multiple ends. Emissions reductions are certainly one of the goals being pursued by some initiatives, but others goals – changing infrastructure, promoting renewables, developing the green economy, emissions trading and carbon markets (as ends in themselves), and revolutionizing IT infrastructure – are also included in the diverse targets pursued by transnational governance. The proximate cause of climate change (i.e. emissions) is not ignored, but looking across the myriad transnational initiatives, it is joined by a focus on the underlying causes – fossil fuel dependence of the energy system and economy. Individual initiatives might very well focus on emissions, but because of the diversity in the population of experiments, a broader, if decentralized, overall focus is clearly observable. Transformation, not just emissions reductions, is the collective goal of transnational governance.

This more holistic understanding does not ignore emissions reductions, but many initiatives treat them more as a side-effect of other action than the ultimate end. Initiatives in the experimental world are focused on "smaller" problems. The Climate Group is working to get LED lighting to be the norm for large municipalities across the world (http://www.theclimategroup.org/). The Voluntary Carbon Standard is working to improve the measurement and accounting of carbon offset credits (http://v-c-s.org/). The C40 group of large cities is working on developing building standards and electric public transportation fleets (www.c40cities.org/). Ironically by beginning with a substantially "larger" perspective – climate change as a product of the modern economy and energy systems – the collective efforts of the transnational governance system are not hampered by the need to devise a single binding emissions reductions goal that directs their action. Moving beyond a focus on emissions reductions, transnational climate governance initiatives undertake a myriad of responses. A recent study of 57 of these initiatives found 10 broad kinds of policy options being pursued and only a few (7) dealing specifically with mandating emissions reductions (Table 1.1).

Table 1.1　Experimental governance activity.

Activity	Number of initiatives
1. Catalogue emissions/undertake inventory	20
2. Set targets/formulate action plan/do risk assessment	32
3. Efficiency measures or offsetting	15
4. Education/information and best practice exchange/regular meetings	49
5. Set certification standards/funding criteria	4
6. Mandate emissions reductions	7
7. Emissions trading	8
8. Monitoring (of implementing actors)	16
9. Enforcement	7
10. Technology development	7

Source: Hoffmann, Matthew J. 2011. *Climate Governance at the Crossroads: Experimenting with a Global Response after Kyoto*. New York: Oxford University Press: chapter 2.

The advantage here is the diversity of tools available. The disadvantage is the lack of a concentrated target and centralized process of monitoring and enforcement. On their own, almost all of the transnational initiatives are small. Scaling up individual initiatives or coordinating multiple initiatives will be no mean task (Abbot 2012). In addition, with the opening and fragmenting of climate governance, actors are able to create and/or join experiments that suit them and their preferences best. They can strategize by asking what is best for me materially and/or ask what is appropriate for my values. This sorting action may be detrimental. Actors may find just the right kind of experiment that suits their needs and values (Hoffmann 2011). The USA was at the forefront in pushing voluntary, small-group multilateralism during the Bush administration because it fit their interests in moving slowly on climate change. Every actor may find an initiative to suit its interests, but this does not necessarily equal an effective response to climate change. The question that remains is whether open sorting into initiatives that match actors' preferences will provide enough of a response. If sorting occurs in the absence of legally binding enforcement of broader climate change goals and activities – enforcement that can likely only be achieved through international treaties and national laws – it may not provide an effective catalyst for climate action.

Transnational climate governance initiatives (collectively) thus go beyond orthodoxy on multiple dimensions because the notion of what kind of problem climate change presents is understood very differently. This changes the political dynamics and the policy tools available. Climate governance becomes the province of multiple actors working towards multiple goals. It defines climate change as a global problem of transformation towards decarbonization. With a broader understanding of the problem comes a wider variety of policy tools, but also the potential pitfalls of a decentralized approach that includes significant fragmentation of the global response to climate change (Biermann *et al.* 2009, 2010; Zelli 2011).

Caught between a Rock and a Hard Place?

Diagnosing the problem of climate change has not been an obstacle to addressing climate change. If anything the issue is that there are too many possible ways to diagnose the problem. The clear consensus on key aspects of climate science has told us that there is a problem and given us a relatively clear picture of its characteristics. What it has not been able to do, however, is tell us what kind of problem it is and how to respond to it. The policy options and responses – governance – that can be and have been generated are dependent upon this latter understanding.

This chapter has demonstrated the wide disparity in the foundational understanding of climate change between the dominant multilateral approach and the nascent transnational one. In concluding I will venture two conjectures on what should be done in climate policy and research given this disparity. First, both policy-makers and climate scholars need to re-imagine climate change as a bigger problem than emissions reductions – we need to align our thinking on what kind of problem climate change is with the collective transnational governance approach. Ironically, conceiving of climate change as a problem of widespread transformation makes it easier to address in interesting ways. It allows for tackling various pieces of the problem – decarbonization in the transportation sector, in the energy system, in agriculture, altering building codes and the built environment, and so forth. Further, it moves climate governance out of the realm of distributing costly emission reduction commitments to the realm of seeking out benefits from transformation. There may be enormous value in shifting understanding of the problem to one where there are incentives to cooperate (i.e. engender coordination effects) rather than conceptions of the problem that create obstacles to cooperation (collective action problems) that must be overcome.

Second, let us not abandon the flailing multilateral approach; rather, we should reconceive its role in climate governance (Sanwal 2007). The two systems of governance – traditional multilateral and transnational – have never been entirely independent of one another. In addition, the annual conferences of the parties (COPs) are about more than negotiations. They serve as a focal point for all kinds of actors (NGOs, corporations, students, interested individuals, and media), enhancing the spotlight on this key problem and communicating the sense of urgency that surrounds it. But the role of treaty-making as the primary governance response must change. Multilateral climate negotiations will not and should not be abandoned. I contend that treaty-making must instead be used to ratify and further developments in the transnational governance system. As climate governance experiments innovate in multiple areas and at multiple scales, multilateral approaches can be used to scale up and coordinate, link and further the dynamics bubbling up from below (Abbot 2012).

For policy-making these suggestions mean combining the advantages of both governance approaches while avoiding their respective disadvantages. Prins and Rayner (2007) are substantially correct when they call for a "buckshot" approach rather than a silver bullet – the global response must have the multiple kinds of activities taking place at many scales to be effective. However, coordination and enhancement of these multiple issues will be necessary to ensure that the transnational or experimental approach constitute an effective global response. The central role of multilateral treaty-making should evolve – it should not be relied upon to drive the process, but rather use multilateral treaty-making to make the transnational process better and more effective. For the research community, this analysis implies that we need to focus our energy on how synergies between diverse kinds of activities can be created and exploited. We need to better understand the nature of multilevel authority relationships and governance dynamics. We need to better understand how transformative pathways (Bernstein and Cashore 2012) can be created.

None of these sets of tasks is small and time is short.

Notes

1 Adaptation to climate change is no less important, but is not the focus of this chapter. Please see Chapter 28 in this volume.
2 It seems likely, though difficult to prove definitively, that the close temporal proximity to the ozone-depletion negotiations influenced the international community's understanding in this dimension (Betsill and Pielke 1998; Hoffmann 2005). The ozone depletion negotiations focused exclusively on reducing the emissions of ozone-depleting substances. This gas-centric focus translated easily to the next environmental problem faced by the international community in that climate change also nominally arises from emissions of particular gases.

References

Abbot, Ken. 2012. "Engaging the Public and the Private in Global Sustainability Governance." *International Affairs*, 88: 543–564.

Andonova, Liliana B., Michele M. Betsill, and Harriet Bulkeley. 2009. "Transnational Climate Governance." *Global Environmental Politics*, 9(2): 52–73.

Barrett, Scott. 1992. *Convention on Climate Change: Economic Aspects of Negotiations*. Paris: OECD.

Barrett, Scott. 2003. *Environment and Statecraft*. Oxford: Oxford University Press.

Bernstein, Steven. 2001. *The Compromise of Liberal Environmentalism*. New York: Columbia University Press.

Bernstein, Steven, Michele Betsill, Matthew Hoffmann, and Matthew Paterson. 2010. "A Tale of Two Copenhagens: Carbon Markets and Climate Governance." *Millennium: Journal of International Studies*, 39(1): 161–173.

Bernstein, Steven and Benjamin Cashore. 2012. "Complex Global Governance and Domestic Pathways: Four Pathways of Influence." *International Affairs*, 88: 585–604.

Betsill, Michele and Harriet Bulkeley. 2004. "Transnational Networks and Global Environmental Governance: The Cities for Climate Protection Program." *International Studies Quarterly*, 48: 471–493.

Betsill, Michele and Matthew Hoffmann. 2011. "The Contours of Cap and Trade." *Review of Policy Research*, 28: 83–106.

Betsill, Michele and Roger Pielke. 1998. "Blurring the Boundaries: Domestic and International Ozone Politics and Lessons for Climate Change." *International Environmental Affairs*, 10: 147–172.

Biermann, Frank, Philipp Pattberg, Harro van Asselt, and Fariborz Zelli. 2009. "The Fragmentation of Global Governance Architectures: A Framework for Analysis." *Global Environmental Politics*, 9: 14–40.

Biermann, Frank, Philipp Pattberg, and Fariborz Zelli, eds. 2010. *Global Climate Governance beyond 2012: Architecture, Agency and Adaptation*. Cambridge: Cambridge University Press.

Bodansky, Daniel. 1994. "Prologue to the Climate Change Convention." In *Negotiating Climate Change: The Inside Story of the Rio Convention*, ed. Irving Mintzer and J.A. Leonard, 45–74. Cambridge: Cambridge University Press.

Bulkeley, Harriet. 2005. "Reconfiguring Environmental Governance: Towards a Politics of Scales and Networks." *Political Geography*, 24: 875–902.

Bulkeley, Harriet, Liliana Andonova, Karin Backstrand *et al.* 2012. "Governing Climate Change Transnationally: Assessing the Evidence from a Database of Sixty Initiatives." *Environment and Planning C*, 30: 591–612.

Bulkeley, Harriet and Kristine Kern. 2006. "Local Government and the Governing of Climate Change in Germany and the UK." *Urban Studies*, 43: 2237–2259.

Bulkeley, Harriet and Peter Newell. 2010. *Governing Climate Change*. Abingdon: Routledge.

The Copenhagen Diagnosis. 2009. *Updating the World on the Latest Climate Science*. I. Allison, N.L. Bindoff, R.A. Bindschadler *et al.* Sydney, Australia: The University of New South Wales Climate Change Research Centre (CCRC), http://www.copenhagendiagnosis.com/ (accessed October 15, 2012).

Depledge, Joanna. 2006. "The Opposite of Learning: Ossification in the Climate Change Regime." *Global Environmental Politics*, 6: 1–22.

Dessler, Andrew and Edward Parson. 2006. *The Science and Politics of Climate Change*. Cambridge: Cambridge University Press.

Engels, Anita. 2006. "Market Creation and Transnational Rule-Making: The Case of CO_2 Emissions Trading." In *Transnational Governance: Institutional Dynamics of Regulation*, ed. M.L. Djelic and K. Sahlin-Anderson, 329–348. Cambridge: Cambridge University Press.

Falkner, Robert, Hannes Stephan, and John Vogler. 2009. "International Climate Policy after Copenhagen: Towards a 'Building Blocks' Approach." *Global Policy*, 1: 252–262.

Gardiner, Stephen M. 2004. "Ethics and Global Climate Change." *Ethics*, 114: 555–600.

Grubb, Michael. 1993. *The Earth Summit Agreements: A Guide and Assessment*. London: Earthscan.

Gupta, Joyeeta. 2000. "North–South Aspects of the Climate Change Issue: Towards a Negotiating Theory and Strategy for Developing Countries." *International Journal of Sustainable Development*, 3: 115–135.

Hajer, Maarten. 2003. "Policy without Polity? Policy Analysis and the Institutional Void." *Policy Sciences*, 36: 175–195.

Hoffmann, Matthew J. 2005. *Ozone Depletion and Climate Change: Constructing a Global Response*. Albany, NY: State University of New York Press.

Hoffmann, Matthew J. 2011. *Climate Governance at the Crossroads: Experimenting with a Global Response after Kyoto*. New York: Oxford University Press.

Houghton, John. 2009. *Global Warming: The Complete Briefing*. Cambridge: Cambridge University Press.

Hulme, Mike. 2009. *Why We Disagree about Climate Change*. Cambridge: Cambridge University Press.

Hulme, Mike. 2011. "You've Been Framed." *The Conversation*, http://theconversation.edu.au/youve-been-framed-six-new-ways-to-understand-climate-change-2119 (accessed October 15, 2012).

Intergovernmental Panel on Climate Change. 2007. *Contribution of Working Groups I, II and III to the Fourth Assessment Report of the Intergovernmental Panel on Climate Change*. Geneva: IPCC, http://www.ipcc.ch/publications_and_data/ar4/syr/en/contents.html (accessed October 15, 2012).

International Energy Agency. 2009. *CO_2 Emissions from Fuel Combustion: Highlights*, 2009 edn. Paris: IEA, http://www.iea.org/co2highlights/co2highlights.pdf (accessed October 15, 2012).

International Energy Agency. 2012. *Tracking Clean Energy Progress: Energy Technology Perspectives 2012 Excerpt as IEA Input to the Clean Energy Ministerial*. Paris: IEA, http://www.iea.org/media/etp/Tracking_Clean_Energy_Progress.pdf (accessed October 15, 2012).

Kahan, Dan, Hank Jenkins-Smith, and Donald Braman. 2010. "Cultural Cognition of Scientific Consensus." *Journal of Risk Research*, 14: 147–174.

Lederer, Markus. 2011. "REDD +Governance." *Wiley Interdisciplinary Reviews: Climate Change*, 3: 107–113.

Litfin, Karen. 1994. *Ozone Discourses*. New York: Columbia University Press.

Maslin, Mark. 2004. *Global Warming: A Very Short Introduction*. New York: Oxford University Press.

National Research Council. 2011. *Advancing the Science of Climate Change*. Washington, DC: National Academy of Science, http://nas-sites.org/americasclimatechoices/samplepage/panel-reports/87-2/ (accessed October 15, 2012).

Newell, Peter, and Matthew Paterson. 2010. *Climate Capitalism: Global Warming and the Transformation of the Global Economy*. Cambridge: Cambridge University Press.

Paterson, M. 2010. "Legitimation and Accumulation in Climate Change Governance." *New Political Economy*, 15: 1–23.

Prins, Gwyn and Steve Rayner. 2007. "The Wrong Trousers: Radically Rethinking Climate Policy." Joint Research Paper of the James Martin Institute for Science and Civilization and the MacKinder Centre for the Study of Long-Wave Events, James Martin Institute, Oxford.

Rabe, Barry G. 2004. *Statehouse and Greenhouse: The Emerging Politics of American Climate Change Policy*. Washington, DC: Brookings Institution Press.

Rabe, Barry G. 2007. "Beyond Kyoto: Climate Change Policy in Multilevel Governance Systems." *Governance: An International Journal of Policy, Administration, and Institutions*, 20: 423–444.

Raufner, Roger and Stephen Feldman. 1987. *Acid Rain and Emissions Trading*. New York: Rowman and Littlefield.

Roberts, J. Timmons and Bradley C. Parks. 2007. *A Climate of Injustice: Global Inequality, North–South Politics, and Climate Policy*. Cambridge, MA: MIT Press.

Rowlands, Ian. 1995. *The Politics of Global Atmospheric Change*. New York: Manchester University Press.

Sanwal, Mukul. 2007. "Evolution of Global Environmental Governance and the United Nations." *Global Environmental Politics*, 7: 1–12.

Selin, Henrik and Stacy D. Van Deveer, eds. 2009. *Changing Climates in North American Politics: Institutions, Policymaking, and Multilevel Governance*. Cambridge, MA: MIT Press.

Sell, Susan. 1996. "North–South Environmental Bargaining: Ozone, Climate Change, and Biodiversity." *Global Governance*, 2: 93–116.

United Nations Environment Programme. 2009. *Climate Change Science Compendium*. New York: UNEP, http://www.unep.org/Documents.Multilingual/Default.asp?DocumentID=596&ArticleID=6326&l=en (accessed October 15, 2012).

United Nations Framework Convention on Climate Change. 2009. "Report of the Conference of the Parties on its Fifteenth Session, Held in Copenhagen from 7 to 19 December 2009," http://unfccc.int/resource/docs/2009/cop15/eng/11a01.pdf (accessed October 15, 2012).

United Nations Framework Convention on Climate Change. 2010. "Report of the Conference of the Parties on its Sixteenth Session, Held in Cancun from 29 November to 10 December 2010," http://unfccc.int/resource/docs/2010/cop16/eng/07a01.pdf (accessed October 15, 2012).

United Nations Framework Convention on Climate Change. 2011. "Report of the Conference of the Parties on its Seventeenth Session, Held in Durban from 28 November to 11 December 2011," http://unfccc.int/documentation/documents/advanced_search/items/3594.php?rec=j&priref=600006771 (accessed October 15, 2012).

Vanderheiden, Steve. 2008. *Atmospheric Justice: A Political Theory of Climate Change*. New York: Oxford University Press.

Victor, David. 2004. *Climate Change: Debating America's Policy Options*. New York: Council on Foreign Relations.

Victor, David. 2011. *Global Warming Gridlock*. Cambridge: Cambridge University Press.

Voß, Jan-Peter. 2007. "Innovation Processes in Governance: The Development of 'Emissions Trading' as a New Policy Instrument." *Science and Public Policy*, 34: 329–343.

Zelli, Fariborz. 2011. "The Fragmentation of Global Climate Governance Architecture." *Wiley Interdisciplinary Reviews: Climate Change*, 2: 255–270.

Chapter 2

Global Water Governance

Joyeeta Gupta

Introduction: Global Water Challenges

About 80% of the global population faces a human security challenge in relation to water (Vörösmarty *et al.* 2010). The amount of water available per person is shrinking because of the growing demands of a consumer society as well as a growing population base. This problem is exacerbated by governance failures in controlling the use and abuse of water while protecting the water system so that it can sustainably provide the wealth of ecosystem services (supporting, provisioning, regulatory, and cultural) that society has always depended upon.

This failure is not only about inappropriate water use and water abuse, but also poor land use management, including deforestation, non-sustainable agriculture, mining, rampant urbanization, and atmospheric pollution, and is connected to almost all sectors of society.

Water is a complex issue. Water can be fresh surface water, groundwater, ocean water, grey (waste) water, green water (water in leaves and plants), and virtual water (water embodied in products). Water is essential for ecosystems and human life; it is used for almost all human activities. Although globally there is enough water, aquatic and other ecosystems are degrading rapidly and water is not always available in the quantities and qualities needed in specific areas to sustain human life. Global wetlands have decreased by half in the last 100 years, and the number of freshwater fish species has decreased by 50% in the last 40 years (WWAP 2009). Although water includes ocean water, this chapter limits itself to fresh water (on the marine environment, see Chapter 4).

Good water management calls for (a) maximizing, and creatively and equitably sharing, the ecosystem services derived from water as it flows; (b) ensuring inclusive and participatory processes; and (c) promoting resilience and flexibility in the

The Handbook of Global Climate and Environment Policy, First Edition. Edited by Robert Falkner.
© 2013 John Wiley & Sons, Ltd. Published 2013 by John Wiley & Sons, Ltd.

governance process so that the impacts of climate change can be taken into account (Postel 2011). This requires (a) an integrated transdisciplinary framework; (b) massive advances in education, training, and learning across institutions; and (c) enhanced communication (Sivakumar 2011). At the international level, the key issues are (a) ensuring sustainability, equitable utilization, and coherence as well as contextual relevance in the management of the 263 transboundary rivers (TFDD 2008) and 273 transboundary aquifers (UNESCO 2009); and (b) ensuring that international rules which are relevant to water management (environment, human rights, trade, and investment) are in line with, and supportive of, an integrative and multilevel water management framework.

From a systemic perspective, local, national, and transboundary water issues can also be seen as global in scope: first, the hydrological system is a unitary system; second, the driving forces that influence water use, abuse, and impacts may be global in nature (e.g. climate change); third, the cumulative effects of local problems/solutions may lead to serious global trends; and finally, qualitative and quantitative changes in water may have global impacts – such as on migratory birds and fish species (Pahl-Wostl *et al.* 2008). Nevertheless the subsidiarity principle should also be applied within the increasingly important context of multilevel governance systems.

Against this briefly sketched background, this chapter provides an overview of global governance on water issues. Water governance has long been on the international agenda. Governance arrangements have evolved along three tracks in the fresh water regime: the development of organizations to manage transboundary and global water issues; the evolution of water (and related) law over the centuries; and the advancement of water policy and management over the last 60 years. To some extent these tracks overlap and match, but – more often than not – progress in the field has been fragmented along policy and law lines. The fragmented nature of water governance has led to a discussion on how water should be organized within the UN arena.

There is no easy way to structure the information around water governance. In the main sections of this chapter I have chosen to discuss the organization of water governance; key policy decisions on water; relevant legal issues; the debate on the likelihood of a water war; and then to draw some conclusions.

The Organization of Water Governance

The Phases of Governance

The recent history of global water governance can be divided into four phases: phase 1: transboundary institutionalization experiments (pre-1960); phase 2: global water policy initiatives (1960–1992); phase 3: hybridization of policy initiatives (1992–2003); and phase 4: attempt at system-wide coherence (2003–2012). These phases are briefly explained below.

Phase 1: Transboundary Institutionalization Experiments (pre-1960)

In the early phase of water governance, pre-1960, interstate treaties and transboundary water commissions were established to govern transboundary water issues. Since

1873, the International Law Association (ILA) (consisting of legal professionals) has been actively engaged in promoting the development of water law. Hundreds of water basin agreements have been concluded over the centuries (TFDD 2008), many establishing transboundary river commissions. The Rhine, for example, has been managed for nearly two centuries – since the 1816 treaty that established the Rhine as a navigable river and defined rules regarding its use. Since then, treaties on the Rhine have been adopted, *inter alia*, to protect salmon and prevent pollution. In 1950, the International Commission for the Protection of the Rhine was set up. Water matters on the Danube can be traced back to the Belgrade Convention of 1948, and currently a Danube Commission manages water issues. The International Joint Commission of 1909 regulates rivers between the USA and Canada; the International Boundary and Water Commission between the USA and Mexico of 1944 builds upon governance relations initiated as early as 1889. In Asia, the Mekong Committee was established in 1957 and the Mekong Commission was established in 1995. The 1960 Indus Commission regulates the Indus. The establishment of boundary water commissions is a key institution of water management today.

These transboundary commissions initially regulated navigation, subsequently water use, then pollution, and more recently they have also taken ecosystem services into account. Increasingly, new commissions have been set up in different parts of the world from the Aral Sea to the Southern Africa Development Community (SADC) region. Based on early experiences and communications, the ILA adopted the *Helsinki Rules on the Uses of International Rivers* in 1966 (ILA 1966), which has had an enduring influence on transboundary water governance since then (see the section in this chapter on "The Law Arena").

Phase 2: Global Water Policy Initiatives (1960–1992)

In the second phase (1960–1992), many intergovernmental agencies that were in one way or another linked to fresh water use undertook individual interventions in the water arena to promote governance. UNESCO launched the International Hydrological Decade (1965–1975) to promote the systematic collection of knowledge about hydrological systems, which led to the establishment of the International Hydrological Programme at UNESCO in 1975. A few attempts were made to create coherence in the water governance field through policy-making via global declarations, beginning with the Stockholm Conference on the Human Environment in 1972 and culminating in detailed policy elaboration in Chapter 18 of Agenda 21 in 1992 (see the section in this chapter on "The Policy Arena").

Phase 3: Hybridization of Policy Initiatives (1992–2003)

In the third phase (1992–2003), the limited effectiveness of – and vacuum in – water governance led to the birth of a number of hybrid organizations willing to take the lead in governance. The International Conference on Water and the Environment in Dublin in 1992 was initiated by certain countries that wished to push the water governance agenda further. Subsequently, growing frustration with the rampant development of large dams (from 5000 to about 50 000 between 1950 and 2011: Postel 2011) and the lack of an authority at the global level to develop policies with respect to these dams, led to the establishment of a hybrid body – the World Commission on Dams – by the World Bank and the International Union for the Conservation of Nature (IUCN) in 1997. This body had the mandate to assess the lessons from completed dams and to draw policy implications from past experiences (see the section in this chapter on "The Policy Arena," especially the subsection covering the years 1992 to 2003).

The global water governance vacuum also created the conditions for the development of the World Water Council in 1996, an international multi-stakeholder forum aiming to stimulate knowledge, awareness, commitment, and action. It does so primarily through the World Water Forum that has occurred once every three years since 1997 in Morocco, Netherlands, Japan, Mexico, Istanbul, and Marseille thus far.

Phase 4: Attempt at System-Wide Coherence (2003–2012)

The fourth phase (2003–2012) was characterized by a perceived need to create coherence in the water field. In 2003, the UN established UN Water as a coordinating mechanism and in 2004 it established the Secretary General's Advisory Board on Water and Sanitation. The former replaced the UN ACC Subcommittee on Water Resources which was expected to implement Agenda 21. UN Water aims to promote system-wide coherence on water issues and improve the visibility and credibility of UN action on water. It has around 30 members from UN agencies and programs, and a growing membership from non-UN bodies. With an annual budget of about US$2 million, it has four programs: a joint monitoring program to oversee progress towards meeting the Millennium Development Goals (see the section in this chapter on "The Policy Arena," especially the subsection covering the years 1992 to 2003); a World Water Assessment Programme which prepares the World Water Development Report; and Programmes on Capacity Building and Advocacy and Communication. Five task forces create greater awareness and coordination of information. Although this body has coordination aspirations, its relatively small size and influence implies that, while it has some impact in achieving awareness-raising and monitoring, its influence falls short of actual coordination of UN-wide activities in the water field (Schubert 2010). The question that now arises is whether a UN Water (Lite) is adequate to address the issue of system-wide coherence.

The Policy Arena

Introduction

In the global policy arena, policies were essentially developed in the second to fourth phase of water institutionalization and correspond to (a) the ad hoc development of individual policy ideas in the UN in the first phase; (b) the ad hoc development of policy ideas within and outside the UN and in hybrid agencies in the second phase; and (c) some attempts at creating coherence in the third phase.

1960–1990

The 1972 UN Conference on the Human Environment adopted a declaration and recommendations (Stockholm Declaration; UN Conference on the Human Environment

2002). Principle 21 of the Stockholm Declaration established that although states were sovereign entities they should not cause harm to other states. Furthermore, Recommendation 51 of the Stockholm Action Plan defined a duty of notification to other states when contemplated domestic water activities would have transboundary effects and laid down that water should be used carefully, pollution should be minimized, and that the benefits of transboundary water regimes should be equitably shared by riparian states.

The first global water conference in Mar del Plata in 1977 led to the adoption of a plan which called on countries to assess their water resources; use water efficiently; ensure regular environmental, health, and pollution control; undertake policies, planning and management efforts; implement measures to deal with natural hazards; educate and train the public while encouraging water-based science; and promote regional and international cooperation (UN Water Conference 1977). As a follow-up to this declaration, the 1980s were declared the International Drinking Water Supply and Sanitation Decade, and supply to rural and urban residents increased by 240% and 150% (totaling 1.3 billion people), while access to sanitation improved for 750 million people. But billions were still without access to water (1.2 billion) and sanitation (1.7 billion) at the end of the decade (Sivakumar 2011: 541).

Following the end of the Cold War, it was expected that greater resources could be devoted to social and environmental issues. Twenty years after Stockholm, the United Nations Conference on Environment and Development adopted a set of 27 principles that could have a bearing on water issues (Rio Declaration; UN Department of Economic and Social Affairs 1992). Furthermore, chapter 18 of Agenda 21 (1992) emphasizes the need for effective integrated management of water. It recognizes the necessity to satisfy basic human needs as a priority (Agenda 21 1992: section 18.8). The other seven programs include water resources assessment, protection of water resources, water quality and aquatic ecosystems, drinking water supply and sanitation, water for sustainable urban development, water for sustainable food production and rural development, and impacts of climate change on water resources. The document included a number of targets to be reached by 2005, but most have not been achieved.

1992–2003

A significant landmark is the International Conference on Water and the Environment and its Dublin Principles (Dublin Declaration 1992). These principles include the recognition of water as a "finite and vulnerable resource, essential to sustain life, development and the environment"; the promotion of a participatory approach concurrently at all levels of governance; taking into account the role of women in the "provision, management and safeguarding of water"; and that water "should be recognized as an economic good." These principles have been very influential in shaping water governance.

In 2000, the World Commission on Dams concluded that although large dams contributed significantly to development, the achievements regularly came at an unacceptable and often unnecessary price in terms of social and environmental costs (WCD 2000). The Commission recommended greater integration of externalized costs and transparency (WCD 2000). It did not, however, deal with dam renewal, interbasin water transfers, or the growing need for dams both to mitigate and adapt to climate change. Since the commission no longer exists, there is no natural home for discussions of dam-related policy.

The World Water Forums (WWF) have also contributed to ideas for water policy. The second meeting in 2000 identified common global water problems such as meeting basic needs and securing food supply, protecting ecosystems, sharing water resources, managing risks, valuing water, and governing water wisely. To address such problems, it called for integrated water resources management, collaboration, and partnership at all levels. The third meeting emphasized that water is a driving force for sustainable development and focused on the role of local authorities and communities, and the need for good governance, capacity-building, and financing. The fourth meeting committed itself to the concept of integrated water resource management and the role of actors at all levels of governance in achieving this. The fifth meeting reached new political heights, as many heads of state attended and committed themselves to taking action based on solidarity, security, and ensuring adaptability. An Istanbul Water Consensus was drawn up and cities were invited to sign up to it if they were willing to develop action plans in accordance with the consensus. The World Water Forums provide a centralized venue to discuss water policy and develop political vision and commitment. But whether this will develop into something more substantial – like legally binding decision-making – remains to be seen (Sivakumar 2011).

In an effort to prioritize global development issues, the UN General Assembly adopted the Millennium Declaration (2000). Focusing on the uneven distribution of the benefits of globalization, the Declaration included a target on water:

> To halve, by the year 2015, the proportion of the world's people whose income is less than one dollar a day and the proportion of people who suffer from hunger and, by the same date, to halve the proportion of people who are unable to reach or to afford safe drinking water (Para III 19).

Two years later, the World Summit on Sustainable Development reiterated the goal of halving the number of people without safe access to drinking water and sanitation by 2015, and the need to develop integrated water resources management and water efficiency plans by 2005 (Johannesburg Declaration; UN Department of Economic and Social Affairs 2002).

The policy process has thus far made three major contributions to water: first, the need for an integrated water resource management approach; second, the emphasis on water as an economic good; third, the gradual emphasis placed on access to water and sanitation. However, there is a lack of long-term planning and ground-breaking ideas, incoherence in the policy process, and no systematic progress towards legally binding policy (Sivakumar 2011; cf. Gleick and Lane 2005).

The Law Arena

Phases

During the first phase of transboundary institutionalization experiments, hundreds of transboundary agreements were adopted. In the second phase (1960–1992) there was a gradual codification of water rules. In the 1960s, the UN General Assembly requested the International Law Commission (ILC) to make a draft treaty on transboundary waters. In 1966 the ILA drafted the Helsinki Rules on the Uses of Waters of International Rivers (ILA 1966). This academic document was very influential in helping countries design their national and transboundary water policies (Bourne 1996). The ILC eventually presented its draft convention to the General Assembly in 1990. The third phase corresponds to the phase in which regional (UNECE 1992) and global-level institutionalization of legal norms through treaties occurs (Watercourses Convention of 1997). It is not clear if the global community is already in a fourth phase in law. This section discusses water law principles, global and regional treaties, the Berlin Rules, supranational and transboundary agreements, zooms in on groundwater issues and the human right to water, and discusses critical court cases and other legal regimes that influence water governance.

Principles

Prominent water principles include the sovereignty principle. While some countries have in the past, and continue till today, to claim absolute territorial sovereignty (the Harmon doctrine, as formulated by Mr Judson Harmon, an Attorney General of the USA), many make concessions in practice (Islam 1987; McCaffrey 2001). In contrast, other states have argued in favor of absolute integrity of state territory, which implies that an upstream country cannot alter the quantities and qualities of water flowing into the downstream country (Max Huber, cited in Berber 1959: 20). As a compromise, some jurists have promoted the concept of community of property in water, which calls on states to treat international rivers in an integrated fashion (Lipper 1967: 38). However, probably the most dominant principle is that of limited territorial sovereignty, which restricts the harm states can cause to other states (*Trail Smelter* Arbitration 1941; Stockholm Declaration 1972; Rio Declaration 1992).

Furthermore, principles regarding the navigational uses of transboundary water include the principle of freedom of navigation and of commerce for riparian states; the freedom of commerce, but not of navigation of non-riparian states; and the duty to consult and settle all matters concerning navigation by common agreement among riparian states.

The UN Watercourses Convention

Governing non-navigational uses of watercourses is more complex. In 1966, the Helsinki Rules (ILA 1966) codified the law on international watercourses (see the subsection on "Transboundary Institutionalization Experiments (pre-1960)" in the section in this chapter on "The Organization of Water Governance"). The Watercourses Convention of 1997 based on the International Law Commission's draft was quite similar in content. This Convention has yet to enter into force. The World Wide Fund for Nature launched a campaign in 2006 to promote its ratification and entry into force.

This Convention focuses on transboundary (surface and ground) watercourses that flow into a common terminus. It elaborates on the rights and responsibilities of states in managing and sharing the watercourse; and what should be undertaken in the event of a dispute. Two critical elements should be pointed out: the duty not to cause harm to other states; and the need to equitably utilize the water body according to six criteria, including natural factors such as geographical or ecological conditions, social and economic needs; the effects of the use of watercourses by one state on another state; existing and potential uses; conservation, protection, development, and economy of the use of water and the costs of measures taken to that end; and the availability of alternatives (McCaffrey 2001).

UNECE Convention

In 1992, the UN Economic Commission for Europe (UNECE 1992) also adopted a Convention on the Protection and Use of Transboundary Watercourses and International Lakes (38 parties), which is in force. This Convention includes surface and groundwaters. It obliges all parties to prevent, reduce, or control transboundary impacts, enforced through a combination of standards, limits on discharges, and monitoring. The use of waters needs to be ecologically sound and embody rational water management, the conservation of water resources, and environmental protection. Reasonable and equitable use is promoted, and the precautionary, polluter pays, and sustainable development principles are included.

The Protocol on Water and Health (25 parties) (UNECE 1999) focuses more on the individual and collective human health aspects, including access to good-quality water and sanitation services. The Protocol includes the precautionary, polluter pays, intergenerational equity, preventive, sovereignty subject to duty, subsidiarity, access to information and public participation, the catchment area, equitable access to water, and protection of vulnerable people principles (UNECE 1999: Article 5). Discussions on whether the UNECE Conventions should become gradually more universal are ongoing.

Berlin Rules

The ILA, in the meanwhile, adopted the Berlin Rules in 2004 (ILA 2004). These aim at a comprehensive water regulatory approach more appropriate to the twenty-first century. These rules, which are not legally binding, cover the management of waters in general and drainage basins in particular. They include rules for transboundary waters and groundwater, discuss the rights of individuals to sufficient, safe, acceptable, physically accessible, and affordable water on a non-discriminatory basis, and elaborate on state responsibility, navigation issues, and the protection of water and water installations in times of armed conflict. Unlike the Helsinki Rules, environmental issues are more explicitly taken into account, the role of stakeholders in management is accounted for, and remedies for damage are suggested.

Supranational and Transboundary Agreements

Supranational policy on water is made primarily by the European Union (EU). Water policy and law within the EU has undergone three phases (Castro 2009): in the first phase (1973–1988), the law coordinated water quality, resulting in directives on drinking water, bathing water, and others; in the second phase (1988–1995), emission and water treatment standards with directives regulating manure, cadmium, and urban waste water were adopted. Since 2000, the focus has been on comprehensiveness with the EU's Water Framework Directive (EC 2000) and the Marine Strategy Framework Directive (EC 2008).

The Water Framework Directive provides a framework for action by all member-states to achieve "good status" for surface- and groundwaters in the EU by 2015. The steps to achieve good status include assessing the pressures and impacts on a river basin to implementing a program of specific measures. This Directive aims at providing a comprehensive approach to water management and integrates past governance efforts that focused on individual issues such as drinking water, urban waste water, and so forth. It sees the river basin as a unit and tries to control water pollution and achieve qualitative objectives for water. Although the quality of the implementation varies in different countries (EC 2007, 2009; Kelly *et al.* 2009), this Directive is extremely important in shaping fresh water policy in the EU. The Marine Strategy Framework Directive aims at a good environmental status for the marine waters of the EU by 2020 and establishes European Marine Regions.

Groundwater Rules

Although groundwater is often seen as comprising 97% of the world's fresh water resources (Brölman 2011), water fluxes are more important than storage (Koutsoyiannis 2011). Annual surface water fluxes are 44,700 km^3 compared to 2200 km^3 for groundwater fluxes to the oceans – a difference of 20 times (Shiklomov and Sokolov 1985, cited in Koutsoyiannis 2011). Around 273 transboundary aquifers exist globally (UNESCO 2009). Thus far, relatively limited attention has been paid to groundwater governance both at the national and international level. The law governing transboundary aquifers draws from the law on shared natural resources as well as the law on shared water resources.

Ideas for groundwater governance emerge from the work of epistemic communities. The Helsinki Rules (ILA 1966) covered groundwater to the extent that it was part of a hydraulic system that included surface water and flowed into a common terminus. Twenty years later the ILA developed rules on confined groundwater in its Seoul Rules (ILA 1986). The 2004 Berlin Rules also dealt with groundwater (ILA 2004).

The first legally binding agreement covering groundwater was the UNECE Water Convention, which includes all kinds of groundwater as long as it interacts with transboundary waters (UNECE 1992). In 2000 the UNECE prepared its Guidelines on Monitoring and Assessment of Transboundary Groundwaters. The International Law Commission's Draft Articles on the Non-Navigational Uses of International Watercourses also covers groundwaters when they are part of a hydrological system that includes surface waters and flows into a common terminus. In 2008, the ILC adopted its Draft Articles on the Law of Transboundary Aquifers (ILC 2008). At regional level a number of agreements have been made recently. The Nubian sandstone aquifer system states concluded agreements in 2000; the North-Western Sahara Aquifer countries reached an agreement in 2002; Niger, Nigeria, and Mali signed a comprehensive agreement on the Iullemeden Aquifer system in 2009; and in 2010 an Agreement on the Guaraní Aquifer was signed by Argentina, Brazil, Paraguay, and Uruguay.

The trends in groundwater governance include a focus on aquifers and not only groundwater; call for protection, preservation, and management of the resource; equitable and reasonable utilization as well as equitable and reasonable sharing of the benefits; the obligation not to cause harm to others; recognition of the sovereignty of the state over its own part of the aquifer; and provisions on limitations on, or notifications of abstractions – but there is still a long way to go in developing groundwater law at transboundary level (Mechlem 2011; Tanzi 2011).

The Human Right to Water

As of 2011, 1–2 billion humans lack access to either potable water or sanitation facilities. This violates human dignity. The discussion on human rights can be traced back to the Human Rights Declaration of 1948 and covenants of 1966. A limited recognition of the human right to water for women (CEDAW 1979) and children (African Charter 1999) was adopted in legal treaties.

Globally economic, social, and cultural rights have progressed relatively slowly. In the area of water, discussions on this right can be traced back to the 1972 Stockholm Declaration, which discusses equitable use of water, and the Mar del Plata Conference Plan (UN Water Conference 1977), which explicitly recognized the need to create the right to drinking water (UN Water Conference 1977: Resolution II, 66). The Rio Declaration (UN Department of Economic and Social Affairs 1992) included 27 principles but no human right to water, although Agenda 21 (1992) emphasized the human need for water and sanitation services. Despite the gradual increase in the momentum as global conferences (Declaration of the International Conference 1994; UN Human Settlements Programme 1996) emphasized the human right to water, the UN Watercourses Convention (UN General Assembly 1997) did not mention this right.

The Millennium Declaration (2000) addressed basic needs issues including water. The UN Committee on Economic, Social and Cultural Rights' General Comment No. 15 on the Right to Water in 2002 and a number of regional conferences (the 2007 Asia Pacific Water Summit and the third South Asian Conference on Sanitation in 2008) paved the way for the UN Human Rights Council to adopt this topic for a three-year study in 2008 and for the UN General Assembly (2010) to adopt a Declaration on the Human Right to Water and Sanitation that was accepted by 122 nations. In 2010, the UN Human Rights Council adopted a resolution on the human right to access safe drinking water and sanitation.

The current status of the human right to water and sanitation is that there is political recognition, and it is arguably legally binding. Such a right creates enforceable rights and responsibilities and in the process may empower the vulnerable. However, it is not self-enforcing and it still requires a legally aware pro bono community and a justice system that can help implement these rights (Gupta *et al.* 2010).

Court Cases on Water

Adjudication has been another source of legal precedent in the water area (Castillo-Laborde 2009).

Key principles of international water law are the notions of freedom of navigation and community of interests of riparian states in a navigable river, arising from the decision of the Permanent Court of International Justice (PCIJ) in the *River Oder* case (1929) between Poland and the downstream countries. It argued that nature gives rise to the transboundary relationship and this leads to obligations of states to cooperatively manage the transboundary river. This principle was recognized also with respect to non-navigational uses in the International Court of Justice's (ICJ) judgment of the *Gabčikovo–Nagymaros* case (1997) between Hungary and Slovakia. Such cooperation is required for all parts of an international river – the tributaries and sub-tributaries, the navigable and non-navigable parts.

The right to equitable use of waters was recognized by the PCIJ in the *Diversion of the Meuse River* case (1937) between Belgium and the Netherlands and once more in the *Gabčikovo–Nagymaros* case (1997). The right to use waters is accompanied by the no-harm principle and requires that states notify others of potentially harmful activities (*Lake Lanoux* Arbitration 1957 between France and Spain) and repair that harm if it occurs (*Gut Dam* case of 1968 between the USA and Canada).

In relation to the Kushk River, a boundary river between Afghanistan, present-day Turkmenistan, and Russia, a commission held in 1893 that the boundary between Afghanistan and Russia was the thalweg in the river. This rule is still often used today in navigational boundary rivers. With respect to the Zarumilla River (between Peru and Ecuador), a Brazilian arbitral award of 1945 held that the thalweg in a canal to be constructed between the islands in the river would be the boundary between the two countries.

On water quality, the 2004 *Protection of the Rhine against Pollution by Chlorides* Arbitration (between the Netherlands and France) held that the concept of "legal community" meant that pollution should be addressed and ordered France to compensate the Netherlands for its excess costs in dealing with the pollution. In the 2010 *Pulp Mills* case between Uruguay and Argentina, the ICJ ruled against Argentina and allowed Uruguay to continue with the mills. Subsequently, the two governments set up a joint commission to address pollution problems in the river.

Other Legal Regimes with an Impact on Water

Three other regimes influence water governance: climate change, investment, and trade regimes.

Environmental agreements (e.g. biological diversity, desertification, wetlands, and climate change) have implications for water. I will briefly focus on climate change. Climate change impacts water through the increased risk of glaciers melting and influencing surface water flows; changing rainfall and evaporation patterns influencing the local availability of water; and the increased likelihood of extreme weather events such as droughts, floods, and cyclones (Intergovernmental Panel on Climate Change 2007). A historical overview of the progress made in terms of implementing the climate convention (Gupta 2010) shows that the problem is far from being addressed and it is now time to climate-proof fresh water agreements and governance worldwide (Cooley and Gleick 2011).

With the recognition of water as an economic good, interest in privatizing some water services has risen. This has led to foreign investment in water. Such foreign investments are generally regulated by investment law, which protects the interests of foreign investors. Investment law includes bilateral and multilateral investment treaties. Such treaties between countries encourage foreign investors to invest in one another's territories in return for protecting their rights. The proliferation of bilateral treaties led the OECD to try and consolidate these treaties in the Multilateral Agreement on Investment, but this failed (Werksman and Santoro 1998) and the proliferation continues. These treaties generally define which investments are covered by them, prescribe equal treatment between nationals and foreigners (national treatment clause) and between investors from different countries, stipulate fair and equitable treatment of the investors, require free transferability of funds into and out of the country and compensation in case of an expropriation or damage to the investment, and establish specific dispute-settlement mechanisms. These mostly call for arbitration under the Convention on the Settlement of Investment Disputes between States and Nationals of Other States (ICSID Convention 1965) or the Rules of the Permanent Court of Arbitration. These treaties often imply that, once the water sector is open to private investment, governments cannot differentiate between local and foreign investors; that the contracts drawn up are confidential and not subject to public control; and that in the event of a dispute, international confidential arbitration may take the subject matter out of the control of the host country. There have been several arbitration cases thus far and their judgments tend to have an impact on water law (Tecco 2008).

Global trade law also influences the water sector. The World Trade Organization (WTO) and regional trade agreements regulate trade, and to the extent that water is a traded commodity, it too falls under this jurisdiction (Barlow and Clarke 2002). When member countries trade water, they need to respect the national treatment principle. This may lead to foreign investors exporting water for profit even if the host state sees water as critical for its domestic interest.

The "Water Wars" Debate

Finally, I reflect briefly on the "water wars" debate. Towards the end of the last century, two emerging schools of thought developed about whether countries and peoples will see water as so critical for their survival and identity that they will be willing to engage in conflict. Water is important, scarce, poorly distributed, and shared, and these four characteristics make it subject to being a source of stress (Gleick 1993; Myers 1993; Kaplan 1994; Villiers 2001; cf. Stalley 2003).

Others argue that water (or environment) is generally only one of a series of variables and it would be difficult to isolate which conflicts could be attributed to water shortage (Levy 1995; Gleditsch 2001). Homer-Dixon (1994) argues that such wars occur only in very exceptional circumstances, for example, following a history of military antagonism. Wolf (1995) argues that the cost of a war far outweighs the costs of desalination plants or other infrastructures to deal with water problems and that makes war unlikely. Based on case-study work, Kalpakian argues that "serious conflict is reserved for matters that touch peoples' identities such as their language, history, heritage and self image" (2004: 1).

What is not disputed is that water scarcities can trigger tensions and, in combination with other factors, may create conditions that could potentially escalate into diplomatic disputes if not outright war. However, there is a much stronger case to argue in favor of the role of water in promoting cooperation between peoples and countries and to ensure that the institutionalization of negotiating and diplomatic processes leads to greater peace globally.

Key Issues in Water Governance for the Twenty-first Century

This chapter's overview raises some questions. First, should water be regulated at the global level? While water issues are clearly often mostly local or fluvial in nature, a systemic approach may call for seeing the hydrological system as one, and globalization has increasingly led to the establishment of production, distribution, and consumption patterns and governance processes that imply that even so-called local issues are often influenced by global demand and supply, processes, and knowledge systems. There seems to be clearly an increasing need to have some degree of global governance on water.

Second, water issues are dispersed throughout the UN and non-UN system. While UN Water tries to harmonize some of the activities of global actors on water issues, it has a relatively small mandate, few resources, and little authority. The competing processes at global level have led to different trajectories for governing water – a policy trajectory that UN Water plays a role in, a law trajectory where legal instruments, arbitral and court awards, and legal epistemic communities shape the debate, and a human rights trajectory that is being pushed in the UN Human Rights Council and the General Assembly.

Third, there are defining moments which have led to the birth of new ideas – the Helsinki Rules in 1966 and its articulation of equitable sharing of water, the Dublin Declaration in 1992 and the birth of integrated water resources management, as well as the notion of water as an economic good, and the UN General Assembly Resolution of 2010 and the coming of age of the human right to water and sanitation.

Fourth, behind the dispersed and competing governance trajectories, confusion exists regarding the discourses that should help shape water governance. The liberalization discourse focuses on private sector participation in water, confidential water contracts, trade, and investment law; the good governance discourse emphasizes transparency, legitimacy, accountability, and participation; the water governance discourse has evolved from the hydraulic mission with its emphasis on optimizing water use through infrastructure development to equitable sharing of water; the water management discourse is shifting from sectoral through integrated water resource management to, possibly, adaptive management; the human rights discourse promotes a focus on the human right to water and sanitation, and indigenous people's rights; environmental discourses are centered on sustainable development, environmental protection, and ecosystem services; and the new scientific framing discourses focus, *inter alia*, on concepts like virtual water trade (Gupta 2009)! Not all of these discourses are reconcilable, and each is being promoted actively at the global level by specific advocacy coalitions. While some see the growing number of dams as redesigning waterscapes and landscapes (Postel 2011), others regard them as necessary to meet water, food, and energy needs in the twenty-first century (Koutsoyiannis 2011). While some see water as an economic good (Dublin Declaration 1992), others argue that it should be seen as a heritage, a human right (Gleick 2003;

Gupta 2010), or a political good (Schouten and Schwartz 2006). There is a lack of critical evaluations of the usefulness and usability of concepts such as sustainable development, integrated water resources management (see Conca 2006), good governance, decentralization vs. centralization, stakeholder participation, private sector participation, the role of bi- and multilateral aid in water policy, even the role of science in water policy. Diffuse policy processes with limited authority or legitimacy and restricted access to quality scientific assessments are unable to generate information and consensus about which of these ideas, norms, and concepts is most likely to be consistent with sustainability and within which specific contexts. Little policy work has been conducted on the conditions for sustainable water transfer from one basin to another, the potential of sea-water desalination and transfer of icebergs, and the sustainability of both small and large dams, among others. Here, too, two schools of thought have emerged. Some advocate a centralized approach to managing water, while others favor a light, coordinating approach.

Fifth, the legal arena is dense with bilateral and regional agreements that either directly relate to water or are in fields that have consequences for water. This has not been integrated into one comprehensive framework. Although this may or may not be necessary, depending on one's perspective on the need to centralize and formalize, a priority now is the need to climate-proof transboundary water agreements in the coming years (Cooley and Gleick 2011).

Sixth, whether one accepts the hypothesis that countries may be willing to go to war on water issues or not, what is clear is that the maldistribution of water is likely to be a source of tension and create human insecurity and calls for better water governance.

Water is a critical resource for countries, not just because of its role in meeting basic needs and its contribution to the national economy, but also because of the significant cultural, religious, and aesthetic function of water. The density of governance on water is both pluralistic and fragmented, embodying competing value systems. The question for the future is whether a harmonized, comprehensive water management system (including an organizational framework and law), is more likely to successfully address the critical water challenges of the twenty-first century? Or, if such an agreement is politically possible, will it merely imply the superimposition of certain values over other values and create greater inequities at the local level? Or should the global community try and prioritize a few issues first and try to regulate those as a priority? Clearly, governance at the global level needs to co-develop with a comprehensive multilevel system of governance that is coherent where possible while being contextually relevant in the diversity of localities where it is to be implemented.

References

African Charter. 1999. *African Charter on the Rights and Welfare of the Child.* Adopted by the Organisation of African Union (OAU), Doc. Cab/Leg/24.9/49 (1990). Entered into force, November 29, 1999.

Agenda 21. 1992. "Agenda 21 Programme for Action," (UN Doc. A/CONF. 151/26/Rev. 1, 14 June 1992, http://www.un.org/esa/dsd/agenda21/res_agenda21_00.shtml (accessed October 15, 2012).

Barlow, Maude and Tony Clarke. 2002. *Blue Gold: The Battle against Corporate Theft of the World's Water*. London: Earthscan.

Berber, Friedrich Joseph. 1959. *Rivers in International Law*. London: London Institute of World Affairs.

Bourne, Charles B. 1996. "The International Law Association's Contribution to International Water Resources Law." *Natural Resources Journal*, 36: 155–216, http://heinonline.org/HOL/Page?handle=hein.journals/narj36&g_sent=1&collection=journals&id=167 (accessed October 15, 2012).

Brölman, Catherine. 2011. "Transboundary Aquifers as a Concern of the International Community." *International Community Law Review*, 13: 189–191. doi:10.1163/187197311X582377.

Castillo-Laborde, Lilian del. 2009. "Case Law on International Watercourses." In *The Evolution of the Law and Politics of Water*, ed. Joseph W. Dellapenna and Joyeeta Gupta, 319–335. Dordrecht: Springer-Verlag.

Castro, Paulo Canelas de. 2009. "European Community Water Policy." In *The Evolution of the Law and Politics of Water*, ed. Joseph W. Dellapenna and Joyeeta Gupta, 227–244. Dordrecht: Springer-Verlag. doi:10.1111/1468-5965.00148.

Conca, Ken. 2006. *Governing Water: Contentious Transnational Politics and Global Institution Building*. Cambridge, MA: MIT Press.

Convention on the Elimination of All Forms of Discrimination against Women (CEDAW). 1979. Adopted by the General Assembly of the United Nations, Resolution 2106 (XX), UN Doc. A/34/46, New York, December 18.

Cooley, Heather and Peter H. Gleick. 2011. "Climate-Proofing Transboundary Water Agreements." *Hydrological Sciences Journal*, 56: 711–718. doi:10.1080/02626667.2011.576651.

Declaration of the International Conference on Population and Development. 1994. "Programme of Action," A/CONF. 171/13/Rev.1, Cairo, September 15, Principle 2.

Delhi Declaration. 2008. Third South Asian Conference on Sanitation (SACOSAN III, November 21).

Dublin Declaration. 1992. The Dublin Statement on Water and Sustainable Development. Report of the International Conference on Water and the Environment, January 31.

EC. 2000. "Directive 2000/60/EC of the European Parliament and of the Council of 23 October 2000 Establishing a Framework for Community Action in the Field of Water Policy." *Official Journal of the European Communities*, L327: 1–72.

EC. 2007. *Towards Sustainable Water Management in the European Union*. Commission Staff working document, accompanying document to the Communication from the commission to the European Parliament and the Council. First stage in the implementation of the Water Framework Directive 2000/60/EC. Brussels. [COM(2007) 128 final], [Sec(2007) 363].

EC. 2008. "Directive 2008/56/EC of the European Parliament and of the Council of 17 June 2008 Establishing a Framework for Community Action in the Field of Marine Environmental Policy," June 17.

EC. 2009. "Report from the Commission to the European Parliament and the Council in Accordance with Article 18.3 of the Water Framework Directive 2000/60/EC on Programmes for Monitoring of Water Status." Brussels. [SEC(2009)415].

Gleditsch, Nils Petter. 2001. "Armed Conflict and the Environment." In *Environmental Conflict*, ed. Paul F. Diehl and Nils Petter Gleditsch, 251–272. Boulder, CO: Westview Press.

Gleick, Peter H. 1993. "Water and Conflict: Fresh Water Resources and International Security." *International Security*, 18: 79–112.

Gleick, Peter H. 2003. "The Water Conflict Chronology," www.worldwater.org/conflict (accessed January 3, 2011).

Gleick, Peter H. and Jon Lane. 2005. "Large International Water Meetings: Time for a Reappraisal." *Water International*, 30: 410–414. doi:10.1080/02508060508691883.

Gupta, Joyeeta. 2009. Driving Forces around Global Fresh Water Governance." In *Water Policy Entrepreneurs: A Research Companion to Water Transitions around the Globe*, ed. D. Huitema and S. Meijerink, 37–57. Cheltenham: Edward Elgar.

Gupta, Joyeeta. 2010. "A History of International Climate Change Policy." *Wiley Interdisciplinary Reviews: Climate Change*, 1: 636–653. doi:10.1002/wcc.67.

Gupta, Joyeeta, Rhodante Ahlers, and Lawal Ahmed. 2010. "The Human Right to Water: Moving toward Consensus in a Fragmented World." *Review of European Community & International Environmental Law*, 19: 294–305.

Homer-Dixon, Thomas. 1994. "Environmental Scarcities and Violent Conflict." *International Security*, 19(1): 5–40.

ICSID Convention. 1965. Washington, DC: International Centre for Settlement of Investment Disputes. Entered into force, October 14, 1966.

Intergovernmental Panel on Climate Change (IPCC). 2007. *Impacts, Adaptation and Vulnerability. Contribution of Working Group II to the Fourth Assessment Report of the Intergovernmental Panel on Climate Change*. Cambridge and New York: Cambridge University Press.

International Law Association (ILA). 1966. *The Helsinki Rules on the Uses of the Waters of International Rivers*. Report of the Fifty-Second Conference, Helsinki. London: ILA.

International Law Association (ILA). 1986. *Rules on International Groundwaters (Seoul Rules)*. Report of the Sixty-Second Conference, Seoul. London: ILA.

International Law Association (ILA). 2004. *Berlin Rules*. Report of the Seventy-First Conference, Berlin. London: ILA.

International Law Commission (ILC). 2008. *Draft Articles on the Law of Transboundary Aquifers*. Report of the International Law Commission on the Work of Its Sixtieth Session, UN GAOR, 62d Sess., Supp. No. 10, at 19, UN Doc. A/63/10 (2008).

Islam, Muhammad Rafiqul. 1987. *Ganges Water Dispute: Its International Legal Aspects*. Dhaka: University Press Limited.

Kalpakian, Jack. 2004. *Identity, Conflict and Cooperation in International River Systems*. Aldershot: Ashgate Publishers.

Kaplan, Robert. 1994. "The Coming Anarchy." *Atlantic Monthly*, 273: 44–76, http://www.theatlantic.com/magazine/archive/1994/02/the-coming-anarchy/4670/ (accessed October 15, 2012).

Kelly, Martyn, Cathy Bennett, Michel Coste *et al.* 2009. "A Comparison of National Approaches to Setting Ecological Status Boundaries in Phytobenthos Assessment for the European Water Framework Directive: Results of an Intercalibration Exercise." *Hydrobiologia*, 621: 169–182. doi:10.1007/s10750-008-9641-4.

Koutsoyiannis, Demetris. 2011. "Scale of Water Resources Development and Sustainability: Small Is Beautiful, Large Is Great." *Hydrological Sciences Journal*, 56: 553–575. doi:10.1080/02626667.2011.579076.

Levy, Marc. 1995. "Is the Environment a National Security Issue?" *International Security*, 20: 35–62.

Lipper, Jerome. 1967. "Equitable Utilisation." In *The Law of International Drainage Basins*, ed. Albert H. Garretson, Robert D. Hayton, and Cecil J. Olmstaed, 15–64. Dobbs Ferry, NY: Oceana Publications.

McCaffrey, Stephen C. 2001. *The Law of International Watercourses: Non-navigational Uses*. Oxford: Oxford University Press.

Mechlem, Kerstin. 2011. "Past, Present and Future of the International Law of Transboundary Aquifers." *International Community Law Review*, 13: 209–222. doi:10.1163/187197311X582278.

Myers, Norman. 1993. *Ultimate Security: The Environmental Basis of Political Stability*. New York: W.W. Norton.

Pahl-Wostl, Claudia, Joyeeta Gupta, and Daniel Petry. 2008. "Governance and the Global Water System: A Theoretical Exploration." *Global Governance*, 14: 419–435.

Postel, Sandra L. 2011. "Foreword: Sharing the Benefits of Water." *Hydrological Sciences Journal*, 56: 529–530. doi:10.1080/02626667.2011.578380.

Schouten, Marco and Klaas Schwartz. 2006. "Water as a Political Good: Implications for Investments." *International Environmental Agreements: Politics, Law and Economics*, 6: 407–421. doi:10.1007/s10784-006-9013-3.

Schubert, Susanne. 2010. "Do the UN Coordination Bodies Fulfill Their Coordination Functions? A Case Study of UN Water, UN Energy and UN EMG." MSc thesis, Hafencity University of Hamburg.

Sivakumar, Bellie. 2011. "Water Crisis: From Conflict to Cooperation. An Overview." *Hydrological Sciences Journal*, 56: 531–552. doi:10.1080/02626667.2011.580747.

Stalley, Phillip. 2003. "Environmental Scarcity and International Conflict." *Conflict Management and Peace Science*, 20: 33–58. doi:10.1177/073889420302000202.

Tanzi, Attila. 2011. "Furthering International Water Law or Making a New Body of Law on Transboundary Aquifers? An Introduction." *International Community Law Review*, 13: 193–208. doi:10.1163/187197311X583240.

Tecco, Nadia. 2008. "Financially Sustainable Investments in Developing Countries Water Sectors: What Conditions Could Promote Private Sector Involvement?" *International Environmental Agreements: Politics, Law and Economics*, 8(2): 129–142.

Transboundary Freshwater Dispute Database (TFDD). 2008. "Case Studies: Water Conflict Resolution." Department of Geosciences, Oregon State University, http://www.trans boundarywaters.orst.edu/research/case_studies/index.html (accessed October 15, 2012).

UN Committee on Economic, Social and Cultural Rights. 2002. "General Comment No. 15. The Right to Water" (Articles 11 and 12 of the International Covenant on Economic, Social and Cultural Rights).

UN Conference on the Human Environment. 1972. Stockholm Declaration, June 16. UN Doc. A/CONF.48/14/Rev.1, 1973.

UN Department of Economic and Social Affairs. 1992. Rio Declaration and Agenda 21. Report on the UN Conference on Environment and Development, Rio de Janeiro, June 3–14, UN doc. A/CONF.151/26/Rev.1 (Vols. I–III).

UN Department of Economic and Social Affairs. 2002. Johannesburg Declaration on Sustainable Development. World Summit on Sustainable Development (A/CONF.199/20), September 4.

UNECE (UN Economic Commission for Europe). 1992. Convention on the Protection and Use of Transboundary Watercourses and International Lakes. Helsinki, March 17. Entered into force, October 6, 1996. UN Reg. Nr. 33207, 1936 UNTS 269.

UNECE (UN Economic Commission for Europe). 1999. London Protocol on Water and Health on the Convention on the Protection and Use of Transboundary Watercourses and International Lakes (25 Parties).

UNESCO. 2009. *Transboundary Aquifers: Managing a Vital Resource*. Paris: UNESCO.

UN General Assembly. 1948. Universal Declaration on Human Rights. UNGA Resolution 217A (III), December 10.

UN General Assembly. 1997. Convention on the Law of the Non-navigational Uses of International Watercourses, New York, May 21, Doc. A/51/869, adopted in Resolution A/RES/51/229.

UN General Assembly. 2000. United Nations Millennium Declaration (A/RES/55/2, September 8), http://www.un.org/millennium/declaration/ares552e.htm (accessed October 15, 2012).

UN General Assembly. 2010. "Resolution on Human Right to Water and Sanitation," A/64/292, 28 July, http://www.un.org/News/Press/docs/2010/ga10967.doc.htm (accessed December 29, 2011).

UN Human Settlements Programme. 1996. "Istanbul Declaration on Human Settlements and the Habitat Agenda. Adopted at the Second United Nations Conference on Human Settlements." Document No. A/CONF.165/14 (Istanbul, June 14), para 11.

UN Water Conference. 1977. "Report of the UN Water Conference. Mar del Plata, Argentina, 14–25 March, 1977." Document No. E/Conf.70/29.

Villiers, Marq de. 2001. *Water: The Fate of Our Most Precious Resource*. Boston, MA: Mariner Books.

Vörösmarty, Charles J., Peter McIntyre, Mark O. Gessner *et al.* 2010. "Global Threats to Human Water Security and River Biodiversity." *Nature*, 467: 555–561.

Werksman, Jake and Claudia Santoro. 1998. "Investing in Sustainable Development: The Potential Interaction between the Kyoto Protocol and a Multilateral Agreement on Investment." In *Global Climate Governance: Inter-linkages between the Kyoto Protocol and other Multilateral Regimes*, ed. Bradnee W. Chambers, 59–76. Tokyo: United Nations University.

Wolf, Aaron T. 1995. *Hydropolitics along the Jordan River: Scarce Water and Its Impact on the Arab-Israeli Conflict*. Tokyo: United Nations University Press.

World Commission on Dams (WCD). 2000. *Dams and Development: A New Framework for Decision-Making*. London: Earthscan.

World Water Assessment Programme (WWAP). 2009. *The United Nations World Water Development Report 3: Water in a Changing World*. Paris: UNESCO Publishing, and London: Earthscan.

Chapter 3

Biodiversity and Conservation

Stuart Harrop

The Importance of Biodiversity and the Current Challenges

The diversity of life within species, habitats, and ecosystems has provided a foundation for human civilizations throughout history and continues to support our contemporary social, economic, and industrial systems and structures. Apart from its intrinsic worth beyond anthropogenic concepts of value, this diversity is of critical importance to human welfare and its notional economic value is vast (Constanza *et al.* 1997). The economic value of biodiversity derives from the services provided to us, among many other things, in sinking carbon, purifying air and water, disposing of our waste, pollinating our crops; and providing food, medicines, and many other raw and naturally refined materials that we take for granted. There are many other indirect benefits. Thus the diversity of life also secures buffers against disease that threaten its own persistence and within a wider matrix operates as a balancing mechanism that secures the dynamic continuance of ecosystems and meta-life systems on Earth (Wilson 1992). Indeed, according to Lovelock, for so long as life has been flourishing on our planet, the Earth has been able to adapt to and withstand the worst of the perturbations that the solar system beyond our biosphere has deigned to inflict upon it (Lovelock 1995).

The importance of biodiversity for the persistence of life systems can often be lost in wider debates. Within the climate change debate discussions, by example, dialogues so often remain in the realm of emissions control or reduce into almost abstract terms such as "carbon" – a term that does not so easily connote the complexity of nature and the systems operating to protect life on Earth. As a result biodiversity protection receives less global attention and is a subsidiary priority to carbon emissions regulation (Gilbert 2010). Climate change is of course devastating to biodiversity (Pounds 1999; Walther *et al.* 2002; Opdam and Wascher 2004;

The Handbook of Global Climate and Environment Policy, First Edition. Edited by Robert Falkner.
© 2013 John Wiley & Sons, Ltd. Published 2013 by John Wiley & Sons, Ltd.

Gregory *et al.* 2009) and even scientific approaches to measuring and monitoring the effects of climate change have been known to have a potentially devastating effect on species (Saraux *et al.* 2011). But climate change and biodiversity loss are closely related if not inextricably and irrevocably entangled. Indeed a vicious circle of reactions and feedbacks manifests when biodiversity begins to decline through the effects of rapid climate change, whereby both the processes of biodiversity decline and exacerbated climate change function together to create an increasing spiral of deleterious effects (Laurance and Williamson 2001; Hoegh-Guldberg *et al.* 2007). Thus the climate change debate cannot be divorced from the parallel discussions that relate to stemming the tide of one of the greatest extinction spasms the world has ever known. Moreover, when the position is clearly understood, no debate at any level in the world convened to seek to resolve complex issues relating to global development or even to resolve relentless waves of economic turbulence (as experienced at the time of writing in early 2012) can sensibly avoid integrating both climate change and biodiversity loss into strategic thinking. Unfortunately, this enlightened attitude is only rarely in evidence; the development of nature conservation law has often been a niche interest, and the subject has only recently become more mainstreamed into wider areas of governance and into wider dialogues.

The pace of development of international law is also slow and there have been very few developments in the last 20 years, consequently most conservation instruments were drafted prior to the more significant understanding of the relationship between biodiversity and climate change that we now have (Trouwborst 2009). Moreover, the challenges faced by biodiversity loss are complex, and effective regulatory systems need to reflect the sophistication of the complex systems that they seek to control (Johannsdottir *et al.* 2010; Underdal 2010). In addition, although conservation science has had significant influence over practical conservation, this has had little effect on international policy priorities (Robinson 2006), and the existing portfolio of instruments does not necessarily reflect conservation exigencies (Rands *et al.* 2010) as they are now understood or indeed the speed at which the problem of diminishing biodiversity is overtaking our ability to respond.

The slow development of international law and policy and its failure to keep pace with contemporary understanding of biodiversity preservation priorities is in sharp contrast to the rate of decline of biodiversity. Indeed, according to the official global institution dealing with biodiversity issues – UNEP – "the biodiversity loss is hundreds of times faster than previously in recorded history and the pace shows no indication of slowing down." This is evidenced by ecosystem loss (UNEP cites 35% of mangrove swamps and 20% of coral reefs as examples) and through extreme rates of species loss (between 150 and 200 species becoming extinct every 24 hours). This relentless rate of decline is caused by human action – habitat encroachment, changes of land use, overexploitation, trade and overuse of species, invasive species carried along human vectors, pollution, and, coming round again in the final part of the circle, climate change. Indeed, it is estimated that "approximately 20–30 per cent of plant and animal species assessed so far are likely to be at greater risk of extinction if increases in global average temperature exceed 1.5 to 2.5 Celsius" – a temperature rise that is now accepted as inevitable by most commentators (Millennium Ecosystem Assessment 2005; UNEP 2010).

Whereas extinction is a natural phenomenon, the current rate of decline is far greater than the replenishment rate of adaptation and evolution, and, therefore, planetary homeostasis (Lovelock 1995) is challenged to the extreme. This challenge has the potential to affect the foundations of biodiversity on our planet in the same manner as previous events in the remote past when one of a number of rare "mass extinctions" occurred that irrevocably and drastically altered the course of the development of life on Earth. For humans the threat of such a mass extinction also represents a severe challenge to our cultures, societies, and economic structures since they are irretrievably embedded in and founded on the matrix and diversity of life on Earth.

The Initial Global Regulatory Response

The idea of creating an instrument designed to protect biological diversity for its own sake rather than endangered or threatened species and habitats – and linking such preservation firmly to climate change and development issues – is a relatively novel and recently developed concept. Prior to the signature of the Convention on Biological Diversity (CBD) in 1992, a number of international conventions and regional instruments had been negotiated which rarely cross-reference one another, only in a few cases deploy consistent and similar terminology, and otherwise have little relationship except where they occasionally overlap in jurisdictions. Instruments such as the Convention on the Conservation of Migratory Species of Wild Animals 1979 (Bonn) and the Convention on the Conservation of European Wildlife and Natural Habitats 1979 (Berne) cover, in part, similar subject matter and, because of their focus on protecting endangered and threatened species and habitats, require parallel and relatively similar legal approaches. And yet they use quite different language, have different levels of detail in provisions, and do not refer to each other. This failure to use consistent language can result in different legal interpretations, inconsistent implementation, and different approaches to protection. This may be best evidenced practically by examining the defenses to illegal taking of species described in Article 9 of Berne and Article II(5) of Bonn, which bear virtually no relationship to each other even where they deal with similar topics. Thus in Berne the defense dealing with reintroduction allows the taking of endangered species where it is for the purpose of "repopulation, reintroduction and for the necessary breeding" and the parallel provision in Bonn permits taking where it is for the purpose of "enhancing the propagation or survival of the affected species." There is also a relatively haphazard approach in the coverage of nature conservation issues by international law irrespective of the importance of the subject matter. The International Whaling Commission established by the International Convention for the Regulation of Whaling (1946) deals exclusively with cetacean species and there is no real parallel in other instruments dealing with the predicament of the many other key marine species that are on the brink of extinction. Moreover the Commission was established to maintain the "orderly development" of the whaling industry and, therefore, was not founded on anything like a contemporary conservation ethic and has been forced to adapt and transmute, through its long history, to deal more with wider and more familiar contemporary issues (Harrop 2003b). The approach to regulation within an instrument can also vary widely from convention to convention. The Convention on International Trade in Endangered Species of Wild Fauna and Flora 1973 (CITES), dealing with international trade, is a comprehensive, detailed, highly focused "lawyer's" legal instrument. It contains technical legal provisions that are largely capable of being transmuted into national law with relative ease and with effective consistency throughout its member-states. Some hard-law instruments, and the CBD is certainly one (the detail will be examined later), are so vaguely drafted and hedge-bound with qualifications that the countries implementing them can operate under widely different policy, law, and interpretations of these conventions' provisions. Some other instruments have very simple, widely drafted, and often weak terms creating limited obligations. The Convention on Wetlands of International Importance, especially as Waterfowl Habitat (1971) (Ramsar) deals with an issue that is just as important as the concerns of CITES. Whereas trade is a key driver of extinction, and hence the importance of CITES, many of the world's most sensitive, biodiverse, and important habitats are wetlands, and that fact lends great importance to Ramsar. However, the convention only requires states to register – and thus substantially protect – one wetland (Article 2(4)), and there are even circumstances where that state can delete the potential site from the list (Article 4(2)). Other criticisms of Ramsar focus on the convention's generalist approach towards the protection of crucial havens of biodiversity in wetland areas, with little detailed provision for states to implement and enforce (de Klemm and Shine 1993).

No doubt there are many reasons for this disparate approach to the design of instruments aimed at biodiversity protection and for the seemingly ad hoc manner in which legislation has evolved. Certainly it may be difficult to engage the attention of politicians and opportunities may have to be grasped when available by the NGOs who occasionally manage to move them into action through their lobbying. However, bearing in mind the short event horizon of politicians (who may expect only a few years in office) and their unwillingness to risk losing votes by compromising the sovereignty of their nations to secure elusive benefits that may not be realizable for a number of generations into the future, it is a miracle that any international instruments to conserve nature exist at all.

One exceptional event which did capture the attention of politicians around the world was the United Nations Conference on Environment and Development in 1992. This conference certainly lived up to its alternate name as the "Earth Summit" in terms of its pageant, apparent visionary nature, and its ability to capture the attention of world leaders. Indeed, in time it may prove to have been the pinnacle of efforts at the global governance level to face up to and assume responsibility for the negative human impact on global life-supporting systems. The summit produced two hard-law instruments: the United Nations Framework Convention on Climate Change and the Convention on Biological Diversity (CBD). Bearing in mind the relentless deforestation taking place particularly in tropical rainforests around the world, it had been hoped by some that a third legal instrument would also have resulted in dealing with the international protection of forests but this was not to be (McConnell 1996). Instead the Forest Principles were agreed, but these were embedded in a soft-law instrument whose name, Non-Legally Binding Authoritative Statement of Principles for a Global Consensus on the Management, Conservation and Sustainable Development of All Types of Forests, made it abundantly clear that no nation would be bound to implement them.

Of course when we reach the pinnacle we have no choice but to descend. Although the Earth Summit, by agreeing to promulgate a convention dealing with climate change on the one hand and biodiversity protection on the other, seemed to have fully appreciated the inseparable nature of the emissions problem and the biodiversity destruction challenge, it may be that since that date we have steadily followed a descending path. Moreover, in terms of the CBD, which is the central subject of this chapter, its wide-ranging and comprehensive but framework stipulations are necessarily lacking in detail and its 20-year history has now demonstrated that those provisions have not been built upon in the manner originally intended.

The rest of this chapter explores the road from Rio until it reaches Nagoya, where at the close of 2010, the CBD set out its strategy for the future. In so doing this analysis also evaluates the nature of this path and attempts to measure its effectiveness.

The Convention on Biological Diversity

The CBD dealt with such a wide range of subject matter that its text is a reflection of extensive compromise. However, it did seek to move on from the approach of previous negotiations of predecessor instruments, whereby texts were dominated by the perspective of "Northern" developed states (for example CITES is generally regarded as representing Northern consumer interests rather than "Southern" producer interests: Hutton and Dickson 2000) by reflecting all global interests and political philosophies through dealing directly with the "North–South" divide. The CBD does deal, to an extent, with aspects of equity between the developed and the then-developing world, requiring, *inter alia*, informed consent and equitable sharing arrangements prior to access to genetic resources. Nevertheless, whether the convention succeeded in reconciling the disparate parts of the globe is measurable only to a degree, and the different stances between biodiversity-range states (usually at the time developing countries) and consumers (the major developed countries) resulted in extensive compromise and contributed to the failure of the USA to ratify the convention – arguably one of the most serious obstacles to the CBD's effectiveness (McConnell 1996; Falkner 2001).

The CBD is a very different instrument from all that preceded it in the field of the conservation of nature. Its general terms could be said to embrace all the subject matter of the existing international legal portfolio and yet there is no reference to specific instruments in its text (beyond a general reference to "the law of the sea" in Article 22), nor was the opportunity taken to coordinate and corral the work of the already existing conservation institutions and convention activities. It was certainly a revolutionary instrument in many ways. Among others it contemplates the full relationship that humans have had with the natural world throughout history (see e.g. Article 8(j)); it does not focus on merely endangered species and habitats but on the whole matrix of species, habitats, and ecosystems, including the non-living environment in which they subsist (Article 2), and it seeks to engage with almost every aspect of international biodiversity conservation in the context of a developing world.

The objectives described in Article 1 comprise: "the conservation of biological diversity, the sustainable use of its components and the fair and equitable sharing of the benefits arising out of the utilization of genetic resources." At first glance, the inclusion of such a wide-ranging and ostensibly contradictory set of objectives would appear to be courageous, especially if the convention succeeded in reconciling the conflicts inherent in promoting use and conservation in one instrument. However, the width of the objectives reflected rather the existence of the conflict rather than its resolution and epitomized the type of compromise forced on the CBD text in negotiations between those who recognized the finite nature of the Earth's natural resources and those who sought development (McConnell 1996). The sheer scope of the objectives also forced the convention to display little detail in its text, but early commentators anticipated that this would be remedied by subsequent, subsidiary legal instruments (Sand 1993). In retrospect, since much of the scope of the CBD has not been dealt with in further international laws, the attempt to cover so many issues has resulted in a text that is spread too thin. Furthermore the CBD is imprecisely drafted and replete with obligations that have lost their edge through a profusion of qualifications littered throughout the text as a result of the many compromises made during the negotiation process (McConnell 1996). Consequently, states can accept its obligations without in many cases implementing laws and policies that have any real uniformity around the world (Harrop and Pritchard 2011) or by simply assuming that existing state laws more or less cover the CBD's scope. In this connection, whereas CITES is implemented by precise laws (in the EU through the current CITES regulation (EC 1996 and subsequent amendments and through resultant national laws), the CBD has been implemented less by new laws and more by policies that may or may not be backed up by normative measures (see for example the UK government's approach: DEFRA 2012).

The CBD contains a facility in Article 28 to create subsequent subsidiary instruments in the form of protocols, and early commentators assumed that the CBD would use this provision to develop the anticipated detailed laws that would build a solid structure over its framework (Glowka *et al.* 1994), but this process has been slow and only a small proportion of the CBD's provisions have been expanded into subsidiary instruments. Furthermore, some of these instruments duplicate existing provisions in the principal text, and may indeed perpetuate some of the damage done by qualifications negotiated into the CBD at its inception. The recently promulgated Nagoya Protocol (discussed later in this chapter) recites what is largely in place already in Articles 8(j), 10(c), and 15 in the parent convention and brings the number of CBD protocols to two (along with the Cartagena Protocol on Biosafety, which relates specifically to only one of the sub-clauses in Article 8) (Harrop 2011; on the Cartagena Protocol, see Chapter 6 in this volume).

Rather than work to produce comprehensive, detailed obligations in the form of protocols, and thereby fill in the many holes left in its framework of provisions, the CBD's recent strategy has veered towards setting voluntary targets for members. This trajectory suggests a softening rather than the anticipated hardening of the CBD's foundational text, and in order to appreciate the potential consequences of this softening approach an examination of both the inherent textual weaknesses of the CBD and the normative nature of the targets is required.

The CBD's Inherent Weaknesses

Instruments of international law are not coherent in quality and the CBD has been described as comprising of "'soft' diffuse obligations" (Braithwaite and Drahos 2000) that are not characteristic of clear law. This is inevitable in instances where there are substantially different perspectives on the subject matter of an instrument – indeed international law can only be as strong as its creators – sovereign nation-states – wish it to be. However, in extreme cases this may result in a convention being reduced to a mere agreement to develop policy and maintain debate tantamount to a declaration of policy. In such circumstances, although such a result may be the best practical achievement (Abbott and Snidal 2000) the outcome may be disappointing in terms of the urgency of the subject matter, where, by example in the CBD's case, we are warned of dire consequences to us all if the rapid rate of biodiversity decline is not halted. However, where there is sufficient interest in a subject and where the international community is convinced that international rather than merely inward-facing national measures are essential, the community is perfectly capable of concluding a reasonably strong and coherent regime of law, at least from a positivist legal perspective. The WTO portfolio is a good example of such a regime and it was founded on general consensus that a strong regulatory system to facilitate open multilateral trade would expand and develop state economies and indeed would, according to the then Director-General of the WTO, "encourage and contribute to sustainable development, raise people's welfare, reduce poverty, and foster peace and stability" (WTO 2012).

Beyond its "soft" obligations, the CBD also reflects a low level of priority extended to biodiversity conservation by the international community and positively provides for other conventions to overrule it. Thus, Article 22 specifically subjugates its provisions to other international rights and obligations "except where the exercise of those rights and obligations would cause a serious damage or threat to biological diversity." The latter provision may appear to be a useful escape clause but, bearing in mind the complexity of ecosystems and the longitudinal studies often required to prove any hypothesis relating to ecosystem dynamics, it would be highly ineffective where evidence was required to prove an impending and "serious damage or threat to biological diversity" within an immediate and pragmatic dispute between two conflicting areas of international law (Glowka et al. 1994). By contrast, the provisions of the World Trade Organization texts function within a closed circle and contain no provisions to compromise the fulfillment of its clear and unequivocal objectives. Indeed the WTO is an international organization managing an "integrated and distinctive legal order" (Lamy 2006) and even operates its own dispute resolution mechanism where it embraces not just the evaluation of trade measures but also their relative impact on nature conservation. In doing so it involves neither NGOs nor international conservation governance institutions in its proceedings (see the *Shrimp–Turtle* case: WTO 1998). Moreover, in order to deal with the interface between trade and existing multilateral environmental agreements, rather than subjugating its authority to these MEAs, the WTO unilaterally assumed the jurisdiction to examine the relationship within its internal Committee on Trade and Environment (WTO 1994a, 1994b, 1994c) and within the remit of the Doha Declaration. Therefore, the WTO is willing to balance its relationship with other international law on its own terms, under control, and without compromising its independent and unassailable legal position. This is in sharp contrast to the CBD's approach to its relationship with other international laws.

Most of the articles of the CBD use terms that have insufficient clarity. Purported obligations may be construed, with little ingenuity, to be optional. Notably, clauses lack the use of auxiliary verbs such as "shall" or "will," which in other agreements serve to reinforce obligations. Similarly, where the obligations are expressed clearly a secondary clause or phrase, such as "subject to national law," "subject to patent law," "as far as possible," effectively destroys both the impact of clarity and weakens the obligation. Whereas reasonable qualification is found in well-drafted and effective law it must remain in the realms of clear limits and it is usually possible to deduce these limits for such qualification from the context and by following ordinary rules of legal construction. In the CBD's context the qualifications go beyond this and can be susceptible to many interpretations. These "obligations" are not for the most part measurable in terms of enactment, implementation, or enforcement, and even subsidiary, soft material generated has been severely criticized for its lack of indicators to facilitate monitoring of impact (Walpole *et al.* 2009).

Article 8(j) deals with a crucial subject and was revolutionary at the time. It provides a useful example of the CBD's approach to drafting. The article endeavors to acknowledge, in terms of legal obligations, the crucial role of traditional practices in conservation, but is hampered from the very beginning of its text by a phrase that makes each member-state's obligations "subject to its national legislation." This is such a wide qualification that it is conceivable that national legislation already enacted that opposes generally accepted norms of human rights could frustrate any implementation of this sub-article. It is also peculiar that a supposed *hard*-law instrument should be expressed to be subject to the very state law that it necessarily must alter to achieve its objectives. Moreover, Article 8(j) also contains an obligation to *respect* traditional conservation practices. The word "respect" presents some unusual problems in terms of monitoring implementation of international law (Harrop 2003a). How, practically, could such an obligation be implemented and enforced in a coherent manner? The word "respect" may be understood on the streets but does not make a useful phrase in the statute books. Designing legislation to enforce "respect" poses extreme challenges even to the most imaginative of legal draftsmen and, if the idea of "respect" is left to be expanded upon by legislators, it is likely that implementation would not only be inconsistent but also incoherent.

It is also interesting to note that deficiencies in drafting and textual design can also be inherited by the CBD's offspring in the form of protocols. Thus the Nagoya Protocol, which in part extends and develops the framework encapsulated in Article 8(j), perpetuates the parent article's weaknesses by, on occasion, using the same language as deployed in the article (Harrop 2011).

The CBD's Regulatory Development

As has been stated already the CBD's strategy was anticipated, as evidenced by observations of experts at the outset, to be expanded through subsidiary protocols which would build substance on the framework provisions within the convention text. Had this happened in a substantial manner the lack of precision and abundant qualifications in the provisions of the CBD could easily have been overlooked as the optimum political achievement at the time of the Earth Summit and thus seen retrospectively as a work in progress. However, in the 20 years since its inception, two CBD protocols have been concluded and together they deal with the subject matter of one full clause and one sub-clause (Article 15 and Article 8(j)) in the case of the Nagoya Protocol and one sub-clause (Article 8(h)) in the case of the Cartagena Protocol. This has hardly made a dent in the objectives of the CBD expressed in Article 1. Moreover, in terms of the priorities that we face today within the scope of the CBD, it could also be said that the needs to control the use of genetically modified organisms (the subject matter of the Cartagena Protocol) and the need to deal with equitable access and benefit sharing (dealt with in the Nagoya Protocol) are not subjects of the utmost priority when we are faced with rapidly diminishing biodiversity on a global scale and resultant critical feedbacks that are themselves exacerbating climate change. The subject matter of these protocols is not placed amongst the key priorities in relevant literature in the field of biodiversity conservation (Sutherland *et al.* 2009).

Although perhaps not the highest priorities in the face of issues causing biodiversity decline, the two protocols are necessarily steps forward in building the detail on the CBD's framework. The Cartagena Protocol is dealt with in Chapter 6 in this volume, but it is appropriate to briefly describe the Nagoya Protocol, promulgated late in 2010 (*Nagoya Protocol on Access to Genetic Resources* 2010).

The protocol focuses on the CBD's objective dealing with "the fair and equitable sharing of the benefits arising out of the utilization of genetic resources" and promotes the preservation of traditional conservation practices (thereby putting detail on the bones of Articles 8(j) and 10(c) in the parent convention as well as extending the provisions in Article 15 of the CBD dealing with "prior informed consent" and access to genetic resources). In terms of conservation value, fulfillment of these objectives should promote stakeholder involvement in conservation initiatives, create local incentives to conserve ecosystems and their components, and provide biodiversity benefits that accrue through preserving long-evolved traditional knowledge.

As has been already mentioned the protocol mirrors the qualities (in part including deficiencies) of the parent convention. This is clear not only in the textual approach but also in respect of specific terms. Thus the protocol, in similar manner to the CBD, subjects its terms to all other international law except where there is a "serious damage or threat to biological diversity."

The protocol deals with some aspects of criticism of the CBD. Thus the access to genetic resource and benefit-sharing provisions in the parent text applied between states but did not directly protect local people's rights in such resources nor guarantee that they would share in any benefits ensuing from the resources' exploitation. The protocol remedies this by directly referring to these local rights in a number of provisions and in a number of different ways. Nevertheless drafting weaknesses derived from the parent text are perpetuated and the new provisions may therefore be implemented in widely differing ways depending on the state's perspective on the matter (Harrop 2011).

A new and useful mechanism, established in Article 11 of the protocol, is the new *Access and Benefit-Sharing Clearing House*. This creates a focal point for sharing key information deriving from the operation of the protocol such as, *inter alia*, measures implementing the protocol text, access permits, model contractual clauses for access, and benefit sharing.

Finally, Article 18 of the Nagoya Protocol requires states to encourage *non-parties* to comply with its provisions. The stipulation provides a diplomatic approach to securing support from parties, such as the USA, which has not ratified the CBD or the Nagoya Protocol.

The CBD's Strategic Deployment of Targets

Since 2002 the CBD has followed a strategy that does not appear to have been envisaged at its inception but may be its only alternative, bearing in mind its legal weaknesses and the lack of willingness of the global community to assume further obligations. This strategy has transformed its identity, from a law-making instrument to an institution whereby it sets voluntary time-bound targets for its members. This approach is a growing trend in some areas of international policy and regulation and can be shown in some instances to have had effective impact (Jolly 2003).

The first set of targets developed by the CBD (endorsed at the World Summit on Sustainable Development in 2002) were designed to achieve a "significant reduction of the current rate of biodiversity at the global, regional, and national level" by 2010 (CBD 2002a). This principal target was heavily criticized for its lack of specificity and measurability.

The targets derived from CBD COP 6 Decision VI/26 (Strategic Plan for the Convention on Biological Diversity) and that decision expressly recognized that the implementation of the convention had been "impeded by many obstacles." These included specifically a "lack of political will"; "legal/juridical impediments"; "lack of appropriate policies and laws"; and a number of other factors described in the appendix to that decision. Bearing in mind these impediments and the clear under-standing expressed in the appendix, it is notable that the targets deriving from the strategic plan still failed to create obligations to set precise, implementable objec-tives. Indeed, the targets did not improve the lack of precision in the mother text but instead perpetuated the problem and in some cases prescribed a lower standard. In other cases targets merely repeated, in paraphrase, existing CBD provisions. A number of commentators raised substantial concerns about the failings of these tar-gets (see generally: Walpole *et al.* 2009; Harrop and Pritchard 2011), and indeed when the strategic plan of the CBD came to be renegotiated in 2010 the Conven-tion's Director-General described the outcome of the targets set in 2002 as a "total disaster" (Vidal 2010).

It is useful to analyze some aspects of the targets in more detail. The umbrella "2010 Biodiversity Target" resembled more a policy aspiration than an instrument designed to fulfill the objectives of a legal instrument. It lacked specificity, and since it described a global aspiration it provided no guidance for implementation by indi-vidual states, since biodiversity, species, and ecosystems do not respect boundaries and in many cases a state could not determine how its individual actions would have supported the target. Indeed, apart from its setting of a deadline (2010), it added little if anything to the already generalized and heavily qualified obligations within the parent convention.

In subsequent decisions of the CBD (COP 7 Decision VII/30 and COP 8 Decision VIII/15) further work was done to elaborate the general target through subsidiary goals. However, these subsidiary targets provided little more detail (and in many cases less detail) than had already been set out in the CBD text. For example: targets 2:1 and 2:2 (see CBD 2002b) aimed to "restore, maintain, or reduce the decline of population species" and to improve the status "of threatened species." How-ever, both of these aspirations are already present in Article 8(f). This article simi-larly describes the restoration of "degraded ecosystems" and the promotion of "the recovery of threatened species, *inter alia*, through the development and implementation of plans and management strategies." The article is more specific than the target, however, in identifying that planning instruments should also be implemented. Other sub-targets are expressed with less precision than in the original text of the parent convention. Thus targets 9:1 and 9:2, dealing with the goal to "maintain sociocultural diversity of indigenous and local communities," sets out summaries of rather more complex and thus more normatively challenging requirements in a number of articles in the CBD including: Article 8(j) (preserving traditional knowledge and benefit sharing), Article 10(c) (protection of customary use of biological resources), and Article 15 (access to genetic resources). Sub-target 4:3 appears to duplicate, in albeit extremely general terms, other international law. Its aim is for "no species of wild flora or fauna [to be] endangered by international trade." The subject of wildlife trade is not specifically dealt with in the CBD text but is dealt with extensively and in much more detail, as already indicated, in the text of CITES. And yet the rather trite statement in the sub-target does not acknowledge the existence of CITES. Bearing in mind the need to create coherent strategies across conservation conventions in order to strongly respond to the challenges we face (Heller and Zavaleta 2009), this omission was certainly disappointing.

Despite the time and resources allocated to this elaboration of the basic 2010 CBD target, the convention was forced to acknowledge that few countries succeeded in establishing national targets "and even fewer have had time to implement them" (UNEP-CBD 2009). In response to this poor performance, the CBD's strategic plan was drastically revised in Nagoya, Japan at the end of 2010 (CBD 2011). The negotiations produced the Aichi Biodiversity Targets (CBD 2011), which are more sophisticated than the predecessor targets and have been in part designed in response to the extensive criticism of the previous CBD goals (Harrop 2011; Harrop and Pritchard 2011). There is now much more precision in the text of the targets but some of the problems remain. Thus the percentages of the Earth that are to be protected appear to be arbitrary and do not assist selection of crucial habitats nor do they set priorities in regions or between habitats. Again, because these percentages are set at the global level they give little guidance to individual states about how a coordinated and thus more effective approach can be made. Similarly some of the detailed targets continue to duplicate the text within the CBD (Harrop 2011). Nevertheless there are many improvements and time is needed for the CBD to develop the detail and to maintain some sort of pressure on states to engage with detailed design. However, the most difficult challenge to the CBD may be its diminishing authority to drive this process.

The CBD's Identity and the Direction of International Biodiversity Preservation Law

The first paragraph of the CBD's new Strategic Plan seeks to

promote effective implementation of the Convention through a strategic approach comprising a shared vision, a mission, strategic goals and targets that will inspire broad-based action by all Parties and stakeholders.

The crucial word that sets the scene is "inspire." This describes the transformed identity of the CBD, not so much as a normative instrument of international law, which may be implemented into law and policy and enforced through pragmatic governance measures, but as a visionary institution from which inspiration may be gained. Although commentators (Abbott and Snidal 2000) have observed how multilateral environmental objectives have been implemented through a varied level of instruments through hard law to soft policy, the CBD appears to have gone one step further. The softening trajectory (Harrop and Pritchard 2011) of the CBD from a hard-law instrument to a mere political inspirer of policy, promoting a vision through rousing rhetoric, appears to be sealed by the new strategic plan and its targets. Questions about the nature of international law and how it may be categorized philosophically (see, e.g., Weil 1983 and Koskenniemi and Leino 2002) have been asked for some time but in this instance a new phenomenon is occurring which may alter the direction of the discourse. We are already aware that the text of the CBD is deficient in its normative quality, and in this regard, as Weil points out, "the capacity of the international legal order to attain the objectives it was set up for will largely depend on the quality of its constituent norms" (Weil 1983). However within the text of the strategic plan there is a distinct avoidance of legal requirements – indeed the plan patently evades the imposition of obligations. Thus paragraph 3, which closes the door firmly on the plan's potential normative function, simply "urges" parties to adhere to it and consequently leaves the success of the convention to both hope and chance in the face of more immediate political expediencies.

Sections V and VI of the Annex to the Strategic Plan specifically focus on implementation and support mechanisms for the plan and it is here that we must search for remnants of normative capacity. This part of the plan responds to criticism of the previous targets whereby it was argued that for targets to function effectively they need to be supported by a "suite of tools" including, among others, specific economic and legal components (Cawardine *et al.* 2009). Section 13 describes the plan as a "flexible framework for the establishment of national and regional targets and national biodiversity strategies" but falls short of prescribing legal obligations. However, the plan does seek to raise the profile of the predicament and importance of biodiversity in a dynamic manner, thus Section 16 seeks to broaden political support by "working to ensure that Heads of State and Government and the parliamentarians of all Parties understand the value of biodiversity and ecosystem services." There are also a number of references to the crucial need to *mainstream* biodiversity issues throughout national, regional, and global governance mechanisms. In tune with the rest of the plan, the language used is aspirational and not prescriptive.

The CBD, coupled with its subsidiary agreements and strategic documents, remains legally weak and imposes largely diffuse obligations on its parties. In leading the planet to this state of affairs the global community may well be inhibited by a short political event horizon that pushes these fundamental problems onto the burden of future generations; however, there are occasional glimmers of an international conscience and within the new strategic plan an understanding of pressing needs becomes discernible.

The CBD's history describes a steady download slide, not just from hard to soft law but also on to a third category whereby, through the expression of mere aspiration, it is becoming little more than a source of exhortation, inspiration, and vision.

Nevertheless, these qualities comprise a component of leadership and, in agreeing to the relevant terms within the strategic plan, there is a suggestion that the international community recognizes that such leadership is required. However, bearing in mind the current dire need for a strong global response to the relentless decline and dilution of biodiversity, much more is required to put inspiration into practice.

There are alternate governance models that may be appropriate, and the global community has to date accepted comparatively strong legal authority in other areas of global regulation such as in the multilateral trade regime, where the operations of a wide-ranging portfolio of agreements and their implementation of them is overseen by a single entity. Thus the CBD's targets could be in part delegated to other, relevant MEAs who would then work to the CBD's guidance within their special areas of expertise (trade, migratory species, marine conservation, wetland conservation, etc.) to fulfill them as priorities – adjusting their own strategies accordingly to correspond with a concerted and coordinated approach. There also appear to be the rudiments of recognition of the need for a unified international approach to conservation in the CBD strategic plan. Section 17 of its Annex seeks to promote partnerships and such relationships are expressed to include *other conventions*. This echoes and strengthens aspects of the provisions in the parent text of the strategic plan, which, among others, includes a statement that the plan "represents a useful flexible framework that is relevant to all biodiversity-related conventions." In addition, paragraph 17(c) of the principal text recognizes the need to create a "coherent implementation of biodiversity-related conventions and agreements."

The other aspect of leadership, deriving from the model described, involves over-seeing national implementation. This requires appropriate links and networks. In this respect the CBD already links into all of its member-states very effectively through its Clearing House Mechanism established pursuant to CBD Article 18(3), whereby it operates as the center for the coordination and communication of all national actions plans and strategies that operate in response to the CBD's strategic plan. Indeed, this is one area where the CBD's provisions may be regarded as operating well. Nevertheless, without coordinated efforts to build on the effectiveness of its other provisions – which itself will require strong leadership – the CBD may ultimately be remembered only for its efficiency in gathering information to simply observe – rather than prevent – the relentless decline of biodiversity.

References

Abbott, K.W. and D. Snidal. 2000. "Hard and Soft Law in International Governance." *International Organization*, 54(3): 421–456.

Braithwaite, J. and P. Drahos. 2000. *Global Business Regulation*. Cambridge: Cambridge University Press.

Carwardine, J., C. Klein, K. Wilson *et al.* 2009. "Hitting the Target and Missing the Point: Target Based Conservation Planning in Context." *Conservation Letters*, 2: 3–10.

CBD (Convention on Biological Diversity). 2002a. "2010 Targets," http://www.cbd.int/2010-target/about.shtml (accessed May 1, 2012).

CBD (Convention on Biological Diversity). 2002b. "Goals and Sub-targets," http://www.cbd.int/2010-target/goals-targets.shtml (accessed May 1, 2012).

CBD (Convention on Biological Diversity). 2011. "Strategic Plan for Biodiversity 2011–2020, Including Aichi," http://www.cbd.int/sp/ (accessed May 1, 2012) (see also: COP 10

Decision X/2X/2. Strategic Plan for Biodiversity 2011–2020, http://www.cbd.int/decision/cop/?id=12268 (accessed May 1, 2012)).

Costanza, R., R. D'Arge, R. de Groot *et al.* 1997. "The Value of the World's Ecosystem Services and Natural Capital." *Nature*, 387: 253–260.

DEFRA. 2012. "UK Implementation of the Convention on Biological Diversity," http://www.defra.gov.uk/environment/natural/biodiversity/internationally/cbd/uk-implementation/ (accessed May 1, 2012).

de Klemm, C. and Shine, C. 1993. *Biological Diversity Conservation and the Law*. Gland, Switzerland and Cambridge: IUCN.

EC. 1996. "Council Regulation No. 338/97 of 9 December on the Protection of Species of Wild Fauna and Flora by Regulating Trade Therein." *Official Journal of the European Union*, L 061, 03/03/1997: 1–69.

Falkner, Robert. 2001. "Business Conflict and U.S. International Environmental Policy: Ozone, Climate, and Biodiversity." In *The Environment, International Relations, and U.S. Foreign Policy*, ed. Paul G. Harris, 157–177. Washington, DC: Georgetown University Press.

Gilbert, N. 2010. "Biodiversity Hope Faces Extinction." *Nature*, 467: 764.

Glowka, L., F. Burhenne-Guilmin, and H. Synge. 1994. *A Guide to the Convention on Biological Diversity*. Global Biodiversity Strategy Environmental Law and Policy Paper No. 30. Bonn: IUCN Environmental Law Centre, IUCN Biodiversity Programme.

Gregory, R.D., S.G. Willis, F. Jiguet *et al.* 2009. "An Indicator of the Impacts of Climate Change on European Bird Populations." *PLoS ONE*, 4: e4678. doi:10.1371/journal.pone.0004678.

Harrop, S.R. 2003a. "Human Diversity and the Diversity of Life: International Regulation of the Role of Indigenous and Rural Human Communities in Conservation." *Malayan Law Journal*, 4: xxxviii–lxxx.

Harrop, S.R. 2003b. "From Cartel to Conservation and on to Compassion: Animal Welfare and the International Whaling Commission." *Journal of International Wildlife Law and Policy*, 6: 79–104.

Harrop, S.R. 2011. "'Living in Harmony with Nature'? Outcomes of the 2010 Nagoya Conference of the Convention on Biological Diversity." *Journal of Environmental Law*, 23(1): 117–128.

Harrop, S.R. and D.J. Pritchard. 2011. "A Hard Instrument Goes Soft: The Implications of the Convention on Biological Diversity's Current Trajectory." *Global Environmental Change*, 21(2): 474–480.

Heller, E.N. and S.E. Zavaleta. 2009. "Biodiversity Management in the Face of Climate Change: A Review of 22 Years of Recommendations." *Biological Conservation*, 142: 14–32.

Hoegh-Guldberg, O., P.J. Mumby, A.J. Hooten *et al.* 2007. "Coral Reefs under Rapid Climate Change and Ocean Acidification." *Science*, 318: 1737.

Hutton, J.M. and B. Dickson. 2000. *Endangered Species, Threatened Convention: The Past, Present and Future of CITES*. London: Earthscan.

Johannsdottir, A., I. Creswell, and P. Bridgwater. 2010. "The Current Framework for International Governance of Biodiversity: Is It Doing More Harm than Good?" *Review of European Community & International Environmental Law*, 19: 139–149.

Jolly, R. 2003. "Global Goals: The UN Experience." United Nations Development Programme Background Paper Human Development Report. New York: UN Human Development Report Office.

Koskenniemi, M. and M. Leino. 2002. "Fragmentation of International Law? Postmodern Anxieties." *Leiden Journal of International Law*, 15: 553–579.

Lamy, P. 2006. "The Place of the WTO and Its Law in the International Legal Order." *European Journal of International Law*, 717(5): 969–984.

Laurance, W.F. and G.B. Williamson. 2001. "Positive Feed-backs among Forest Fragmentation, Drought, and Climate Change in the Amazon." *Conservation Biology*, 15: 1529–1535.

Lovelock, J. 1995. *Ages of Gaia*. Oxford: Oxford University Press.

McConnell, F. 1996. *The Biodiversity Convention: A Negotiating History*. London: Kluwer Law International.

Millennium Ecosystem Assessment. 2005. *Ecosystems and Human Well-Being*. Washington, DC: Island Press.

Nagoya Protocol on Access to Genetic Resources and the Fair and Equitable Sharing of Benefits Arising from Their Utilization to the Convention on Biological Diversity. 2010. Conference of the Parties to the Convention on Biological Diversity 10th Meeting, Nagoya, Japan, October 18–29, 2010, Agenda Item 3, http://www.cbd.int/abs/text/ (accessed November 9, 2012).

Opdam, P. and D. Wascher. 2004. "Climate Change Meets Habitat Fragmentation: Linking Landscape and Biogeographical Scale Levels in Research and Conservation." *Biological Conservation*, 117(3): 285–297.

Pounds, J.A. 1999. "Biological Response to Climate Change on a Tropical Mountain." *Nature*, 398: 611–615.

Rands, M.R.W., W.M. Adams, L. Bennum *et al*. 2010. "Biodiversity Conservation: Challenges beyond 2010." *Science*, 329: 1298–1303.

Robinson, J.G. 2006. "Conservation Biology and Real-World Conservation." *Conservation Biology*, 20(3): 658–669.

Sand, P. 1993. "International Law after Rio." *European Journal of International Law*, 4: 377–389.

Saraux, C., C. Le Bohec, J.M. Durant *et al*. 2011. "Reliability of Flipper-Banded Penguins as Indicators of Climate Change." *Nature*, 469: 203–206.

Sutherland, W.J., W.M. Adams, R.B. Aronson *et al*. 2009. "One Hundred Questions of Importance to the Conservation of Global Biological Diversity." *Conservation Biology*, 23(3): 557–567.

Trouwborst, A. 2009. "International Nature Conservation Law and the Adaptation of Biodiversity to Climate Change: A Mismatch?" *Journal of Environmental Law*, 21(3): 419–442.

Underdal, A. 2010. "Complexity and Challenge of Long-Term Environmental Governments." *Global Environmental Change*, 20: 386–393.

UNEP (United Nations Environmental Programme). 2010. *The State of the Planet's Biodiversity*, http://www.unep.org/wed/2010/english/biodiversity.asp#trends (accessed May 1, 2012).

UNEP/CBD/COP/9/14/Add.1. 2009. "Updating and Revision of the Strategic Plan: Note by the Executive Secretary," www.cbd.int/doc/meetings/cop/cop-09/.../cop-09-14-add1-en.doc (accessed May 1, 2012).

Vidal, J. 2010. "The Real Butterfly Effect: Destroying Nature Will Ruin Economies and Cultures, Pleads UN." *The Guardian* (London) August 17, 4.

Walpole, M., R.E.A. Almond, C. Besançon *et al*. 2009. "Tracking Progress toward the 2010 Biodiversity Target and Beyond." *Science*, 325: 1503.

Walther, G., E. Post, P. Convey *et al*. 2002. "Ecological Responses to Recent Climate Change." *Nature*, 416: 389–395.

Weil, P. 1983. "Towards Relative Normativity in International Law?" *American Journal of International Law*, 11: 413–437.

Wilson, E.O. 1992. *The Diversity of Life*. London: Penguin Books.

WTO (World Trade Organization). 1994a. "The Decision on Trade and Environment Adopted by Ministers at the Meeting of the Uruguay Round Trade Negotiations Committee in Marrakesh on April 14."

WTO (World Trade Organization). 1994b. "Understanding on Rules and Procedures Governing the Settlement of Disputes adopted by Adopted by Ministers at the Meeting of the Uruguay Round Trade Negotiations Committee in Marrakesh on April 14."

WTO (World Trade Organization). 1994c. "Agreement Establishing the World Trade Organization Adopted by Ministers at the Meeting of the Uruguay Round Trade Negotiations Committee in Marrakesh on April 14."

WTO (World Trade Organization). 1998. "United States: Import Prohibition of Certain Shrimp and Shrimp Products," WT/DS58/AB/R.

WTO (World Trade Organization.) 2012. "About the WTO: A Statement by the Director-General," http://www.wto.org/english/thewto_e/whatis_e/wto_dg_stat_e.htm (accessed May 1, 2012).

Chapter 4

Marine Environment Protection

Markus Salomon

Introduction

Oceans and seas have long been recognized as huge, unknown, and unexploited natural habitats which can provide endless resources, especially as food. Today we know that our marine resources are not infinite; this is true even for surface areas. Furthermore, human activities are responsible for severe pressures and impacts on the marine environment. The most important direct threats stem from fisheries, shipping, the oil and gas industry, mineral extraction, mariculture, maritime construction work, and tourism, as well as land-based activities such as agriculture. All the activities mentioned are a threat to marine biodiversity. Anthropogenic impacts and pressures lead to eutrophication, accumulation of toxic substances in marine animals at the end of the food chain, degradation of benthic habitats, and marked decline in number and size of exploited large marine animals. Climate change further adds to the mentioned diverse pressures, altering water temperatures, sea levels, and the pH levels of marine waters.

It is already documented that about 90% of the biomass of exploited large fish species and mammals has been lost in comparison to their historic levels. Human activities have already eliminated approximately 65% of seagrass and wetland habitats in temperate zones (Census of Marine Life International Secretariat: Consortium for Ocean Leadership 2011). The coral reefs are at high risk of extinction. The main reasons are the emission of carbon dioxide and the effects of acidification (Pandolfi *et al.* 2003).

There is no doubt that the protection and sustainable management of the oceans and seas are becoming more and more important, especially with the increase in the exploitation and uses of marine resources and therefore of pressures and impacts. Shipping and fisheries are two of the main sectors that are causing harm to the

The Handbook of Global Climate and Environment Policy, First Edition. Edited by Robert Falkner.
© 2013 John Wiley & Sons, Ltd. Published 2013 by John Wiley & Sons, Ltd.

marine environment. This chapter offers an overview of international policies and regulations for the protection of the marine environment from shipping and fishing activities. The European fisheries policy is brought up as an example for a regional approach in managing fish stocks.

The Marine Environment

It is common knowledge that we live on a blue planet. Seventy-one percent of the Earth's surface is covered by salt water. There are on the one hand the five oceans – Pacific, Atlantic, Indian, Antarctic, and Arctic – and on the other hand the marginal seas like the Mediterranean, Baltic, and Bering. The oceans and most of the seas are connected to one another. More than half of the oceans' area is over 3000 m deep.

The oceans are an important part of the biosphere, but only about 5% of them have been systematically explored. Approximately 250 000 marine species excluding microbes have been described by scientists to date, but it is suggested that the total number of species could be at least four times higher (Census of Marine Life International Secretariat: Consortium for Ocean Leadership 2011). The marine environment is usually divided into the benthic (of the sea bottom) and pelagic (open water) zone, which can be further subdivided.

The greater part of human activities in connection with the marine environment take place near the coast or, better said, on the continental shelf. In these shallow waters of up to 200 m in depth, the productivity is generally higher than in offshore waters because of the permanent nutrient input from land and recycling runoff. High productivity means also high biomass of marine species. This is one reason why most fishery activities take place near the coast. Furthermore, these areas are also the ones with the highest biodiversity (Tittensor *et al.* 2010) (see Figure 4.1). Habitats of high ecological value are found in these shallow waters, such as mangrove forests, which are closely interlinked with freshwater and terrestrial ecosystems. The coastal seas can be highly diverse, for example about 133 different marine and coastal habitat types have been described for the Baltic Sea (IOW 2003).

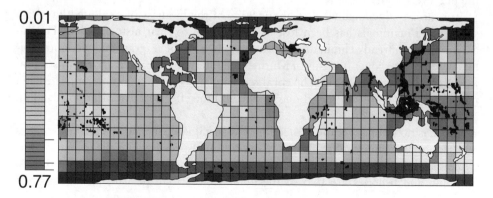

Figure 4.1   Global marine species richness for all taxa.
Source: Tittensor, D.P., C. Mora, W. Jetz *et al.* 2010. "Global Patterns and Predictors of Marine Biodiversity across Taxa." *Nature*, 466: Figure 2b.

The natural conditions for coral reefs in the tropics are different; with a tremendous variety of life, they are one of the most fascinating ecosystems of the world. Reefs are perhaps the most diverse and species-rich areas in the marine environment. This is remarkable because warm-water coral reefs appear in tropical waters that are extremely poor in nutrients. Outside of the reefs productivity is very low. One of the main reasons why reefs are very productive might be that they are able to hold all nutrients in the system and cycle them in a very efficient way, similar to rainforests. Coral reefs are very sensitive to changing conditions, for example in temperature, pH level, salinity, physical disturbance, and sedimentation (Nybakken 1993).

The far regions of the oceans are often called blue deserts because primary production as well as number of species are often very low. These areas are characterized by limited nutrition. The oceanic zone can be divided into the very thin top layer, which receives light and where primary production (growth of microalgae) occurs, and the aphotic zone, the permanently dark water masses. Approximately 90% of the volume of the ocean constitutes the dark, cold area which is called the deep sea (Nybakken 1993). The deep sea is the largest and least known ecosystem on the planet. It harbors a high and fascinating biodiversity which is adapted to extreme conditions like high pressure, low food availability, and permanent darkness.

The blue desert is interrupted by little ocean oases, the seamounts. These are undersea mountains which are usually volcanic in origin. It is estimated that about 1000 seamounts exist in the Atlantic Ocean and 30 000 in the Pacific Ocean. Seamounts are quite important because they are hotspots of marine life. The upwelling of nutrient-rich deepwater at the seamounts results in high productivity. High productivity means a high biomass of phyto- and zooplankton, which attracts a lot of fish. This is why many top predators such as marine mammals and large fish species congregate over seamounts. However, because of the high biomass of fish there, seamounts are therefore also attractive to the fishing industry. Seamounts have been targeted by fishermen since the middle of the twentieth century. The problem with this exploitation is that not much is known about the targeted fish species and stocks. They are often, like the orange roughy (*Hoplostethusatlanticus*), long-lived, slow–growing, and late-maturing species with a low productive potential. Despite their aggregation at seamounts, their overall abundance is actually very low. Therefore, they are quickly fished out, as already documented, and will likely need decades for recovery (Morato n.d.).

The marine ecosystems provide a huge number of goods and services to human beings. Benefits people obtain from ecosystems are generally called ecosystem services (ES). In the Millennium Ecosystem Assessment ecosystem services are divided into the following four categories (Reid *et al.* 2005):

1. provisioning services: goods and benefits to people with a clear monetary value;
2. regulating services: for example regulation of climate and control of local rainfall;
3. cultural services: which provide no direct material benefit, like aesthetic beauty of coastal formations;
4. supporting services: which are of no direct benefit to people but essential for the ecosystems, like formation of sediments or nutrition cycling.

Important provisioning services from oceans and seas are food, energy, transportation routes, minerals, recreational areas, and biomedical products. The marine ecosystem also have other services that are of great value: oceans, for instance, are the most important natural sink for anthropogenic carbon dioxide (CO_2) (Sabine *et al.* 2004). Some of the marine ES create direct economic benefits, while the value of others, especially of regulating and supporting services, are generally very difficult to estimate. This is true for example for climate regulation, water purification, and coastal protection.

International Legal Background for the Use of the Marine Environment

The oceans have been subject to the freedom of the seas doctrine since the seventeenth century. This principle limits national rights and jurisdiction over the oceans and seas to a narrow area surrounding a nation's coastline. The remainder of the oceans is free to all and belongs to none. The doctrine was slightly restricted by different conventions and measures, falling within the competences of different global international organizations for marine matters. Important organizations in this respect include the Food and Agriculture Organization of the United Nations (FAO), the International Maritime Organization (IMO), and the United Nations Environment Programme (UNEP), among others.

The most important convention regulating access to the resources of the oceans is the 1982 United Nations Convention on the Law of the Sea (UNCLOS) (United Nations Office for Legal Affairs 2011). UNCLOS entered into force in 1994 and is an important attempt by the international community to regulate all aspects of resources and use of the oceans and the seas. In this way, UNCLOS is more or less delivering a framework with limited concrete obligations but it calls for implementation through other global or regional treaties (Kachel 2006).

The main motivation for the convention was to end rising tensions between nations over conflicting claims to ocean space and resources. The hope was that with more stable governance, conflicts among nations over the use of the different resources of the oceans could be avoided in the future.

The most important aspects of the Law of the Sea treaty are: navigational rights, such as the right of innocent passage through the territorial sea, territorial sea limits, economic jurisdiction, legal status of resources on the seabed beyond the limits of national jurisdiction, conservation and management of living marine resources, a marine research regime, a binding procedure for the settlement of disputes between states, and the protection of the marine environment.

States are in principle free to enforce any law, regulate any use, and exploit any resources in coastal waters up to 12 nautical miles from the national coastline. This sovereignty extends to the air space over the territorial sea as well as to its bed and subsoil. The introduction of the exclusive economic zone (EEZ) was an innovative step made under the Law of the Sea treaty. The EEZ is a seazone over which a state has special rights for the exploration of marine resources. It stretches from the seaward edge of the state's territorial sea out to 200 nautical miles from its coast or even to the continental shelf beyond the 200-mile limit. In the EEZ third states enjoy freedom of navigation and have the right of other uses of the sea that are in line with international agreements. The reason for establishing the EEZ was the increasing interests in exploiting offshore resources like fish, oil, or gas but also concerns over the threat of pollution and wastes from shipping, especially from oil tankers (United Nations Office for Legal Affairs 2011).

International Shipping

Transportation of goods and products is essential for international trade. It is the bloodstream of the global economy. And as all countries are more or less surrounded by the sea, most of the goods are transported by ship. Furthermore, the costs for transporting goods and bulk materials on ships are the lowest compared to all other forms of commercial transportation and they produce the lowest CO_2 emissions. The establishment of a global system of trade was highly dependent on the development of shipping. About 90% of global trade today is done by sea. There was a steady and considerable increase in goods loaded aboard ship between the 1970s and 2008; total cargoes loaded on ship went up from 2.6 billion t. to 8.2 billion t. (Figure 4.2). This development was temporarily hampered by the financial crisis in 2009 that was accompanied by a decline in the volume of global merchandise trade (IMO 2012).

Impacts of Shipping on the Marine Environment

Even though shipping is the most climate-friendly form of commercial transport it is not at all environmentally friendly. Shipping is responsible for a number of pressures

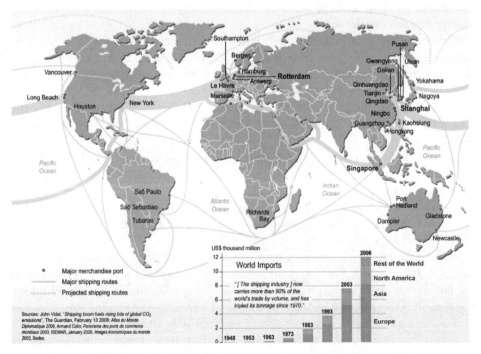

Figure 4.2   The boom in shipping trade.
Source: UNEP/GRID-Arendal. 2009. "The Boom in Shipping Trade." UNEP/GRID-Arendal Maps and Graphics Library 2, http://maps.grida.no/go/graphic/the-boom-in-shipping-trade1.

and risks to the seas and oceans: discharges of hazardous substances such as oil, atmospheric emissions, noise pollution, discharge and disposal of sewage and litter, introduction of non-indigenous species, and risk of oil spills. Most of the threats from shipping are not well monitored.

The release of airborne pollutants by ships has harmful impacts on the ecosystem and the atmosphere. Of key importance for the marine environment are sulfur dioxide (SO_2) and nitrogen oxide (NO_x) emissions that result in the acidification and eutrophication of sea water. Heavy oil or bunker oil are used as fuel in ships. These oils have a high sulphur content of around 3%, which leads to high SO_2 emissions. Another problem is the emission of particulate matter, which not only affects the climate but also human health. The most important greenhouse gas emitted by shipping is CO_2. It is estimated that shipping is responsible for 3.3% of global emissions of CO_2 (Buhaug *et al.* 2009). A large proportion of emissions are released near the coast. As a consequence, atmospheric emissions from ships have created a serious air-quality problem in some port cities (EMSA n.d.).

Shipping is an important source of oil inputs in the sea. Oil enters the marine environment via incidental, operational, illegal discharges and accidental oil spills (OSPAR Commission 2010). Oils and their components can damage the marine ecosystems in a variety of ways, such as toxic effects on organisms that ingest oil components. An important indicator for chronic oil pollution is the occurrence of oiled seabirds at beaches. Major sources for oil pollution are not accidental oil spills but illegal discharges, especially of residues from fuel preparation and tank-wash water. In 2009, for example, 173 oil spills were observed in the North Sea and Baltic Sea through aerial surveillance, with a high density of occurrence near the main shipping routes (Figure 4.3).

Other harmful substances in the marine environment stemming from shipping are antifouling agents. A well-known example is tributyltin (TBT), a highly toxic substance that affects the endocrine system of mollusk species in particular. With the global ban on TBT the release of this antifouling agent from ships' hulls is expected to cease, while such agents as copper are expected to increase. It is nevertheless still necessary to foster the development of less toxic substitutes (OSPAR Commission 2010).

It is acknowledged that shipping is a major source of marine litter (UNEP 2009). The illegal disposal of waste, especially plastic, is a threat to marine life. The ingestion of all kinds of plastic debris by water-surface-feeding seabirds such as the fulmar that leads to a variety of negative health effects is well documented (Mallory 2008). Illegal sewage discharges near the coast pose a problem for water quality and are related to the eutrophication of the seas.

Ship traffic has been shown to be a major source of low-frequency noise in the sea, especially in coastal waters. A strong increase in shipping noise was estimated in the last decades with the growing ship traffic. Noise pollution can cause behavioral changes in cetaceans, for instance abandoning breeding and feeding areas, and in extreme cases can lead to physical damage or stranding and death (IFAW 2008).

Furthermore, shipping is the main vector for introducing non-indigenous species in the seas and oceans. They are usually transported in the ballast water. The main problem is the transportation of species over physical barriers that separate communities and ecosystems. These non-indigenous species can harm native species and

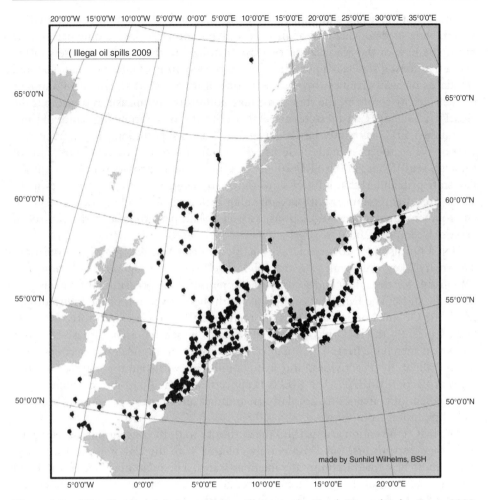

Figure 4.3   Oil spills observed via aerial surveillance in the North Sea and Baltic Sea in 2009. Source: Bonn Agreement. 2010. *Bonn Agreement Aerial Surveillance Programme: Annual Report on Aerial Surveillance for 2009.* London: Bonn Agreement Secretariat, http://www .bonnagreement.org/eng/doc/2009_report_on_aerial_surveillance.pdf.

ecosystems as predators, competitors, parasites, and pathogens. The main negative ecological effect is the loss of region-specific species in the habitats (Gollasch n.d.). However, economic damages have also been documented (Paalvast and van der Velde 2011).

Regulating International Shipping

UNCLOS, among others, provides a framework for the management of shipping worldwide (Khee-Jin Tan 2006). The Law of the Sea treaty is based on the principle that shipping standards concerning construction, equipment, seaworthiness, and manning of ships are primarily the concern of the respective flag states. Ships are subject to the exclusive jurisdiction of the state whose flag they are flying. If requested by another state, a flag state must investigate any and all violations committed by a ship flying its flag. Measures against ships flying foreign flags are only permissible under the Law of the Sea treaty if they are restricted to the territorial waters of the respective coastal states. If a port state detects the violation of international rules and standards on seaworthiness by a vessel in one of its ports that threatens to damage the marine environment, the state must take administrative measures to prevent the vessel from sailing. Coastal states also have authority to monitor compliance of ships within their EEZs with international regulations against pollution. However, coastal states have in the past made little use of this authority to prosecute pollution of the high seas (König 2002; SRU 2004). One of the main problems linked with the flag-state principle is that a lot of ships today are flagged-out to "convenient" flags. These are often countries with lax manning standards and rules, poor safety and environmental protection regulations, as well as bad performance in the control of vessels (DeSombre 2006).

UNCLOS also lays down general obligations for marine environment protection. The treaty's key organization is the IMO. The IMO is the specialized UN agency responsible for the safety and security of shipping and the prevention of marine pollution by ships. The IMO was established with the Convention of the International Maritime Organization, which was adopted in 1948 and came into force 10 years later. Currently the IMO has 170 member-states which represent almost 100% of the world merchant fleet. The organization consists of an assembly, a council for the coordination of activities, and five committees, with one for marine environment protection (MEPC). The MEPC is concerned with adoption, amendment and enforcement of conventions, regulations, and measures for the prevention and control of pollution from ships.

The IMO convention grants IMO the authority to draft conventions, agreements, and other instruments. But IMO is only charged with the elaboration of a treaty, which must be agreed upon by the member-states. The main objective of the IMO is the preparation of the draft instruments and serving as a forum for the states and international organizations during their committee meetings and conferences (Anianova 2006). Besides member-states, non-governmental organizations and intergovernmental organizations are also involved in the activities of the IMO; but only member-states have voting rights.

The main convention for protecting the marine environment from shipping pollution (which IMO is responsible for) is the 1973/1978 International Convention for the Prevention of Pollution from Ships (MARPOL) (Nauke and Holland 1992). The MARPOL convention with its thematic Annexes I to VI regulates all kinds of marine pollution from ships: the emission, release, and discharge of oil; noxious liquid substances carried in bulk; harmful substances carried by sea in packaged form; sewage; garbage; and air pollution.

One recent success story for IMO's work was the amendment of Annex VI of the MARPOL convention for the prevention of air pollution in 2008 to further reduce harmful emissions from ships. This welcomed revision of the annex was surprisingly far reaching. It reduced the global cap on sulfur content in heavy oil used in shipping from 4.5% to 3.5% from January 2012 and to 0.5% by January 2025 at the latest. The sulfur limit in Emission Control Areas is also to be reduced in two steps from the current 1.5% to 0.1% from January 2015. Under the revised Annex VI the use of alternative technologies like exhaust gas cleaning systems is also allowed if they are able to achieve the relevant emission reductions. Another aspect of the revision was the tightening of NO_x emission limits for engines onboard ship. It is expected that the new international limit for sulfur content of marine fuels will significantly reduce emissions of SO_2 (EMSA n.d.).

Spatial rules are an important element in marine environmental protection. The IMO recognized that certain marine areas require a stricter regulatory regime against pollution from ships. For this reason Special Areas and Particularly Sensitive Sea Areas (PSSAs) were introduced under the work of the IMO by a resolution of the MEPC in 1991. PSSAs are areas that need special protection for ecological, socio-economic, or scientific reasons and which may be vulnerable to damage by international maritime activities. One ecological criterion is to be a unique or rare ecosystem. When an area is approved as a PSSA, specific measures can be used to control maritime activities in that area, such as strict application of MARPOL discharge and equipment requirements for ships. To date IMO has adopted 13 PSSAs.

In Special Areas, the adoption of special mandatory methods for the prevention of sea pollution is required due to technical reasons related to their oceanographical and ecological condition and to their sea traffic. Today for example the North Sea and Baltic Sea are designated as SO_x Emission Control Areas (SECAs) where sulfur content in fuel used by ships must be lower than that set under the global cap. The expansion of PSSAs and Special Areas to other marine regions is one option for increasing the protection of highly polluted coastal seas from shipping pressures (Kachel 2008).

Another successful step towards reducing pollution from shipping at international level was the adoption of the Convention on the Prevention of Marine Pollution by Dumping of Wastes and Other Matter (London Convention, LC) from 1972 and of the 1996 London Protocol of the convention (Nauke and Holland 1992). Both convention and protocol in general ban dumping of waste and other matters into the sea, with some exceptions, for example dredged materials and sewage sludge (IMO n.d.).

An important part of IMO's work deals with the safety of ships, which is highly relevant for the protection of the marine environment, particularly the prevention of harmful substance release due to shipping accidents. Some important IMO conventions in this context are: the 1978 International Convention on Standards of Training, Certification, and Watchkeeping for Seafarers (STCW) as amended, the 1972 Convention on the International Regulations for Preventing Collisions at Sea (COLREG), and the 1965 International Convention Relating to Intervention on the High Seas in Cases of Oil Pollution Casualties (INTERVENTION). In this respect Port State Control and the classification of ships are very important issues. Port State Control is the inspection of foreign ships carried out at national ports to verify that the condition of the ship and its equipment as well as its handling comply with the requirements of international regulations.

The IMO has encouraged the establishment of regional Port State Control organizations and agreements, for example the Paris Memorandum of Understanding on Port State Control (Paris MoU), which is applicable to Europe. This Memorandum signed by 27 states requires shipping authorities to inspect ships entering their ports. In 2010 alone, over 24 000 inspections had been performed in European ports (Paris

MoU 2011). In May 2011, the Paris MoU adopted a new inspection regime (Paris MoU 2010). The goal of this welcomed revision is to implement a more risk-based targeting mechanism for ship inspection. With the new inspection regime, the obligation for an inspection is dependent on the risk profile of the ship and its last inspection in one of the Paris MoU member-states. This system intends to reward quality shipping on the one hand and to concentrate efforts on high-risk ships on the other hand. Furthermore, flag states are categorized according to their performance, which influences the profiling of ships under their flags. The Paris MoU published annually for a number of years so-called White/Grey/Black Lists of flag states. The flag-state administrations are ranked according to their ships' performance in Port State Controls over the last three years. On the White List are flag states with a very good performance and on the Black List are flag states with a very bad performance. This instrument works according to the principle of "blaming and shaming" and seems to be quite successful. In recent years, there was a steady increase in ships on the White List and a steady decrease of ships on the Black List (Paris MoU 2011).

Some 50 international conventions and numerous soft-law instruments have until recently been initiated by and adopted under the IMO. There are some success stories for clean shipping, aside from the one already mentioned, as well as the phasing out of single-hull tankers and TBT as an antifouling agent. Some developments are still very slow, such as reducing air pollution from ships, particularly nitrogen oxides, CO_2 and particulate matter. One of the main critiques of the work done under IMO is that the law-making process is very slow (Anianova 2006). But this is not really the fault of the institution; rather it is a weakness in the international decision-making procedure. Generally, the IMO members try to make decisions by consensus, and it is also a challenge to adapt instruments or conventions unanimously. For example, the debate about making double hulls for tankers mandatory took up to 20 years; the convention for this task was adopted in 1992 and the complete global phasing out of single-hull tankers will go on until 2015.

Another problem is the enforcement of international standards. The effective implementation and enforcement of international standards for clean shipping are primarily the responsibility of the flag states. There are still numerous low-standard vessels sailing the oceans. One reason is that many states are not willing or are not able to guarantee compliance with international standards. Reasons for the latter are the lack of financial capacity or technical expertise, and conflicts between national and international regulations (Knudsen and Hassler 2011).

In general the chance of single states regulating shipping is low given the international nature of shipping. Action is restricted to a few measures in territorial waters. The priority given to the flag-state principle in the Law of the Sea treaty is another reason that has prevented regional decision-making bodies from setting regional protection standards (SRU 2004: paragraph 506). Nevertheless, Europe and the United States have acted as drivers of global developments of more stringent shipping provisions in the past (SRU 2004; EPA 2011).

Fisheries

Fish are on the one hand an important food resource but on the other hand a central part of the marine ecosystem. Fishing is one of the oldest human activities exploiting living resources. It is an important industry and source of income and livelihood for millions of people, especially coastal communities, to date. While there is still a constant increase in employment in the fishing industry in some parts of the world, it is clearly a downward trend in industrialized continents like Europe and North America. Several factors are responsible for the latter; one is the decrease in catches. In 2009 about 80 million t. of fish was captured in the seas and oceans. Global production of marine-capture fisheries reached a peak of 86.3 million t. in 1996 and then declined slightly to 79.9 million t. in 2009 (FAO 2010).

Impacts of Fisheries on the Marine Environment

Fishing is responsible for considerable pressures on the marine environment. The main challenges of fishing are at present the overexploitation of fish stocks, the discard problem, and the destruction of benthic habitats. The consequences are depletion of main predator and prey species, threat to sea birds and marine mammal populations, impacts on the marine food web, loss of sensitive benthic communities, but also, let us not forget, loss of income and the destruction of important food resources.

According to the last status report on world fisheries and aquaculture from the FAO, 32% of the world fish stocks were overexploited, depleted, or recovering from overexploitation (FAO 2010) (Figure 4.4). Since the mid-1970s, there has been a steady increase in fish stocks, which have suffered from excessive fishing pressure. During the same period, a steady decrease in stocks considered as underexploited or moderately exploited was observed. In 2008 this was true for only 18% of the monitored fish stocks and 53% of the stocks were fully exploited, meaning that no room for further expansion was available.

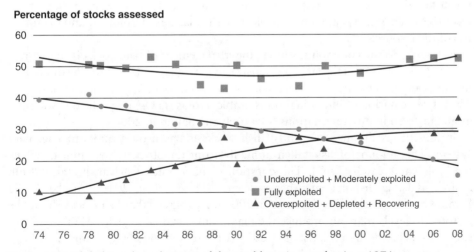

Figure 4.4 Global trends in the state of the world marine stocks since 1974.
Source: FAO (Food and Agriculture Organization of the United Nations). 2010. *The State of World Fisheries and Aquaculture 2010*. Rome: FAO.

Another problem of present fishing practices is the high quantities of fish and other species which are caught unintentionally (bycatch) and discarded, and consequently wasted. Fish are thrown overboard for several reasons: some fish have very low commercial value, fishermen lack quotas for the species, or there are length constraints. Most of the discarded fish die because they are badly injured, which becomes a loss for stock productivity and the ecosystem. It is estimated that the worldwide weighted discard rate of fish is at 8% (Kelleher 2005), which seems to be pretty low. But discard rates in some fisheries, particularly bottom trawling of flatfish, is much higher, at up to 90% of the caught species (STECF 2010). It is important to note that it is generally very difficult to monitor discard. Bycatch and discard can therefore be a direct threat for species diversity. For example, the bycatch of harbor porpoises in gillnets in the Baltic Sea is a direct threat to the population, which is at high risk of extinction (HELCOM 2010). However, there is also evidence of reduction in discards in recent years due to the increased utilization of catches, more selective fishing gears, and introduction of bycatch and discard regulations (Kelleher 2005).

Fishing activities can also harm the marine ecosystem directly. Bottom trawls are responsible for causing severe damage to the benthic habitats. For example tickler chains used in beam trawling dig several centimeters into the sea-bed and plow through the sediment or benthic communities on hard substrates. Sessile organisms or coral reefs in particular are very sensitive to bottom trawling. Recovery from one bottom trawling over a cold-water coral reef can take up to hundreds of years (Institute for Marine Research in Bergen n.d.).

Lost fishing gear can also harm the marine environment, as seen in ghost fishing or the entangling of marine species in lost nets or net fragments.

International and Regional Activities and Regulations for Fisheries

Instruments and Measures   There exist a number of instruments and measures for sustainable fisheries management. The first step is to limit the exploitation of fish resources to an extent that is compatible with the marine environment and the future use of the resource.

Fish are a classic common resource; the management of fish stocks is therefore a special challenge. Often the lack of exclusive access rights is blamed for the difficulties in implementing a fisheries management system that is focused on long-term interests. The implementation of individual transferable quotas (ITQs) might be one solution, as already seen in different nations (Hentrich and Salomon 2006).

In most cases the management of fish stocks is based on a quota-management system. The amount of fish taken annually from the stock is determined by the total allowable catch (TAC). The question of how much fish can be taken from a stock is being disputed in fisheries science (Froese *et al.* 2010). At the World Summit on Sustainable Development in 2002 it was agreed that fish stocks should be restored to levels that can produce the maximum sustainable yield (MSY), and this should be achieved by no later than 2015 (United Nations 2002). Simplified, MSY is theoretically the largest yield or catch that can be taken from a fish stock over an indefinite period. The concept of MSY aims to maintain the fish stock size at the point of maximum growth rate by harvesting only the surplus production. Fish stocks adapt to fishing pressures through an increase in productivity. This increase in production in comparison to natural conditions, called surplus production, can be fished without affecting the size of the stock. MSY is a rather old concept that is criticized for several reasons, for example: it focuses solely on a particular species in question; it ignores the size, age, and reproductive status of the caught individuals, and also ignores a lot of the natural factors influencing the development of a fish stock (Larkin 1977). The concept was permanently modified as a result of the weaknesses mentioned. It is therefore important to take the precautionary approach into account by applying MSY, especially given the many uncertainties caused by incomplete data on fish stocks (SRU 2011). Furthermore, fish species are often dependent on one another, for example in predator–prey or competitive relationships and should, therefore, be managed using multi-species approaches.

Besides proper management of commercial fish stocks, measures are necessary for protecting non-target species and the marine ecosystems from fishing activities, such as reducing bycatch. The strongest incentive for fishermen to take efforts to reduce bycatch is a discard ban, that is, the obligation to land the entire catch. Other options are economic incentives, guidelines, or obligatory standards for the use of less destructive fishing technologies. Important instruments for the protection of marine biodiversity, especially sensitive ecosystems, from impacts of fisheries are marine protected areas (MPAs).

Regulations and Standards Sea fisheries have international characteristics because fishing activities are not restricted to marine areas under national jurisdiction, although most fish worldwide are captured in coastal waters or in the EEZ. As mentioned earlier, the Law of the Sea treaty gives coastal states exclusive authority over living marine resources in coastal waters and the EEZ. But the states are obliged to adopt proper conservation and management measures to ensure sustainable use of the fish stocks and to avoid overexploitation of the living resources. Therefore, best available scientific evidence should be taken into account. The target mentioned in UNCLOS for the management of the fish stocks is MSY.

Article 62 of UNCLOS contains some conservation measures and instruments – for example licensing of fishermen and vessels, fisheries research programs, catch quotas, temporal and spatial fishing bans, gear restrictions, vessel position reporting, enforcement procedures, and minimum fish sizes – which coastal states may establish. If a coastal state does not have the capacity to harvest the total allowable catch, it could give other states access to the surplus of the allowable catch. But these states must comply with the conservation measures and with the other terms and conditions established in the laws and regulations of the coastal state.

Non-target species and the protection of the marine ecosystem are also addressed in UNCLOS. According to Article 61(4) with respect to conservation and management measures, "the coastal states shall take into consideration the effects on species associated with or dependent upon harvested species." There are also general obligations to protect the marine environment, as well as rare and fragile ecosystems, and the habitats of depleted, threatened, or endangered species; the latter applies for example to marine mammals.

Fish stocks are not confined within national borders. A lot of stocks are highly mobile, which means they migrate long distances and are abundant, at least temporarily, in the high seas. For the regulation of the management of these living resources, the Straddling Fish Stocks Agreement (SFSA, http://www.un.org/depts/los/convention_agreements/convention_overview_fish_stocks.htm) was adopted in 1995. The agreement is closely linked to UNCLOS and its objective is to ensure the long-term conservation and sustainable use of straddling and highly migratory fish stocks. The SFSA provides a framework for cooperation in the conservation and management of those resources to ensure that measures taken in areas under national jurisdiction and in the adjacent high seas are compatible and coherent. In some instances the agreement goes beyond the UNCLOS provisions. For example, the conservation and management of relevant fish stocks should be based on the precautionary approach. Other aspects agreed upon by the signatory states are: assessment of impacts of fishing on target stocks and the affected marine environment, adoption of conservation and management measures for non-target species, minimization of bycatches, protection of marine biodiversity, and performance of effective monitoring and controls. Coastal states are obliged to set precautionary reference points to maintain populations at levels that can produce MSY and to take conservation and management measures to prevent fish stocks from falling below a limit reference point or to facilitate stock recovery. The SFSA was signed by 59 nations as of September 2011.

Another approach in regulating the use of marine biological resources in international waters is the introduction of Regional Fisheries Bodies (RFBs) (FAO 2009). RFBs are international bodies formed by countries with fishing interests in the same area to foster cooperation towards conservation, management, and development of the living marine resources. Some of them manage all fish stocks abundant in a specific area, while others focus on specific highly migratory species, like tuna, through vast geographical areas. The first regional fisheries body had in fact already been introduced in 1948 with the Indo-Pacific Fisheries Council. RFBs are based on an agreement or an arrangement between different states. The function and the mandate of the different RFBs vary to a great extent. Some RFBs have a purely advisory mandate; their decisions are consequently not binding for their members. Other bodies adopt fisheries conservation and management measures that have binding characteristics. Their duties are therefore different. In some cases it is just the collection, analysis, and dissemination of information and data; in other cases concrete decisions for stock management and conservation are made. If they have a clear management mandate, they are called Regional Fisheries Management Organizations (RFMOs). Currently, 44 RFBs exist worldwide, and 20 of them are management bodies. The status of tuna stocks shows quite clearly that RFBs are not as successful as they should be, even though several attempts had been made in the past to improve their performance. Reasons for this are: implementation lag of management decisions, egoism of national states hindering good fisheries governance, members' unwillingness to fund research, difficulties in consensus decision-making, and focus on crisis management instead of long-term management (FAO 2009).

The FAO adopted a Code of Conduct for Responsible Fisheries in 1995 to meet international management obligations and in response to the increasing problems in world fisheries (http://www.fao.org/docrep/005/v9878e/v9878e00.htm#7). This recommendation is not legally binding. However, it contains a range of principles and provisions for the management of fish stocks and the protection of the marine environment from fishing activities, including long-term sustainability of fishery resources, conservation and protection of aquatic biodiversity and endangered species, recovery of depleted fish stocks, and minimization of adverse impacts from fisheries on the target stocks and the remaining ecosystem. In the meantime a number of further FAO guidelines on fisheries such as the 2008 International Guidelines for the Management of Deep-Sea Fisheries in the High Seas have been adopted (FAO 2009).

Overall, the international law of fisheries provides a number of obligations and provisions for sustainable fisheries management and the protection of marine life from fishing activities. The major problem to date is the absence and incoherent implementation and enforcement of international or regional standards at the national level. The consequences are inadequate control regimes, management systems that force fishermen to compete for their share instead of supporting sustainable fish stocks management, unsustainable subsidies, and defiance of scientific advice (Markowski 2009; Mora *et al.* 2009; SRU 2011).

Europe as an Example of a Regional Approach   The European nations started to cooperate in the field of fisheries back in the 1970s. The main concern in those days was to prevent conflicts between nations by exploring living marine resources. It was also the time when a lot of coastal states were extending their territorial waters and before EEZs were introduced. The Common Fisheries Policy (CFP) was only formally created in 1984 (Markus 2009).

Today, the EU has the exclusive competence for the management and conservation of marine living resources in sovereign waters (within the 12-mile zone) and in common waters under national jurisdiction (EEZ). But the Community has redelegated some power to the member-states, for example they are allowed to take emergency measures for the protection of living aquatic resources (Markus 2009). The aims of the CFP are to give member-states equal access to common waters, excluding coastal waters in the 12-mile zone; to increase fisheries productivity; to provide acceptable livelihood for fisheries-dependent communities; and to stabilize the market and secure supply. The four central elements of the CFP are:

1. rules for the management and conservation of fisheries resources;
2. common structural policy;
3. common organization of the market for fisheries products;
4. the external dimension of the CFP or fishing agreements with third countries.

One objective of the CFP is the sustainable exploitation of living aquatic resources. Therefore, the community is limiting fishing opportunities by adopting total allowable catches (TACs), effort limitations, and technical measures. By requiring the use of less destructive fishing gear and prohibiting fishing in sensitive marine areas, environmental concerns are therefore taken into consideration.

The core management instrument for the regulation of fishing effort is the regular setting of TACs for the entire EU, which must be based particularly on the scientific recommendations of the International Council for the Exploration of the Sea (ICES). One of the main shortcomings of the CFP was that the scientific advice of the ICES had been regularly ignored. Accordingly, the TACs exceeded the scientific advice on average by more than 40% (European Commission 2011). Consequently, two-thirds of the European fish stocks are still overexploited. The responsibility for setting the TACs and most of the other decisions under the CFP lies in the hands of the council of the fisheries minister of the member-states. In order to improve the management of the fish stocks, the EU has most recently started to draw up multiannual management plans. These management plans have helped to improve the status of some important stocks (SRU 2011).

An important aspect of sustainable fishing management is the adaptation of fishing fleets to the sustainable use of the available resources. This is the intention of the structural policy of the CFP. Different instruments had therefore been implemented in the past, for example an entry–exit regime (new entries into the fleet must be compensated for by the exit of 1.35 times the amount of capacity), scrapping premiums, and withdrawal of subsidies for vessel modernization. Unfortunately, the measures taken had only minor successes. One of the reasons was the unwillingness of some member-states to reduce their fishing fleet. Another was the existence of ongoing direct and indirect subsidies supporting overcapacities (SRU 2011).

The common organization of the European markets in fishery and aquaculture encompasses quality and marketing standards, as well as requirements on minimum level of consumer information. Further, it establishes producer organizations in order to improve the sale conditions of members' products. It also includes intervention mechanisms for guaranteeing a minimum price for fish products. Another aspect is the building up of a trade system with third countries (Markus 2009).

The EU has bilateral and multilateral fisheries agreements with third countries. Today, more than a quarter of the fish caught by the European fleet comes from waters outside of the EU. Partnership agreements with developing countries in particular are under criticism for being responsible for the overexploitation of marine living resources in external waters.

The CFP is currently under reform to improve its sustainability (SRU 2011).

Conclusions: Special Challenges in International Marine Protecting Policies

Only if we are able to protect marine biodiversity and use marine resources in a sustainable way will future generations have the chance to profit from all services provided by the marine ecosystems. The example of shipping and fisheries shows clearly that the implementation of marine environmental protection instruments and measures in relevant sectors is still a great challenge. Although there are already some successful international initiatives for the greening of shipping, developments at the international level to protect marine waters from impacts of sea transport are still very slow. For example, ambitious emission-control standards for particulate matter and CO_2 and effective control systems to prevent inputs of litter and oil residues are urgently needed. One possibility of speeding up these processes at the international level lies in leadership initiatives taken by important actors, such as the USA or Europe.

International obligations for sustainable fisheries management exist, but they are too weak to enforce national implementation. Some coastal states such as Iceland and Norway are successful in managing their marine living resources; Europe is on its way to improving its fisheries policy. These approaches might be good examples for others. One prime reason for the poor compliance of national regulations with international marine protection standards is the focus on short-term economic and political interests. In order to break this dominance, it is necessary to give the consumers and the public, who own the marine ecosystems and the living resources, more say in the decision process.

One way of enforcing the implementation of internationally agreed decisions might be for coastal states or regions to develop an integrated maritime policy with ambitious targets and goals for the conservation and protection of their marine waters. Strategies for the protection of the marine ecosystems are an important part of this policy approach. This is one reason why the EU adopted a marine strategy framework directive (MSRL) in 2008 as the environmental pillar of the European maritime policy. With the MSRL the European coastal states are for the first time obliged to take actions in all fields of marine environmental protection. An important part of this obligation is the setting of targets for good environmental status of the marine waters concerned and developing programs for measures (Salomon 2009). Important instruments that can assist this process are marine spatial planning and MPAs, as well as concepts for the implementation of the ecosystem approach in the management of marine resources. However, the basic requirement is the improvement of our knowledge of marine ecosystems and the threats human activities present.

Finally, it would be highly desirable to have a more in-depth public discussion and understanding of our oceans and seas, their sustainable use as well as their marine ecosystem services.

References

Anianova, E. 2006. "The International Maritme Organization (IMO): Tanker or Speedboat?" In *International Maritime Organisations and their Contribution towards a Sustainable Marine Development*, ed. P. Ehlers and R. Lagoni, 77–104. Hamburg: Transaction Publishers.

Bonn Agreement. 2010. *Bonn Agreement Aerial Surveillance Programme: Annual Report on Aerial Surveillance for 2009*. London: Bonn Agreement Secretariat, http://www.bonnagreement.org/eng/doc/2009_report_on_aerial_surveillance.pdf (accessed October 16, 2012).

Buhaug, Ø., J.J. Corbett, Ø. Endresen *et al.* 2009. *Second IMO GHG Study 2009*. London: International Maritime Organization.

Census of Marine Life International Secretariat: Consortium for Ocean Leadership. 2011. *Scientific Results to Support the Sustainable Use and Conservation of Marine Life: A Summary of the Census of Marine Life for Decision Makers*. New York: Census of Marine Life International Secretariat.

DeSombre, E.R. 2006. *Flagging Standards: Globilization and Environmental, Safety, and Labor Regulations at Sea*. Cambridge, MA: MIT Press.

EMSA (European Maritime Safety Agency). n.d. "Air Emissions: General Background," http://emsa.europa.eu/implementation-tasks/environment/air-pollution/149-air-pollution/532-air-emissions-general-background.html (accessed December 1, 2011).

EPA (US Environmental Protection Agency). 2011. "Oil Pollution Act Overview," http://www.epa.gov/osweroe1/content/lawsregs/opaover.htm (accessed December 7, 2011).

European Commission. 2011. "Communication from the Commission Concerning a Consultation on Fishing Opportunities." Brussels: European Commission.

FAO (Food and Agriculture Organization of the United Nations). 2009. *The State of World Fisheries and Aquaculture*. Rome: FAO.

FAO (Food and Agriculture Organization of the United Nations). 2010. *The State of World Fisheries and Aquaculture 2010*. Rome: FAO.

Froése, R., T.A. Branch, A. Proelß *et al.* 2010. "Generic Harvest Control Rules for European Fisheries." *Fish and Fisheries*, 12(3): 340–351.

Gollasch, S. n.d. "Aliens Invade the Sea," http://www.ices.dk/marineworld/alienspecies.asp (accessed October 16, 2012).

HELCOM (Helsinki Commission). 2010. *Maritime Activities in the Baltic Sea. An Integrated Thematic Assessment on Maritime Activities and Response to Pollution at Sea in the Baltic Sea Region, vol. 123, Baltic Sea Environment Proceedings*. Helsinki: HELCOM.

Hentrich, Steffen and Markus Salomon. 2006. "Flexible Management of Fishing Rights and A Sustainable Fisheries Industry in Europe." *Marine Policy*, 30(6): 712–720.

IFAW (International Fund for Animal Welfare). 2008. *Ocean Noise: Turn It Down: A Report on Ocean Noise Pollution*. Yarmouth Port, MA: IFAW, http://www.ifaw.org/sites/default/files/Ocean%20Noise%20Pollution%20Report.pdf (accessed October 16, 2012).

IMO (International Maritime Organization). 2012. *International Shipping Facts and Figures: Information Resources on Trade, Safety, Security, and the Environment*. London: IMO Maritime Knowledge Centre.

IMO (International Maritime Organization). n.d. "Convention on the Prevention of Marine Pollution by Dumping of Wastes and Other Matter," http://www.imo.org/About/Conventions/ListOfConventions/Pages/Convention-on-the-Prevention-of-Marine-Pollution-by-Dumping-of-Wastes-and-Other-Matter.aspx (accessed December 9, 2011).

Institute for Marine Research in Bergen. n.d. "Coral Reefs in Norway: Fishery Impact," http://www.imr.no/coral/fishery_impact.php (accessed December 12, 2011).

IOW (Baltic Sea Research Institute Warnemünde). 2003. *Baltic Sea System Study. Final Report*, http://www2008.io-warnemuende.de/Projects/Basys/reports/final/en_home.htm (accessed October 16, 2012).

Kachel, M.J. 2006. "Competencies of International Maritime Organisaitions to Establish Rules and Standards." In *International Maritime Organisations and their Contribution towards a Sustainable Marine Development*, ed. P. Ehler and R. Lagoni, 21–52. Hamburg: Transaction Publishers.

Kachel, M.J. 2008. *Particularly Sensitive Sea Areas: The IMO's role in Protecting Vulnerable Marine Areas*. Berlin and Heidelberg: Springer-Verlag.

Kelleher, K. 2005. *Discards in the World's Marine Fisheries: An Update*. FAO Fisheries Technical Paper 470. Rome: FAO.

Khee-Jin Tan, A. 2006. *Vessel-Source Marine Pollution: The Law and Politics of International Regulation*. New York: Cambridge University Press.

Knudsen, Olav F. and Björn Hassler. 2011. "IMO Legislation and Its Implementation: Accident Risk, Vessel Deficiencies and National Administrative Practices." *Marine Policy*, 35(2): 201–207.

König, D. 2002. "Port State Control: An Assessment of Europena Practice." In *Marine Issues from a Scientific Political and Legal Perspective*, ed. P. Ehlers, E. Mann-Borgese, and R. Wolfrum, 37–54. The Hague: Kluwer Law International.

Larkin, P.A. 1977. "An Epitaph for the Concept of Maximum Sustained Yield." *Transactions of the American Fisheries Society*, 106(1): 1–11.

Mallory, M.L. 2008. "Marine Plastic Debris in Northern Fulmars from the Canadian High Arctic." *Marine Pollution Bulletin*, 56: 1486–1512.

Markowski, M. 2009. "The International Legal Standard for Sustainable EEZ Fisheries Management." In *Towards Sustainable Fisheries Law*, ed. G. Winter, 3–28. Bonn: International Union for Conservation of Nature and Natural Resources.

Markus, Till. 2009. *European Fisheries Law: From Promotion to Management*. Groningen and Amsterdam: Europa Law Publishing.

Mora, Camilo, Ransom A. Myers, Marta Coll *et al.* 2009. "Management Effectiveness of the World's Marine Fisheries." *PLoS Biology*, 7(6): e1000131.

Morato, T. n.d. "Seamounts: Hotspots of Marine Life," http://www.ices.dk/marine world/seamounts.asp (accessed October 16, 2012).

Nauke, M. and G.L. Holland. 1992. "The Role and Development of Global Marine Conventions: Two Case Histories." *Marine Pollution Bulletin*, 25(14): 74–79.

Nybakken, J.W. 1993. *Marine Biology: An Ecological Approach*, 3rd edn. New York: Harper-Collins College Publishers.

OSPAR Commission. 2010. *Quality Status Report 2010*. London: OSPAR Commission.

Paalvast, Peter and Gerard van der Velde. 2011. "New Threats of an Old Enemy: The Distribution of the Shipworm *Teredo navalis L*. (Bivalvia: *Teredinidae*) Related to Climate Change in the Port of Rotterdam Area, the Netherlands." *Marine Pollution Bulletin*, 62(8): 1822–1829.

Pandolfi, J.M., R.H. Bradbury, E. Sala *et al.* 2003. "Global Trajectories of the Long-Term Decline of Coral Reef Ecosystems." *Science*, 301: 955–958.

Paris MoU (The Paris Memorandum of Understanding on Port State Control). 2010. "Paris MoU: New Inspection Regime," http://www.google.de/url?sa=t&rct=j&q=new %20inspection%20regime%20nir%20paris%20mou&source=web&cd=5&ved=0CE0 QFjAE&url=http%3A%2F%2Fwww.rws.nl%2Fimages%2FNew%2520Inspection% 2520Regime%25202010_tcm174-289603.pdf&ei=GcbgTovLEYTj4QSQupzSBg&usg= AFQjCNH872tonOkmcCkKSkT1TvC4BA0TYQ (accessed October 20, 2012).

Paris MoU (The Paris Memorandum of Understanding on Port State Control). 2011. "Port State Control: Voyage Completed, a New Horizon Ahead." In *Annual Report 2010*. Paris: MoU.

Reid, Walter V., Harold A. Mooney, Angela Cropper *et al.* 2005. *Ecosystems and Human Wellbeing: Synthesis. A Report of the Millennium Ecosystem Assessment*. Washington, DC: Island Press.

Sabine, C.L., R.A. Feely, N. Gruber *et al.* 2004. "The Oceanic Sink for Anthropogenic CO_2." *Science*, 305(5682): 367–371.

Salomon, Markus. 2009. "Recent European Initiatives in Marine Protection Policy: Towards Lasting Protection for Europe's Seas?" *Environmental Science & Policy*, 12(3): 359–366.

SRU (German Advisory Council on the Environment). 2004. *Marine Environment Protection for the North and Baltic Seas*. Berlin: Nomos-Verlag.

SRU (German Advisory Council on the Environment). 2011. *Sustainable Managment of Fish Stocks: Reforming the Common Fisheries Policy*. Berlin: SRU.

STECF (Scientific Technical and Economic Committee for Fisheries). 2010. *Report of the SGMOS-08-01 Working Group on the Reduction of Discarding Practices*. European Commission Joint Research Centre, Institute for the Protection and Security of the Citizen. Luxembourg: Office for Official Publications of the European Communities.

Tittensor, D.P., C. Mora, W. Jetz *et al.* 2010. "Global Patterns and Predictors of Marine Biodiversity across Taxa." *Nature*, 466: 1098–1101. doi:10.1038/nature09329.

UNEP (United Nation Environmental Programme). 2009. *Marine Litter: A Global Challenge*. Nairobi: UNEP.

UNEP/GRID-Arendal. 2009. "The Boom in Shipping Trade." UNEP/GRID-Arendal Maps and Graphics Library 2, http://maps.grida.no/go/graphic/the-boom-in-shipping-trade1 (accessed November 9, 2011).

United Nations. 2002. *World Summit on Sustainable Development: Plan of Implementation of the World Summit on Sustainable Development*. n.p.: United Nations.

United Nations Office for Legal Affairs. 2011. "The United Nations Convention on the Law of the Sea (A Historical Perspective)," http://www.un.org/depts/los/convention_ agreements/convention_historical_perspective.htm (accessed November 30, 2011).

Chapter 5

Deforestation

David Humphreys

Introduction

Forests play a key role in global climate regulation as a major sink for carbon dioxide, one of the main greenhouse gasses. Forests currently absorb carbon dioxide, thus promoting tree growth and slowing atmospheric warming. However, photosynthesis becomes difficult for tree species as temperatures warm and enzyme molecules start to break down. Above a certain temperature threshold (the exact temperature will vary between ecosystems and species) forests will start to leach carbon dioxide back into the atmosphere leading to further warming (Melillo *et al.* 2011). Additional warming would increase the risks of forest fires, which could release huge amounts of carbon dioxide into the atmosphere. Such loss of tree cover could have a deleterious effect on the world's climate. A holistic policy response to climate change thus needs to pay full attention to the iterative relationship between climate change and changes to the world's forest cover. Yet it is only since the turn of the millennium that there has been a sustained international policy focus on the role of forests in climate change. Prior to that climate change and deforestation were dealt with on largely separate international policy tracks.

After briefly considering how deforestation may be imagined as a global political issue this chapter will provide a historical overview of international cooperation on forests. The chapter concludes by examining the policy idea known as Reducing Emissions from Deforestation and forest Degradation (REDD), in the process noting some criticisms that this idea has attracted as well as some unresolved policy issues.

Defining Forests as a Political Problem

The climate regulation function of forests is just one of the public goods that forests provide. Forests provide a broad range of public goods at different spatial scales,

The Handbook of Global Climate and Environment Policy, First Edition. Edited by Robert Falkner.
© 2013 John Wiley & Sons, Ltd. Published 2013 by John Wiley & Sons, Ltd.

from the local level to the global. Public goods are non-excludable, in that no one can be prevented from benefiting from the goods that forest provide, and non-rival, in that consumption of a public good by one person does not affect what is left for others. At a global scale everyone benefits from the climate regulation services that forests provide and no one can be excluded from the benefits of a stable climate. Similarly, forests provide local public goods. They provide watershed services to local ecosystems and communities and they may replenish natural aquifers. Deforestation alters local hydrological regimes, changing the water in the soil and the moisture in the atmosphere, sometimes leading to the drying up of streams and rivers and local climatic change. Forests also provide local soil conservation functions. They may serve as cultural spaces and provide recreational and spiritual functions for local communities and indigenous peoples. They are the habitat for a huge range of biological diversity, both flora and fauna, and thus help maintain the diverse gene pool that is necessary for adaptable and resilient ecological systems and species. Forests, therefore, are shared between the world's people in the sense that they host or provide a range of public goods with life-preserving functions for both proximate and distant users (Perrings and Gadgil 2003; Humphreys 2006).

The political problem of deforestation is in part contention between different actors over the various public goods that forests provide. A forest valued solely for its carbon function would be managed very differently from one valued solely for its biodiversity, with a preponderance of fast-growing tree species that can provide rapid uptake of carbon dioxide in the case of the former and a wide range of different species in the case of the latter. But deforestation is in large part a political issue because as well as providing public goods forests also provide private goods, those that are excludable in that the owner of a good has the right, legally at least, to prevent others from using the good, and rival in that the more one actor consumes a given good, the less is left for others to enjoy. Private goods include timber, nuts, fruits, and rubber.

The harvesting of forest private goods is not necessarily incompatible with forest public good maintenance providing the harvesting of private goods does not lead to the depletion of the resource base. Often, however, unsustainable harvesting techniques are used, such as clear felling an area of forest for timber without replanting. A further problem is that while standing forests provide private goods, so too may the land on which forests stand. Deforestation may take place, for example, to clear land for agriculture, urban settlements, and oil prospecting (Barraclough and Ghimire 2000; Grainger 2009). Forest land itself may be privately owned, with different owners having very different approaches to forest management: long-term conservation in some cases and revenue maximization from short-term exploitation in others. Forests may be said to be sustainably managed when maximizing the yields of the private goods that forests provide does not lead to the degradation of any forest public good (Humphreys 2006).

Political conflicts over forests are due to contention between different actors over the various public and private goods that forests provide. This contention is played out at different spatial levels, from the local level to the global, with different actors making different claims to forest territory. Indigenous peoples and local community groups in tropical countries, often supported by civil society organizations such as international NGOs, argue that local communities are best placed to conserve the forests, and that deforestation results when customary and traditional forms of forest ownership are undermined by outsiders who wish to exploit the forest for short-term gain. Indigenous peoples and local community groups argue that forests are, first and foremost, local commons (May 1992; Colchester 1994; Chhatre and Agrawal 2008). However, under international law forests, along with other natural resources, are a sovereign national resource of the state.

The two claims are not incompatible: national law may recognize and uphold local land claims and when this is the case the ability of local groups to resist incursions onto their ancestral lands is greatly enhanced. Many local groups in the tropics argue that they are best placed to conserve and sustainably use their forests when their traditional rights are recognized by the state and when they have secure and legally enforceable tenure rights. However, community groups dispute both state-owned and private forms of property in which customary and traditional rights are not recognized. Finally, and as noted, forests service a global public good, namely the atmospheric commons. The tension between these various scales – local, national, global – defines deforestation as an international policy issue.

From Vienna to Rio: A Century of International Cooperation on Forests (1892–1992)

Forest management first emerged as an international scientific issue in 1892 when the International Union of Forest Research Organizations (IUFRO) was established in Vienna following the recommendation to create an international forest science research body at the 1890 Congress of Agriculture and Forestry. The IUFRO has since emerged as the most visible and significant international forest science network. In 1945 the UN Food and Agriculture Organization (FAO) was established with responsibility for dealing with global food and agricultural issues. The FAO was also given responsibility for international forest issues, although only approximately 4% of the organization's budget is dedicated to forests, and for its first 40 years the FAO confined itself primarily to technical management issues. However, in the mid-1980s deforestation emerged as a global political issue.

International Tropical Timber Organization

Accelerating tropical deforestation in the 1960s and 1970s – in the Congo Basin, Southeast Asia, and Latin America, particularly Brazil, where the military government viewed the Amazon as a frontier to be rolled back and developed (a "land without people for people without land") – had led many governments, environmental NGOs, development banks, and international organizations to conclude that concerted international action was needed to address the problem. In 1985 the International Tropical Timber Organization (ITTO) was created with the mandate of promoting the expansion and diversification of the international trade in tropical timber (the ITTO has no mandate on national-level trade within countries) and encouraging the sustainable utilization and conservation of tropical forests. Despite this latter mandate the ITTO is not a resource conservation organization (Colchester 1990). It is primarily an international commodity organization, one of a series that were created by the United Nations Conference on Trade and Development following concerns expressed by developing countries in the 1960s and 1970s that they were not receiving a fair return on their natural resources and that they should have an organized voice in international politics. The ITTO encourages cooperation between producers and consumer governments and promotes market transparency, with some collaboration on development projects. As of 2011 ITTO had funded some 800 projects totalling more than US$300 million (ITTO 2011). The main donors to project funding are the governments of Japan, the Netherlands, Switzerland and the USA, and the European Union.

In 1988 a study commissioned by the ITTO concluded that less than 1% of the international tropical timber trade (namely from Queensland, Australia) came from sustainable sources (Poore *et al.* 1989). These findings catalyzed the adoption by the ITTO in 1990 of the Year 2000 Objective, namely the target that by the end of the twentieth century the international trade in tropical timber should come from sustainably managed sources. Although many countries had adopted policies that supported the objective, these policies had not always been translated into action on the ground and the target was not met (Poore and Chiew 2000). The ITTO's two objectives – of promoting the international timber trade while conserving the forest resource base – have always sat uneasily at the heart of the organization (Humphreys 1996).

Tropical Forestry Action Plan

The same year that the ITTO was created, the FAO, in collaboration with the World Bank, the United Nations Development Programme, and the Washington-based World Resources Institute, launched an ambitious plan, the Tropical Forestry Action Plan, intended to address tropical deforestation and organized around five program areas. The program areas were forestry in land use (at the interface between forestry and agriculture), forest-based industrial development, fuelwood and energy, the conservation of ecosystems, and the action program on institutions. The intention of the plan was that individual tropical forest countries would initiate national forestry action programs structured around these five program areas (FAO *et al.* 1987).

By 1990, 79 countries had initiated, or expressed an interest in initiating, a national forestry action program. But TFAP was attracting widespread criticism, both from NGOs and from an FAO-initiated independent review. Significantly, the TFAP had not succeeded in slowing tropical deforestation. The independent review noted that "the rate of deforestation appears to have accelerated in spite of the TFAP" (Ullsten *et al.* 1990: 12). Other criticisms that TFAP attracted were that it was overly focused on the forest sector thus ignoring causes of deforestation outside the forest sector, and that national forest action programs were donor-driven, reflecting the interests of donors rather than of forest communities. Reflecting this last point, critics argued that national forest action plans were focused predominantly on forest-based industrial development, to the exclusion of other action programs. As a result of this criticism the idea of a global plan receded and following the 1992 United Nations Conference on Environment and Development the emphasis changed to national-level action.

The United Nations Conference on Environment and Development

At the 1992 UNCED two conventions were opened for signature: the UN Framework Convention on Climate Change (UNFCCC) and the Convention on Biological

Diversity. But from the outset of the conference's preparatory negotiations there was contention over the measures for addressing deforestation due to some deep-rooted disagreements between developed and developing countries. While dichotomies between developed and developing countries are often overstated in analyses of international politics, in the case of the UNCED forest negotiations the distinction is a valid one. All the main developed countries – the EU (and its precursor, the European Community), Canada, the USA, Japan – argued for a forest convention for the conservation and sustainable management of the world's forests. However, almost all developing countries, including the main tropical forest countries in South America, Central Africa, and Asia, backed by China, opposed a convention which, it was suggested, could limit the sovereignty of states over their forest resources (Johnson 1993; Humphreys 1996).

One area of disagreement was on the causes of deforestation. Through the Group of 77 developing countries (G77), tropical countries argued that many of the causes of deforestation were to be found in global economic inequities. The G77 argued that unsustainable patterns of consumption in the developed countries, including high demand for tropical timber and agricultural produce farmed on deforested land, were drivers of tropical deforestation. Furthermore, it was claimed, high levels of international debt meant that developing countries paid more in debt interest and repayments than they received in international aid, leading many tropical countries to export timber and other natural resources to earn hard currency for debt repayment. The G77 argued that if tropical forests were to be conserved there should be an agreed package of international debt relief (Humphreys 1996).

The negotiations thus saw the developing countries using their forests to bargain for economic concessions from the developed countries. On this view the UNCED negotiations may be seen as a price negotiation: the tropical countries argued that if the developed countries wanted tropical forests conserved they would have to pay for it (Davenport 2005). The G77 introduced into the negotiations the concept of compensation for opportunity cost foregone: if tropical countries were to conserve their forests rather than exploit them they would forgo a major revenue stream and should expect to be compensated for this by those countries that wanted tropical forests conserved. While the developed countries did agree to some modest increased in development assistance during the UNCED negotiations these were insufficient to meet the aspirations of the developing countries.

In effect, the developing countries created a bargaining issue linkage between forests and other issues and in so doing they raised the price of forest conservation, a price that developed countries were unwilling to pay. There was a clear power-based dimension to the negotiations. In terms of possession of economic power the developed countries possess more capabilities than the developing countries on all important indicators, such as share of global GDP, GDP per capita, level of technological development, and influence in international economic and financial institutions. The developing countries have a history of staking a claim to the finance and technology of their developed counterparts, but prior to the advent of international environmental issues they had little they could bargain with in return. Now cognizant of the value attached to their forests by other countries the developing countries, recognizing their improved bargaining leverage, negotiated from a perceived position of increased strength, seeking to translate the concerns of developed governments and citizens on tropical deforestation into economic and political gain.

This illustrates an important point from negotiating theory. The mere possession of a capability is not on its own sufficient for an actor to maximize its bargaining leverage. A cognitive factor enters the equation in that the actor in question must appreciate the value attached by other actors to its capabilities if it is to bargain to maximum advantage and secure the best returns in any negotiated deal (Raiffa 1982; Fisher and Ury 1992; Goldman and Rojot 2003). However, the developed countries were unwilling to meet the demands of the developing countries because they wished to protect their underlying economic power position. For example, to have agreed to technology transfer on preferential and concessional terms would have resulted in lower returns on research and development costs for businesses based in developed countries. Debt relief would have harmed the international banking sector or the tax-payer in developed countries. Agreeing to the demands of developed countries, therefore, would have eroded the North's relative advantages in international trade and finance. Ultimately no agreement was possible as neither side was prepared to meet the terms of the other (Humphreys 1996; Davenport 2005). The outcome from the negotiations was the non-legally binding statement of principles on forests, a declaration which, at the insistence of the G77, reaffirms in its first paragraph the principle that states have sovereignty over their natural resources (United Nations 1992).

International Forest Politics since Rio

Following the UNCED in Rio there was a three-year hiatus in international forest politics. The disagreements that had surfaced during the negotiations, along with the criticism of the Tropical Forestry Action Plan, contributed to a chilling effect in international forest politics. It was not until 1995 when, following a joint initiative from the Canadian and Malaysian governments, a temporary UN forum, the Inter-governmental Panel on Forests, was created under the auspices of the Commission for Sustainable Development (CSD) with a two-year life span.

One of the main political outputs from the Intergovernmental Panel on Forests (1995–1997) was the recommendation that all states should form "national forest and land use programmes." The emphasis on "land use" reflected a consensus born out of the criticism that the Tropical Forestry Action Plan had encountered, namely focusing solely on forests and forestry would not address the causes of deforestation, which invariably lie outside the forest sector. Indeed, the Panel spent considerable time deliberating the causes of deforestation, a subject on which there has been extensive academic debate (Brown and Pearce 1994; Jepma 1995; Spray and Moran 2006; Spillsbury 2009; Boucher et al. 2011).

There is no "universal" theory of deforestation, and given how the causes of deforestation vary over time and space such a theory is likely to prove elusive. Monocausal explanations blame deforestation on single factors, such as population pressure, the high demand for tropical timber, or clearance for alternative land uses. Such explanations have been criticized for failing to take into account the often complex causes of deforestation and the variegated ways in which they may interact to produce deforestation in one space rather than another. But if monocausal explanations are unsatisfactory then so too is the view that the causes of deforestation are impenetrably complex with no clear causal patterns evident (Geist and Lambin 2002).

A viewpoint that lies between that of monocausality and impenetrable complexity, one that has attracted a measure of consensus from scholars and which found support at the Intergovernmental Panel on Forests, is that there are different interactions between multiple causal factors, with different synergies of causation apparent in different places at different times. Many analyses now distinguish between direct causes (sometimes referred to as proximate causes) and underlying causes. To Geist and Lambin (2002: 143) proximate causes are "human activities or immediate actions at the local level, such as agricultural expansion, that originate from intended land use and directly impact forest cover." Direct (or proximate) causes involve forest conversion to other land uses and the deliberate modification of forests at the local level; the felling of a tree in a particular space is, after all, an essentially local act.

Underlying causes, in distinction, are "fundamental social processes, such as human population dynamics or agricultural policies, that underpin the proximate causes" (Geist and Lambin 2002: 143). Underlying causes relate to the social forces and pressures that shape actors' behavior and incentivize those actors who fell trees to do what they do. They may operate locally, but often operate from a distance. So, for example, the increasing international demand for tropical timber for furniture is an underlying cause of deforestation, while tree felling in tropical countries to feed that industry would constitute a direct, or proximate, cause. Other underlying causes of deforestation include the international demand for agricultural produce such as beef, soybeans, and palm oil (Boucher *et al.* 2011).

The Intergovernmental Panel on Forests was the first intergovernmental organization to adopt and work with the distinction between direct and underlying causes of deforestation. It was clear after the UNCED negotiations that if the Panel was to try to reach conclusions on the causes of deforestation, even at a very generalized level, then there would be political disagreement. Instead the Panel opted to develop a diagnostic framework that would enable individual countries to trace the causal chains that affect their forests (Table 5.1). No order of importance was implied in the framework. The Panel noted that the correlation between underlying and direct causes of deforestation is not always straightforward and the relative values assigned to forests and the alternative uses of forested land will change over time (United Nations 1996).

Eight types of underlying cause were identified by the Panel (United Nations 1996). The first type is economic and market distortions, in particular the valuing of private goods such as timber that can be bought and sold and the undervaluing of the public goods values of forests. Second, policy distortions include building roads into forested areas that enable migration from those seeking to exploit the forests for commercial gain as well as from the rural landless poor. Other policy distortions included providing subsidies to actors to convert forests to other land uses and promoting forest colonization. Third, insecurity of tenure refers to unclear property rights so that ownership of areas of forest is unclear, promoting open access and incursions from outsiders. Fourth, lack of livelihood opportunities refers to poverty and the lack of life opportunities that may lead the poor to exploit forests unsustainably, catering to short-term needs rather than the long-term viability of the resource base. Fifth, government deficiencies include lack of enforcement capacity resulting in

Table 5.1 Diagnostic framework: relationships between selected direct and underlying causes of deforestation and forest degradation.

Direct causes	Underlying causes							
	1	2	3	4	5	6	7	8
Replacement:								
By commercial plantations	X					X	X	
Planned agricultural expansion	X	X				X		
Pasture expansion	X	X				X	X	
Spontaneous colonization		X	X	X		X		X
New infrastructure						X	X	
Shifting agriculture			X	X				X
Modification:								
Timber harvesting damage	X		X		X		X	
Overgrazing			X		X			
Overcutting for fuel			X		X			
Excessive burning				X	X			
Pests or diseases					X			
Industrial pollution					X		X	

Source: United Nations. 1996. "Intergovernmental Panel on Forests, Programme Element I.2, Underlying Causes of Deforestation and Forest Degradation." E/CN.17/IPF/1996/2.
Key:
1 Economic and market distortions
2 Policy distortions, particularly inducements for unsustainable exploitation and land speculation
3 Insecurity of tenure or lack of clear property rights
4 Lack of livelihood opportunities
5 Government failures or deficiencies in intervention or enforcement
6 Infrastructural, industrial or communications developments
7 New technologies
8 Population pressures causing land hunger

limited compliance with laws and regulations, with transgressors often unpunished. Sixth, infrastructural, industrial, or communications developments include shifts in the global prices of products which may lead to forest clearance (for example, a rise in the price of agricultural produce leading to increased demand for agricultural land) and pressures for new land for urban expansion. Seventh, new technologies have accelerated land clearance. The invention of the chainsaw and its application to forestry in the early twentieth century revolutionized forestry and led to accelerated rates of tree felling. New technologies such as biofuels and genetically modified trees may also increase pressure on forest space. But technology is not necessarily a malign force in forests. New technologies may reduce wastage in wood processing, leading to reduced pressure for tree felling. Finally, demographic factors may affect forest use. While population increases in developing countries need not necessarily translate into deforestation, population hotspots in forested areas, perhaps due to colonization or road building, will increase pressure on forests.

The diagnostic framework was offered as a tool for countries to identify those causes of deforestation relevant for their national context. It was stressed that the framework was illustrative and that countries should add to and adapt the framework in line with national circumstances. The framework, it was suggested, could be used to identify those underlying causes associated with particular direct causes so that appropriate remedial policies could be designed. For example if a country is experiencing deforestation due to an increase in commercial plantations, then according to Table 5.1 the underlying causes may be economic and market distortions; infrastructural, industrial, or communications developments; new technologies; or any combination of these factors.

The Intergovernmental Panel on Forests was succeeded by another temporary forum of the CSD, the Intergovernmental Forum on Forests, which had a three-year life span and was, to all intents and purposes, the same forum as its predecessor. Between them the Intergovernmental Panel on Forests and the Intergovernmental Forum on Forests produced some 270 proposals for action, namely suggestions on possible measures that countries can take when designing and implementing national forest programs.

When the mandate for the Intergovernmental Forum on Forests expired in 2000 it was replaced by a new body, the United Nations Forum on Forests (UNFF), which reports direct to the UN Economic and Social Council. In 2007 the UNFF negotiated a Non-Legally Binding Instrument on All Types of Forests which added to the growing body of soft law on forests. This instrument established four global objectives, namely reversing the loss of forest cover worldwide, enhancing forest-based benefits, increasing the areas of protected forests and sustainably managed forests, and increasing development assistance and financial resources for sustainable forest management (United Nations 2008). Despite the pledge to increase development assistance no new monies were pledged when the instrument was agreed.

By now many policy-makers were shifting attention away from UN forest institutions, which concentrated primarily on the negotiation of soft international law, and focusing instead on more practice-oriented measures to reduce deforestation. A major development has been the creation of non-state market-based forest certification schemes. Forest certification is the process by which an independent third party certifies that a forest management process conforms to agreed management standards. The first global scheme was the Forest Stewardship Council (FSC) created in 1993. As a market-based scheme the FSC relies for its success on demand from environmentally discriminating consumers who wish to purchase timber from well-managed sources, and forest managers and retailers who are prepared to meet this demand (Cashore *et al.* 2004; Gulbrandsen 2004; Pattberg 2005; Auld and Gulbrandsen 2010). The creation of the FSC subsequently led to the creation of a number of competitor schemes, most of which are now consolidated under the Programme for the Endorsement of Forest Certification schemes (PEFC).

The role of the market is also prominent in a recent policy innovation to reduce deforestation in order to tackle climate change. This policy has become known as Reducing Emissions from Deforestation and forest Degradation, or REDD.

REDD: An Evolving Concept

The fourth assessment report of the Intergovernmental Panel on Climate Change (IPCC) found that despite widespread deforestation over the last century there is still more carbon in the world's forests than in the atmosphere. It estimated that approximately 17% of greenhouse gas emissions are caused by deforestation and forest degradation (IPCC 2007). Most of these emissions take place in tropical countries that are subject to severe deforestation pressures, such as Brazil, the countries of the Congo Basin, and the forested countries of Southeast Asia, in particular Indonesia. Reducing these emissions is now seen as an important dimension of international action to tackle climate change.

Following a proposal from the governments of Costa Rica and Papua New Guinea to the eleventh conference of parties to the UN Framework Convention on Climate Change (UNFCCC) in 2005, there is now an international consensus that governments should actively seek to reduce greenhouse gas emissions from forests. Subsequent deliberations within the UNFCCC and the United Nations have developed and refined the principle that governments and communities that take measures to prevent deforestation that would otherwise have occurred should receive compensation. This idea has evolved significantly since 2005. Initially known as "avoided deforestation" (AD) the idea then became "Reducing Emissions from Deforestation" (RED). In 2007 the Stern Review of the economics of climate change endorsed the idea of reducing carbon emissions from deforestation, concluding that "A substantial body of evidence suggests that action to prevent further deforestation would be relatively cheap compared with other types of mitigation, if the right policies and institutional structures are put in place" (Stern 2007: xiii).

In 2006 during discussions on RED at the UNFCCC it was pointed out that there are significant emissions from forest degradation in addition to deforestation (Griffiths 2007). Deforestation is the complete removal of forest canopy from an area, while forest degradation is the partial removal of the forest canopy. So forest degradation is a qualitative change in forest cover while deforestation is a quantitative reduction. Degradation may take the form of a temporary disturbance of the forest canopy that can be repaired, either naturally or through planting of new saplings or careful management. But because degradation may be a precursor to broader changes leading in time to full deforestation, the concept of RED was broadened in 2007 to become Reducing Emissions from Deforestation and forest Degradation (REDD) (Kanninen et al. 2007).

Although the basic principle that countries taking action to reduce deforestation and forest degradation should be paid was now attracting support from a growing coalition of governments, it was pointed out that REDD privileged one particular forest good, namely carbon sequestration, above all others. So the principle was further broadened to REDD+, the plus sign denoting that forests provide a range of public and private goods in addition to carbon sequestration and that REDD+ should pay a poverty-alleviation role. While the REDD+ acronym is now widely used, within the UN system REDD remains the accepted acronym.

REDD has the potential to transform international climate and forest politics. When emissions from deforestation and forest degradation are taken into account (as opposed to just energy-based emissions) then some of the largest emitters of carbon dioxide are tropical forest countries, such as Brazil and Indonesia. The basic idea underpinning REDD is the "opportunity cost forgone" argument made by the G77 during the UNCED forest negotiations. Forests are cut down for one of two reasons: because they are worth more as timber than they are standing, or because alternative land uses are worth more than standing forest. A forest owner will conserve its forests if it can earn at least as much from conserving an area of forest as it would from clearing it. The rationale is that REDD schemes will generate sufficient financial resources to incentivize those who make decisions that will generate forest degradation or deforestation, to change their behavior so that forests that would have been lost are conserved. The Stern Review estimated that the opportunity cost of protecting forests in the eight countries that emit more than 70% of non-energy-related emissions would be around US$5 billion per annum, rising over time (Stern 2007: xxvi).

In order for reduced deforestation and forest degradation to be measured there needs to be a baseline, in other words the background rate of forest loss that would have taken place in the absence of REDD. Any difference between observed deforestation and the baseline is, it is assumed, due to changed behavior from the forest owner, who is thus entitled to REDD payments in line with the area of forest that has been conserved. Strictly speaking the baseline rate of deforestation should be determined scientifically from historical data, such as satellite imagery and ground observations. However, agreement on baselines is likely to involve an element of political bargaining. For example, some countries may negotiate for generous baselines – in other words baselines that tend to overestimate historical deforestation so that reduced deforestation is easier to achieve – before agreeing to participate in a REDD scheme (Humphreys 2008).

Parties to the UNFCCC endorsed in principle the idea of REDD at the Bali conference of parties in 2007. However, there is as yet no agreement on the legal principles that should govern REDD and what role, if any, it should play in a successor to the Kyoto Protocol. REDD has evolved as a governing principle and idea with no single multilateral, institutional focus. It is a broad term used for the various options for paying a government or other actor, such as a private forest owner, to conserve forests and their carbon stocks in order to slow anthropogenic climate change. REDD payments may come from a range of actors, including international development agencies such as the World Bank, donor governments, conservation NGOs, and businesses interested in investing in offsets.

Donors may have one of two motives for financing national REDD schemes. The first is altruism, the desire to make a contribution to stabilization of the world's climate. The second concerns offsets, in other words compensating for greenhouse gas emission made elsewhere. Actors criticized for high emissions may wish to demonstrate that they are taking action to offset their emissions through carbon sink enhancement activities. Criticism for high emissions may come from the electorate in the case of governments, shareholders in the case of businesses, and, for both, NGOs. Offsets are permitted under the Clean Development Mechanism of the Kyoto Protocol (for example, when a high-emitting Annex I country plants trees to absorb a share of its emissions) and under the EU's Emissions Trading Scheme. Individual companies and other actors may also buy offsets outside these schemes in the smaller voluntary carbon offsets market.

REDD funding can take place through bilateral arrangements between a forested country and a donor. Under such arrangements forested countries would negotiate bilaterally with donors to agree the forest area that should be conserved and the period of time over which it should be conserved in exchange for an agreed sum of money. Bilateral REDD arrangements are most likely to appeal to those countries with a significant expanse of forest cover and which are adept at leveraging development assistance from donors. Brazil, the country with the world's largest expanse of tropical forest, favors such an approach and is a supporter of bilateral REDD funding.

An alternative is a market-based approach under which participating forested countries would sell carbon credits which would be bought and sold on the international offset market, bringing together those demanding offsets and those supplying carbon credits. An international market-based approach could be incorporated in a post-Kyoto protocol. However, a market-based approach to REDD will work only if the market price is high enough to compensate forest owners for the forgone opportunity cost of alternative land uses. If the revenue per hectare that a forest owner can earn from REDD is less than can be earned from deforestation and conversion to, say, planting soybeans then (and ignoring the transaction costs that would arise from changing land use) the rational forest owner seeking to maximize the revenue available from his land would plant soybeans. However, a price for REDD credits that conserves forests today need not necessarily do so in the future. Continuing with the example of soybeans: the REDD market would need to track the price of soybeans so that the per hectare revenues the forest owner can earn from REDD continue to remain just ahead of those that can be earned from soybeans. Furthermore, global commodity prices are dynamic and changing, and the forest owner may be tempted to convert to another land use, say palm oil or cattle farming, if the per hectare revenue earnings were to outstrip those from both REDD and soybeans. In short, REDD prices will only continue to incentivize reduced deforestation and forest degradation if they rise to track the returns offered by the best opportunity cost forgone by the forest owner. Because the opportunity cost forgone will vary from place to place, a REDD price may be high enough to reduce deforestation in one forest, yet insufficiently high to do so in another (Costenbader 2011: 36–37).

This discussion reveals a problem with the idea of an international market for REDD credits: there is no basis in market theory for concluding that a price in one product will rise to track those of competitor products. A REDD price may lag behind the price for alternative uses of forest land when the demand for REDD credits is low relative to alternative land uses, or when the REDD market is flooded with supplies of REDD credits so that, again, the price is suppressed relative to other land uses. And if, for whatever reason, the price of REDD credits is insufficiently high to prevent deforestation, what then will happen? Either deforestation will take place, or it will be avoided only if the shortfall between willingness to pay and the opportunity cost forgone is met so that forest owners are incentivized to continue maintaining their forest cover. The most obvious way that such a shortfall can be met is through international public finance (Karsenty *et al.* 2013).

To make financial sense REDD payments would go to those areas of forest where the money would make the most difference, in other words to those forests that are most at risk and where the investment will thus conserve the most carbon. A criticism that REDD has attracted is that it is therefore at risk of being exploited by unscrupulous forest owners (Griffiths 2007). An owner with no prior intention of deforesting his land could threaten to do so in order to claim REDD payments. Similarly, an investor may buy up an area of pristine land and threaten to clear it if he calculated that the cost of purchasing the land would be justified by the REDD payments that would be received. Ironically, therefore, forest communities who have conserved their forests for centuries and who have no intention of deforesting their land would be unlikely to benefit from REDD payments. There is a thin dividing line between rewarding those who maintain an environmental service and those who threaten to destroy it, and it can be argued that REDD would favor the latter over the former.

REDD is redefining the idea of carbon offsets. The original idea of the carbon offset was to compensate for emissions of carbon dioxide in one space by reducing emissions or increasing carbon sink capacity in other space. So, for example, under the UNFCCC's Kyoto flexibility mechanisms offsets may take the form of an actor funding additional tree planting to offset its emissions. The problem of offsets is more problematic with REDD, where there would be no increase in forest cover. An actor could offset its carbon emissions in one space not through establishing additional forest cover elsewhere, but by maintaining forest cover that already exists. Proponents of REDD claim that such an offset model is acceptable if the emissions that have taken place are offset against emissions that have demonstrably been avoided in an area suffering a high background rate of deforestation in relation to an agreed baseline. Critics argue that merely maintaining forest cover, while certainly desirable in its own right, should not be a pretext for additional emissions from fossil-fuel burning. Given that two major biophysical processes have led to anthropogenic climate change, namely deforestation and the excavation and burning of fossil fuels, the solution to the problem lies in reforestation and emissions reduction. On this view the notion of REDD offsets is a flawed idea: while it is clearly desirable to conserve existing sink capacity through forest conservation, it is far more problematic in climate stabilization terms to offset such activities against additional carbon dioxide emissions (Humphreys 2008).

Another politically contentious issue is the participation of local communities in REDD schemes (Cotula and Mayers 2009; von Scheliha *et al.* 2009). Many indigenous peoples' groups have been suspicious of REDD due its focus on the global public good of climate regulation rather than the welfare of local communities, arguing that REDD will institutionalize global and national control over forests at the expense of customary local commons regimes, with most of the benefits flowing to national treasuries rather than to local communities (Griffiths 2007). A study from the Forests Peoples' Programme on the implementation of REDD in Peru concluded that REDD policies are "undermining rights of indigenous people and are likely to lead conflicts over land and resources" (Espinoza Llanos and Feather 2011: 6). One commentator has argued that

> REDD is not inherently pro-poor and could be anti-poor. Market-based REDD could end up compensating wealthy developers who are threatening to cut down the rainforest rather than communities that have conserved forests for centuries (Richards 2008).

While local people will be the most affected by REDD they will have a low level of influence on the design and implementation of REDD schemes unless donors, host governments, and forest owners insist upon it.

The Kyoto Protocol established legally binding emissions targets only for the Annex I countries (the countries of Europe plus Canada, the USA, Japan, Australia, and New Zealand). The seventeenth conference of parties to the UNFCCC in Durban in 2011 agreed that an international legal agreement on greenhouse gas emissions reductions should be agreed as soon as possible, but no later than 2015, to come into effect by 2020 (on climate negotiations, see Chapter 20 in this volume). If such an agreement can be reached it is very likely to include emissions reductions from forests and would thus provide a firm legal footing for REDD. Until then emissions reductions from forests will be voluntary and non-legally binding, and REDD will lack a single international institutional focus.

To date three international REDD initiatives have emerged. The UN-REDD program was established in 2008 to help developing countries develop national REDD strategies. The UN agencies involved are the FAO, the United Nations Environment Programme, and the United Nations Development Programme. The program also works closely with the United Nations Forum on Forests, the Global Environment Facility, and the UNFCCC secretariat. As of 2011 14 countries were receiving support to help build capacity for implementing a REDD scheme (so-called "REDD readiness" programs). The countries are Bolivia, Democratic Republic of Congo, Ecuador, Indonesia, Nigeria, Panama, Papua New Guinea, Paraguay, the Philippines, the Solomon Islands, Tanzania, Vietnam, and Zambia, none of which had yet been paid for any reduced emissions from their forests (UN-REDD 2011). The three largest donor governments to the UN-REDD program are Norway, followed by Denmark and Spain.

The second initiative, the Forest Carbon Partnership, is a partnership of governments with donors also including The Nature Conservancy and BP. Established in 2008, the Forest Carbon Partnership helps tropical and subtropical countries develop REDD readiness programs including developing national monitoring systems, management systems, and stakeholder consultation arrangements. The World Bank is a trustee for the Forest Carbon Partnership.

The third multilateral REDD initiative is the Forest Investment Program, which sits within the Strategic Climate Fund, one of two Climate Investment Funds set up to promote the objectives of the FCCC (the other being the Clean Technology Fund). The two funds aim to promote low-carbon development through scaled-up funding channelled through multilateral development banks (including the World Bank, which administers the Forest Investment Program) and other sources, including the private sector. Established in 2009, the Forest Investment Program aims to finance efforts that will address those underlying causes of deforestation identified in national REDD readiness programs. It is possible that these three multilateral REDD initiatives could become consolidated under the UNFCCC. For now they exist independently, although with very similar aims and overlapping membership in terms of recipients and donors.

Conclusions

While international scientific cooperation on forestry is now well over a century old, deforestation's emergence as an international political issue dates only to the 1980s. Early attempts to address deforestation focused on international organizations, plans, and negotiations, all of which have failed in one way or another. While the International Tropical Timber Organization continues to exist it remains primarily focused on international trade issues rather than forest conservation. The

Tropical Forestry Action Plan failed, and the UNCED forest negotiations failed to produce a global forest convention.

Since then there have been a succession of international forest institutions, the latest version of which, the United Nations Forum on Forests, continues to exist, meeting every two years. However, the focus of international forest cooperation has progressively narrowed. The Intergovernmental Panel on Forests agreed that the emphasis should be on national forest programs. Since then international agreements on forests have been soft and non-legally binding.

The degradation of forest public goods is due to the overexploitation of forest private goods, in particular timber, and the clearance of forests to free land for alternative uses, notable agriculture. REDD, which operates at the interface between climate and forest policy, seeks to place a value on the public goods dimension of forests, thus providing an economic incentive for forest conservation. But REDD represents a further narrowing of forest policy, valuing as it does the carbon sequestration function of forests. REDD has been criticized for this, and if REDD policies are to be effective at reducing deforestation they will have to overcome both political resistance and technical challenges, such as agreeing baselines and measuring forest carbon.

For now REDD initiatives are spread over a range of international institutions. However, REDD does offer the prospect that governments could, for the first time, agree a legally binding international policy on forests; not a global forest convention (an option still supported by many governments) but a forest policy embedded within a post-Kyoto climate change agreement. In this respect the future of REDD is now tied to a comprehensive agreement limiting emissions of carbon dioxide and other greenhouse gasses not covered by the Montreal Protocol. Finalizing such an instrument is certain to see some hard political bargaining, and the agreement and ratification of such an instrument should certainly not be taken for granted.

References

Auld, Graeme and Lars H. Gulbrandsen. 2010. "Transparency in Nonstate Certification: Consequences for Accountancy and Legitimacy." *Global Environmental Politics*, 10(3): 97–119.

Barraclough, Solon L. and Krishna B. Ghimire. 2000. *Agricultural Expansion and Tropical Deforestation: Poverty, International Trade and Land Use*. London: Earthscan.

Boucher, Doug, Pipa Elias, Katherine Lininger *et al.* 2011. *The Root of the Problem: What's Driving Tropical Deforestation Today?* Cambridge, MA: UCS (Union of Concerned Scientists) Publications.

Brown, Katrina and David Pearce, eds. 1994. *The Causes of Tropical Deforestation: The Economic and Statistical Analysis of Factors Giving Rise to the Loss of Tropical Forests*. London: UCL Press.

Cashore, Benjamin, Graeme Auld, and Deanna Newsom. 2004. *Governing through Markets: Forest Certification and the Emergence of Non-state Authority*. New Haven, CT: Yale University Press.

Chhatre, Ashwini and Arun Agrawal. 2008. "Forest Commons and Local Enforcement." *Proceedings of the National Academy of Sciences*, 105(36): 13286–13291.

Colchester, Marcus. 1990. "The International Tropical Timber Organization: Kill or Cure for the Rainforest?" *The Ecologist*, 20(5): 166–173.

Colchester, Marcus. 1994. *Salvaging Nature. Indigenous Peoples, Protected Areas and Bio-diversity Conservation*. UNRISD Discussion Paper 55. Geneva: United Nations Research Institute for Social Development.

Costenbader, John. 2011. "REDD+Benefit Sharing: A Comparative Assessment of Three National Approaches." UN-REDD Programme, June 1, http://www.theredddesk.org/resources/reports/redd_benefit_sharing_a_comparative_assessment_of_three_national_policy_approaches (accessed February 14, 2011).

Cotula, Lorenzo and James Mayers. 2009. *Tenure in REDD: Start-Point or Afterthought?* Natural Resources Issues No. 15. London: International Institute for Environment and Development.

Davenport, Deborah S. 2005. "An Alternative Explanation for the Failure of the UNCED Forest Negotiations." *Global Environmental Politics*, 5(1): 105–130.

Espinoza Llanos, Roberto and Conrad Feather. 2011. *The Reality of REDD+ in Peru: Between Theory and Practice*. Moreton-in-Marsh, UK: Forest Peoples Programme.

FAO, World Bank, World Resources Institute, and United Nations Development Programme. 1987. *Tropical Forestry Action Plan*. Rome: FAO.

Fisher, R. and W. Ury. 1982. *Getting to Yes: Negotiating an Agreement without Giving In*, 2nd edn. London: Random House.

Geist, Helmut J. and Eric F. Lambin. 2002. "Proximate Causes and Underlying Driving Forces of Tropical Deforestation." *Bioscience*, 52(2): 143–150.

Goldman, Alvin L. and Jacques Rojot. 2003. *Negotiation: Theory and Practice*. The Hague: Kluwer Law International.

Grainger, Alan. 2009. *Controlling Tropical Deforestation: 14 (Natural Resource Management Set)*. London: Earthscan.

Griffiths, Tom. 2007. *Seeing "RED"? "Avoided Deforestation" and the Rights of Indigenous Peoples and Local Communities*. Moreton-in-Marsh, UK: Forest Peoples Programme.

Gulbrandsen, Lars. 2004. "Overlapping Public and Private Governance: Can Forest Certification Fill Gaps in the Global Forest Regime?" *Global Environmental Politics*, 4(2): 75–99.

Humphreys, David. 1996. *Forest Politics: The Evolution of International Cooperation*. London: Earthscan.

Humphreys, David. 2006. *Logjam: Deforestation and the Crisis of Global Governance*. London: Earthscan.

Humphreys, David. 2008. "The Politics of 'Avoided Deforestation': Historical Context and Contemporary Issues." *International Forestry Review*, 10(3): 433–442.

Intergovernmental Panel on Climate Change (IPCC). 2007. *Climate Change 2007: The Physical Science Basis: Summary for Policymakers. Contribution of the Working Group I to the Fourth Assessment Report of the Intergovernmental Panel on Climate Change*. Cambridge: Cambridge University Press.

International Tropical Timber Organization (ITTO). 2011. "About ITTO," http://www.itto.int/about_itto/ (accessed December 8, 2011).

Jepma, C.J. 1995. *Tropical Deforestation: A Socio-economic Approach*. London: Earthscan.

Johnson, Stanley. 1993. *The Earth Summit: The United Nations Conference on Environment and Development (UNCED)*. London: Graham and Trotman.

Kanninen, Markku, Daniel Murdiyarso, Frances Seymour *et al.* 2007. *Do Trees Grow on Money? The Implications of Deforestation Research for Policies to Promote REDD*. Bogor, Indonesia: Centre for International Forestry Research.

Karsenty, Alain, Aurélie Vogel, and Frédéric Castel. 2013. "'Carbon Rights,' REDD+ and PES." *Environmental Science & Policy*.

May, P. 1992. "Common Property Resources in the Neotropics: Theory, Management Progress and an Action Agenda." In *Conservation of Neotropic Forests: Working from Traditional Resource Use*, ed. K. Redford and C. Padoch, 359–378. New York: Columbia University Press.

Melillo, J., S. Butler, J. Johnson *et al.* 2011. "Soil Warming, Carbon–Nitrogen Interactions, and Forest Carbon Budgets." *Proceedings of the National Academy of Sciences*, 108(23): 9508–9512.

Pattberg, Philipp H. 2005. "The Forest Stewardship Council: Risks and Potential of Private Forest Governance." *Journal of Environment & Development*, 14(3): 356–374.

Perrings, Charles and Madhac Gadgil. 2003. "Conserving Biodiversity: Reconciling Local and Global Public Benefits." In *Providing Global Public Goods: Managing Globalization*, ed. Inge Kaul, Pedro Conceição, Katell Le Goulven and Ronald U. Mendoza, 532–555. Oxford: Oxford University Press.

Poore, Duncan, Peter Burgess, John Palmer *et al.* 1989. *No Timber without Trees: Sustainability in the Tropical Forest*. London: Earthscan.

Poore, Duncan and Thang Hooi Chiew. 2000. "Review of Progress towards the Year 2000 Objective." Twenty-Eighth Session of the International Tropical Timber Council, ITTC(XXVIII)/9/Rev.2, November 5.

Raiffa, H. 1982. *The Art and Science of Negotiation*. Cambridge, MA: Harvard University Press.

Richards, Michael. 2008 "REDD, the Last Chance for Tropical Forests?" August. Policy Brief of FRR, IDL Group Ltd, http://www.theidlgroup.com/documents/PolicyBrief-REDDLastChanceforTropicalForestsAugust2008FINAL.pdf (accessed December 8, 2011).

Spillsbury, Richard. 2009. *Deforestation: Development or Destruction?* East Sussex: Wayland.

Spray, Sharon L. and Matthew David Moran, eds. 2006. *Tropical Deforestation. Exploring Environmental Challenges Series*. Lanham, MD: Rowman and Littlefield.

Stern, Nicholas. 2007. *The Economics of Climate Change: The Stern Review*. Cambridge: Cambridge University Press.

Ullsten, Ola, Salleh Mohammed, and Montague Yudelman. 1990. *The Tropical Forestry Action Plan: Report of the Independent Review*. Kuala Lumpur: FAO.

United Nations. 1992. "Non-legally Binding Authoritative Statement of Principles for a Global Consensus on the Management, Conservation and Sustainable Development of All Types of Forests." A/CONF.151/6/Rev.1, Rio de Janeiro, Brazil.

United Nations. 1996. "Intergovernmental Panel on Forests, Programme Element I.2, Underlying Causes of Deforestation and Forest Degradation." E/CN.17/IPF/1996/2, February 13.

United Nations. 2008. "Non-legally Binding Instrument on All Types of Forest." A/RES/62/98, General Assembly, January 31.

UN-REDD. 2011. "FAQs," http://www.un-redd.org/AboutUNREDDProgramme/FAQs/tabid/586/Default.aspx (accessed November 12, 2011).

von Scheliha, Stefanie, Björn Hecht, and Tim Christopherson. 2009. *Biodiversity and Livelihoods: REDD Benefits*. Eschborn, Germany: Deutsche Gesellschaft für Technische Zusammenarbeit.

Chapter 6

Biotechnology and Biosafety

Aarti Gupta

Biosafety: An Anticipatory Governance Challenge

Use of the techniques of modern biotechnology in agriculture and food production has given rise to impassioned debates over the last two decades about the benefits versus the risks posed by genetically modified organisms (GMOs) and products thereof. So-called transgenic varieties now constitute significant percentages of important globally traded commodity crops, such as maize, canola, soybean, and cotton (James 2011).

Governance of such products and ensuring their biosafety (i.e. safe uptake and use) remains a quintessentially *anticipatory* challenge, one where the very existence and nature of risk and harm remains scientifically and normatively contested (Gupta 2001; see also Guston 2010). Its anticipatory nature is related to the existence of "epistemological uncertainty" in this domain, whereby uncertainty and outright unknowability "lies at the core of a problem" (Funtowicz and Ravetz 1992: 259). Such uncertainty complicates the process of devising appropriate, adaptable, and stable biosafety governance arrangements. Anticipatory governance has to *co-evolve* with rapid socio-technical and environmental change, the contours of which are not easily discernible.

Global governance of GMOs has thus been shaped by the contested nature of concerns over potential environmental, human health, and socio-economic risks posed by GMOs and the resultant diverse framings of the nature of the governance challenge. This chapter reviews current global policy approaches to biosafety, including their central elements as well as potential conflicts between the global institutions wherein they take shape. It also reviews the national ramifications of existing global biosafety policy approaches. In so doing, it assesses diverse scholarly perspectives

The Handbook of Global Climate and Environment Policy, First Edition. Edited by Robert Falkner.
© 2013 John Wiley & Sons, Ltd. Published 2013 by John Wiley & Sons, Ltd.

on the suitability and effectiveness of these approaches, with implications for future policy directions and scholarship.

The (Contested) Problem of Biosafety

Modern biotechnology can be defined as

> the application of (i) *in vitro* nucleic acid techniques, including recombinant deoxyribonucleic acid (DNA) and direct injection of nucleic acid into cells or organelles, or (ii) fusion of cells beyond the taxonomic family that overcome natural physiological reproductive or recombination barriers (WHO 2005: 1).

Use of such technologies can result in introduction of novel traits into plants, animals, and microorganisms that go beyond what would be feasible with traditional breeding techniques. It is this crossing of species barriers and the novel genetic material introduced into food and feed crops that raises a variety of ecological, health, and safety concerns.

Ecological concerns include, for example, potential adverse impacts on biodiversity from novel gene flow from engineered plants, such as creation of herbicide-tolerant weeds; or adverse impacts on non-target organisms or development of pest resistance to toxins engineered in plants (e.g. Rissler and Mellon 1996; Wolfenbarger and Phifer 2000). Human health concerns include potential allergenicity or toxicity from consuming genetically modified ingredients in food (McHughen 2000). Going beyond this, various socio-economic and ethical concerns include potential adverse impacts of relying on high-tech, capital-intensive interventions such as genetic engineering for local and subsistence food production systems; the creation of patentable monopolies and concentration of ownership in seed and plant varieties of vital food and commodity crops; or the moral (un)acceptability of modifying nature (Nuffield Council on Bioethics 2003; Kleinman and Kinchy 2007).

Such claims of adverse ecological, health-related, or socio-economic impacts are countered by those who emphasize, instead, the multiple benefits to be derived from transgenic crops. For some scholars and practitioners, such benefits include, *inter alia*, enhanced food production, reduced use of synthetic pesticides, and improved food security. According to such a perspective, the central risk to be guarded against is development of overly stringent biosafety regulations that will impede spread of this technology globally, to the detriment of the world's poorest (Herring 2007; Paarlberg 2008; for a detailed critique of "pro-poor" biotechnology perspectives, see Jansen and Gupta 2009).

A central feature of global biosafety governance is thus the lack of societal consensus on the existence, nature, and extent of risks or benefits posed by use of modern biotechnology in agriculture; whether these risks and benefits are likely to materialize; and how they will be distributed within and across societies. As a result, norms underpinning global governance remain contested as well, and diverse global governance forums become sites of conflict to negotiate contested meanings of biosafety. Furthermore, the role of science and expertise in shaping biosafety policy choices comes to the fore. Devising appropriate mechanisms by which to provide scientific input into governance of such new technologies is then a central challenge in biosafety governance. These features underpin and shape current global policy approaches in this domain, as discussed next.

Global Policy Approaches

Two dominant policy approaches to ensuring biosafety currently coexist at the global level. The first is the *science-based harmonization* of national biosafety decisions encouraged by the global trade regime of the World Trade Organization (WTO), so as to facilitate transfers of transgenic products worldwide. This is the approach promoted by the WTO's Agreement on the Application of Sanitary and Phytosanitary Standards (henceforth SPS Agreement) to govern GMO trade. The second is *mandatory disclosure* by GMO producers of biosafety information and the intention to export GMOs, as a way to facilitate *informed choice* about import of transgenic products in diverse national contexts. This is the approach adopted by the multilaterally negotiated Cartagena Protocol on Biosafety (CPB) under the United Nations Convention on Biological Diversity, concluded in 2000.

Both approaches have unleashed multifaceted scholarly debates, of which four strands are discussed in this chapter. The first is the *appropriate role of science* in anticipatory risk governance, given a (global) push for science-based decision-making in this area. A second is whether *harmonization or a privileging of diversity* is promoted by existing global policy approaches, particularly in the current context of globalization. A third relates to potential *normative conflicts* between global biosafety policy approaches and associated *institutional conflicts* between international trade and environmental regimes. And a fourth relates to the promise versus perils of relying on *information disclosure* in anticipatory risk governance. Each of these areas of scholarly debates is discussed in turn.

Global Biosafety Governance: What Kind of Science–Society Contract?

An oft-deployed lens for assessing the suitability and effectiveness of global biosafety policy approaches is scrutinizing the role of science in this contested domain of global policy. The focus on science is stimulated, in the first instance, by the WTO-SPS Agreement's call for national sanitary and phytosanitary measures (relating to animal and plant health, thus also including biosafety measures) to be based on scientifically sound evidence of harm, so as to prevent unnecessary restrictions on trade, and avoid protectionism masquerading as risk avoidance, both key concerns of the global trade regime (SPS Agreement 1994; see also Christoforou 2000; Gupta 2002; Eckersley 2004).

The WTO-SPS Agreement also, however, allows for legitimate context-specific differences in judgments of appropriate levels of safety. This balancing act is reflected in the Agreement's preamble, which states that

> no Member should be prevented from adopting or enforcing measures necessary to protect human, animal or plant life or health, subject to the requirement that these measures are not applied in a manner which would constitute an arbitrary or unjustifiable discrimination between members where the same conditions prevail or a disguised restriction on international trade (SPS Agreement 1994: Preamble).

While this attempted balance between permitting legitimate difference while preventing illegitimate discrimination is laudable in principle, in practice it has been complicated to interpret and institutionalize. In particular, science is increasingly implicated in such determinations, with the WTO-SPS Agreement specifying that national (trade restrictive) decisions should be based upon scientifically sound assessments in order to be compatible with trade rules. Thus the Agreement states that nationally appropriate measures should be "applied only to the extent necessary to protect human, animal or plant life or health, [be] based on scientific principles and ... not [be] maintained without sufficient scientific evidence" (SPS Agreement 1994: Article 2). At the heart of the WTO-SPS Agreement, then, is a requirement that higher national standards must have clear scientific justification. Given the inevitable existence of scientific uncertainties in certain instances, the agreement permits recourse to trade-restrictive precautionary measures, as long as these are strictly time-bound and documentable efforts are ongoing to generate concrete scientific evidence of harm (SPS Agreement 1994: Article 5.7).

While analyses of the precautionary principle in environmental and biosafety governance abound (e.g. O'Riordan and Jordan 1995; Foster *et al.* 2000; Paarlberg 2001; Sunstein 2005) the essential debate turns on how precaution is to be interpreted and institutionalized, and whether it is part of or goes beyond a "sound-science" decision-calculus. This is related to whether biosafety governance itself transcends (or should transcend) science-based decision-making, even one that permits precautionary actions in the face of scientific uncertainty (Gupta 2002). Should biosafety governance, in other words, be *socially precautionary* as well?

These questions occupied center stage in negotiation and subsequent interpretation of the obligations of the Cartagena Protocol on Biosafety, seen by many as an important counter to the WTO SPS-Agreement's science-based harmonization imperative (for detailed histories of these negotiations, see Gupta 2000; Bail *et al.* 2002). When concluded in 2000, the Cartagena Protocol was hailed by many as the first to institutionalize a precautionary approach to multilateral environmental and biosafety governance (Falkner 2000). The Protocol was demanded by developing countries and confers upon a potential importing country the right to give its "advance informed agreement" prior to trade in certain GMOs (Cartagena Protocol on Biosafety 2000: Article 15). However, such agreement is also to be based upon an assessment of scientific harm. The Protocol nonetheless permits precautionary restrictions in the face of scientific uncertainty. In this, it appears to give more flexibility to countries than the WTO-SPS Agreement, since it does not specify a time frame within which precautionary decisions must be reviewed (Gupta 2002; Millstone and Van Zwanenberg 2003).

Analyses of how the Protocol's inclusion of precaution is to be interpreted vary greatly, ranging from arguments that it is much more far-reaching than the WTO, thus providing an essential counterweight to the global trade regime (Isaac and Kerr 2003) to a view that the Protocol's language on precaution is largely aligned with the WTO-SPS agreement's push for science-based decision-making even if it does permit greater consideration of scientific uncertainties (e.g. Gupta 2002) to claims that its excessive emphasis on precaution hampers global spread of transgenic crops, particularly to the poor in Africa (e.g. Morris 2008).

These global policy debates and developments are grounded in early efforts in key OECD countries in the 1970s and 1980s – particularly in the United States and Europe – to develop biosafety regulatory frameworks to address risks posed by use of modern biotechnology (for early histories, see Wright 1994; Gottweis 1998). These efforts have resulted in the well-known transatlantic divide in GMO regulatory approaches between the USA and the EU over the last two decades (Jasanoff 1995; Levidow 2007; Pollack and Shaffer 2009; Cho 2010).

The United States has evolved a product-based approach to biosafety and biotechnology regulation since the late 1980s, based on the principle of substantial equivalence (e.g. Jasanoff 2005). According to this principle, products of genetic engineering do not require regulation if judged to be substantially equivalent to a non-genetically modified product, regardless of whether genetic engineering techniques were used in the production process (Millstone *et al.* 1999; Bernauer 2003). This focus on substantial equivalence has long prompted impassioned debate, not least because the United States remains the main producer and exporter of GM seeds, food, and crops (James 2011), ensuring that its own regulatory approach has consequences for global biosafety governance and for diffusion of biosafety regulatory frameworks to other parts of the world (Murphy and Levidow 2006).

In an influential early critique of substantial equivalence, Millstone and colleagues questioned both the adequacy of tests relied upon to establish equivalence, and the underlying assumption that equivalence of transgenic with conventional foods could even be established. Their conclusion was that substantial equivalence was a "pseudo-scientific concept because it is a commercial and political judgment masquerading as if it were scientific." They noted, for example, that it did not require biochemical or toxicological tests and hence served to "discourage and inhibit informative scientific research" (Millstone *et al.* 1999: 526).

In contrast to the USA, the EU has consistently strengthened its regional GMO governance architecture through its directives on deliberate release, traceability, and labeling of GMOs and their products (Pollack and Shaffer 2005; Levidow 2007). At the heart of the EU approach is a focus on the production process. Thus *use* of techniques of genetic engineering is sufficient to trigger regulation, regardless of the characteristics of the resultant product. Furthermore, the EU's approach institutionalizes reliance on precaution, insofar as (trade) restrictive actions can be used to avert potentially serious harm, even given lack of clear scientific evidence of harm (Murphy and Levidow 2006; Falkner 2007).

Both the USA and the EU have sought in the last decades to export their respective approaches to GMO regulation to the global arena, and bilaterally to developing countries. In its broad contours, the WTO-SPS Agreement's science-based harmonization approach is consistent with a US sound-science based regulatory approach; while the EU's precautionary approach is perceived by many to have been institutionalized within the Cartagena Protocol. Beyond this, the consequences of the transatlantic divide in GMO regulatory approaches, in particular for developing countries, are much debated. Some scholars argue that the transatlantic EU–USA conflict constrains developing country biosafety policy choices, insofar as it forces choice between the two (e.g. Bernauer and Aerni 2007), while others argue that developing countries do not necessarily have to ally themselves with the EU or US biosafety model, but can, to greater or lesser extent, forge their own path in GMO regulation (e.g. Falkner and Gupta 2009). This debate relates directly to another strand of scholarly literature assessing global GMO policy, namely whether such policy promotes *harmonization or diversity* in national-level biosafety governance approaches, to which I turn next.

Global Biosafety Governance: Privileging Harmonization or Diversity?

With globalization as a catchword of the 1990s and beyond, a debate permeating scholarly analysis of global biosafety policy approaches relates to whether economic globalization and, in particular, globalized systems of food production and trade promote a *harmonization* of national policy choices or whether policy diversity persists. Harmonization is seen as desirable in a global trade context as a way to promote a level playing-field in the stringency and scope of national environmental policies, and hence to facilitate easy transfers of transgenic crops (Bernauer 2003; Drezner 2007; but see also Jasanoff 1998 for a different conceptualization of harmonization).

In the case of agricultural biotechnology and biosafety, the debate has focused on whether global policy will result in a "trading up," that is, an upward harmonization of national biosafety policies towards the more stringent level of the EU (e.g. Prakash and Kollman 2003; Young 2003). Evidence for such trading up remains inconsistent, however, and in recent years, attention has shifted to the persisting regulatory polarization between the USA and the EU and its consequences for national policy choices, particularly in the developing world (Bernauer 2003; Pollack and Shaffer 2009). Several prominent studies point to negative consequences of this transatlantic GMO conflict for developing country policy choices, arguing either that countries will be forced to choose one or the other pathway, that is, that there will be regulatory harmonization in the South along two nodes (Drezner 2007), or that the transatlantic conflict will have varied adverse impacts in the South, including the development of inconsistent regulatory approaches; impediments to public and private investment in agricultural biotechnology; and/or lack of public support (Paarlberg 2001; Bernauer 2003).

Others, however, provide alternative interpretations of whether globalization fuels harmonization, and with what implications for policy choices. In line with a long-standing literature that questions a view of globalization as a homogenizing force (e.g. Appadurai 1996), this strand of writing emphasizes the persistence of regulatory diversity. This is also because of the important mediating influence of domestic institutions and priorities, which fuel diverse national responses to globalization, also in the biosafety realm (Millstone and Van Zwanenburg 2003; Falkner and Gupta 2009; Pollack and Shaffer 2009). Recent studies propose typologies of domestic biosafety regulatory approaches that go beyond the EU–US regulatory dichotomy. For example, Kleinman *et al.* (2009) identify three regulatory models – which they term "liberal science-based"; "precautionary science-based"; and "social values-based" – that they claim different countries blend in distinctive ways to determine their specific policy approach.

In this, the role of the Cartagena Protocol is also noted. Analysts have focused on how the Cartagena Protocol's privileging of importing country choice in relation to the GMO trade permits context-specific differences to persist in biosafety regulatory choices, in contrast to the harmonization imperative of the global trade regime (Jaffe 2005; Gupta and Falkner 2006). Millstone and Van Zwanenberg (2003) note, for example, that rather than a convergence of policies across jurisdictions, the persisting scientific conflicts over GMO safety provide leeway to countries in the South to pursue divergent policy choices, and in this they are bolstered by the Cartagena Protocol's privileging of importer choice and precaution. Others concur with the existence of diverse rather than harmonized biosafety policies in the South, yet attribute this diversity to conflicting trade imperatives rather than to scientific uncertainty over GMO safety, arguing that policy diversity is related to differential needs to access US and/or EU markets (Clapp 2006). Yet others highlight a variety of domestic political imperatives, extending beyond scientifically assessable harm, in shaping policy choices (Jaffe 2005; Falkner 2006; Gupta and Falkner 2006; Falkner and Gupta 2009).

This body of work has also highlighted that domestic biosafety policy choices transcend science-based decisions. Analyses of various national-level biosafety rules reveal that domestic biosafety decisions are based on myriad criteria that include, but go beyond, scientifically assessable ecological and human health risks. The influence of so-called "socio-economic considerations" in GMO decision-making was an important axis of conflict during the global negotiations of the Cartagena Protocol as well. Developing countries pushed to have such considerations included as a legitimate basis for national GMO regulatory choices, even those which would have resulted in a restriction of trade. The Protocol allows, as a result, for countries to "take into account, consistent with their international obligations, socioeconomic considerations arising from the impact of [genetically modified organisms] on the conservation and sustainable use of biological diversity" (Cartagena Protocol 2000: Article 26).

This, however, can be interpreted in a manner that is similar to the WTO-SPS Agreement's (limited) provisions on relevant socio-economic factors to be taken into account in domestic risk decisions. The WTO-SPS Agreement permits consideration of "relevant economic factors" in a risk assessment that is to form the basis for national decisions. Such considerations include

the potential damage in terms of loss of production or sales in the event of the entry, establishment or spread of a pest or disease; the costs of control or eradication in the territory of the importing Member; and the relative cost-effectiveness of alternative approaches to limiting risk (SPS Agreement 1994: Article 5.3).

Thus, socio-economic factors that can be considered under this Agreement relate to potential economic damages *arising from* sanitary or phytosanitary harm.

The Cartagena Protocol's provisions on socio-economic considerations can also be interpreted as restricted to those impacts arising from adverse impacts on biodiversity. Furthermore, the explicit stipulation that such considerations be "consistent with international obligations" implies deference to the WTO-SPS Agreement's provisions, in particular its call for science-based national decisions and (limited) inclusion of "relevant" economic factors. This appears to exclude considerations such as disruption to traditional livelihoods or undue dependence on patented seed as legitimate socio-economic bases for restricting GMO trade, concerns voiced by

developing countries during negotiation of the Cartagena Protocol. While such ratio-
nales to restrict trade appear to clearly run afoul of WTO rules, a more open question
is whether public acceptability or consumer opposition to transgenic crops; lack of
capacity to segregate modified from non-modified varieties of key crops; and/or lack
of capacity to monitor safe handling and use in diverse contexts should be seen
as legitimate socio-economic considerations influencing domestic biosafety policy
choices (see e.g. Stabinsky 2000; Gupta 2002; Kleinman and Kinchy 2007).

As with most instances of global policy implementation, the key test lies in how
criteria for domestic biosafety choices – whether those advanced by the Cartagena
Protocol or the WTO-SPS Agreement – are interpreted and applied in diverse national
contexts. A recent analysis of how "socio-economic considerations" permitted by
the Protocol are being transposed into domestic legislation in developing countries
shows that relatively few countries formally and systematically include such considerations in their biosafety assessment and approval processes (Falck-Zepeda and
Zambrano 2011). Furthermore, what "socio-economic factors" include is open to
multiple interpretations, with implications for what constitutes legitimate biosafety
governance outcomes.

Whether and how to systematically include socio-economic factors in domestic
decision-making is related, more broadly, to debates about democratic modes of risk
governance as a way to lend legitimacy to contested risk and safety decisions (Jasanoff
2004, 2005; Lövbrand *et al.* 2011). In a recent analysis of the evolving domestic
biosafety regime in India, Gupta (2011), for example, identifies two contrasting
(and contradictory) sources of legitimacy that the Indian governance architecture
has relied upon in reaching contested biosafety decisions: objective science versus
democratic deliberation.

India has so far approved only genetically modified varieties of cotton for
commercial use. This looked set to change in 2010/2011, when a nation-wide
controversy erupted over granting commercial approval to a variety of eggplant (or
brinjal, as it known as in India) modified to be pest resistant. Brinjal is a widely
consumed vegetable in India and such approval would have resulted in the first
transgenic food crop being approved for commercial use in India. In this instance,
the then Minister of Environment and Forests launched an innovative experiment in
deliberative democracy as a means to reach *and* legitimize a decision. He did this by
organizing a nation-wide series of open public meetings to debate the risks and benefits of this particular transgenic crop. Following this, he announced a moratorium on
approval of the modified brinjal variety, yet simultaneously held out the hope that
better scientific information about risks and benefits could, in the future, help guide
decisions on these contested issues. This was notwithstanding the fact that his own
decision to impose a moratorium evoked concerns going well beyond scientifically
assessable harm.

This paradoxical outcome – namely relying on democratic deliberation as a way
to legitimize risk-related decisions, but nonetheless calling on objective science to
serve as ultimate arbiter of conflicting views – highlights the continued deference
to science as a way to legitimize contested risk decisions. Yet, as Gupta (2011)
concludes, the brinjal experience makes clear that biosafety governance in India
will *perforce* have to engage with (potentially messy) democratic decision-making
processes, even if these are seen as ad hoc in a global context. The challenge lies in

how to institutionalize deliberative processes so as to rely on them in a systematic manner in making risk-related societal choices.

In line with this, research has also examined the role of global regimes, such as the Cartagena Protocol, in contributing to a democratization of debate and broadening of domestic biosafety policy agendas. Global obligations under the Protocol, for example, have been embraced by local and/or translocal protest movements as a means to contest global neoliberal or expert-driven discourses and practices of risk governance. Scoones (2008), for example, focuses on how anti-GMO mobilization and protest movements in countries such as India, South Africa, and Brazil shape domestic biosafety policy choices. He documents how such local and translocal mobilizations make "an important contribution to democratic debate in context(s) where, because of the forces of neo-liberalism, alternatives have little space" (Scoones 2008: 317). Nonetheless, the policy space for domestic biosafety regulatory choices is not without limits. Newell (2007) highlights, for example, the "bounded autonomy" of developing countries, particularly least developed countries, in selecting preferred biosafety regulatory pathways, given multiple global-national-local nexus of influences, both public and private, that need to be negotiated in making such policy choices (see also Pollack and Shaffer 2009).

Notwithstanding this, a relocalization of globally negotiated norms in given domestic contexts appears inevitable in practice. Global governance arenas such as the Cartagena Protocol are best viewed, then, as sites for negotiating shared understanding of concepts such as biosafety, rather than as vehicles for the global diffusion of consensual or universally valid decision criteria (Jasanoff 1998; Gupta 2004; see also Alemanno 2011). Such negotiated understandings are apt to rely, furthermore, on ambiguity and interpretative flexibility, rather than standardization on the basis of technical precision or scientific objectivity. If so, global policy arenas such as the Cartagena Protocol are themselves arenas for the generation of local – that is, context-specific – knowledge, which then inevitably requires relocalization to be relevant to other (local) contexts (Gupta 2004).

Global Biosafety Governance: Norm Conflicts or Collusion?

If global policy forums can be conceptualized as sites for deliberating and generating shared understandings of contested concepts, such a lens can also be brought to bear on analyzing *inter-linkages between global forums*, such as the Cartagena Protocol and world trade agreements. Analyses of such linkages have been another key strand of scholarly research in global environmental and biosafety governance (for a comprehensive overview, see Young *et al.* 2008). A concern with so-called "regime interplay" dates back to the mid-1990s, with a proliferation ever since of typologies of different types, sources, and consequences of regime interplay (e.g. Gehring and Oberthür 2006). Much scholarly attention has focused on whether there are synergies or conflicts among global policy forums. Capping this is a recent interest in interplay management (Oberthür and Stokke 2011).

Inter-linkages between the WTO-SPS Agreement and the Cartagena Protocol on Biosafety remain one of the most analyzed examples of such phenomena (e.g. Safrin 2002; Eckersley 2004; Oberthür and Gehring 2006; Gupta 2008). One concern has been with the existence and nature of (what are mostly assumed to be undesirable) *conflicts* across these global regimes. A related focus is on the practice of so-called "forum-shopping" by key actors, who search for a favorable international institutional environment to further their specific political interests and globalize their preferred GMO regulatory approaches (for a detailed analysis, see Pollack and Shaffer 2009). Much literature on inter-linkages has also noted the consequences of what Eckersley (2004) has termed the "big chill" or the shadow cast by the WTO over negotiation and evolution of multilateral environmental agreements that address trade-related environmental issues.

In a recent analysis that compares institutional interactions across global climate, biodiversity and trade regimes, Zelli *et al.* (2012) argue that, in addition to analyzing dyadic relations between two regimes, the broader normative context shaping inter-regime interactions is also crucial to understand. They point, for example, to the overarching normative dominance of market liberalism in shaping regime interactions, even as this dominance is incomplete and contested in specific instances by key actors. Market liberalism privileges economic efficiency, unfettered markets, deregulation, and privatization in governance, a configuration that Steven Bernstein has labeled the "compromise of liberal environmentalism" (Bernstein 2001). Building on this, Zelli *et al.* (2012) argue that liberal environmental norms not only shape the provisions and practices of individual global regimes, but also their interactions with each other. In such a view, if the WTO casts a shadow over global environmental regimes, it is because the overarching normative context supports such an outcome. As such, these authors suggest that institutional interactions in the global environmental domain might be characterized more by a problematic normative *collusion* (that is, a homogeneity or similarity) rather than the more frequently assumed *conflict* between regimes.

Whether there is normative conflict between the two regimes in practice remains open to interpretation. Political conflicts over the appropriate relationship between the Cartagena Protocol and the multilateral trade regime led to the near-collapse of protocol negotiations in 1999, before an agreement was finally concluded in 2000. Reaching an agreement hinged on the inclusion and formulation of a so-called "savings clause" in the Protocol, a legal term for a provision that makes explicit its relationship to a prior and related treaty, in this case WTO Agreements (Gupta 2000; Bail *et al.* 2002).

The preamble of the Protocol contains this savings clause, which begins with a shared ideal that "trade and environment agreements should be mutually supportive." It goes on to state, however, that "this Protocol shall not be interpreted as implying a change in the rights and obligations of a Party under any existing international agreements," a sentence that was inserted at the insistence of GMO producer countries and seen as vital to securing agreement of the USA and other GMO producers to a protocol. A final sentence, insisted on by the EU as a way to counter the preceding one, notes that "the above recital is not intended to subordinate this Protocol to other international agreements" (Cartagena Protocol 2000: Preamble). The outcome is that these two almost contradictory statements can be differently interpreted to suit specific needs.

With the Protocol in force since 2003, scholarly and practitioner attention has shifted to how the relationship between it and global trade agreements is, for example, being interpreted in judicial decisions and dispute settlement processes. Here, a 2003 WTO a case brought by the United States against the EU's GMO regula-tory approach has proved instructive. In this case, the USA, supported by Australia and Canada, asserted that the EU was violating its WTO-SPS obligations, given its restrictive approach to GMO imports and the *de facto* moratorium in place on new approvals of transgenic crops (WTO 2003; see also Isaac and Kerr 2003). In its defense, the EU invoked, *inter alia*, its rights and obligations under the Cartagena Protocol on Biosafety. As a result, the highly anticipated outcome of the dispute was expected to shed light on how the WTO interpreted its relationship to the Cartagena Protocol (Isaac and Kerr 2003).

The WTO ruling, when it finally came in May 2006, found largely in favor of the United States, arguing that the European Union was partially in violation of its WTO-SPS obligations. It is noteworthy, however, that the panel did not rule on the substantive elements of EU GMO regulation, including the reliance on precaution therein. Instead, the finding highlighted the failure of the EU to conduct the needed risk assessments prior to imposing restrictions on trade. In reaching this conclusion, moreover, the dispute settlement panel did not take global obligations under the Cartagena Protocol into account (Lieberman and Gray 2008). A rationale was that the two regimes do not share similar (relevant) membership, given that the USA – as the major GMO producer and exporter – is not a party to the Cartagena Protocol. More importantly, the Cartagena Protocol was not yet in force when the EU GMO restrictions were introduced.

This WTO panel finding has been interpreted in multiple ways, with one issue being whether the transatlantic GMO conflict is amenable to juridical resolution. Alemanno (2011: 2) asserts, for example, that the WTO case proved important not so much in what it decided but "in constraining the conflict by channelling it into a legal process, so as to deflect pressure within the US to retaliate aggressively against the EU." Cho (2010: 12) argues, in contrast, that resorting to a WTO dispute settlement mechanism aggravates conflict. As he puts it, "the adversarial legalism entrenched in the WTO adjudicative mechanism tends to judicialize science" with "duelling experts" (Pollack and Schaffer 2009: 172) seeking to mediate between uncertainties and conflicts over whose science is sound. A conclusion stressed by these authors, then, is that the WTO dispute settlement's procedural emphasis on "reason-giving and deliberation" among parties may be a better vehicle for generating shared understandings than (misguided) attempts to adjudicate whose science is valid or better aligned with WTO obligations.

A procedural focus on transparency and deliberation also underpins the final strand of debates about global policy examined here, reliance on information disclo-sure as key to a right to know and choose in anticipatory risk governance.

Global Biosafety Governance: The Trials and Triumphs of Disclosure?

A more recent strand of research on global policy approaches to biosafety has empha-sized the focus on information disclosure as an anticipatory risk governance strategy in a global context. The Cartagena Protocol's call for "advance informed agree-ment" can, in this view, be seen as an example of "governance by disclosure," whereby disclosure of information is relied upon to further normative, procedural, and substantive governance aims. These include furthering a right to know and the exercise of choice by GMO importing countries, so as to facilitate risk mitigation (Gupta 2010a, 2010b).

Advance informed agreement as a way to govern GMO trade derives from the longer-established notion of prior informed consent (PIC). PIC has its origins in medical ethics as a risk–benefit balancing strategy for individuals participating in potentially risky clinical trials. Since the mid-1980s, prior informed consent has also underpinned voluntary guidelines and globally negotiated legally binding regimes governing trade in hazardous waste and restricted chemicals (Mehri 1988; Wolf 2000). The choice of prior informed consent as a governance mechanism is a compromise between the two extremes of an outright *ban* on trade in potentially hazardous substances, versus *caveat emptor* ("let the buyer beware"), whereby it is left to the market to govern potentially risky trade (Mehri 1988; Gupta 2010b). This compromise is reflected in the Cartagena Protocol as well, which privileges importing country *choice* based on mandatory information disclosure relating to GMOs in trade, rather than either calling for restrictions on such trade itself or leaving disclosure requirements or restrictions solely to the dictates of markets.

The current limited disclosure obligations institutionalized within this global regime, however, ensure that a norm of *caveat emptor* prevails in practice. This holds at least for GMO varieties contained in the bulk agricultural commodity trade of crops such as maize, canola, or cotton (Gupta 2010b). In documentation accompanying such trade, exporters are required only to disclose that shipments "may contain" GMOs. This requirement does not require existing market practices, such as current lack of segregation between GM and non-GM crop varieties, to change. As a result, governance by disclosure in this case is market following rather than market forcing. Furthermore, in order for importing countries to put the limited disclosed information to use in risk decisions, the onus remains on them to undertake sampling, testing, and verification of such disclosed information.

Debates in the global context of the Cartagena Protocol have now shifted to standardization of sampling criteria, appropriate detection methods, and availability of testing protocols, where again divergent EU–US approaches to such issues come to the fore. The Protocol's disclosure obligations for the GMO commodity trade are thus another global arena where the transatlantic conflict plays out (see Gupta 2010b for a detailed analysis of these dynamics). Given the many hurdles (capacity- and cost-related) to implementing such a disclosure, testing, verification, and labeling approach to domestic biosafety governance, it is noteworthy that some smaller or poorer developing countries are now imposing bans or moratoria on entry of GMOs, notwithstanding the existence of the Cartagena Protocol and its privileging of importer choice (Falck-Zepeda 2006).

A market-following rather than market-shaping outcome of disclosure is also discernible in the Protocol's requirement that parties disclose certain risk and biosafety-related information to its online Biosafety Clearing House (BCH). Information to be disclosed to the BCH includes the genetically modified varieties approved for commercialization in GMO producer countries (as demanded by potential importers); but also domestic biosafety laws and contact persons responsible for import decisions in GMO importing countries (CBD 2008). Given, however, that most GMO producer countries have not ratified the Protocol, the burden of disclosure has fallen, ironically, on importing countries who are parties to the Protocol. Disclosure of information such as contact persons for GMO import decisions can, however, be trade facilitating, given that such information is necessary for trade to occur (Gupta 2010a).

In sum, as this line of research suggests, it is not clear that a biosafety governance approach based on disclosure and choice (and an associated need for segregation, labeling, and extensive infrastructures for sampling, detection, and verification) is an appropriate governance pathway for all countries. Given persisting political conflicts, technical complexities, and high costs of such routes to governance, it is noteworthy that bans and moratoria are becoming a fallback option for some of the poorest countries faced with the challenge of managing entry and safe use of GMOs within their borders. This is likely to remain the case as long as the veil of unknowability continues to hang over future normative and political developments in (global) GMO use, trade, and governance.

Conclusions and Global Policy Implications

This chapter has highlighted normative disagreements over the nature of the biosafety problem, and the implications of such disagreements for evolution and implementation of multilevel biosafety policy approaches. Various strands of research suggest that biosafety policy is shaped by an overall market-liberal bent to (global) environmental governance. This is evident, for example, in how a market-enabling global policy environment appears to have curtailed the evolution of stringent disclosure obligations in the Cartagena Protocol. It is also reflected in the rise and spread of corporate voluntary biosafety approaches in recent years, and their influence in shaping domestic biosafety trajectories (see e.g. Newell 2003; Glover and Newell 2004; Clapp 2007).

Yet a dominance of market liberalism in shaping biosafety trajectories is neither predetermined nor static, but is consistently challenged at multiple levels and by multiple actors (Zelli *et al.* 2012). This is evident, not least, from the policy evolution continually underway within the global context of the Cartagena Protocol. Two aspects now on the global biosafety agenda that may shift the normative balance include, first, a recently concluded agreement on GMO-related liability (Jungcurt and Schabus 2010) and, second, a much-awaited strengthening of trade-related disclosure obligations on GMO exporters, still to be negotiated.

Notwithstanding how these developments will proceed, a key issue remains how to conceptualize *and* institutionalize socially appropriate and democratically legitimate pathways for GMO governance that vary across contexts. Jasanoff (2005) analyzes, for example, the distinctive evolution of biotechnology and biosafety trajectories across OECD countries, and the role of distinct risk rationalities and political cultures therein. Building on this, a future global biosafety research agenda requires further delineation of what democratizing risk governance *can mean* in diverse global and national contexts (Jasanoff 2004, 2006; Gupta 2011; Lövbrand *et al.* 2011) and how it might be linked to imperatives of sovereignty and markets that shape technological trajectories.

Such a research agenda also has to engage with the anticipatory nature of the biosafety governance challenge and the veil of uncertainty over how markets for GM and non-GM crops will develop, which crops will be approved and win acceptance (or not) in key markets, and how understandings of risks will evolve. The challenge remains selecting appropriate governance pathways in the absence of such knowledge and in the presence of epistemological uncertainties relating to risk and harm. The economic, political, and normative stakes are high for all, but particularly for the poorest countries, who might be recipients of (unwanted) GMOs (on "unplanned exposure" to transgenic crops in the South, see Clapp 2006).

Ultimately, anticipatory governance of biosafety remains contested because different governance pathways implicate different (and contested) visions of a desirable future (Jansen and Gupta 2009). Pro-poor biotechnology writings evoke a vision of a future *without* biotechnology as threatening and a future *with* biotechnology as beckoning and promising (on this, see also de Wilde 2000). Either way, the message is that if we do not embrace agricultural biotechnology and are too circumspect about safety, we will run out of time, and the future (at least for some) will be bleak. However, not all share such depictions of the future. The anticipatory governance challenge requires engaging with such diverse views in a legitimate manner; and in democratically agreeing upon appropriate decision parameters and sources of (policy) legitimacy.

References

Alemanno, Alberto. 2011. "Review of Mark A. Pollack and Gregory C. Shaffer, *When Cooperation Fails: The International Law and Politics of Genetically Modified Foods*, Oxford University Press, 2009." *European Law Journal*, 17(3): 1–4.

Appadurai, Arjun. 1996. *Modernity at Large: Cultural Dimensions of Globalization*. Minneapolis: University of Minnesota Press.

Bail, Christoph, Robert Falkner, and Helen Marquard, eds. 2002. *The Cartagena Protocol on Biosafety: Reconciling Trade in Biotechnology with Environment and Development?* London: RIIA/Earthscan.

Bernauer, Thomas. 2003. *Genes, Trade, and Regulation: The Seeds of Conflict in Food Biotechnology*. Princeton, NJ: Princeton University Press.

Bernauer, Thomas and Philipp Aerni. 2007. "Competition for Public Trust: Causes and Consequences of Extending the Transatlantic Biotech Conflict to Developing Countries." In *The International Politics of Genetically Modified Food: Diplomacy, Trade and Law*, ed. Robert Falkner, 138–154. Basingstoke: Palgrave Macmillan.

Bernstein, Steven. 2001. *The Compromise of Liberal Environmentalism*. New York: Columbia University Press.

Cartagena Protocol on Biosafety. 2000. *Cartagena Protocol on Biosafety to the Convention on Biological Diversity: Text and Annexes*. Montreal: Secretariat of the Convention on Biological Diversity.

CBD (Convention on Biological Diversity). 2008. "Operation and Activities of the Biosafety Clearing House." Note by Executive Secretary UNEP/CBD/BS/COP-MOP/4/3, 29.

Cho, Sungjoon. 2010. "Failure of Cooperation or Cooperation of Failure? Transatlantic Regulation of Genetically Modified Foods." *American Journal of International Law*, 104: 324.

Christoforou, Theofanis. 2000. "Settlement of Science-Based Trade Disputes in the WTO: A Critical Review of the Developing Case Law in the Face of Scientific Uncertainty." *New York University Environmental Law Journal*, 8: 622–648.

Clapp, Jennifer. 2006. "Unplanned Exposure to Genetically Modified Organisms: Divergent Responses in the Global South." *Journal of Environment & Development*, 15(1): 3–21.

Clapp, Jennifer. 2007. "Illegal GMO Releases and Corporate Responsibility: Questioning the Effectiveness of Voluntary Measures." *Ecological Economics*, 66(2–3): 348–358.

de Wilde, Rein. 2000. *De voorspellers: eenkritiek op de toekomstindustrie* [The Prophesiers: A Critique of the Futures Industry]. Amsterdam: De Balie Publishers.

Drezner, Daniel W. 2007. *All Politics is Global: Explaining International Regulatory Regimes.* Princeton, NJ: Princeton University Press.

Eckersley, Robyn. 2004. "The Big Chill: The WTO and Multilateral Environmental Agreements." *Global Environmental Politics*, 4(2): 24–50.

Falck-Zepeda, Jose. 2006. "Co-existence, Genetically Modified Biotechnologies and Biosafety: Implications for Developing Countries." *American Journal of Agricultural Economics*, 88(5): 1200–1208.

Falck-Zepeda, Jose and Patricia Zambrano. 2011. "Socioeconomic Considerations in Biosafety and Biotechnology Decision-Making: The Cartagena Protocol and National Biosafety Frameworks." *Review of Policy Research*, 28(2): 171–195.

Falkner, Robert. 2000. "Regulating Biotech Trade: The Cartagena Protocol on Biosafety." *International Affairs*, 76(2): 299–313.

Falkner, Robert. 2006. "International Sources of Environmental Policy Change in China: The Case of Genetically Modified Food." *Pacific Review*, 19(4): 473–494.

Falkner, Robert. 2007. "The Political Economy of 'Normative Power' in Europe: EU Environmental Leadership in International Biotechnology Regulation." *Journal of European Public Policy*, 14(4): 507–526.

Falkner, Robert and Aarti Gupta. 2009. "Limits of Regulatory Convergence: Globalization and GMO Politics in the South." *International Environmental Agreements: Politics, Law and Economics*, 9(2): 113–133.

Foster, Kenneth R., Paolo Vecchia, and Michael H. Repacholi. 2000. "Science and the Precautionary Principle." *Science*, 288 (May 12): 979–981.

Funtowicz, Silvio O. and Jerome R. Ravetz. 1992. "Three Types of Risk Assessment and the Emergence of Post-Normal Science." In *Social Theories of Risk*, ed. Sheldon Krimsky and Dominic Golding, 251–271. Westport, CT: Praeger.

Gehring, Thomas and Sebastian Oberthür. 2006. "Comparative Empirical Analysis and Ideal Types of Institutional Interaction." In *Institutional Interaction in Global Environmental Governance: Synergy and Conflict among International and EU Policies*, ed. Sebastian Oberthür and Thomas Gehring, 307–371. Cambridge, MA: MIT Press.

Glover, Dominic and Peter Newell. 2004. "Business and Biotechnology: Regulation and the Politics of Influence." In *Agribusiness and Society: Corporate Responses to Environmentalism, Market Opportunities and Public Regulation*, ed. Kees Jansen and Sietze Vellema, 200–231. London: Zed Books.

Gottweis, Herbert. 1998. *Governing Molecules. The Discursive Politics of Genetic Engineering in Europe and the United States.* Cambridge, MA: MIT Press.

Gupta, Aarti. 2000. "Governing Trade in Genetically Modified Organisms: The Cartagena Protocol on Biosafety." *Environment*, 42(4): 23–33.

Gupta, Aarti. 2001. "Searching for Shared Norms: Global Governance of Biosafety." PhD dissertation, Yale University Graduate School of Arts and Sciences (UMI No. 3030777).

Gupta, Aarti. 2002. "Advance Informed Agreement: A Shared Basis to Govern Trade in Genetically Modified Organisms?" *Indiana Journal of Global Legal Studies*, 9(1): 265–281.

Gupta, Aarti. 2004. "When Global is Local: Negotiating Safe Use of Biotechnology." In *Earthly Politics: Local and Global in Environmental Governance*, ed. Sheila Jasanoff and Marybeth Long-Martello, 127–148. Cambridge, MA: MIT Press.

Gupta, Aarti. 2008. "Global Biosafety Governance: Emergence and Evolution." In *Institutional Interplay: Biosafety and Trade*, ed. Oran R. Young, W. Bradnee Chambers, Joy Kim, and Claudia ten Have, 19–46. Tokyo: United Nations University Press.

Gupta, Aarti. 2010a. "Transparency to What End? Governing by Disclosure through the Biosafety Clearing House." *Environment and Planning C: Government and Policy*, 28(2): 128–144.

Gupta, Aarti. 2010b. "Transparency as Contested Political Terrain: Who Knows What about the Global GMO Trade and Why Does It Matter?" *Global Environmental Politics*, 10(3): 32–52.

Gupta, Aarti. 2011. "An Evolving Science–Society Contract in India: The Search for Legitimacy in Anticipatory Risk Governance." *Food Policy*, 36(6): 736–741.

Gupta, Aarti and Robert Falkner. 2006. "The Influence of the Cartagena Protocol on Biosafety: Comparing Mexico, China and South Africa." *Global Environmental Politics*, 6(4): 23–55.

Guston, David H. 2010. "The Anticipatory Governance of Emerging Technologies." *Journal of the Korean Vacuum Society*, 19(6): 432–441.

Herring, Ronald. 2007. "Stealth Seeds: Bioproperty, Biosafety, Biopolitics." *Journal of Development Studies*, 43: 130–157.

Isaac, Grant and William Kerr. 2003. "Genetically Modified Organisms at the World Trade Organization: A Harvest of Trouble." *Journal of World Trade*, 37(6): 1083–1095.

Jaffe, Gregory. 2005. "Implementing the Cartagena Biosafety Protocol through National Biosafety Regulatory Systems: An Analysis of Key Unresolved Issues." *Journal of Public Affairs*, 5: 299–311.

James, Clive. 2011. *Global Status of Commercialized Biotech/GM Crops: 2010*. Executive Summary, ISAAA Brief No. 42. Ithaca, NY: ISAAA.

Jansen, Kees and Aarti Gupta. 2009. "Anticipating the Future: 'Biotechnology for the Poor' as Unrealized Promise?" *Futures*, 41(7): 436–445.

Jasanoff, Sheila. 1995. "Product, Process or Programme: Three Cultures and the Regulation of Biotechnology." In *Resistance to New Technology*, ed. Martin Bauer, 311–331. Cambridge: Cambridge University Press.

Jasanoff, Sheila. 1998. "Harmonization: The Politics of Reasoning Together." In *The Politics of Chemical Risk*, ed. Roland Bal and Willem Halffman, 173–194. Dordrecht: Kluwer Academic Publishers.

Jasanoff, Sheila, ed. 2004. *States of Knowledge: The Co-production of Science and Social Order*. New York and London: Routledge.

Jasanoff, Sheila. 2005. *Designs on Nature: Science and Democracy in Europe and the United States*. Princeton, NJ: Princeton University Press.

Jasanoff, Sheila. 2006. "Biotechnology and Empire: The Global Power of Seeds and Science." *OSIRIS*, 21: 273–292.

Jungcurt, Stefan and Nicole Schabus. 2010. "Liability and Redress in the Context of the Cartagena Protocol on Biosafety." *Review of European Community & International Environmental Law*, 19(2): 197–206.

Kleinman, Daniel Lee and Abby Kinchy. 2007. "Against the Neoliberal Steamroller? The Biosafety Protocol and the Social Regulation of Agricultural Biotechnology." *Agriculture and Human Values*, 24(2): 195–206.

Kleinman, Daniel Lee, Abby Kinchy, and Robyn Autry. 2009. "Local Variation or Global Convergence in Agricultural Biotechnology Policy? A Comparative Analysis." *Science and Public Policy*, 36(5): 361–371.

Levidow, Les. 2007. "The Transatlantic Agbiotech Conflict as a Problem and Opportunity for EU Regulatory Policies." In *The International Politics of Genetically Modified Food: Diplomacy, Trade and Law*, ed. Robert Falkner, 118–137. Basingstoke: Palgrave Macmillan.

Lieberman, Sarah and Tim Gray. 2008. "The World Trade Organization's Report on the EU's Moratorium on Biotech Products: The Wisdom of the US Challenge to the EU in the WTO." *Global Environmental Politics*, 8(1): 33–52.

Lövbrand, Eva, Roger Pielke Jr, and Silke Beck. 2011. "A Democracy Paradox in Studies of Science and Technology." *Science, Technology, and Human Values*, 36: 474–496.

McHughen, Alan. 2000. *Pandora's Picnic Basket: The Potential and Hazards of Genetically Modified Foods*. Oxford: Oxford University Press.

Mehri, Cyrus. 1988. "Prior Informed Consent: An Emerging Compromise for Hazardous Exports." *Cornell University Law Journal*, 21: 365–389.

Millstone, Erik, Eric Brunner, and Sue Mayer. 1999. "Beyond 'Substantial Equivalence'." *Nature*, 401 (October 7): 525–526.

Millstone, Erik and Van Zwanenberg, Patrick. 2003. "Food and Agricultural Biotechnology Policy: How Much Autonomy Can Developing Countries Exercise?" *Development Policy Review*, 21(5–6): 655–667.

Morris, Jane. 2008. "The Cartagena Protocol: Implications for Regional Trade and Technological Development in Africa." *Development Policy Review*, 26(1): 29–57.

Murphy, Joseph and Les Levidow. 2006. *Governing the Transatlantic Conflict over Agricultural Biotechnology: Contending Coalitions, Trade Liberalisation and Standard Setting.* Abingdon: Routledge.

Newell, Peter. 2003. "Globalization and the Governance of Biotechnology." *Global Environmental Politics*, 3(2): 56–71.

Newell, Peter. 2007. "Corporate Power and 'Bounded Autonomy' in the Global Politics of Biotechnology." In *The International Politics of Genetically Modified Food: Diplomacy, Trade and Law*, ed. Robert Falkner, 67–84. Basingstoke: Palgrave Macmillan.

Nuffield Council on Bioethics. 2003. *The Use of Genetically Modified Crops in Developing Countries: A Follow-up Discussion Paper.* London: Nuffield Council on Bioethics.

Oberthür, Sebastian and Thomas Gehring. 2006. "Institutional Interaction in Global Environmental Governance: The Case of the Cartagena Protocol and the World Trade Organization." *Global Environmental Politics*, 6(2): 1–31.

Oberthür, Sebastian and Olav Schram Stokke, eds. 2011. *Institutional Interplay and Global Environmental Change. Interplay Management and Institutional Complexes.* Cambridge, MA: MIT Press.

O'Riordan, Timothy and Andrew Jordan. 1995. "The Precautionary Principle in Contemporary Environmental Politics." *Environmental Values*, 4(3): 191–212.

Paarlberg, Robert. 2001. *The Politics of Precaution: Genetically Modified Crops in Developing Countries.* Baltimore, MD: Johns Hopkins University Press.

Paarlberg, Robert. 2008. *Starved for Science: How Biotechnology is Being Kept out of Africa.* Cambridge, MA: Harvard University Press.

Pollack, Mark A. and Gregory C. Shaffer. 2005. "Biotechnology Policy." In *Policy-Making in the European Union*, ed. Helen Wallace, William Wallace, and Mark A. Pollack, 329–351. Oxford: Oxford University Press.

Pollack, Mark A. and Gregory C. Shaffer. 2009. *When Cooperation Fails. The International Law and Politics of Genetically Modified Foods.* Oxford: Oxford University Press.

Prakash, Aseem and Kelly Kollman. 2003. "Biopolitics in the EU and the U.S.: A Race to the Bottom or Convergence to the Top?" *International Studies Quarterly*, 47(4): 617–641.

Rissler, Jane and Margaret Mellon. 1996. *Ecological Risks of Engineered Crops.* Cambridge, MA: MIT Press.

Safrin, Sabrina. 2002. "Treaties in Collision? The Biosafety Protocol and the World Trade Organization Agreements." *American Journal of International Law*, 96: 606.

Scoones, Ian. 2008. "Mobilizing against GM Crops in India, South Africa and Brazil." *Journal of Agrarian Change*, 8(2–3): 315–344.

SPS Agreement. 1994. Agreement on the Application of Sanitary and Phytosanitary Measures. Annex 1A to the Final Act Embodying the Results of the Uruguay Round of Multilateral Trade Negotiations. Marrakesh, April 15.

Stabinsky, Doreen. 2000. "Bringing Social Analysis into a Multilateral Environmental Agreement: Social Impact Assessment and the Biosafety Protocol." *Journal of Environment & Development*, 9: 260–283.

Sunstein, Cass. 2005. *Laws of Fear: Beyond the Precautionary Principle.* Cambridge: Cambridge University Press.

WHO (World Health Organization). 2005. *Modern Food Biotechnology, Human Health and Development: An Evidence-Based Study.* Geneva: WHO.

Wolf, Amanda. 2000. "Informed Consent: A Negotiated Formula for Trade in Risky Organisms and Chemicals." *International Negotiation*, 5(3): 485–521.

Wolfenbarger, LaReesa and Paul R. Phifer. 2000. "The Ecological Risks and Benefits of Genetically Engineered Plants." *Science*, 290 (December 15): 2088–2093.

Wright, Susan. 1994. *Molecular Politics: Developing American and British Regulatory Policy for Genetic Engineering*, 1972–1982. Chicago: University of Chicago Press.

WTO (World Trade Organization). 2003. "European Communities: Measures Affecting the Approval and Marketing of Biotech Products: Request for Establishment of a Panel by the United States," WT/DS291/23, August 8.

Young, Alasdair R. 2003. "Political Transfer and 'Trading Up'? Transatlantic Trade in Genetically Modified Food and U.S. Politics." *World Politics*, 55: 457–484.

Young, Oran R., W. Bradnee Chambers, Joy Kim, and Claudia ten Have, eds. 2008. *Institutional Interplay: Biosafety and Trade*. Tokyo: United Nations University Press.

Zelli, Fariborz, Aarti Gupta, and Harro van Asselt. 2012. "Horizontal Institutional Interlinkages." In *Global Environmental Governance Reconsidered*, ed. Frank Biermann and Philipp Pattberg, 175–198. Cambridge, MA: MIT Press.

Chapter 7

Global Chemicals Politics and Policy

Henrik Selin

Introduction

All over the world, products of the modern chemicals revolution are used to improve human standards of living in numerous ways, including by increasing yields of major cash crops, protecting public health, and producing countless industrial and consumer goods. Contemporary reliance on a multitude of pesticides and industrial chemicals, however, has also resulted in significant environmental and human health problems, ranging all the way from minor skin rashes to alarmingly high rates of cancer-related fatalities (Mancini *et al.* 2005; Langman 2007). At the World Summit on Sustainable Development (WSSD) in Johannesburg in 2002, countries agreed that chemicals worldwide should be "used and produced in ways that lead to the minimization of significant adverse effects on human health and the environment" no later than the year 2020 (WSSD 2002: paragraph 22). While societies now are better at recognizing chemical risks than only a few decades ago, countries are not on track to meet the 2020 goal. Thus, chemicals management is both a critical environmental and human health issue and a significant topic in global environmental politics.

As countries cooperate on a wide range of political, scientific, and technical chemicals abatement issues, they work closely with a host of IGOs and NGOs (Lönngren 1992; Selin 2010; Wexler *et al.* 2012). The institutional framework for managing chemicals is structurally different from many other major environmental regimes. Rather than organizing cooperation under an overarching framework convention, as in for example the cases of climate change, ozone depletion, and biodiversity, international legal and political efforts to address problems of hazardous chemicals are structured around a diverse set of legally independent treaties and programs. This structuring of policy-making across formally independent but functionally linked

The Handbook of Global Climate and Environment Policy, First Edition. Edited by Robert Falkner.
© 2013 John Wiley & Sons, Ltd. Published 2013 by John Wiley & Sons, Ltd.

forums creates both governance opportunities and challenges. Regarding opportunities, it provides states, IGOs, and NGOs with a wide range of policy instruments to address the multifaceted aspects of chemicals management. With respect to challenges, it creates particular needs for coordinated decision-making and implementation to ensure the overall effectiveness of those policy measures that have been taken.

In part because of the institutionally fragmented nature of the chemicals regime, this is an area of global environmental politics and policy-making where regime participants have engaged in relatively long-standing cooperation about ways to promote policy coordination and capture synergies across different agreements and management efforts. As part of these efforts, states and different stakeholder groups in 2006 adopted the Strategic Approach to International Chemicals Management (SAICM), which operates as an umbrella mechanism promoting sound chemical management and harmonization of controls and activities across major agreements and programs. While SAICM is not comparable to a framework convention – it is a voluntary program and not based on a legally binding agreement requiring ratification – it is an important institutional part of continuing work on synergies and treaty implementation. The institutional complexity of the chemicals regime creates several legal, political, and management challenges as national governments and other stakeholder groups seek to improve environmental and human health protection from hazardous chemicals.

Most major institutional parts of the global chemicals regime have been developed by states, IGOs, and NGOs since the 1980s (but with disparate actions on hazardous substances also taken much earlier). The core of the institutionally diverse chemicals regime is structured around four treaties. Three of these are global: the 1989 Basel Convention on the Control of Transboundary Movements of Hazardous Wastes and Their Disposal; the 1998 Rotterdam Convention on the Prior Informed Consent Procedure for Certain Hazardous Chemicals and Pesticides in International Trade; and the 2001 Stockholm Convention on Persistent Organic Pollutants (POPs). The fourth one is regional: the 1998 Protocol on Persistent Organic Pollutants to the Convention on Long-Range Transboundary Air Pollution (CLRTAP). These four treaties cover different but partially overlapping parts of the chemicals life cycle of production, use, emissions, trade, and disposal. Many hazardous chemicals are also covered by two or more treaties. The four treaties are further connected through overlaps in membership and stakeholder engagement.

This chapter addresses global chemicals politics and policy, as a major part of international environmental cooperation and governance. The subsequent section outlines the nature of the chemicals problem, including different ways in which international cooperation has the potential to support better environmental and human health protection from hazardous chemicals. This is continued by a presentation of the chemicals regime and its main policy responses in the form of central treaties and programs as well as some of the major IGOs and NGOs involved in international chemicals management. This section also identifies important linkages between agreements and policy efforts. The next section examines major issues for improving regime effectiveness. The final section identifies areas where there is a need for more empirical research and analysis as well as discussing policy needs towards one day achieving the goal of safe production and use of chemicals.

The Nature of the Chemicals Issue

The chemical industry consists of firms that produce chemicals from raw materials (mainly petroleum) as well as those that alter or blend individual substances into different mixtures. It is unclear how many chemicals are currently in use worldwide, but estimates are in the 60 000 to 100 000 range. Production volumes of individual chemicals range from millions of tonnes per year to quantities of much less than 1000 tonnes annually. Global sales of chemicals grew almost ninefold between 1970 and 2000. Total chemicals production (excluding pharmaceuticals) is currently worth over US$2 trillion, where 45% of this value is traded internationally (including intra-firm trade). Asia (mainly Japan, China, and India) is the world's leading chemicals-producing region in monetary terms, followed by the European Union (EU) and the United States (CEFIC 2006). While firms in industrialized countries in the short term will continue to dominate the market in specialty and life science chemicals – whose production requires advanced technology and educated workers – there is a rapid growth in the production and use of more basic chemicals in many developing countries.

There are no exact global data on chemicals contamination and poisoning, but IGOs and researchers have produced estimates based on (much-debated) extrapolations from local information from different parts of the world. The World Health Organization (WHO) proposed in 1990 that 3 million people were hospitalized every year as a result of pesticide poisoning, resulting in 220 000 deaths (Jeyaratnam 1990). More recent calculations put annual fatality figures closer to 300 000, 99% of which occur in developing countries (Srinivas Rao *et al.* 2005). In addition to the deadly effects of high-dose exposure, environmental risks from low-dose exposure include estrogenic effects, disruption of endocrine functions impairing immune system functions, functional and physiological effects on reproduction capabilities, and reduced survival and growth of offspring (AMAP 2010). Human low-dose exposure has been linked to carcinogenic and tumorigenic effects as well as endocrine disruptions (AMAP 2009). Many national health authorities recommend that pregnant women and small children limit their dietary intake of certain fish containing high levels of chemicals to reduce risks.

Hazardous chemicals are released through agricultural use, common industrial and manufacturing practices, combustion processes, leakages from wastes, and accidents. Many problems are related to the persistence, toxicity, bioaccumulation, and biomagnifications of hazardous substances. Persistence refers to how long a chemical remains in the environment before it is biodegraded. Toxicity refers to the negative effects a chemical may have on an organism or part of an organism (organ, tissue, or cell). Bioaccumulation is an essential biological process that takes place in all living organisms to obtain necessary nutrients but one where problems can arise when hazardous substances are accumulated through the same mechanism, allowing them to build up in fatty tissues over time. Biomagnification is a biological process related to bioaccumulation as hazardous chemicals that have bioaccumulated in a large number of organisms at a lower trophic level are concentrated further by an organism at a higher trophic level as those chemical concentrations are passed up food-webs (AMAP 2009, 2010).

Management of hazardous substances involves balancing benefits and risks of chemicals use, as different interests and needs are considered. Since at least the 1960s, national authorities and international organizations have struggled to design effective mechanisms for risk assessments and regulations. These management efforts have been fueled by many high-profile chemical accidents and contamination scandals, including those occurring in Love Canal, USA; Seveso, Italy; and Bhopal, India (Selin 2010). Although all the world's countries struggle to design effective chemicals policy, many management challenges are particularly difficult in developing countries (which are then also the countries that are experiencing some of the most serious environmental and human health problems). For example, national implementation plans developed under the Stockholm Convention reveal many challenges to effective policy-making and implementation. For instance, in a report submitted by Tanzania, the government states that there are only 15 pesticide inspectors the whole country (Government of Tanzania 2005). Similar low-capacity issues plague many other countries also.

There are several international dimensions to the chemicals problem (Krueger and Selin 2002). One important aspect relates to the transboundary transport of emissions of POPs and other substances. This means that hazardous chemicals can be transported over long distances, cross national borders, and thus cause contamination problems far from where they were originally released. The connection between environmental and human health risks and international trade is also very strong. Many farmers and workers – particularly in developing countries – are exposed to risks from imported pesticides and industrial chemicals, including the rapidly growing trade in electronics wastes (e-wastes). In addition to the serious risks to the people who directly handle hazardous substances, there can also be risks to consumers worldwide. Even if a hazardous substance is banned for direct use in one country but used elsewhere, it may still enter the country in which it is prohibited in the form of residues in imported vegetables and fruits, or as parts of many consumer goods (Emory 2001).

An important aspect of international cooperation is that it can help in diffusing scientific and socio-economic knowledge about the chemicals problem. In many cases, people who regularly use or are exposed to hazardous chemicals are unaware of the risks and are not trained to take even the most basic protective measures. Furthermore countries that recognize widespread domestic problems with hazardous chemicals sometimes have difficulties mustering adequate technical, financial, and/or human resources to initiate more effective risk-reduction measures. This is again particularly true for many developing countries. Ideally, international legal and political activities can function as important catalysts for the dispersion of resources that may enable better domestic actions to reduce environmental and human health effects stemming from the unsafe handling and use of hazardous chemicals. Management improvements are badly needed, as data from all over the world demonstrate that societies have a long way to go to achieve safe production and use of chemicals (Mancini *et al.* 2005; Langman 2007; Liu 2010; Harrison 2011).

The Chemicals Regime

The global community has expanded the chemicals regime to include regulations on the full life cycle of production, use, trade, and disposal of industrial chemicals and pesticides as well as emission controls on by-products of production and combustion processes. The regime also contains provisions and management programs designed to regulate additional chemicals, increase and harmonize information about commercial and discarded chemicals traded across countries, generate more scientific and socio-economic data for risk assessment, and enhance regional and local management capacities. Despite the fact that these many legal and political responses have been developed incrementally and are formally independent, there are both cognitive and practical reasons to regard them as part of a regime (Selin 2010). Cognitively, states, IGOs, and NGOs perceive major chemicals issues to be connected and formulate policy responses and management efforts based on these connections. Practically, countries are parties to multiple agreements and the same chemicals are regulated under more than one treaty and through similar control mechanisms where policy-making under one agreement can greatly shape debates and outcomes in other forums.

The *Basel Convention* addresses the generation and transboundary transport of hazardous wastes. By 2012, 177 countries and the EU were parties. The Basel Convention covers wastes containing hazardous substances and discarded chemicals can also be classified as hazardous wastes. Levels of hazardous wastes have increased sharply since the 1960s. Most waste trade takes place between industrialized countries (O'Neill 2000). However, it was the growing waste trade between industrialized and developing countries that provided the political impetus to the Basel Convention. This included several high-profile cases of illegal dumping of hazardous wastes in developing countries (Kummer 1995; Krueger 1999; Clapp 2001). Responding to political pressure from mainly developing countries, the Governing Council of the United Nations Environment Programme (UNEP) approved the so-called Cairo Guidelines in 1987, setting the first voluntary trade standard. The Cairo Guidelines introduced a prior informed consent (PIC) procedure: export of hazardous wastes from one firm to another could take place only after the national government of the country where the importing firm was located had given explicit permission to go ahead with the trade.

Many developing countries, a few Nordic countries, and several environmental advocacy groups, however, believed that a voluntary system was not strong enough to control unwanted imports and unlawful dumping. In response, countries adopted the Basel Convention in 1989, which prohibits the export of hazardous wastes to Antarctica and to parties that have taken domestic legal measures to ban such imports. Permitted waste transfers to other parties are subject to a legally binding PIC procedure: a party cannot permit export of hazardous wastes to another party without first receiving explicit consent of the importing state to proceed with the transfer. Waste exports to non-parties are prohibited, unless they are subject to an agreement between the exporter and importer that is at least as stringent as the requirements under the Basel Convention. The trade in old or discarded chemicals is subject to controls by the Basel Convention if they are categorized as hazardous wastes under the convention. Even if the Basel Convention provides some legal protection, there remain serious problems with the legal and illegal trade in hazardous wastes (Iles 2004; Pellow 2007).

The issue of even stricter trade restrictions stayed on the agenda. Following some political gains during the first two conferences of the parties (COPs), a coalition led by many African countries convinced the parties in 1995 to adopt the Ban Amendment, which prohibits the export of hazardous wastes for final disposal and recycling from countries listed in Annex VII (parties that are members of the Organisation for Economic Co-operation and Development (OECD) and the EU as well as Liechtenstein) to all other parties (i.e. developing countries). However due to opposition from some industrialized countries and also slow ratification by many developing countries seeking to profit from the waste trade, the Ban Amendment has not yet entered into force. Furthermore, the parties in 1999 adopted the Basel Protocol on Liability and Compensation, which identifies who is financially responsible in the event of an incident during the transboundary movement of hazardous wastes. This protocol, however, has also not yet entered into force. In addition, parties focus on funding, capacity-building, and technology transfer issues. Related to all of these, parties have approved the establishment of 14 regional centers to support regional and local management (Selin 2012a).

The *Rotterdam Convention* also deals with trade, but focuses on commercial substances. By 2012, 145 countries and the EU had ratified the Rotterdam Convention. Similar to the Basel Convention, it was the largely unregulated North–South trade that acted as the main stimulus for policy developments. Following political discussions and initiatives pushed by developing countries dating back to the 1970s, UNEP Governing Council in 1989 adopted the first global voluntary PIC procedure in the so-called Amended London Guidelines. This was similar in operation to that under the Basel Convention. In 1989, the Council of the Food and Agricultural Organization (FAO) also adjusted its Code of Conduct to include a compatible system. The PIC scheme was managed jointly by the FAO for pesticides and by UNEP Chemicals for industrial chemicals (Paarlberg 1993; Victor 1998). In the 1990s, political pressure from mainly developing countries increased to convert the voluntary scheme into a treaty to strengthen the position of importers and provide better environmental and human health protection. Subsequently, the Rotterdam Convention was adopted in 1998 (Kummer 1999; Emory 2001; McDorman 2004).

The Rotterdam Convention PIC procedure stipulates that the government of a potentially importing party can respond in three different ways after receiving a formal request to accept import of a particular chemical on the PIC list. First, the government can declare that it consents to receive the import of the chemical and any other shipments within the same calendar year; second, the government may reject the request; or third, the government may consent to import, but only if specific conditions are met by the exporting party. The government of the potential exporter must abide by any decision made by the potentially receiving country. The Secretariat, which is divided between UNEP Chemicals and FAO, acts as facilitator throughout this process and distributes all the responses between the parties. National governments are in turn responsible for communicating all information and decisions from the other party to all relevant domestic firms. Still, as under the Basel Convention, both legal and illegal trade of commercial chemicals continues to cause problems (Collins 2010).

The Rotterdam Convention stipulates that a party that has banned or severely restricted the use of a chemical of a category covered by the treaty is required to notify the Secretariat. When the Secretariat has received notification from parties from at least two different geographical regions – or a single party that is a developing country or a country with an economy in transition experiencing domestic problems – it forwards all related information to the Chemical Review Committee, which conducts an evaluation and submits a recommendation to the COP making the final decision regarding inclusion on the PIC list (Kohler 2006). As of early 2012, the Rotterdam Convention covered 43 chemicals. Several hundred other chemicals are being lined up for review by the Chemical Review Committee, and it is expected that the parties will continue to expand the PIC list. Operating alongside the Rotterdam Convention, the Globally Harmonized System for the Classification and Labelling of Chemicals is designed to make it easier to identity specific chemicals transported between different countries (Selin 2010).

The *CLRTAP POPs Protocol* was the first multilateral agreement targeting POPs as a separate category of particularly hazardous chemicals (even if many of the substances covered by the protocol were already regulated by earlier regional agreements). By 2012, 30 countries and the EU were parties. The protocol covers the production, use, emissions, and disposal of POPs and operates under the auspices of the United Nations Economic Commission for Europe (UNECE), which comprises North America and Europe as far east as Russia and Kazakhstan. This agreement was largely born out of North American and Northern European concerns with the long-range transport of emissions of hazardous substances to northern latitudes, in particular the Arctic. In Canada more than any other country, the POPs issue became integrated with broader scientific and political concerns about Arctic environmental contamination and health risks, particularly of indigenous peoples (Downie and Fenge 2003; Thrift *et al.* 2009). Indigenous groups were also active participants in Canadian and circumpolar research programs and policy forums, as they lobbied for strong regulatory action under CLRTAP (Selin and Selin 2008).

The CLRTAP assessments led by Canada and Sweden in the early 1990s identified a set of priority POPs that were subject to extensive long-range transport and measured throughout the northern hemisphere, leading to political negotiations on possible control options (Selin 2003). Consequently, the CLRTAP POPs Protocol is designed to reduce the release and long-range transport of POPs emissions. To this end, regulated chemicals are divided into three annexes. The production and use of pesticides and industrial chemicals listed Annex I are banned. Annex II lists pesticides and industrial chemicals for which there are listed use exemptions. POPs by-products of industrial and combustion processes, controlled through applications of best available techniques and best environmental practices, are listed in Annex III. The agreement also set standards for the environmentally sound transport and disposal of discarded POPs, intended to be consistent with stipulations under the Basel Convention. In many ways, the development of this regulatory approach shaped subsequent global assessments and negotiations leading to the Stockholm Convention (Selin 2003, 2010).

The CLRTAP POPs Protocol originally covered 16 chemicals. Similar to the evaluation mechanism set up under the Rotterdam Convention (and also the Stockholm Convention, discussed in the following paragraph), countries designed the agreement so that additional chemicals can be assessed and possibly regulated under any of the three annexes. In 2009, the parties in many cases led by the EU added 9 more chemicals, so that a total of 25 chemicals were covered by the CLRTAP POPs Protocol by 2012. Parties have also nominated other chemicals for evaluation, and it is likely that additional substances will be regulated in the future. As the parties to this regional agreement move forward, there are many institutional linkages connecting it to the global Rotterdam and Stockholm Conventions. Furthermore, legal, political, and management linkages between these agreements are growing, as the number of overlapping parties increases and countries continue to expand the lists of chemicals regulated simultaneously under two or three of these major chemicals treaties.

The *Stockholm Convention* sets global controls on the production, use, emissions, trade, and disposal of POPs. By 2012, 175 countries and the EU had ratified the Stockholm Convention. In 1995, the UNEP Governing Council called for global assessments of 12 POPs ("the dirty dozen"). Based on these assessments, the UNEP Governing Council in 1997 initiated treaty negotiations on the 12 POPs (Downie and Fenge 2003; Selin and Eckley 2003). These substances were generally recognized by both industrialized and developing countries as harmful and demonstrated to pose significant environmental and human health risks (and European and North American countries had already reached consensus on this under CLRTAP and now wanted to initiate global controls). Many countries by the late 1990s had already banned the production and use of the 10 commercial pesticides and industrial chemicals under negotiation, or severely restricted their application. Arctic conditions continued to loom large, but many negotiation issues concerned the interests and situations of developing countries with respect to chemicals use and ways to support better local management, in part because these countries were not part of the earlier CLRTAP work on POPs.

The Stockholm Convention started out regulating 12 POPs (all of which at the time were also controlled by the CLRTAP POPs Protocol and some were also covered by the Rotterdam Convention). Substances are divided into three annexes, similar in structure to the CLRTAP POPs Protocol. Annex A lists pesticides and industrial chemicals whose use and production are prohibited, but parties can apply for country-specific and time-limited exemptions. Annex B lists POPs whose use and production are still permitted, but are subject to specific production and use provisions. Annex C lists by-products covered by the convention, and also outlines general guidelines on best available techniques and best environmental practices for their minimization. In addition, the Stockholm Convention includes a detailed mechanism for evaluating and possibly including additional chemicals under the treaty (again similar to the mechanisms under the CLRTAP POPs Protocol and the Rotterdam Convention). These evaluations are carried out by a designated POPs Review Committee, which reports to the COPs making all final regulatory decisions (Kohler 2006).

Following the entry into force of the Stockholm Convention, the parties have added 10 more chemicals, making it a total of 22 controlled POPs. More substances are likely to be added in the future, as the convention-based assessment work progresses. The import and export of POPs are permitted only for substances subject to use exemptions or for their environmentally sound management and disposal.

On these issues, the Stockholm Convention is designed to be compatible with the Rotterdam and Basel Conventions. In addition, work on formulating technical guidelines for the environmentally sound management of stockpiles and wastes is carried out in collaboration with the Basel Convention, as the Basel parties are simultaneously formulating technical guidance documents on POPs wastes. Parties are also developing guidelines for best available techniques and best environmental practices for controlling by-products. In addition, parties have established 15 Stockholm Convention regional centers to aid capacity-building and technology transfer, mainly in developing countries (Selin 2012a).

In addition to the four main chemicals agreements, countries around the globe are parties to many regional agreements controlling somewhat similar sets of hazardous substances in different ways (Selin 2010). All of these contribute to the institutional complexity of the chemicals regime, shaping policy-making and implementation. Several initiatives have been created under the UNEP's Regional Seas Programme since the 1970s; currently covering 13 actions plans involving over 140 countries. There are also long-standing agreements covering other shared bodies of water outside the UNEP program, including for the Northeast Atlantic, the Baltic Sea, and the Great Lakes. Furthermore, many of the world's transboundary rivers are covered by more or less strict pollution-related legal provisions. In addition, there are several regional hazardous waste treaties covering discarded hazardous chemicals as well as wastes containing them. Some of these waste agreements have been established by developing countries to gain additional legal means and protection to prevent unwanted imports and dumping beyond what is afforded by the Basel Convention.

Many IGOs work on chemicals. As discussed earlier, UNEP helped develop global and regional agreements, and also established the International Register of Potentially Toxic Chemicals in 1976. The International Labour Organization (ILO) addresses chemicals in work places. The OECD coordinates testing requirements and establishes guidelines for data generation and sharing. The FAO and the WHO collaborate in the Codex Alimentarius Commission to establish maximum acceptable levels of pesticide residues in foods. Alongside SAICM, the Inter-Organization Programme for the Sound Management of Chemicals (IOMC) is a mechanism for coordinating action towards the WSSD 2020 goal, involving nine participating organizations: FAO, ILO, OECD, UNEP, WHO, the World Bank, the United Nations Institute for Training and Research (UNITAR), the United Nations Development Programme (UNDP), and the United Nations Industrial Development Organization (UNIDO).

A wide range of environmental NGOs have long focused on international and local problems with hazardous chemicals. Highly active NGOs include not only traditional organizations such as the WWF and Greenpeace, but also more issue-specific ones. These include the Basel Action Network, the Pesticide Action Network, the International POPs Elimination Network, and the International Chemical Secretariat. Indigenous peoples' groups have also been a major presence in the development of international chemicals policy since the 1990s, not least the Inuit Circumpolar Council. In addition, major industry associations and multinational firms have participated in international chemicals politics. For example, the International Council of Chemical Associations, the American Chemistry Council, and the European Chemical Industry Council, together with many firms such as BASF,

DuPont, and Dow Chemical regularly attend international scientific and political meetings. All these different kinds of non-state actors are also in frequent contact with state officials and IGO staff, as they try to shape policy-making in many different ways.

Improving Regime Effectiveness

In a global multilevel governance system such as the one that is under development for the management of hazardous chemicals, states, IGOs, and NGOs interact within and across forums that are formally independent but functionally interdependent. These forums are also located at different governance scales. In this respect, multilevel governance studies overlap with the analysis of how institutional linkages shape politics, policy-making, and implementation across jurisdictions. In an institutionally dense issue area such as that of chemicals management, an important aspect of improving environmental and human health protection from hazardous chemicals is better coordinated policy-making and implementation. In addition to focusing on important agreement-specific issues, regime participants need to address horizontal linkages within governance scales (for example, harmonize decisions on the same or related chemicals under two or more global treaties) as well as vertical linkages across governance scale (for example, consider how global policy-making on a particular chemical relates to regional agreements, influences national management, and affects local communities).

Institutional linkages can have both positive and negative effects on collective problem-solving and regime effectiveness (Selin and VanDeveer 2003; Oberthür and Gehring 2006; Selin 2010; Oberthür and Stokke 2011). Supportive linkages across policy forums can facilitate policy-making and implementation, which may allow regime participants to capture important regulatory and management synergies. Ideally, actions on, for example, PCBs under the Basel and Stockholm Conventions should be complementary and collectively enhance parties' abilities to address PCBs. However, regime participants may also act in ways that hinder policy developments in response to linkages between policy forums. In such cases, controversy between groups of state and non-state actors on, for example, regulations of a particular chemical in one policy forum spills over into another one, causing stalemate. Such an impasse may be more difficult to break because the fact that issues are linked across multiple policy forums raises the overall political stakes. In these respects, institutional density influences linkage politics – the strategic use of institutional linkages to shape policy processes and achieve desired outcomes (Selin 2010).

Sometimes, basic policy coordination and standardization is facilitated by the fact that regime participants share a basic interest in harmonizing principles, norms, rules, and decision-making procedures across forums so that they are not faced with conflicting or contradictory commitments and requirements. This interest in harmonization has been a driving force behind many of the political debates and policy actions on efforts to capture synergies across the main chemicals agreements, dating back to the 1990s (Krueger and Selin 2002). However, not all collaborative efforts on policy coordination and diffusion are uncontroversial. Regime participants may at times want to standardize in different ways and at differing levels of stringency, as they seek varying types of preferred outcomes. In such instances, political disagreement in one forum can spill over into others, hindering decision-making across the regime. An important aspect of such linkage politics is that actors may strategically engage in forum shopping and scale shopping, as they look for political venues where they think they can best advance their interests (Selin 2012b).

Among the many legal, political, and organizational steps that the international community has been taking to improve policy coordination to better capture synergies across forums (based on the cognitive and practical linkages discussed earlier), the adoption of SAICM in 2006 is one of the more important ones. Created by states, IGOs, and a wide range of non-state actors, SAICM is designed to act as an overarching policy framework, promoting policy coordination and the capturing of management synergies across different agreements and programs. One of its main purposes is supporting efforts towards the fulfillment of the WSSD 2020 goal on the safe production and use of chemicals. SAICM objectives are grouped under five main themes: risk reduction; knowledge and information; governance; capacity-building and technical cooperation; and illegal international traffic. Measures under each of these themes are intended to complement treaty-specific activities. While this may be the case sometimes, SAICM together with the IOMC also adds to the institutional complexity of the chemicals regime.

Furthermore, parties to the three global conventions have taken steps to coordinate the COPs – the supreme policy-making bodies of each agreement. The three COPs gain their legal authority directly from each convention and can make only decisions pertaining to their respective treaty. However, in 2010 a simultaneous extraordinary meeting of the COPs to the Basel, Rotterdam, and Stockholm Conventions was held to move forward the coordination and synergies agenda. Issues discussed and acted upon included organizational structures and managerial functions. All three global conventions started out with their own independent secretariats. However, as parties were pushing to better coordinate policy-making and management, states saw benefits in merging secretariat activities and synchronizing budget cycles. To this end, the three COPs in 2011 established the position of Executive Secretary as a joint administrative function for the Basel, Rotterdam, and Stockholm Conventions secretariats (that is, one person is overseeing the activities of all three). In addition, the secretariats among other things have taken steps to create joint legal, financial, and administrative services.

As regime participants continue to consider different ways to improve multilevel governance, there is widespread agreement among countries about the need to better implement existing controls as well as expand the number of regulated chemicals through the introduction of appropriate life cycle controls (even if states and stakeholder groups do not always agree on which chemicals should be regulated and through what means). Studies and reports demonstrate that there are many substances currently not regulated under any of the four main chemicals agreements that pose environmental and human health risks (and these may or may not be covered by much domestic legislation) (Langman 2007; AMAP 2010; PANAP 2010). Importantly, future assessments and policy decisions are linked across treaties, as they are an important part of linkage politics. That is, any decision to regulate or not regulate a specific chemical may directly shape debates and outcomes in other forums. As a result, supporters of regulations seek to expand controls in multiple policy venues while opponents of regulations attempt to block or minimize controls in these same forums.

Another political challenge under the institutionally fragmented chemicals regime is to establish more comprehensive monitoring and compliance mechanisms, as these are central to efforts to improve data collection, implementation, and effectiveness. Such mechanisms could also operate across multiple treaties, as part of the broader effort on policy and implementation coordination. This could help ensure consistency in treaty implementation across the regime as well as reduce the burden on parties to compile and submit separate reports to multiple secretariats. The development of such mechanisms depends heavily on the political interest and will of parties. Both industrialized and developing countries, however, are reluctant to give up sovereignty and approve the design of independent monitoring and compliance mechanisms. At the same time, many developing countries, in particular, are struggling to find human and financial resources to meet expanded data-gathering and reporting requirements. As such, monitoring and compliance issues are closely related to debates and efforts on capacity-building and technology transfer.

While all countries struggle to better address environmental and human health problems stemming from hazardous substances, it is clear that many developing countries are facing particular management challenges. Working through IGOs and NGOs to enhance management capacities in developing countries is an important environmental justice issue. This is also an area where there are growing coordination efforts across treaties and forums (including under SAICM). Activities in this area involve the establishment and operation of regional centers under the Basel and Stockholm Conventions. By 2012, there were 14 Basel Convention regional centers and 15 Stockholm Convention regional centers. Of the 14 Basel Convention regional centers, six also function as regional centers under the Stockholm Convention. Their overall impact is still unclear, but the regional centers have initiated activities in three broad areas important to capacity-building, technology transfer, and treaty implementation: raising awareness, strengthening administrative ability, and diffusing scientific and technical assistance and information (Selin 2012a).

Of course, expanded capacity-building and technology transfer are dependent on not only shoring up the necessary political will, but also the availability of greater resources. Under the chemicals regime – as in many other environmental issue areas – resource debates are part of broader North–South politics on responsibilities and funding. Developing country efforts to establish mandatory funding mechanisms under the main chemicals conventions (such as the multilateral fund under the ozone regime) have been rejected by industrialized countries. Instead, industrialized countries have been willing to commit only to voluntary mechanisms, while the Global Environment Facility (GEF) is connected to some chemicals work (mainly on POPs). However, developing countries argue that not enough funds are made available and that the GEF project procedure is too bureaucratic. Developing countries, together with countries with economies in transition, also argue that they require more resources for capacity-building and technology transfer as part of any effort to strengthen monitoring and compliance mechanisms, to address situations where non-compliance is due to a lack of resources.

As many hazardous substances continue to pose significant environmental and human health problems, the adoption of more proactive and precautionary policies and management approaches is needed (Selin 2010). To date, most international and national regulatory efforts have focused on the management of known or suspected hazardous chemicals, rather than actively promoting the development of less harmful chemicals or non-chemical alternatives. In traditional management procedures, the burden of proof is also on regulators to prove that a chemical is not safe, rather than the producer and/or seller having to produce data demonstrating that a substance is not likely to cause adverse environmental and human health effects. It is, however, only through the development and application of quicker and more proactive procedures for assessment and regulation that the main chemicals treaties can become truly effective. In the end, the best way to protect human health and the environment from hazardous chemicals is to avoid using them in the first place.

The promotion of green chemistry – the utilization of principles that reduce or eliminate the use or generation of hazardous substances in the design, manufacture, and application of chemicals – is an effort to more effectively take environment and health concerns into consideration when synthesizing new chemicals (Anastas and Warner 1998). Green chemistry proponents stress the importance of creating substances with little or no environmental toxicity. Chemicals should also be designed so that at the end of their functional lives they break down into innocuous degradation products, as part of a broader effort to create a more sustainable use of materials (Geiser 2001). In addition to having many significant and much-needed environmental and human health benefits, a more proactive policy approach to chemicals management reduces costs of cleaning up areas contaminated by toxic substances. There are no reliable global data on the costs of dealing with contaminated areas, but these have been – and will continue to be – significant in both industrialized and developing countries.

Future Research and Policy Needs

The chemicals regime offers several opportunities for continued empirical research and analysis. One area involves further research into multilevel governance, institutional linkages, and the design and operation of effective governance structures and bodies. Many international environmental issue areas suffer from implementation and compliance problems and the chemicals regime is no exception. With respect to the protection of the environment and human health from hazardous chemicals, there is a significant gap between stated policy goals and on-the-ground realities all over the world. This raises critical questions about how to create good governance structures. Because the chemicals regime is institutionally diverse and built around a number of formally free-standing agreements and programs, addressing and trying to benefit from institutional linkages becomes an important part of multilevel governance. These kinds of multilevel governance issues have, however, received relatively little scholarly attention.

Furthermore, institutional linkages are set to grow in importance under the chemicals regime. For example, as parties take steps to organizationally link different treaties (through the synchronizing of COPs, combining of secretariat functions, etc.), they become closer connected (even as they remain legally independent). Also, actions such as the continued expansion of regulation of the same pesticides, industrial chemicals, and by-products covered by two or more agreements, as well as the

development of overlapping technical guidelines on the environmentally sound management and disposal of chemicals, further increase the institutional complexity of
the chemicals regime. In addition, recent political developments on mercury pollution
with the intent of adopting a global mercury convention in 2013 will create different
kinds of linkages with other treaties, including when it comes to the operation of the
regional centers under the Basel and Stockholm Conventions. Further study of these
and other kinds of institutional linkages can contribute to the literature on global
environmental politics.

In institutional analysis, there is a need to focus more on actor-based linkages and
how they shape policy-making and implementation. The chemicals regime is a prime
area for more research into the characteristics and implications of linkage politics.
As there is a growth in institutional linkages, this affects the interests and strategies
of policy actors in multiple ways. Regime participants operating in situations of
a high degree of institutional density will not only seek to advance their interests
with respect to a particular policy issue within a single forum, but also engage in
linkage politics as they consider how choices they make in one forum will influence
their interests and policy outcomes in other venues. This, in turn, impacts regime
creation, implementation, and effectiveness. More such in-depth empirical research
into the chemicals regime can be used to further analyze characteristics and implications of linkage politics, where such insights would also be highly relevant to other
environmental issues areas becoming more institutionally complex.

Looking forward, there are several policy needs to meet the goal of safe production and use of chemicals. These go well beyond 2020, where many actors play
critical roles. Ultimately, it is necessary to fundamentally re-evaluate the way in
which chemicals are developed, how they are put on the market, and where they can
subsequently be used. In this, there is shared responsibility by regulatory authorities
and firms that produce and use chemicals. With respect to creating more precautionary systems for assessments and controls, there are ongoing changes in several
places, most notably in the EU. In 2007, the EU passed a regulation on the registration, evaluation, and authorization and restriction of chemicals (REACH). This
aims to improve environmental and health protection through better risk assessment
and earlier identification of hazardous chemicals based on their intrinsic properties
and taking quicker and more comprehensive regulatory action (Selin 2007). Some
countries outside the EU are also taking similar steps, but this is just the beginning
of a long policy process and outcomes are uncertain (Selin 2013).

Alongside taking more aggressive action to address the way new chemicals are
developed, there is much need for continued international assessments and policy-
making on existing chemicals. Global environmental and human health goals can
only be met if industrialized and developing countries continue to work together
under different agreements to introduce appropriate life-cycle regulations on a larger
set of hazardous substances. This also involves better linking of international policy-
making with regional, national, and local management needs and efforts to create
sustainable livelihoods. To this end, IGOs, donor countries, and NGOs play important roles funding and supporting capacity-building and technology transfer. The
ability to better assess progress and target shortcomings in treaty implementation is
dependent on the willingness of national governments to expand collective mechanisms for monitoring and addressing compliance problems. However, it remains to

be seen if the international community has the ability and political will to take all these essential actions.

References

AMAP (Arctic Monitoring and Assessment Programme). 2009. *AMAP Assessment 2009: Human Health in the Arctic*. Oslo: AMAP.

AMAP (Arctic Monitoring and Assessment Programme). 2010. "AMAP Assessment 2009: Persistent Organic Pollutants in the Arctic." *Science of the Total Environment*, 408: 2851–3051.

Anastas, Paul T. and John C. Warner. 1998. *Green Chemistry: Theory and Practice*. Oxford: Oxford University Press.

CEFIC (European Chemical Industry Council). 2006. *Facts and Figures: The European Chemical Industry in a Worldwide Perspective*. December. Brussels: CEFIC.

Clapp, Jennifer. 2001. *Toxic Exports: The Transfer of Hazardous Wastes from Rich to Poor Countries*. Ithaca, NY: Cornell University Press.

Collins, Craig. 2010. *Toxic Loopholes: Failures and Future Prospects for Environmental Law*. Cambridge: Cambridge University Press.

Downie, David L. and Terry Fenge, eds. 2003. *Northern Lights against POPs: Combatting Toxic Threats in the Arctic*. Montreal: McGill-Queen's University Press.

Emory, Richard W. 2001. "Probing the Protections in the Rotterdam Convention on Prior Informed Consent." *Colorado Journal of International Environmental Law and Policy*, 12: 47–69.

Geiser, Kenneth. 2001. *Materials Matter: Towards a Sustainable Materials Policy*. Cambridge, MA: MIT Press.

Government of Tanzania. 2005. "National Implementation Plan (NIP) for the Stockholm Convention on Persistent Organic Pollutants (POPs)." Report prepared for the Stockholm Convention secretariat.

Harrison, Jill Lindsey. 2011. *Pesticide Drift and the Pursuit of Environmental Justice*. Cambridge, MA: MIT Press.

Iles, Alastair. 2004. "Mapping Environmental Justice in Technology Flows: Computer Waste Impacts in Asia." *Global Environmental Politics*, 4(4): 76–107.

Jeyaratnam, J. 1990. "Acute Pesticide Poisoning: A Major Global Health Problem." *World Health Statistics Quarterly*, 43(3): 139–144.

Kohler, Pia M. 2006. "Science, PIC and POPs: Negotiating the Membership of Chemical Review Committees under the Stockholm and Rotterdam Conventions." *Review of European Community & International Environmental Law*, 15(3): 293–303.

Krueger, Jonathan. 1999. *International Trade and the Basel Convention*. London: Royal Institute for International Affairs.

Krueger, Jonathan and Henrik Selin. 2002. "Governance for Sound Chemicals Management: The Need for a More Comprehensive Global Strategy." *Global Governance*, 8(3): 323–342.

Kummer, Katharina. 1995. *International Management of Hazardous Wastes: The Basel Convention and Related Legal Rules*. Oxford: Clarendon Press.

Kummer, Katharina. 1999. "Prior Informed Consent for Chemicals in International Trade: The 1998 Rotterdam Convention." *Review of European Community & International Environmental Law*, 8(3): 323–330.

Langman, James. 2007. "Regional Concern about Pesticides on Rise." *EcoAméricas.com*, October 2007, http://www.ecoamericas.com/en/story.aspx?id=864 (accessed October 16, 2012).

Liu, Lee. 2010. "Made in China: Cancer Villages." *Environment*, 52 (2): 8–21.

Lönngren, Rune. 1992. *International Approaches to Chemicals Control: A Historical Overview*. Sweden: National Chemicals Inspectorate.

McDorman, Ted L. 2004. "The Rotterdam Convention on the Prior Informed Consent Procedure for Certain Hazardous Chemicals and Pesticides in International Trade: Some Legal Notes." *Review of European Community & International Environmental Law*, 13(2): 187–200.

Mancini, Francesca, Ariena H.C. Van Bruggen, Janice L.S. Jiggins *et al.* 2005. "Acute Pesticide Poisoning among Female and Male Cotton Growers in India." *International Journal of Occupational and Environmental Health*, 11(3): 221–232.

Oberthür, Sebastian and Thomas Gehring, eds. 2006. *Institutional Interaction in Global Environmental Governance: Synergy and Conflict among International and EU Policies.* Cambridge, MA: MIT Press.

Oberthür, Sebastian and Olav Schram Stokke, eds. 2011. *Managing Institutional Complexity: Regime Interplay and Global Environmental Change.* Cambridge, MA: MIT Press.

O'Neill, Kate. 2000. *Waste Trading among Rich Nations: Building a New Theory of Environmental Regulation.* Cambridge, MA: MIT Press.

Paarlberg, Robert L. 1993. "Managing Pesticide Use in Developing Countries." In *Institutions for the Earth: Sources of Effective International Environmental Protection*, ed. Peter M. Haas, Robert O. Keohane, and Marc A. Levy, 309–350. Cambridge, MA: MIT Press.

PANAP (Pesticide Action Network Asia Pacific). 2010. *Communities in Peril: Global Report on Health Impacts of Pesticide Use in Agriculture.* Penang: Pesticide Action Network Asia Pacific.

Pellow, David Naguib. 2007. *Resisting Global Toxics: Transnational Movements for Environmental Justice.* Cambridge, MA: MIT Press.

Selin, Henrik. 2003. "Regional POPs Policy: The UNECE/CLRTAP POPs Agreement." In *Northern Lights against POPs: Combatting Toxic Threats in the Arctic*, ed. D.L. Downie and T. Fenge, 111–132. Montreal: McGill-Queen's University Press.

Selin, Henrik. 2007. "Coalition Politics and Chemicals Management in a Regulatory Ambitious Europe." *Global Environmental Politics*, 7(3): 63–93.

Selin, Henrik. 2010. *Global Governance of Hazardous Chemicals: Challenges of Multilevel Management.* Cambridge, MA: MIT Press.

Selin, Henrik. 2012a. "Global Environmental Governance and Regional Centers." *Global Environmental Politics*, 12(3): 18–37.

Selin, Henrik. 2012b. "Global Multilevel Governance of Hazardous Chemicals." In *Handbook of Global Environmental Politics*, 2nd edn, ed. P. Dauvergne, 210–221. Cheltenham: Edward Elgar.

Selin, Henrik. 2013. "Minervian Politics and International Chemicals Policy." In *Leadership in Global Institution Building: Minerva's Rule*, ed. Y. Tiberghien, 193–212. New York: Palgrave Macmillan.

Selin, Henrik and Noelle Eckley. 2003. "Science, Politics, and Persistent Organic Pollutants: Scientific Assessments and Their Role in International Environmental Negotiations." *International Environmental Agreements: Politics, Law and Economics*, 3(1): 17–42.

Selin, Henrik and Noelle Eckley Selin. 2008. "Indigenous Peoples in International Environmental Cooperation: Arctic Management of Hazardous Substances." *Review of European Community & International Environmental Law*, 17(1): 72–83.

Selin, Henrik and Stacy D. VanDeveer. 2003. "Mapping Institutional Linkages in European Air Pollution Politics." *Global Environmental Politics*, 3(3): 14–46.

Srinivas Rao, Ch., V. Venkateswarlu, T. Surender *et al.* 2005. "Pesticide Poisoning in South India: Opportunities for Prevention and Improved Medical Treatment." *Tropical Medicine and International Health*, 10(6): 581–588.

Thrift, Charles, Ken Wilkening, Heather Myers, and Renata Raina. 2009. "The Influence of Science on Canada's Foreign Policy on Persistent Organic Pollutants (1985–2001)." *Environmental Science & Policy*, 12(7): 981–993.

Victor, David G. 1998. "'Learning by Doing' in the Nonbinding International Regime to Manage Trade in Hazardous Chemicals and Pesticides." In *The Implementation and*

Effectiveness of International Environmental Commitments: Theory and Practice, ed. David G. Victor, Kal Raustiala, and Eugene B. Skolnikoff, 221–281. Cambridge, MA: MIT Press.

Wexler, Philip, Jan van der Kolk, Asish Mohapatra, and Ravi Agarwal, eds. 2012. *Chemicals, Environment, Health: A Global Management Perspective*. Boca Raton, FL: CRC Press.

WSSD (World Summit on Sustainable Development). 2002. "Plan of Implementation of the World Summit on Sustainable Development." Johannesburg, South Africa, August 26–September 4.

Part II   Concepts and Approaches

The Handbook of Global Climate and Environment Policy, First Edition. Edited by Robert Falkner.
© 2013 John Wiley & Sons, Ltd. Published 2013 by John Wiley & Sons, Ltd.

Chapter 8

Global Environmental Norms

Steven Bernstein

Introduction

Governance and policy norms define the fundamental substance of how policy actors and political communities understand the appropriate purposes of environmental policy. Thus, attention to norms should arguably be a core concern among scholars and practitioners. Such attention seems especially important in a policy arena driven largely by a perceived imperative to alter human behavior and its interaction with the natural environment, and where complex social and economic dynamics have produced intense and often fraught debates over the proper course of action given the challenge of reconciling competing values, needs, and conditions among communities affected by such policies.

Yet, early global environmental policy scholarship largely ignored norms in its preoccupation with explaining and promoting any form of cooperation on a set of problems that had previously received very little international attention. The subsequent explosion of research and critical attention to the form and substance of environmental policy in the wake of an increasing sense of crisis has not only corrected that neglect, but has led to insights on norms that are now at the cutting edge of debates on their influence and impact on global politics more broadly.

The chapter begins by defining norms and reviewing antecedents in the literature that provided a foundation for later research that corrected that early neglect. It then identifies the broad normative patterns in global environmental policy and extends the early literature on the evolution of global environmental norms to take account of the increased contestation evident in the global negotiations for the 2012 United Nations Conference on Sustainable Development (Rio+20). It then reviews current debates over how to explain and understand the emergence, influence, and impact of those norms internationally and domestically, and the cutting edge of norms

The Handbook of Global Climate and Environment Policy, First Edition. Edited by Robert Falkner.
© 2013 John Wiley & Sons, Ltd. Published 2013 by John Wiley & Sons, Ltd.

scholarship that attempts to capture the dynamic and contested terrain of global environmental norms. The chapter concludes with the implications of increased contestation – and the tendency of global negotiations to mask that contestation – for future action on the most serious global environmental problems, including climate change.

The Conceptual Landscape: Definitions and Antecedents

Norms define and regulate appropriate behavior for actors with a given identity (e.g. Finnemore and Sikkink 1998: 891), assign rights and responsibilities regarding the issue in question, and are publicly or collectively understood as such. This broad definition corresponds to the constitutive, regulative, and deontic function of norms. Norms constitute identities and meanings by defining who may act, in what context they may act, and what their actions mean in that particular context. They regulate by pre/proscribing how actors should behave in defined contexts (Dessler 1989: 456). Finally, norms serve a deontic function when they express values that create new rights and responsibilities (Onuf 1997; Ruggie 1998: 21). Norms perform these functions simultaneously, but to varying degrees.

Norms do not necessarily identify actual behavior; rather they identify notions of what appropriate behavior ought to be. Norms are intersubjective, or shared, but only in the sense of being irreducible to individual beliefs. The importance of norms in global politics comes from their *institutionalization*, which makes them "collective" or part of social structure. Institutionalization concerns the perceived legitimacy of the norm as embodied in law, institutions, or discourse even if all relevant actors do not follow it (Onuf 1997: 17). Legitimacy matters because the question is not whether the norm exists, but the political authority the norm enjoys. A claim of legitimacy does not necessarily mean it adheres to a deeper notion of justice – though that is one source of disagreement in the literature. Rather, legitimacy in this context refers to the basis of obligation being rooted in the acceptance and justification of the norm as defining appropriate behaviour because of agreement or recognition by relevant communities (Franck 1990: 16, 38; Florini 1996: 364–365; Bernstein 2005, 2011). The degree of institutionalization is important because it indicates how strongly challenges to the norm are likely to be contested and ultimately the ability of the norm to (re)define state or other key actors' interests.

Being collectively held, norms are "discrete positivities" and thus can be operationalized more straightforwardly than often portrayed (Onuf 1997: 32). Most international norms are stated explicitly in treaties, resolutions, declarations (including the "soft" declaratory law that has served as a basis for international environmental law and institutions: Dupuy 1991), and in rules and standards established by international organizations. Hence, norms leave behavioral traces in the form of treaty commitments, action programs, practices, policies, and so on.

Norms did not initially attract much attention when the field of global environmental politics began to coalesce in the 1990s. This, despite the almost simultaneous rise of the "social constructivist" research program in International Relations in the late 1980s and early 1990s, where norms figured prominently in its core ontology that rested on the co-constitution of agents and structures. Although English School scholars had long paid attention to the societal aspects of international systems in contrast to American realism, constructivists much more explicitly placed norms at the center of their research program to show that ideas and specifically "inter-subjective" knowledge had a profound impact on the nature and functioning of international relations.

However, environmental scholars at the time were preoccupied with international environmental negotiations and the search for explanations of international coop-eration on environmental problems (e.g. Young 1989; Mintzer and Leonard 1994; Sprinz and Vaahtoranta 1994). Works on how and why environmental cooperation mattered, focused on implementation and effectiveness of international rules, quickly followed (Underdal 1992; Chayes and Chayes 1995; Victor *et al.* 1998; Young and Levy 1999). Yet, even pioneering work that focused on the role of ideas in influencing these outcomes – some of which originated in the environmental politics literature – paid little attention to the specific norms those ideas promoted. Instead, work such as Peter Haas's (1989, 1992) application of the "epistemic communities" concept to the Mediterranean Action Plan or agreements to combat ozone depletion, on the role of experts motivated and empowered by causal and principled beliefs, focused on how scientific ideas could define an interest in cooperating or create a focal point around which cooperation might converge.

Even work focused more explicitly on the constitutive basis of global environmen-tal politics, such as the way deeper sovereignty norms delimited it or how sovereignty could be redefined by it (e.g. contributions to Conca and Lipschutz 1993) or that took a critical stance on values being promoted in the name of environment (e.g. Chaterjee and Finger 1994) did so without much explicit reference or analysis of norms. Still, contributions such as Conca and Lipschutz's pioneering volume drew attention to the constitutive basis and deep structures of global environmental pol-itics that provided a foundation for later work on norms, including not only how structures define and constrain action, but how norm contestation or the introduc-tion of new norms can enable new actors and identities that potentially challenge long-standing practices that were harmful to the environment. Similarly, the liter-ature on compliance and effectiveness began to pay attention to not only material incentives that shifted the costs or benefits of complying with an environmental agreement but also to the force of norms and discourse in social learning, redefining interests, facilitating compliance, and legitimating new practices in compliance with environmental goals.

Meanwhile, international environmental legal scholarship had long paid atten-tion to the substance of environmental norms, but paid less attention to their causal or constitutive effects. That scholarship became a source of interest for social sci-entists perhaps because much of international environmental law emerged initially through statements of broad principles such as the 1972 Stockholm Declaration and antecedents in various UN declarations such as on sovereignty over natural resources.[1] Thus, the study of norms and "soft law" became an important dimen-sion of international legal scholarship (Dupuy 1991; Kirton and Trebilcock 2004), especially since from the legal perspective the line between norms and rule, in terms of their authority and effects, can be extremely blurry (Bodansky 2010).

Drawing these various strands together, the next section traces the evolution of environmental norms before addressing current trends and controversies in the literature.

The Evolution of Global Environmental Norms

Global environmental norms evolved out of a series of North–South compromises, but also owing to an ideational shift in how the international community framed environmental issues and responses to them over the last 40 years.[2] It is easy to forget that current formulations of the environmental problematique differ substantially from the dominant views held when the first concerted efforts at wide-scale global responses to environmental problems began in the late 1960s and early 1970s. From the perspective of those earlier efforts, which focused on the negative environmental consequences of unregulated industrial development, suspicions of economic growth, and planetary consciousness and "loyalty to the Earth," the shift in environmental governance is a remarkable and a largely unforeseen departure. Table 8.1 summarizes the key norms of environmental governance over the last 40 years. It is arranged to identify "norm-complexes" – a set of norms that govern practices in a particular issue area – at key junctures (Bernstein 2001). A norm-complex need not be stated explicitly, or even internally consistent, but can be inferred from specific norms institutionalized at a particular time and can be used to assess the significance of changes.

A logical starting place is the 26 principles of the Declaration on the Human Environment, the main statement of governing norms from the 1972 Stockholm Conference. The political debate in the lead-up to the conference – the first large-scale multilateral gathering explicitly focused on the full range of international environmental issues – quickly began to reflect underlying tensions between North and South. Preparatory meetings especially highlighted concerns in the South over perceived lifeboat ethics, and an unwillingness to give up state sovereignty over resources and policy.

At Stockholm, developing countries succeeded in placing concerns about economic growth on the agenda, but ideas to link environment and development had not yet been formulated. As a result, a weak compromise focusing on *environmental protection*[3] prevailed, consistent with the view of Western environmentalists that development and environmental protection are different, often competing tasks. Attempts to further institutionalize environmental governance concentrated on ways to reconcile competing sets of environment and development norms introduced at Stockholm.

"Sustainable development" emerged in the 1980s as that breakthrough idea, becoming the dominant conceptual framework for responses to international environmental problems and capturing the imagination of world opinion. As promoted by the World Commission on Environment and Development (WCED 1987), known as the Brundtland Commission, the concept aimed to legitimate economic growth in the context of environmental protection – a major shift in framing environmental problems since Stockholm. The second column of Table 8.1 highlights the norms promoted in the Brundtland report.

By 1992, a further shift had occurred along one pathway enabled by the sustainable development concept. The UN Conference on Environment and Development or Rio Earth Summit institutionalized the view that liberalization in trade and finance is consistent with, and even necessary for, international environmental protection, and that both are compatible with the overarching goal of sustained economic growth.

Table 8.1 The evolution of international environmental norms: 1972–2012.

	Stockholm 1972	WCED 1987	UNCED 1992	UNCSD 2012
State Sovereignty and Liability	1. Sovereignty over resources and environmental protection within state borders. Responsibility for pollution beyond state borders. (Principles 21–23).	1. Unchanged.	1. Unchanged (Principles 2, 13, 14) except: (a) advanced notification of potential environmental harm (Principles 18,19); (b) state right to exploit resources is to be pursuant to *development* in addition to environment policies.	1. Unchanged in terms of rights and obligations, but greater recognition of role of business and stakeholders in governance and implementation.
Political Economy of Environment and Development	2. Developed and developing countries differ on sources of and solutions to environmental problems. (Principles 11–13).	2. States have the following *common* responsibilities: (a) revive global growth; (b) share responses to global environmental problems.	2. Common but differentiated responsibility (Principles 3, 7, 11). Development takes precedence if costs of environmental protection too high (Principle 11).	2. Unchanged, but increased North–South contestation over the meaning of differentiation; pressure to universalize sustainable development goals.
	3. Balance free trade with commodity price stability. (Principle 10).	3. Free trade plus an emphasis on global growth balanced with managed interventions and commodity price stability.	3. Free trade and liberal markets. Environment and free markets compatible. (Principle 12).	3. Unchanged, but increased tensions on intellectual property (tech transfer), fossil fuel and "green" subsidies, access to resources, markets for environment goods.

(*continued*)

Table 8.1 (Continued)

	Stockholm 1972	WCED 1987	UNCED 1992	UNCSD 2012
	4. Environmental protection requires technology and resource transfers to developing countries. (Principles 9, 20).	4. Unchanged plus specific proposals such as a tax on use of the global commons.	4. Transfers left primarily to market mechanisms, except for least developed countries.	4. Transfers left primarily to market mechanisms, but contestation over intellectual property rules, e.g. compulsory licensing model and technology clearinghouses.
	5. States should cooperate to conserve and enhance global resource base. (Principles 1–7 and 24). Multilateralism.	5. Multilateral cooperation for global economic growth as necessary for other goals.	5. Same as WCED plus human centered development. (Principles 1, 7, and 27).	5. Contested multilateralism; complex governance.
Environ-mental Management	6. Command-and-control methods of regulation favored over market allocation in national and international planning. (Principles 13 and 14).	6. Mix of command-and-control and market mechanisms. Polluter Pays Principle (PPP) endorsed.	6. Market mechanisms favored. PPP and Precautionary Principle. (Principles 16 and 15).	6. Market mechanisms still favored, increased role for the private sector and public–private partnerships, pressure for greater role of state in regulation. Growth of civic environmentalism.
Norm-Complex	Environmental Protection.	Managed Sustainable Growth.	Liberal Environmentalism.	Contested Sustainable Development.

Source: Adapted from Bernstein, Steven. 2001. *The Compromise of Liberal Environmentalism*. New York: Columbia University Press: 109. © 2001 Columbia University Press. Reprinted with permission of the publisher.

Thus, the Earth Summit embraced, and perhaps even catalyzed, the new economic orthodoxy then sweeping through the developing world (Biersteker 1992). These norms are embodied most explicitly in the Rio Declaration on Environment and Development – "the one 'product' of UNCED designed precisely to embody rules and principles of a general and universal nature to govern the future conduct and cooperation of States" (Pallemaerts 1994: 1).

Notably, proponents of the concept of sustainable development meant it to incorporate *three* pillars: environment, economic, *and* social. However, although the Rio Declaration mentions social goals such as poverty eradication, participation of major groups in decision-making, and recognition of the contribution of indigenous peoples and knowledge to sustainable development, the word "social" appears only once in the Declaration and any goals that could be broadly construed as social appear in the context of "development" more generally. Norms of individual rights and equity, employment, or access to resources are notably absent from the Declaration, with the exception of a general call for equity in meeting environmental and developmental needs (Principle 3). Not surprisingly, 20 years later, a central institutional goal of the 2012 UN Conference on Sustainable Development (Rio+20) remained greater coherence among the three sets of goals in the UN system.

Instead, the main elements of the normative compromise institutionalized at Rio 1992 include state sovereignty over resources (and environment and development policies) within a particular state's borders on the political side, the promotion of global free trade and open markets on the economic side, and the polluter pays principle (and its implicit support of market instruments over strict regulatory mechanisms) and the precautionary principle on the management side. For example, according to Principle 12: "States should cooperate to promote a supportive and open international economic system that would lead to economic growth and sustainable development in all countries, to better address the problems of environmental degradation." The polluter pays principle (Rio Principle 16) refers to the idea that the polluting firm ought to shoulder the costs of pollution or environmental damage by including it in the price of a product. This principle thus favors market mechanisms (such as tradable emission permits or privatization of the commons) since they operate by institutionalizing schemes that incorporate environmental costs into prices. It also promotes an end to market-distorting subsidies. The precautionary principle essentially says that a risk of serious environmental harm warrants a precautionary stance even under conditions of uncertainty. Notably, however, its institutionalization remains limited, in part because it provokes contestation whenever it appears to bump up against liberal trade norms.

The one norm that most directly implies that any obligations toward the environment might operate in anything but a liberal market context is Principle 7, which recognizes the "common but differentiated responsibility" of developed and developing states. It harkens back to long-standing demands for differential obligations of the North and South and hence some possible interference in what might be the most economically efficient means of dealing with global environmental problems (it has also faced significant contestation in recent years, discussed further below).

This principle can also be found in Articles 3(1) and 4(1) of the UN Framework Convention on Climate Change (UNFCCC) and is a fundamental element of the implementation of the treaty, which creates different obligations for developed and developing states. It is repeated in Article 10 of the 1997 Kyoto Protocol. The main operative provisions of the UNFCCC deserve mention in this regard since "common but differentiated responsibility" still appears to fit with using or creating markets and liberal economic norms more generally. In line with common but differentiated responsibilities, Article 4(2)(a) obligates developed states to "tak[e] the lead" in modifying their greenhouse gas emissions, but to do so while recognizing, *inter alia*, "the need to maintain strong and sustainable economic growth." It further states that, "Parties may implement such policies and measures jointly with other Parties." This idea of "joint implementation," along with emission trading and the Clean Development Mechanism, became one side of the compromise that produced agreement on the Kyoto Protocol. The compromise linked quantitative reductions or limits in greenhouse gas emissions in developed countries to the inclusion of these three market mechanisms, which were justified as a way to achieve emissions reductions cost-effectively. Ironically, emission trading, promoted most heavily by the United States in negotiations, has proliferated largely outside of the Protocol, but the norms supporting market mechanisms remain relatively strong, despite a number of criticisms of specific instruments.

The discourse of compatibility between the trade regime and the climate regime has also been an important part of the latter's legitimation (Eckerseley 2009). Language in the UNFCCC thus closely mirrors Principle 12 of Rio Declaration: Article 3.5 states parties should "promote ... [an] open international economic system that would lead to sustainable economic growth" enabling them better to address the problems of climate change, and includes General Agreement on Tariffs and Trade (GATT)/WTO language on non-discrimination. Whether or not future trade measures that might result from the international climate regime or national climate policies could be justified under WTO rules in practice is a matter of some controversy (Eckerseley 2009; Hufbauer *et al.* 2009). The point here is that to the degree that policies, such as border tax adjustments on imports not subject to rules limiting emissions (to prevent carbon "leakage," for example), reveal contradictions within that legitimating discourse or lead the WTO to rule against such a measure, it poses a serious challenge to the climate regime. This risk underlines the enormous normative pressure on UNFCCC, Kyoto Protocol, or any successor agreement to be compatible with international trade rules.

Table 8.1, column three, summarizes the norms institutionalized at UNCED, with principles in parentheses referring to the Rio Declaration. Elsewhere, I have characterized the norm-complex institutionalized as "liberal environmentalism," which predicates environmental protection on the promotion of a liberal economic order.

While space limitations prevent a full account of the institutionalization of these norms in specific treaties, practices, and policies, it is worth noting that states subsequently reaffirmed these norms at Rio anniversary global conferences, including the 1997 UN General Assembly Special Session to review the implementation of Agenda 21 and again at the 2002 World Summit on Sustainable Development (WSSD) in Johannesburg – or Rio+10. The latter is notable for further reinforcing global liberalism, the importance of the private sector, and the declining hope for multilateral management. It thereby reflected underlying structural conditions of freer and accelerated transaction flows, globalizing markets, and the fragmentation of political authority. Rio provided the normative foundations for environmental governance to adapt to such conditions. WSSD also heralded the legitimation of another trend consistent with the pattern of working with the market and private sector: public–private partnerships for sustainable development. This move broadened the location of environmental activity, but without deepening core commitments by states or improving multilateral coordination efforts. The proliferation of "corporate social responsibility" initiatives – which vary widely in terms of their scope, authority, and effectiveness – and NGO-led "certification" systems that attempt directly to regulate environmental and social practices in the marketplace, also emerged out of this compromise. Many emerged in reaction to inadequate or missing multilateral responses.

Ten years on, the process of global negotiations to implement the Rio goals persists. The Rio+20 Conference (or UN Conference on Sustainable Development) in June 2012 is just the latest round, following not only the 2002 WSSD, but also the Financing for Development initiative that emerged out of the 2002 Monterrey Consensus and the 2008 Doha Declaration on Financing for Development. These efforts are notable for the way they reflect a rapprochement between traditional development concerns such as aid and poverty alleviation with the Bretton Woods institutions' focus on liberalization (Pauly 2007). Specifically, the 2008 Doha Declaration on Financing for Development identifies two mechanisms aimed at building macroeconomic coherence by linking the finance and trade regimes – the Enhanced Integrated Framework (EIF) and Aid for Trade (AfT) initiatives, which mostly focused on trade facilitation – as the means of fulfilling the Monterrey Consensus. The latest manifestation of the compromise is the "Green Economy" agenda, one of two conference themes (the other being the reform of the institutional framework for sustainable development) of Rio+20.

Still, the outcome document, following the pattern of Rio+5 and Rio+10, "reaffirms" the core principles in the 1992 Rio Declaration and reflects the universal consensus not to reopen negotiations on norms (United Nations 2012: para. 7). That does not mean debate is closed, however. Contestation continued to bubble through the surface of this latest round of global negotiations. Debate persists, for example, on the meanings of some key norms – not only "common but differentiated responsibility," but also polluter pays, which implies internalizing costs for some but responsibility of industrialized countries "to pay" for their historical pollution for others. More broadly, the Green Economy concept is explicitly linked to sustainable development in a way that highlights still-sharp disagreements about what sustainable development means in practice. Notably, the outcome document identifies this theme as "Green Economy in the context of sustainable development and poverty eradication." This formulation reflects the suspicion articulated by developing country governments in their submissions that the concept may tilt policy too far towards an emphasis on environment and "green" jobs and investment at the expense of poverty alleviation or more general economic growth and social stability concerns. Some developing countries also articulated opposition as a concern over green protectionism and that they will be unable to benefit from such a transition owing to their lack of access to technology, expertise, or investment, thus leaving them even worse off. In many ways, the same conflicts over aid, development financing, and technology transfer that have characterized North–South bargaining persist even as this latest articulation of sustainable development suggests a compromise that attempts a correction from too "liberal" an environmentalism, or, more positively, a more fundamental transformation to a greener, less carbon intensive, capitalism (Newell and Paterson 2010). This contestation signals stress on liberal environmentalism on the one hand, but on the other hand signals that the market-based compromise remains resilient in lieu of a clearly articulated alternative.

Moreover, owing to that ambiguity in interpretation and history of conflict in implementation, the compromise remains weak in policy terms because it masks differences rather than confronts or resolves them. In practice, this has meant that institutions with specific mandates to address parts of the compromise have continued to emphasize their primary missions while using the rhetoric of sustainable development. While there have been some serious efforts to integrate the concept of sustainable development into policy, especially in UN institutions, the ambiguity and lack of precision has contributed to the limited implementation of the integration of environmental and social concerns into core policies and practices of the key financial and trade institutions with greater legal, financial, and political weight in development policy. Thus, although the fundamental compromise appears to remain legitimate, the final column of Table 8.1 reflects increased contestation on a number of fronts.

Explaining Norm Emergence and Evolution

Environmental social movements, norm entrepreneurs, the "teaching of norms" (Finnemore 1996) from IOs such as the United Nations Environment Programme (UNEP), and especially epistemic communities of scientific experts who brought environmental cause–effect relationships to the attention of governments (Haas 1989, 1992) have all contributed to the promotion of international environmental norms. The role of "agency" is thus clearly important, consistent with the first wave of constructivist scholarship on norms (e.g. Haas 1992; Finnemore 1996; Finnemore and Sikkink 1998). The focus on agency is also useful for research on how norms influence domestic policies because actors carry, support, and utilize norms to wield influence. However, agency alone cannot easily explain the evolution toward liberal environmentalism, which is not only a general trend, but is pervasive in a number of policy contexts. For example, scholars have noticed similar normative underpinnings in climate change generally (Eckersley 2009; Newell and Paterson 2010) and specifically in transnational and experimental forms of climate change governance largely outside of the formal multilateral regime (Hoffmann 2011), as well as in forest governance (Humphreys 2006) and water governance (Conca 2005).

Although these authors come from various theoretical perspectives, they all pay attention to the way in which the wider structural environment has shaped responses to global environmental problems. In my own work (Bernstein 2001, 2011) I have emphasized that norms emerge through the interaction of new ideas with broader social structure, an argument based in sociological institutionalism and the idea that that the *social* fitness of proposals for new norms with extant social structure better explains why some norms are selected, while others fall by the wayside (see also Florini 1996). On this view, social structure is composed of global norms and institutions. It serves a constitutive function by defining what appropriate authority is, where it can be located, and on what basis it can be justified. It also serves a regulative function by prescribing and proscribing the boundaries of governance activities. A number of Constructivist IR scholars employ such a notion of social structure under various formulations including an "environment" in which organizations operate, "normative structure," and "social structure" (Finnemore 1996; Meyer *et al.* 1997; Ruggie 1998: 22–25; Reus-Smit 1999; Barnett and Coleman 2005). Their basic insight is that already institutionalized norms define appropriate and inappropriate courses of action, legitimate and delegitimate institutional forms, create a context in which cost–benefit analysis occurs, or, put more generally, make the purposes, goals, or rationale of an institution understandable and justifiable to the relevant audience in society (Weber 1994: 7).

Not surprisingly, there are considerable theoretical differences in the way different authors understand this wider environment, especially in terms of how they understand structure and its social and material bases. My own position is that the evolution of environmental norms occurred in a context of wider shifts in the norms of the international political economy: liberal environmentalism tapped into an evolving set of neoliberal norms around global economic governance, a consensus from which it drew legitimacy for a growth-oriented, privatized, and market-based orientation that favored working with the market to solve social problems.

Another trend in broader social structure worth highlighting is the growing normative consensus since the end of the Cold War on the need to "democratize" global governance. These norms include demands for democratic reform and improved public accountability of international institutions to states and/or broader affected publics (e.g. Payne and Samhut 2004; Held and Koenig-Archibugi 2005), increased transparency (e.g. Florini 2010; Gupta 2010), as well as "stakeholder democracy" that calls for collaboration and truer deliberation among states, business, and civil society (Bäckstrand 2006). Such normative pressure is especially prevalent in international environmental institutions, treaties, and declaratory law, which have been at the forefront of promoting increased public participation and transparency at all levels of governance since the 1972 Stockholm Conference (Mori 2004). Examples of codification include Rio Declaration Principle 10 (which states that environmental issues are best handled with participation from all "concerned citizens at the relevant level") and the Aarhus Convention on Access to Information, Public Participation in Decision-Making and Access to Justice in Environmental Matters, which came into force in 2001.

The focus on social structure opens up a deeper theoretical debate that is the new frontier in research on norms of global governance, namely the shift from research on the static influence of norms on behavior to an examination of power and contestation to understand and explain norm dynamics.

Norms and Power: Two Views

Hoffmann (2010) draws a useful distinction in the current wave of norms literature between those who view agents as reasoning *about* norms and agents who reason *through* norms. These two views coincide with different ways to understand the relationship between norms and power.

The reasoning "about norms" perspective tends to dominate work on norm compliance and socialization – what I would call the "influence" literature. On this view, norms are sources of power and influence because they are persuasive, that is, result from good arguments or related forms of communicative action (e.g. Risse 2000), or because they are attached to rules that carry sanctions, either legally or because they mobilize political action such as protests or boycotts with material consequences. Norms can also expose norm-violators to shaming or other harm to their reputation.

The reasoning "through norms" perspective focuses on contestation within a normative community, that is, where intersubjectively held norms may be taken for granted, but where interpretations of the norm can vary and clash. This view raises questions of normative change and transformation as well as the power/knowledge nexus, where norms may reflect dominant discourses. As Wiener (2004: 203) has argued:

> [T]he interpretation of the meaning of norms, in particular, the meaning of generic socio-cultural norms, cannot be assumed as stable and uncontested. On the contrary, discursive interventions contribute to challenging the meaning of norms and subsequently actors are likely to reverse previously supported political positions.

One way to think about contestation is the "gap between general rules and specific situations" (Sandholtz 2008: 121; Hoffmann 2010: 10).

However, contestation may also result owing to changes in background knowledge or social structure that produce changed understanding of identities, situations, or relationships. If one takes the sociological approach to normative evolution outlined above, for example, underlying social structure evolves, which conditions and constrains as much as it enables particular norms (Bernstein 2001; Adler and Bernstein 2005). Similarly, Epstein (2012) takes a more explicitly Bourdieuian perspective in her work on the anti-whaling norm as a form of structural power, or what Bourdieu calls "symbolic domination." Understood in this way, norms do not merely persuade or even simply define interests, they evacuate "other ways of acting or talking about the issue."

For example, anti-whaling states stripped Iceland of its voting rights in the International Whaling Commission when it sought to rejoin in 2001 (after leaving in 1992) by voting on its membership, which is not normally subject to a vote (Epstein 2012: 168–169). This move was made possible, in part, because the *nomos* shifted from a "whaling order" to an "anti-whaling" order owing to the successful recategorizing of whaling as commercial, but specifically defining aboriginal subsistence as outside of that order. Bourdieu's "*nomos*" refers to the "the underlying matrix of norms regulating the practices, or ways of doing and seeing, pertaining to a particular field" (Epstein 2012: 171). The redrawing of the field allowed the norm of anti-whaling to gain traction where scientific evidence had failed to produce sufficient momentum for a moratorium, but it did so through an anti-whaling discourse that cast commercial whaling as bad, but aboriginal whaling as acceptable.

Adler and Bernstein (2005) similarly draw attention to the "background knowledge" or "episteme" as productive of a normative order. An episteme is the "intersubjective knowledge that adopts the form of human dispositions and practices" or the "bubble" within which "people happen to live, the way people construe their reality, their basic understanding of the causes of things, their normative beliefs, and their identity, the understanding of self in terms of others."

These authors draw in different ways from Foucault and Bourdieu in an effort to capture how norms are part of deeper understandings of the world. As part of a background knowledge, norms are a form of productive power in the sense of defining the order of things or distinguishing normal from abnormal, and producing both social relations and the institutional forms that reflect this background knowledge. In turn, norms can empower some actors over others depending on how they stand in relation to that background knowledge. Such a view of norms draws attention to the perspective of norm "takers" or "targets" who in their resistance may be confronting underlying power relations (Epstein 2012).

These approaches highlight not only the deontological function of norms, but also that the defining of good and bad is infused with power relations. This perspective stands in some tension to the subtext of much of the literature on norms that suggests that norms gain traction through communicative action, which is a form of truth-seeking (Risse 2000).

Taken a step further, some scholarship has explicitly examined the negative consequences of environmental norms. Dimitrov (2005), for example, argues that the norm of "environmental multilateralism" helps explain the creation of a multilateral institution to address forestry – the UN Forum on Forests (UNFF) – with universal participation. However, given other norms and interests actually driving forest policy, the creation of the institution served only to produce a hollow institution "designed to be idle" and "preempt governance" (Dimitrov 2005: 4).

What Do Norms Do? Influence and Impact

More or less coinciding with these two views of power, norms affect outcomes in global politics in two ways. First, they may have direct influence over choices which reflects a "reasoning about norms" approach. A large literature compares the influence of norms to material interests, judging norms based on whether actors respond to normative prescriptions or because they provide guidance in the absence of clearly defined interests. This literature often asks whether behavior in a particular situation is driven by a logic of consequences, which would suggest norms do little or no work apart from the underlying interests of actors, as opposed to a logic of appropriateness, where behavior reflects judgments about what is socially acceptable, legitimate, or the "right" thing to do in a particular circumstance (March and Olsen 1998).

In this regard, Keck and Sikkink outline a series of strategies that transnational actors can undertake to encourage states to follow norms – the politics of information, symbolism, leverage, and accountability (Keck and Sikkink 1998). This literature debates the degree to which domestic policy-making institutions and networks, culture, or ideology can influence the uptake of norms (e.g. Risse *et al.* 1999), and many studies find that some "fit" with domestic factors is important (e.g. Cortell and Davis 1996). More recently, Acharya (2004) has emphasized the ability of local actors to reconstruct international norms to fit with local norms or to reinforce local beliefs or institutions. Others have shown the importance of learning gained through interaction in transnational networks, explicit efforts at dialogue, and/or participation in formal and informal international gatherings or conferences and transgovernmental networks (e.g. Holzinger *et al.* 2008; Bernstein and Cashore

2012) as important mechanisms for the dissemination of, and possible transformation of, norms.

Recent studies of norms promoted by powerful international organizations such as the World Bank and IMF similarly highlight that the apparent stabilization of norms within those organizations does not necessarily fix the impact of norms "on the ground," since the implementation of policies interacts with domestic circumstances and processes (Park and Vetterlein 2010). This interaction can also be a source of norm contestation that feeds back into the life cycle of the norm since it can be a factor in contestation that then occurs in the IO.

A number of studies on the influence of environmental norms on domestic and firm behavior highlight norms influence along these lines. For example, Falkner (2006) attributes China's decision to reverse policy and halt the authorization of new genetically modified crops to the influence of transnational networks promoting environmental norms as well as China's shift in preferences owing to economic globalization.

In the case of forestry, Keskitalo *et al.* (2009) find that the implementation of international environmental and indigenous rights norms promoted by forest certification systems differed in Sweden, Finland, and Russia depending on national infrastructure and market characteristics. Similarly the widespread diffusion and implementation of norms around "sustainable forest management," and some variance in interpretation, have depended in large part on linking with other domestic processes, especially learning processes, links with domestic and transnational networks in support of the norm, and resonance with domestic laws and practices (Bernstein and Cashore 2012). In addition, the norm of transparency, also prevalent in forestry, has had effects, for example in Central Africa, where the raising of awareness and reporting of corruption by international NGOs such as Transparency International, Global Witness, and Resource Extraction Monitoring have been key drivers, though that influence is largely limited to the formal forest sector and not the much larger informal sector (Eba'a Atyi *et al.* 2008: 24).

Haufler has similarly documented transparency's increasing influence in the extractive (oil, gas, and mining) sector, through the Extractive Industries Transparency Initiative (EITI). Its relative success stems both from powerful support through the agency of Tony Blair, but also its intersection with transnational networks with complementary global norms, which facilitated the construction of transparency as a solution for management of resource revenues. Similarly, Florini (2010) found more generally that the domestic uptake of transparency varies based on the norms (e.g. views on democratization, privatization, and regulatory policy) and capacities of countries. Together, these findings suggest different domestic factors and transnational interactions can affect the impact of international norms, but also can feed back into the norm life cycle.

Constructivists have also stressed a second way norms matter that coincides closely with the "reasoning through norms" approach: social structure or background knowledge may define interests and identities in the first place, making it difficult to disentangle interests and norms in practice. Moreover, new policy ideas themselves may stem from the interaction of deeper, sometimes taken-for-granted norms and their interactions with "experienced events" (Hurd 2008: 303). This understanding of norms means that changing ideas and behavior at once reflect some interpretation or understanding of underlying norms but also reproduce or alter those norms in interaction with new circumstances. This can be both unconscious and conscious, for example when states attempt to reconstruct rules to condone their behavior (Hurd 2007).

An example in practice is the active contestation over the norm of common but differentiated responsibility (CBDR). Developing states invoke the norm to justify deferring or differentiating their commitments as a group on environmental action. A number of developed states contest that interpretation. Some, for example, interpret the norm as consistent with differentiation within the group of developing countries. Others have introduced language to emphasize common and universal obligations, even as most still accept some differentiation on commitments based on level of development and capacity, an argument also accepted by some developing countries including Egypt and Bangladesh (Brunnée 2010). Notably, outright rejection of the norm has not been tolerated; rather, debate is around its interpretation. For example, climate negotiators debate how much to weigh historical emissions in calculating obligations, whether to look at per capita versus national level of commitments, or, more broadly, whether all "major" economies should have commitments to reduce their projected growth of greenhouse gas emissions. The latter became the basis for the 2010 Cancun Agreements, although that bargain came at the expense of legally binding commitments.[4]

By addressing contestation, the latest wave of constructivist literature is more careful to step outside of a strict idea of co-constitution. At the same time constructivists do not wish to cede ground to the rationalists who simply juxtapose a logic of consequences and logic of appropriateness as more or less alternative explanations. The middle ground is gained by focusing on "background knowledge" rooted in "practices" or competent performances (Adler and Pouliot 2011: 7, 17). On this view, practices – say of diplomacy or North–South bargaining – take on a common-sense character that is conscious, but largely taken for granted. Moreover, norms, even when followed or promoted, may be "practiced" competently or incompetently as this background knowledge is enacted and reified within relevant communities.

Whether one finds the notion of background knowledge useful, the idea that norms must be enacted and that there is some room for reasoning about them means they are also subject to contestation and should be thought about dynamically.

Contestation and the Future of Global Environmental Norms

The recent turn in the constructivist literature on norms towards a focus on contestation tells us that the apparent stability reflected in a political consensus to not "reopen" discussion of norms may mask underlying contradictions and tensions to its own detriment. The framers of sustainable development norms built the concept around the idea that no significant trade-offs needed to be made to achieve environment and development goals. The norms promoted a win-win (or win-win-win if the social is included) discourse, which evolved to reflect the compatibility of the market with all other social purposes. In practice, however, liberal environmentalism does involve compromises, on principles and substance, of policies sufficient to address sustainable development goals effectively. If formulated in a way that denies that compromises or concessions are necessary, ongoing bargaining on substantive problems becomes more difficult to the degree that it reveals the contradictions previously glossed over.

The Kyoto Protocol and subsequent climate change negotiations are prime examples that reveal these contradictions. When it came time to commit to the compromises the Kyoto Protocol embodied, the United States balked at legal commitments to reduce emissions on one hand, while developing countries refused to budge from a principled commitment to unfettered growth and differential responsibility on the other. The current round of negotiations has made explicit the need to reconcile developing countries' recognition that they are likely to suffer most from the consequences of climate change with the understanding that the North is unwilling to make all the trade-offs necessary to decarbonize capitalism, the ultimate and necessary goal of a green economy. What appears left is an ongoing mutual commitment to, or faith in, the market while multilateralism continues to come under pressure as the appropriate institutional form to work out these differences.

Meanwhile, pressures in deeper social structures – whether reinterpretations of "background knowledge," shifts in the global distribution of power, the proliferation of new actors with legitimate claims to be heard in a shifting and more complex governance environment, or new forms of resistance to global liberalism in light of contradictions revealed in the aftermath of the 2008 financial crisis – suggest that contestation over the meaning and legitimacy of many existing norms will continue to reflect reasonable differences in social purposes on a global scale. The tendency to mask these differences bodes ill for moving forward on the need to transform the global economy sufficiently to avoid dangerous climate change and related planetary crises.

Notes

1 For example, Sohn (1973). State sovereignty over resources is widely considered the foundational norm of international environmental law, existing in various forms in legal decisions and documents such as the UN Charter, but stated explicitly beginning with UN General Assembly Resolution 1803/62 (1962) on Permanent Sovereignty over Natural Resources, and later Principle 21 of the Stockholm Declaration on the Human Environment and Principle 2 of the Rio Declaration, which expanded it to include a sovereign right to exploit resources pursuant to a state's own environment and development policies.

2 This section draws liberally from Bernstein (2001) and Bernstein (2012).

3 The term admittedly does not capture the uneasy mix of conservation, economic development, sovereignty, and state responsibility norms that characterized Stockholm outcomes, but is consistent with common understandings of what Stockholm institutionalized.

4 The agreements can be downloaded at http://unfccc.int/meetings/cancun_nov_2010/items/6005.php. The lack of legally binding commitments may also be temporary, but so far only the European Union, Australia, and a handful of smaller states have agreed to legally binding commitments in a second phase of the Kyoto Protocol.

References

Acharya, Amitav. 2004. "How Ideas Spread: Whose Norms Matter?" *International Organization*, 58 (2): 239–275.

Adler, Emanuel and Steven Bernstein. 2005. "Knowledge in Power: The Epistemic Construction of Global Governance." In *Power in Global Governance*, ed. Michael Barnett and Raymond Duvall, 294–318. Cambridge: Cambridge University Press.

Adler, Emanuel and Vincent Pouliot. 2011. "International Practices." *International Theory*, 3(1): 1–36.

Bäckstrand, Karin. 2006. "Democratizing Global Environmental Governance? Stakeholder Democracy after the World Summit on Sustainable Development." *European Journal of International Relations*, 12(4): 467–498.

Barnett, Michael and Liv Coleman. 2005. "Designing Police: Interpol and the Study of Change in International Organizations." *International Studies Quarterly*, 49(4): 593–619.

Bernstein, Steven. 2001. *The Compromise of Liberal Environmentalism*. New York: Columbia University Press.

Bernstein, Steven. 2005. "Legitimacy in Global Environmental Governance." *Journal of International Law and International Relations*, 1(1–2): 139–166.

Bernstein, Steven. 2011. "Legitimacy in Intergovernmental and Non-state Global Governance." *Review of International Political Economy*, 18(1): 17–51.

Bernstein, Steven. 2012. "Grand Compromises in Global Governance." *Government and Opposition*, 47(3): 368–394.

Bernstein, Steven and Benjamin Cashore. 2012. "Complex Global Governance and Domestic Policies: Four Pathways of Influence." *International Affairs*, 88(3): 585–604.

Biersteker, Thomas. 1992. "The 'Triumph' of Neoclassical Economics in the Developing World: Policy Convergence and Bases of Governance in the International Economic Order." In *Governance without Government: Order and Change in World Politics*, ed. James Rosenau and Ernst-Otto Czempiel, 102–131. Cambridge: Cambridge University Press.

Bodansky, Daniel. 2010. *The Art and Craft of Environmental Law*. Cambridge, MA: Harvard University Press.

Brunnée, Jutta. 2010. "From Bali to Copenhagen: Towards a Shared Vision for a Post-2012 Climate Regime?" *Maryland Journal of International Law*, 25: 86–108.

Chatterjee, Pratap and Matthias Finger. 1994. *The Earth Brokers*. New York: Routledge.

Chayes, Abram and Antonia Chayes. 1995. *The New Sovereignty: Compliance with International Regulatory Agreements*. Cambridge, MA: Harvard University Press.

Conca, Ken. 2005. *Governing Water: Contentious Transnational Politics and Global Institution Building*. Cambridge, MA: MIT Press.

Conca, Ken and Ronnie D. Lipschutz, eds. 1993. *The State and Social Power in Global Environmental Politics*. New York: Columbia University Press.

Cortell, Andrew P. and James W. Davis. 1996. "How Do International Institutions Matter? The Domestic Impact of International Rules and Norms." *International Studies Quarterly*, 40(4): 451–478.

Dessler, David. 1989. "What's at Stake in the Agent Structure Debate?" *International Organization*, 43: 441–474.

Dimitrov, Radoslav. 2005. "Hostage to Norms: States, Institutions and Global Forest Politics." *Global Environmental Politics*, 5(4): 1–24.

Dupuy, Pierre-Marie. 1991. "Soft Law and the International Law of the Environment." *Michigan Journal of International Law*, 12: 420–435.

Eba'a Atyi, Richard, Didier Devers, Carlos de Wasseige, and Fiona Maisels. 2008. "State of the Forests of Central Africa: Regional Synthesis." In *The Forests of the Congo Basin: State of the Forest 2008*, ed. Carlos de Wasseige, Didier Devers, Paya de Marcken *et al.*, 15–41. Luxembourg: Publications Office of the European Union.

Eckersley, Robyn. 2009 "Understanding the Interplay between the Climate Regime and the Trade Regime." In *Climate and Trade Policies in a Post-2012 World*, 11–18. Geneva: United Nations Environment Programme.

Epstein, Charlotte. 2012. "Norms: Bourdieu's *Nomos*, or the Structural Power of Norms." In *Bourdieu in International Relations*, ed. Rebecca Adler-Nissen, 165–178. New York and London: Routledge.

Falkner, Robert. 2006. "International Sources of Environmental Policy Change in China: The Case of Genetically Modified Food." *Pacific Review*, 19(4): 473–494.

Finnemore, Martha. 1996. *National Interests in International Society*. Ithaca, NY: Cornell University Press.

Finnemore, Martha and Kathryn Sikkink. 1998. "International Norm Dynamics and Political Change." *International Organization*, 52(4): 887–918.

Florini, Ann. 1996. "The Evolution of International Norms." *International Studies Quarterly*, 40: 363–389.

Florini, Ann. 2010. "The National Context for Transparency-Based Global Environmental Governance." *Global Environmental Politics*, 10(3): 120–131.

Franck, Thomas. 1990. *The Power of Legitimacy among Nations*. New York: Oxford University Press.

Gupta, Aarti. 2010. "Transparency in Global Environmental Governance: A Coming of Age?" *Global Environmental Politics*, 10: 1–9.

Haas, Peter M. 1989. "Do Regimes Matter? Epistemic Communities and Mediterranean Pollution Control." *International Organization*, 43(3): 377–403.

Haas, Peter M. 1992. "Introduction: Epistemic Communities and International Policy Coordination." *International Organization*, 46: 1–35.

Held, David and Mathias Koenig-Archibugi, eds. 2005. *Global Governance and Public Accountability*. Oxford: Blackwell.

Hoffmann, Matthew J. 2010. "Norms and Social Constructivism in International Relations." In *International Studies Online*, ed. Robert A Denemark, http://www.isacompendium.com/subscriber/tocnode?id=g9781444336597_chunk_g978144433659714_ss1-8 (accessed November 5, 2012).

Hoffmann, Matthew J. 2011. *Climate Governance at the Crossroads: Experimenting with a Global Response after Kyoto*. New York: Oxford University Press.

Holzinger, Katharina, Christoph Knill, and Thomas Sommerer. 2008. "Environmental Policy Convergence: The Impact of International Harmonization, Transnational Communication, and Regulatory Competition." *International Organization*, 62(3): 553–587.

Hufbauer, Gary C., Steve Charnovitz, and J. Kim. 2009. *Global Warming and the World Trading System*. Washington, DC: Peterson Institute for International Economics.

Humphreys, David. 2006. *Logjam: Deforestation and the Crisis of Global Governance*. London and Sterling, VA: Earthscan.

Hurd, Ian. 2007. "Breaking and Making Norms: American Revisionism and the Crises of Legitimacy." *International Politics*, 44: 194–213.

Hurd, Ian. 2008. "Constructivism." In *Oxford Handbook of International Relations*, ed. Christopher Reus-Smit and Duncan Snidal, 298–316. Oxford: Oxford University Press.

Keck, Margaret and Kathryn Sikkink. 1998. *Activists beyond Borders: Transnational Advocacy Networks in International Politics*. Ithaca, NY: Cornell University Press.

Keskitalo, Carina, Camilla Sandström, Maria Tysiachniouk, and Johanna Johansson. 2009. "Local Consequences of Applying International Norms: Differences in the Application of Forest Certification in Northern Sweden, Northern Finland, and Northwest Russia." *Ecology and Society*, 14(2): 1–14.

Kirton, John and Michael Trebilcock, eds. 2004. *Hard Choices, Soft Law: Combining Trade, Environment, and Social Cohesion in Global Governance*. Aldershot: Ashgate.

March, James G. and Johan P. Olsen. 1998. "The Institutional Dynamics of International Political Orders." *International Organization*, 52(4): 943–969.

Meyer, John W., D.J. Frank, A. Hironaka *et al.* 1997. "The Structuring of a World Environmental Regime, 1870–1990." *International Organization*, 51: 623–651.

Mintzer, Irving M. and J.A. Leonard, eds. 1994. *Negotiating Climate Change: The Inside Story of the Rio Convention*. Cambridge: Cambridge University Press.

Mori, Satoko. 2004. "Institutionalization of NGO Involvement in Policy Functions for Global Environmental Governance." In *Emerging Forces in Global Environmental Governance*, ed. Norichika Kanie and Peter M. Haas, 157–175. Tokyo: United Nations University Press.

Newell, Peter and Matthew Paterson. 2010. *Climate Capitalism*. Cambridge: Cambridge University Press.

Onuf, Nicholas G. 1997. "How Things Get Normative," revised version of paper presented to a conference on International Norms, Hebrew University of Jerusalem, May 26–27.

Pallemaerts, Marc. 1994. "International Environmental Law from Stockholm to Rio: Back to the Future?" In *Greening International Law*, ed. Philippe Sands, 1–19. New York: The New Press.

Park, Susan and Antje Vetterlein, eds. 2010. *Owning Development: Creating Policy Norms in the IMF and the World Bank*. New York: Cambridge University Press.

Pauly, Louis W. 2007. "The United Nations in a Changing Global Economy." In *Global Liberalism and Political Order*, ed. Steven Bernstein and Louis W. Pauly, 91–108. Albany, NY: State University of New York Press.

Payne, Rodger and Nayef Samhat. 2004. *Democratizing Global Politics*. Albany, NY: State University of New York Press.

Reus-Smit, Christian. 1999. *The Moral Purpose of the State: Culture, Social Identity, and Institutional Rationality in International Relations*. Princeton, NJ: Princeton University Press.

Risse, Thomas. 2000. "Let's Argue! Communicative Action in World Politics." *International Organization*, 54(1): 1–39.

Risse, Thomas, Stephen C. Ropp, and Kathryn Sikkink, eds. 1999. *The Power of Human Rights: International Norms and Domestic Change*. Cambridge: Cambridge University Press.

Ruggie, John G. 1998. *Constructing the World Polity*. London: Routledge.

Sandholtz, W. 2008. "Dynamics of International Norm Change: Rules against Wartime Plunder." *European Journal of International Relations*, 14(1): 101–131.

Sohn, Louis B. 1973. "The Stockholm Declaration on the Human Environment." *Harvard International Law Journal*, 14: 423–515.

Sprinz, Detlef and Tapani Vaahtoranta. 1994. "The Interest-Based Explanation of International Environmental Policy." *International Organization*, 48(1): 77–105.

Underdal, Arild. 1992. "The Concept of Regime 'Effectiveness'." *Cooperation and Conflict*, 27(3): 227–240.

United Nations. 2012. "'The Future We Want.' Outcome of the UN Conference on Sustainable Development." A/CONF.216/L.1∗ (reissued June 22). New York: United Nations.

Victor, David, Kal Raustiala, and Eugene Skolnikoff, eds. 1998. *The Implementation and Effectiveness of International Environmental Commitments: Theory and Practice*. Cambridge, MA: MIT Press.

WCED (World Commission on Environment and Development). 1987. *Our Common Future*. Oxford: Oxford University Press.

Weber, Steve. 1994. "Origins of the European Bank for Reconstruction and Development." *International Organization*, 48: 1–38.

Wiener, Antje. 2004. "Contested Compliance: Interventions on the Normative Structure of World Politics." *European Journal of International Relations*, 10(2): 189–234.

Young, Oran. 1989. "The Politics of International Regime Formation: Managing Natural Resources and the Environment." *International Organization*, 43: 349–375.

Young, Oran and Marc Levy, eds. 1999. *The Effectiveness of International Environmental Regimes: Causal Connections and Behavioral Mechanisms*. Cambridge, MA: MIT Press.

Chapter 9

Global Governance

Johannes Stripple and Hannes Stephan

Introduction

The uptake of new concepts is often a slow process. It usually takes time for scholars and practitioners to adopt a new word, to agree on its meaning and appreciate the value of using it. But every now and then, conceptual change happens overnight. Few concepts have as swiftly entered academic and policy discussions and become *the* organizing concept as "global governance." A historical parallel might be the way "national security" in the 1940s became the "commanding idea" – a comprehensive label for a range of phenomena, previously discussed as war, defense, military and foreign policy.[1] The historical context for the concept of global governance was the fall of the Berlin Wall and the waning of the Cold War order. At this point in time, the demand was high for new concepts that could capture a rapidly changing world. Global governance is a suggestive term that quickly became established as a key concept in international relations, particularly in UN circles (Jönsson 2010). Weiss (2011: 9) reads the conceptual history of the term as a "shotgun wedding between academic theory and practical policy in the 1990s." The publication of *Governance without Government* (Rosenau and Czempiel 1992) coincided with Sweden's launch of the policy-oriented Commission on Global Governance, and the publication of the commission's report *Our Global Neighborhood* in 1995 coincided with the first issue of the journal *Global Governance: A Review of Multilateralism and International Organization*. Probably because the concept seemed to make immediate sense for capturing and responding to a rapidly unfolding world politics, "global governance" has, rather unfortunately, come to be understood *both* as a new empirical phenomenon and a theoretical term for analyzing it.

Hewson and Sinclair (1999) noted that separate literatures have emerged on global governance – one as a broad theoretical approach and the other as a normative

The Handbook of Global Climate and Environment Policy, First Edition. Edited by Robert Falkner.
© 2013 John Wiley & Sons, Ltd. Published 2013 by John Wiley & Sons, Ltd.

program for the better management of common resources. Among those who deploy an empirical definition of global governance, the actual interpretations diverge considerably. There are those who use it in a relatively narrow sense to rethink regimes as enmeshed in broader systems of governance instead of delimited issue areas. Other scholars employ "global governance" in a broader sense as a vantage point for understanding the sources and political implications of global change. This means that global governance offers a more comprehensive and integral perspective than do other approaches, such as "globalization" or "neo-medievalism." Furthermore, the concept of governance articulates an account of political order that relies on neither international "anarchy" nor on the hierarchical authority of the state (Hurrell 2007: 95). By drawing attention to the rise of hybrid, non-hierarchical, and network-like modes of governing on the global stage, global governance is therefore more than a theory "about" international relations – rather it is what international relations are about. The overarching narrative of the discipline is thus changing from one of anarchy in a system of states to governance within a global society (Barnett and Sikkink 2008). Finally, Dingwerth and Pattberg (2010) observe that scholars tend to overestimate the orderly coordination in world affairs when they capture certain areas, such as health and the environment, as global governance.

Within the field of international environmental politics, the global governance of the environment (i.e. "global environmental governance") became an organizing concept in the mid- to late 1990s, subsuming previous discussions on International Governmental Organizations (IGOs), regime theory, the implementation of environmental agreements at the national level, private (business or civil) self-regulation, social movements, questions about the transparency and legitimacy of international negotiations, legal obligations, and other forms of steering such as codes of conduct or standards. The strength of the concept is its capacity to "convey a sense of an overarching set of arrangements beyond the specificities of individual issue areas or thematic concerns that encompasses a broad range of political foci" (Paterson et al. 2003: 1). Global environmental governance can thus be read as the practical answer to one of the discipline's defining questions:

> Can a fragmented and often highly conflictual political system made up of 170 sovereign states and numerous other actors achieve the high (and historically unprecedented) levels of co-operation and policy co-ordination needed to manage environmental problems on a global scale (Hurrell and Kingsbury 1992: 1)?

Many writings on global environmental governance have drawn attention to a new set of actors in global politics. Governance, understood as "the capacity to get things done without the legal competence to command that they be done" (Czempiel 1992: 250), encourages scholars to look for agency beyond the nation-state and instead among, for example, social movements, multinational companies, and scientific networks. Many scholars (e.g. Biermann 2006; Okereke et al. 2009) argue that the added value of the global governance concept (vis-à-vis state-centric, international approaches) is its ability to account for the increasing participation of non-state actors in the governing of collective affairs.

This brief chapter outlines the main characteristics of global governance as an academic approach to global climate and environmental policy. We trace its intellectual origin in the mid-twentieth century and present an overview of contemporary interpretations of global governance. The second part of the chapter explores global environmental governance as practice. We particularly look at four different sets of practices through which the climate and the environment are being governed. The conclusion summarizes our reasoning, addresses areas in need of research, and tables some ideas about how to improve the system of global environmental governance, particularly with reference to climate change.

Global Governance as an Academic Approach

The concept of "governance" was introduced in the 1940s and used systematically in areas such as business organization, economics, and neo-corporatism since the 1960s (Pierre 2000). Cajvaneanu (2011) shows how governance historically coincided with the representation of society as highly complex and functionally differentiated. The concept emerged as a reflection on areas of "private government" (corporations, labor unions, and, later on, the university) as "autonomous, self-governing units within society that share the governing of individuals with the public government" (Cajvaneanu 2011: 70). This early articulation of governance as referring to a system of rules that organized the functioning of entities in both the public and private spheres (Merriam 1944; Eells 1960) helps us to understand where contemporary ideas about "global governance" originated. In the 1960s and 1970s, the concept of governance was shaped through developments within cybernetics and systems theory (control in the context of high complexity); in the development of transaction cost/neo-institutional economics (markets and networks as alternative governance structures that can enhance economic efficiency); and, finally, through the debates on the crises of the welfare state and the inefficiency of centralized policy-making. These later "areas of thought" laid the foundation for the normative "good governance" discourse (transparency, accountability, rule of law, participation) that organizations like the World Bank, IMF, and OECD have promoted (Cajvaneanu 2011) and which frames many contemporary discussions on global environmental governance.

While global governance as an academic approach emerged in the early 1990s, its main precursor in IR was the literature on transnational environmental actors in the 1970s. Keohane and Nye's (1972) edited volume *Transnational Relations and World Politics* is one of those few publications that instantaneously demarcate a new field of study. The book tried to account for what was understood as a new phenomenon in world politics – the influence of non-state actors (mostly multinational companies) on state behavior. Shortly afterwards, Keohane and Nye (1977) formalized the arguments into a theoretical model, "complex interdependence," which captured a world where transnational activity affects states' capacity to act. Military force was seen as ineffective in a world characterized by new interdependencies (of which the environment was one) and, hence, the distinction between "high politics" (security) and "low politics" (trade) appeared obsolete. Moreover, the mutual independence of states and other actors and the multiplicity of intersocial contacts were believed to lead to the breakdown of the state-centric nature of international politics.[2] Already at this point, environmental politics was an empirical terrain that inspired important research for the discipline at large. Thompson-Feraru's (1974) classic "Transnational Political Interests and the Global Environment" theorized non-state activities at the

United Nations Conference of the Human Environment (UNCHE). Henceforth, the literature on environmental NGOs in international politics has had its own lineage through publications such as Willets (1982), Chatterjee and Finger (1994), and Keck and Sikkink (1998) (see also Chapter 16 in this volume).

While Keohane and Nye's original intention was about measuring changes in state power and influence, a key contribution of the literature on global governance has, in fact, been a return to questions of authority in world politics. Borrowing from earlier writings on governance throughout the social sciences, Rosenau (1992: 4–5) argued that governance

> embraces governmental institutions, but it also subsumes informal, non-governmental mechanisms... Governance is a system of rule that is dependent on intersubjective meanings as on formally sanctioned constitutions and charters.

And, in a formulation that has become the trademark of global governance, Rosenau (1992: 5) stated that "it is possible to conceive of governance without government – of regulatory mechanisms in a sphere of activity which function effectively even though they are not endowed with formal authority." The emergence of such new authority structures led Rosenau to identify two (separate) political worlds, one "state-centric" consisting of "sovereignty-bound states" and the other "multi-centric" consisting of "sovereignty-free" actors. In *Turbulence in World Politics*, Rosenau (1990) elaborated the elements, parameters, and evolution of these two separate worlds, while the sequel, *Along the Domestic–Foreign Frontier* (Rosenau 1997), accounted for non-state actors as more generic "spheres of authority" which govern within their respective and often overlapping domains.[3] Rosenau's pathbreaking usage of the global governance concept therefore consists of three trends that are significant for global change: the relocation of authority in multiple directions (rescaling and new spheres of authority), the emergence of a global civil society, and a restructuring of the global political economy (Hewson and Sinclair 1999).

Global Environmental Governance

The academic study of international environmental politics (IEP) is informed by a dominant narrative that commands widespread support. Because environmental degradation does not respect territorial borders, environmental problems are located within the realm of the international – a political space that has certain characteristics (sovereign authority, territoriality), which turns IEP into a question of how to regulate the polluting activities of states in the absence of a world government. The catchphrase of the Brundtland Commission "the Earth is one but the world is not" (WCED 1987: 28) captures this predicament. The need for collective action leads to the creation of international environmental regimes, which play a significant role in forging cooperation between states. Regimes are seen as the key institutions for tempering or overcoming the fundamental condition of international anarchy.

Soon after Rosenau's pioneering work in the early 1990s, "global environmental governance" emerged as a key theme in IEP. Books such as *Global Governance: Drawing Insights from the Environmental Experience* (Young 1997), *Global Civil Society and Global Environmental Governance: The Politics of Nature from Place to Planet* (Lipschutz and Mayer 1996), and *Environmental Governance: The Global Challenge* (Hempel 1996) explored various dimensions of the concept. Young (1997) provided a seminal contribution, but the concept of governance used was narrower than, for example, Lipschutz and Mayer's (1996) multi-scalar and civil society-oriented approach, which argues that governance through some form of non-state-based social relations, rather than hierarchy or markets, is likely to be the most effective means of protecting nature. The units of governance will be defined by function and social meanings, anchored to particular places but linked globally through networks of knowledge-based relations (1996: 254). Jagers and Stripple (2003), for instance, built a case around the activities of the insurance industry's attempt to govern climate change "beyond the state." They argued that global climate governance could be conceived as comprising "all purposeful mechanisms and measures aimed at steering social systems towards preventing, mitigating, or adapting to the risks posed by climate change" (2003: 385).

Paterson *et al.* (2003) draw attention to how the inaugural issue of the journal *Global Environmental Politics* displays rather different interpretations of global environmental governance. One set of articles articulated a green version of the debates surrounding the Commission on Global Governance. Global environmental governance (GEG) here is taken to mean a programmatic, reformist orientation to institutional arrangements in global politics, principally the UN system (Paterson *et al.* 2003: 2). Another set of articles explored globalization and resistance and here "GEG can be seen as a product of two phenomena: the pursuit of neoliberal forms of globalization; and the resistance to such centralization of power." The differences in the interpretations of each of the words "global," "environmental," and "governance" have led scholars to imagine global environmental governance in very different ways. Scholars with a background in neoliberal institutionalism have reconceptualized regimes as enmeshed in broader systems of governance instead of issue areas (Stokke 1997). GEG becomes the sum of the overlapping networks of interstate regimes on environmental issues (Paterson *et al.* 2003: 4). Biermann (2006), who also writes within a liberal institutionalist framework, delineates how a new system of global environmental governance departs from *international* politics. Biermann tries to seize the middle ground between the rebranded, but still state-centric, regime theory and Rosenau-inspired, rather vague definitions where almost anything can be labeled global governance. Biermann argues, on empirical grounds, that global governance is defined by a number of new phenomena in world politics. Hence, the concept of global governance should be restricted to denote those features that make "the world of today different from what it used to be in the 1950s" (2006: 241). Biermann points to (1) the increased participation of non-state actors (e.g. networks of experts, environmentalists, multinational corporations but also new agencies, intergovernmental organizations, and international courts); (2) new forms of cooperation beyond the traditional negotiation of international law (partnerships, networks, practices of standard-setting); (3) a new segmentation of policy-making, both vertically (multilevel governance) and horizontally (multipolar governance) (2006: 243–247).

However, it seems that the empirical focus on patterns of global governance has come at the expense of a sustained analysis of power dimensions. The processes through which global activities are directed and world orders are produced ultimately require an analysis of the workings of power (Barnett and Duvall 2005: 2). The limits, silences, and unwanted legitimations of the concept have not been properly understood and there is an urgent need to develop a self-consciously critical governance approach. Barnett and Sikkink (2008: 79) urge scholars to consider multiple dimensions of power – from "compulsory power" and "institutional power" to power inherent in the constitution of subject's capacities ("structural power") and in the discursive production of subjectivity in world politics ("productive power"). Furthermore, Douglas (1999) contends that global governance scholars misread the history of the modern state and the genealogy of modern power. In his view, globalization extends, rather than fragments, state power. The modern project of government cannot be confined to actions of state institutions, and thus decentralization and diffusion "beyond the state" imply an extension of the modern project of government. Neumann and Sending (2010) describe global governance as a particular way of governing, a neoliberal governmental rationality characterized by a drive to govern *more* in the sense of covering more and more geographical and functional domains, but also in the sense of governing *less* – governing through "indirect rule," that is, through the freedom of subjects to govern themselves in various areas. There is also a large literature on the political economy of global environmental governance (Paterson 2000; Newell 2005) which criticizes the existing literature for neglecting the "root causes" of environmental change (the way in which trade, production, and finance are organized in a globalized world economy) and for failing to account for the substantive outcomes of emerging governing mechanisms. On this view, global environmental governance will necessarily fail if the underlying growth dynamic of capitalism is not confronted.

Besides such diverse conceptual approaches to global environmental governance, how is governing actually achieved in these arenas in the absence of the formal authority of the state? In the next section, we briefly review and illustrate what might be called the practices of global environmental governance. By this we mean to highlight different processes of governing. We will cast the net wide and draw attention to four different sets of practices through which environmental and climate issues are being governed.

Global Environmental Governance as Practice

Our typology of practices covers norm creation, informational governance, standard-setting, and capacity-building and implementation. It is important to note that this represents a set of simplified distinctions. More often than not, any particular governance mechanism involves more than one practice and the third and fourth elements of this list are effectively reliant on the existence of governing norms and informational resources. While seemingly inferior to international, state-based, and "hard" legally grounded mechanisms for altering the behavior of actors and achieving formal compliance, the above list of practices reflects a wide variety of "soft" instruments of global governance (Bulkeley and Newell 2010: 56). Soft instruments work through moral persuasion or economic incentives and they encourage self-binding. In the medium term, they may still achieve a measure of transnational governance, if governance is understood broadly as "authoritatively allocating resources and exercising control and coordination" (Andonova *et al.* 2009: 55). Furthermore, transnational governance is to be distinguished from mere ad hoc cooperation between different actors by the emergence of "institutional arrangements that structure and direct actors' behavior in an issue-specific area" (Falkner 2003: 72, 73).

Norm Creation

A prime example of "soft" practices of global governance is the creation of norms that stipulate overarching ethical principles or prescribe certain forms of political or commercial conduct. Although norms should be seen as an essential part of any governance activity, practices of norm creation are usually wrapped up in broader, more diffuse strategies and campaigns by civil society actors, international organizations, and national governments. One of the rare instances of norm creation as an explicit objective was the World Commission on Dams (WCD). Set up in 1997 at a meeting of critics and proponents of large hydroelectric dams, this cross-sectoral initiative consisted of 12 official Commissioners, who represented the whole spectrum of opinions, and of a forum with 70 members to facilitate discussions with a wide variety of stakeholders. The WCD was not a spontaneous development, but drew on decades of civil society mobilization around large dams and their often destructive impact on livelihoods and the environment. Its ultimate aim was to provide for a transparent and inclusive process of fact-finding and discussion to enable mutual learning and the articulation of normative guidelines for future projects (Khagram and Ali 2008).

The WCD's final report was published in November 2000, but it was not universally welcomed. For instance, the Chinese delegation had already withdrawn by this time and the Indian government rejected the legitimacy of the initiative. Nor could the WCD report be used as a blueprint for decision-making, as it reflected a rather complex collection of overarching values, strategic priorities, and policy guidelines. However, regardless of criticism or support, it was very significant that this norm-creating process could not be ignored by any of the major policy actors. Subsequent developments confirmed that the values, norms, and guidelines projected by the report began to spread to other global governance institutions. The World Bank recognized the normative framework, UNEP set up a Dams and Development Project to ensure follow-up, and development policy-making around the world was, to some degree, influenced by the WCD's results. As Dubash (2011: 207) concludes, it may therefore be best to view the WCD as a "norm-changing process" with long-term effects.

Frequently, of course, normative shifts happen largely outside institutional venues through political campaigns and processes of global socio-cultural change, as for example with the increasing interest in "climate justice" and the practice of carbon offsetting in the late 2000s (Pattberg and Stripple 2008). At the same time, normative dynamics on their own do not necessarily suffice to bring about new governance practices. Strategic economic self-interest on the part of nation-states and MNCs may play an important role, even if these interests can only be understood within a broader normative context – such as concerns over a company's reputation among consumers and investors. But before such definable interests come into play, there must be a sufficient degree of relevant knowledge and other informational resources that help to make an issue "governable" in a technical sense.

Informational Governance

At its most basic level, meaningful information has to be generated before it can be diffused. The UNFCCC, for instance, provides detailed guidelines for the national accounting of greenhouse gas emissions. In the transnational realm as well, efforts are underway to increase transparency with regard to the environmental and social impacts of multinational corporations. The Global Reporting Initiative (GRI), established in 1997 by the US-based non-profit organization CERES and UNEP, is a cross-sectoral initiative which – with the help of stakeholder committees – defines and disseminates one of the most widely used, voluntary standards for sustainability reporting, often referred to as triple bottom line standards. Equivalent goals of measurement, standardization, and comparability are pursued by the 1998 "Greenhouse Gas Protocol," an organization that runs along similar lines under the aegis of a think tank (the World Resources Institute) and a business NGO (the World Business Council on Sustainable Development). Its own web site describes it as "the most widely used international accounting tool for government and business leaders to understand, quantify, and manage greenhouse gas emissions."[4] The GHG Protocol is now working to develop new accounting standards for agriculture and food products, for cities, and for national-level climate mitigation policies (Ranganathan 2011).

Initiatives like the GRI and the GHG Protocol do not only hope to influence the activities of other transnational organizations and of national governments, but are also based on the assumption that rendering environmental damage measurable makes it more likely that companies will begin to factor it into product design, procurement, and marketing. Some programs go a step further by actively encouraging the disclosure of environment-related information and thus exposing the measured emissions to the public at large. For example, by collecting and publishing data on carbon emissions from participating companies in around 60 countries, the Carbon Disclosure Project (CDP), which assembles over 500 institutional investors with a total of US$71 trillion in assets, seeks to influence public and private investors' decisions by highlighting companies' performance on energy use and climate mitigation. The peer pressure that commonly operates in investor–company relationships has allowed it to gather an increasing number of data submissions from large companies – over 3000 in 2010 and achieving a 74% disclosure rate from the 500 largest companies (measured by market capitalization). The literature suggests that disclosure of carbon intensity/emissions and carbon reduction measures can improve a company's relationship with government, lower the reputational risk of negative campaigns by NGOs, and may lead to increasing interest by investors (Ziegler et al. 2011). Altogether, though such practices of information sharing do not amount to targeted regulatory efforts, they "perform crucial governance functions by framing issues, setting agendas, defining what counts as responsible for effective action, offering inspiration, and providing a means of benchmarking achievements" (Bulkeley and Newell 2010: 56).

A second category of informational governance is the generation and dissemination of knowledge – for instance about the scientific foundations of climate change and the effectiveness of environmental or climate policy. Regarding the science of climate change, the Intergovernmental Panel on Climate Change (IPCC) was created in 1988 by national governments and its published assessments are, to some degree, subject to negotiation. Yet, the IPCC also draws on the services of a large number of scientists from around the world and can therefore be regarded as a transnational epistemic community.[5] Notwithstanding significant uncertainties within the natural science of climate change, the vast majority of specialists agree on the basic processes and ramifications of anthropogenic global warming. Because of the perceived legitimacy and authoritative nature of this form of knowledge, there is considerable cognitive pressure on policy-makers to acknowledge the problem and devise appropriate policy responses.

This cognitive background applies to all levels of policy-making, and where political actors concur on the main causes of climate change, they may seek to generate policy-relevant knowledge that can be diffused and applied in many different settings. The required degree of consensus has been easier to achieve at the local level than in international forums, and several global transnational municipal networks are actively employing knowledge-related governance mechanisms. First, the Climate Alliance (assembling over 1500 European cities and indigenous rainforest peoples), founded in 1990, focuses on local climate mitigation through projects on energy efficiency and renewable energy. Second, the Cities for Climate Protection Program (hosted by the International Council for Local Environment Initiatives) originated in the same year and works on both climate mitigation and adaptation. Third, the Large Cities Climate Leadership Group (dubbed C-40) emerged in 2005 and proclaims a special responsibility of the biggest global cities, given that over 80% of global GHG emissions are produced within city boundaries. Fourth, World Mayors Council on Climate Change (established in December 2005) has over 70 members. They engage in political advocacy but their regular interaction also provides the benefits that these types of initiatives have traditionally delivered, namely, the dissemination of knowledge related to policy formulation and implementation, for instance through "showcasing" best practices and successful climate policies with the potential to be replicated in other municipal settings (Bulkeley and Newell 2010). Some scholars refer to the "technical leadership" (Toly 2008) exerted by cities in demonstrating how the necessary transition in the energy and transport sectors could be accomplished in practical terms. In sum, therefore, the actions of global transnational municipal networks may well affect the political dynamics within nation-states and international regimes, but their chief contribution arguably lies in the realm of policy learning.

Standard-Setting

Norms, information, and knowledge are not merely used in indirect ways to influence the decisions of governments and economic actors. At times, they equally provide the basis for the formulation of well-defined standards, which, although still voluntary, may acquire considerable importance in the global marketplace (see also Chapter 23 in this volume). When initiated by civil society actors, the mechanisms of choice tend to be certification and labeling schemes. These can be understood as "deliberative and adaptive governance institutions designed to embed social and environmental norms... that derive authority directly from interested audiences, including those they seek to regulate, [... and] not from sovereign states" (Bernstein and Cashore 2007: 348). In short, they represent "a combination of normative and market mechanisms" (Andonova *et al.* 2009: 61).

In the climate change arena, transnational standard-setting is a relatively recent practice. The Verified Carbon Standard (VCS), the Climate, Community, and Bio-diversity Standard (CCB), the Carbon Fix Standard (CFS), and the Gold Standard represent initiatives that have established particular standards for voluntary carbon markets. They all have slightly different orientations, with the CFS aiming at carbon sequestration from forestry projects and putting somewhat less emphasis on sustainable development objectives than the CCB, but more than the VCS. Perhaps the most interesting scheme is the Gold Standard, which straddles the Kyoto Protocol's Clean Development Mechanism (CDM) and voluntary carbon markets. Established in 2003, it seeks to remedy some of the flaws of the existing CDM process which generates carbon credits that can be bought by countries (and sometimes firms) to meet their climate reduction targets. The Gold Standard strives to ensure the environmental integrity of projects while delivering sustainable development benefits. It does so through a more rigorous assessment and certifies only well-designed and -managed projects in the areas of renewable energy and energy efficiency. The certified credits are attractive to countries and organizations wanting to maximize the climate-mitigating effects of their "offsets" as well as to buyers in the voluntary markets concerned about the reputational risks of ethically questionable projects (Levin *et al.* 2009). For now, it remains to be seen whether the Gold Standard will stagnate in the future or whether it will gain an ever-greater market share, for instance by being adopted as the minimum standard of emergent national, regional, or global carbon markets.

Such projections are critical for assessing the overall effectiveness of transnational standard-setting, but they are easier to attempt when it comes to more long-standing initiatives. One prominent example is the Forest Stewardship Council (FSC), a multi-stakeholder organization created in 1993 to remedy the absence of an international forestry regime and to "promote environmentally appropriate, socially beneficial, and economically viable management of the world's forests."[6] Accordingly, its main decision-making organ, the General Assembly, consists of three chambers of environmental, social, and economic organizations – subject to an agreed North–South allocation. The FSC label is granted to a variety of forestry products and guarantees their provenance from certified forests that are managed in line with the 10 Principles and 56 Criteria for Forest Stewardship. Working through its nationally tailored standards and with the help of independent auditing firms, by December 2011 the FSC had certified over 147 million hectares of forest in 80 countries. There is some evidence that the legitimacy and credibility of the FSC label – particularly in the OECD world but also beyond – has influenced retailers' management of supply chains, consumer purchasing decisions, and national forest policies. Nevertheless, the FSC's global market share, as measured in certified forest area, has remained under 10%, and the overwhelming majority of certified forests are located in the global North. Moreover, an increasing number of environmental NGOs argue that the FSC's continued growth has been achieved only through the watering down of its standards (Gulbrandsen 2010). The relative success of the FSC in setting and monitoring normative standards for forestry products, however, is coming under threat from the proliferation of competing, less stringent standards, such as the

Sustainable Forestry Initiative (SFI) or the Programme for the Endorsement of Forest Certification (PEFC).

If particular producer groups feel squeezed by an erosion of conventional markets or the cost of certification and regular auditing, then the same fate is likely to befall similar standard-setting activities initiated by progressive corporations and civil society organizations. In this respect, it is worth briefly considering the Marine Stewardship Council (MSC), set up in 1996 by the World Wide Fund for Nature and Unilever. The MSC is organized along similar lines to the FSC, but its Stakeholder Council only has an advisory function. With certification covering more than 7% of the world's fisheries, the most visible success of the MSC has been the generation of consumer awareness, brand recognition, and thus the delivery of clear marketing advantages for certified products. However, this influence remains largely confined to Europe and North America and has barely affected fast-growing Asian markets (Hale 2011). The experience of individual countries nonetheless suggests that the MSC might be more likely to simultaneously shape retailers' and consumers' purchasing decisions, on the one hand, and retain the cooperation of major producer groups, on the other. For instance, in the Netherlands the growing importance of MSC labels has compelled fishermen and NGOs to seek enhanced institutional and personal interactions. At the national, though not necessarily global, scale this process may over time generate more trust between stakeholders and ensure a stable market environment for sustainable fishery products (de Vos and Bush 2011).

Capacity-Building and Implementation

A final practice associated with transnational actors are coordination services and direct interventions "on the ground." Early examples included financial transfers in return for increased environmental conservation, as intended by the debt-for-nature swaps pioneered by the NGO Conservation International in the mid-1980s. For the desired outcome to be attained, however, the active cooperation of two governments (debtor and indebted) is also required and this pattern of public–private mixity has become a hallmark of many practices in global environmental governance. Activities such as capacity-building, implementation, financing, and coordination assume a general acceptance of core norms, standards, and policy objectives as well as the possession of considerable informational and material resources. These requirements often lead to the formation of hybrid global public policy networks and public–private partnerships whose functional advantage lies in the ability to "leverage transnationally the resources and skills of multiple actors from different levels of governance and sectors in society" (Andonova *et al.* 2009: 65).

The Gold Standard certification for carbon credits, described in the previous section, is one illustration of a hybrid governance instrument, as it represents a private initiative interacting with the public enterprise of the Kyoto Protocol's carbon market. Hybrid public–private policy networks often become formalized and instituted as partnership arrangements. So-called "type 2" public–private partnerships constituted the principal outcome of the 2002 World Summit on Sustainable Development (WSSD). By December 2011, 348 such partnerships had been registered with the UN Commission on Sustainable Development (CSD). Although they could be regarded as modest substitutes for the failure of legally binding international agreements, partnerships may also represent "flexible cooperation mechanisms" that deliver at least three important functional and political benefits: "[1] learning by doing, [2] building coalitions of the willing, and [3] dividing a complex governance problem into smaller components" (Andonova 2009: 197).

Selected "type 2" partnerships can, moreover, demonstrate the practical, outcome-oriented nature of cooperative ventures which may often encompass governmental actors, international organizations, and private actors. Szulecki *et al.* (2011) have analyzed the effectiveness of a number of partnerships in the field of sustainable energy and have recorded their functions as including services such as knowledge dissemination, technology transfer, technical implementation, training, planning, and capacity-building. For example, one of the largest CSD partnerships, the Renewable Energy and Energy Efficiency Partnership (REEEP), which is linked to a network of over 250 other organizations, does not merely list services related to information-sharing and professional advice, but also funds and promotes small-scale projects as well as strategic mechanisms for the further development of sustainable energy.[7] REEEP's greatest impact is likely to derive from its contribution to capacity-building and market development in over 50 countries (Parthan *et al.* 2010).

It is important to recognize that practices of implementation and capacity-building are not necessarily more politically neutral than the preceding three categories of transnational governance (norms, information/knowledge, and standards). As Bulkeley and Newell (2010: 58) put it:

> [B]y being able to determine the criteria on which funding is distributed, to set the rules of the game, and by providing certain forms of advice or access to particular sorts of technologies or information, networks can find themselves in powerful positions.

With regard to capacity-building, clear differences can, for instance, be discerned between the World Bank's Prototype Carbon Fund (PCF) and the Global Environment Facility's Small Grants Programme (SGP). Bäckstrand (2008: 85) describes the PCF as an "implementation partnership between multinational firms and governments to promote Kyoto carbon markets." The PCF, established in 1999, claims to offer a "learning-by-doing" opportunity for all the parties involved (17 MNCs and 6 OECD governments) and invests their contributions in projects that deliver both sustainable development and emission credits. This form of institutional capacity-building is, in one way, merely a pragmatic service, but it also represents a platform for launching a new "product" (i.e. carbon credits) and familiarizing potential buyers with the mechanics of using the emerging carbon market.

By contrast, the Global Environment Facility (GEF) designed the SGP to favor more bottom-up forms of capacity-building. Since 1995, it has awarded 12 000 small grants (of up to US$50 000) in 122 countries, usually directly allocated to local communities and NGOs that seek to reconcile socio-economic and environmental goals, for instance through projects on biodiversity or climate change. The SGP's objectives are to simultaneously increase the effectiveness of implementing global environmental objectives and ensure greater local "ownership" of these overarching agendas (Andonova 2009: 211). Thus, proclaimed as the "people's GEF" on the SGP web site,[8] the program is clearly driven by a strategic motivation. But its grass-roots structure and selection of projects partners allow for considerable leeway, which marks a real difference from the more prescriptive, tightly organized PCF.

Evidently, both the nature and the performance of the above governance practices have been critically investigated by many scholars. There is now a growing literature examining the legitimacy of transnational governance, with a focus on key components such as representativeness/participation, transparency, accountability, and effectiveness. While hybrid public–private initiatives, such as the Clean Development Mechanism, score most highly due to multiple accountability features (Bäckstrand 2008), there remain serious deficiencies which, at the very least, "complicate a simultaneous attainment of procedural quality and problem-solving effectiveness" (Lövbrand *et al.* 2009: 76). The deliberative potential of multi-actor transnational governance schemes may be further questioned because of unequal participation of Northern and Southern representatives. Examining three of the initiatives surveyed above (WCD, GRI, and FSC), Dingwerth (2008) concludes that only the FSC has institutionalized North–South parity among its stakeholders and that, in general, Southern actors tend to be underrepresented when it comes to the knowledge-related aspects of decision-making. Notwithstanding the proliferation of transnational governance practices, it thus remains to be seen whether sufficiently legitimate compromises can be found in a context marked by inequalities of political and economic power/resources and divergent interpretations of sustainable development.

Conclusion

This chapter has shown how, since the early 1990s, the suggestive concept of global governance has inspired (and subsumed) a vast amount on research on the environment. Global governance, as an academic approach, has pioneered the study of modes of governing where states do not occupy the pole position. Instead, the emergence of non-state and network-like forms of governance in various issue areas (e.g. forestry, fisheries, biodiversity, climate) have been approached as instances of "global environmental governance." Rather than focusing on the wide range of political actors in this domain – which has been a staple of the global governance literature – we have reviewed four different sets of practices through which environmental and climate issues are being governed: norm creation, informational governance, standard-setting, and capacity-building and implementation. Our ambition has been to explore how governing is performed in the absence of the formal authority of the state. However, as the various examples have demonstrated, the jury is still out on the extent to which the new practices of environmental and climate governance are making a genuine difference.

For an issue like climate change, where the intergovernmental struggle to construct a comprehensive legal architecture is progressing at a snail's pace, all eyes are naturally turned toward the possibility (and capability) of governing "beyond the state." Biermann (2010: 287) points out that it is too early to judge if the current experimentation with various climate governance initiatives is indicative of a fundamental incapability of the modern state to deal with the complexity of the global climate or if this is just a temporary phenomenon. The latter would imply that non-state networks and institutions might lose their influence and significance once an intergovernmental consensus on the key parameters of a strong global climate regime emerges. Yet, it is also possible to understand global climate and environmental governance as part of a broader shift. The institutionalization of environmental governance "beyond the state" resembles what Ruggie has called the reconstitution of a global public domain. As a domain, it does not replace states but embeds systems of governance in broader global frameworks of social capacity and agency that did not exist previously (Ruggie 2004: 519). Therefore, it is possible that the activities "beyond the state" that we have captured in this chapter indicate, in Ruggie's words,

> the arrival on the global stage of a distinctive public domain – thinner, more partial, and more fragile than its domestic counterpart, to be sure, but existing and taking root apart from the sphere of interstate relations (Ruggie 2004: 522).

In our opinion, instead of the endless waiting for – or imaginative design of – a "global climate deal," the cutting of the Gordian knot that will settle matters for years to come with a brief stroke of a pen, scholars and practitioners should try to grasp the emerging climate order in its entirety. The overall *global climate governance complex*, the state *and* the non-state in a single analytical framework, is not very well understood, although Keohane and Victor (2011) have made an inspiring start. Here awaits a potentially fruitful area of scholarship on the norms, rules, and practices it embodies and on the ways in which it is shaping the subjects of governance, such as states, communities, and individuals. To view environmental governance as a "governance complex" might deliver a small seed of hope in times of despair.

Notes

1 Yergin (1977: 195) notes that the concept of "national security" entered with such force that it "seemed always to have been with us."
2 However, Keohane (1984) moved a few years later away from the idea of providing a quite separate perspective on non-state actors in world politics and instead constructed a functional theory of regimes that could account for patterns of international cooperation. This crucial move made Realist and Liberal schools of thoughts united in a shared "rationalist" research program, premised on the condition of anarchy in the international system and oriented towards investigating international cooperation generally, and specifically when, where, and how regimes and institutions make a difference. This agenda still inspires considerable research in international environmental politics.
3 Rosenau (2003: 295) includes a list of illustrations: "An SOA can be an issue regime, a professional society, an epistemic community, a neighborhood, a network of the like-minded, a truth commission, a corporation, business subscribers to codes of conduct (e.g., the Sullivan principles), a social movement, a local or provincial government, a diaspora, a regional association, a loose confederation of NGOs, a transnational advocacy group, a paramilitary force, a credit-rating agency, a strategic partnership, a transnational network, a terrorist organization, and so on across all the diverse collectivities that have become sources of decisional authority in the ever more complex multi-centric world."
4 http://www.ghgprotocol.org (accessed October 20, 2012).
5 Haas (1992: 3) defines epistemic communities as networks of professionals with "recognised expertise and competence in a particular domain and an authoritative claim to policy relevant knowledge within that domain or issue-area."
6 See http://www.fsc.org/vision_mission.html (accessed October 20, 2012).
7 See http://www.reeep.org/48/about-reeep.htm (accessed October 20, 2012).
8 See http://sgp.undp.org (accessed October 20, 2012).

References

Andonova, Liliana B. 2009. "International Organizations as Entrepreneurs of Environmental Partnerships." In *International Organizations in Global Environmental Governance*, ed. Frank Biermann, Bernd Siebenhüner, and Anna Schreyögg, 195–222. Abingdon: Routledge.

Andonova, Liliana B., Michele M. Betsill, and Harriet Bulkeley. 2009. "Transnational Climate Governance." *Global Environmental Politics*, 9: 52–73.

Bäckstrand, Karin. 2008. "Accountability of Networked Climate Governance: The Rise of Transnational Climate Partnerships." *Global Environmental Politics*, 8: 74–102.

Barnett, Michael N. and Raymond Duvall. 2005. "Power in Global Governance." In *Power in Global Governance*, ed. Michael N. Barnett and Raymond Duvall, 1–32. Cambridge: Cambridge University Press.

Barnett, Michael N. and Kathryn Sikkink. 2008. "From International Relations to Global Society." In *The Oxford Handbook of International Relations*, ed. Christian Reus-Smit and Duncan Snidal, 62–83. Oxford: Oxford University Press.

Bernstein, Steven and Benjamin Cashore. 2007. "Can Non-state Global Governance Be Legitimate? An Analytical Framework." *Regulation and Governance*, 1: 347–371.

Biermann, Frank. 2006. "Global Governance and the Environment." In *Palgrave Advances in International Environmental Politics*, ed. Michele M. Betsill, Kathryn Hochstetler, and Dimitris Stevis, 237–261. New York: Palgrave Macmillan.

Biermann, Frank. 2010. "Beyond the Intergovernmental Regime: Recent Trends in Global Carbon Governance." *Current Opinion in Environmental Sustainability*, 2: 284–288.

Bulkeley, Harriet and Peter Newell. 2010. *Governing Climate Change*. Abingdon: Routledge.

Cajvaneanu, Doina. 2011. *A Genealogy of Government. On Governance, Transparency and Partnership in the European Union*. Trento: University of Trento, School of International Studies.

Chatterjee, Pratap and Matthias Finger. 1994. *The Earth Brokers*. London: Routledge.

Czempiel, Ernst O. 1992. "Governance and Democratization." In *Governance without Government: Order and Change in World Politics*, ed. James N. Rosenau and Ernst O. Czempiel, 250–271. Cambridge: Cambridge University Press.

de Vos, Birgit I. and Simon R. Bush. 2011. "Far More than Market-Based: Rethinking the Impact of the Dutch Viswijzer (Good Fish Guide) on Fisheries' Governance." *Sociologica Ruralis*, 51: 284–303.

Dingwerth, Klaus. 2008. "Private Transnational Governance and the Developing World: A Comparative Perspective." *International Studies Quarterly*, 52: 607–634.

Dingwerth, Klaus and Philipp Pattberg. 2010."How Global and Why Governance? Blind Spots and Ambivalences of the Global Governance Concept." *International Studies Review*, 12: 696–719.

Douglas, Ian. 1999. "Globalization as Governance: Toward an Archeaology of Contemporary Political Reason." In *Globalization and Governance*, ed. Jeffrey A. Hart and Aseem Prakash, 134–160. London: Routledge.

Dubash, Navroz K. 2011. "World Commission on Dams." In *Handbook of Transnational Governance*, ed. Thomas Hale and David Held, 202–210. Cambridge: Polity.

Eells, Richard Sedric Fox. 1960. *The Meaning of Modern Business: An Introduction to the Philosophy of Large Corporate Enterprise*. New York: Columbia University Press.

Falkner, Robert. 2003. "Private Environmental Governance and International Relations: Exploring the Links." *Global Environmental Politics*, 3: 72–87.

Gulbrandsen, Lars H. 2010. *Transnational Environmental Governance: The Emergence and Effects of the Certification of Forests and Fisheries*. Cheltenham: Edward Elgar.

Haas, Peter. 1992. "Introduction: Epistemic Communities and International Policy Coordination." *International Organization*, 46: 1–36.

Hale, Thomas. 2011. "Marine Stewardship Council." In *Handbook of Transnational Governance*, ed. Thomas Hale and David Held, 308–313. Cambridge: Polity.

Hempel, Lamont. 1996. *Environmental Governance. The Global Challenge.* Washington, DC: Island Press.

Hewson, Martin and Timothy J. Sinclair. 1999. "The Emergence of Global Governance Theory." In *Approaches to Global Governance Theory*, ed. by Martin Hewson and Timothy J. Sinclair, 3–22. Albany, NY: State University of New York Press.

Hurrell, Andrew. 2007. *On Global Order: Power, Values, and the Constitution of International Society.* Oxford: Oxford University Press.

Hurrell, Andrew and Benedict Kingsbury, eds. 1992. *The International Politics of the Environment: Actors, Interests, and Institutions.* Oxford: Clarendon Press.

Jagers, Sverker and Johannes Stripple. 2003. "Climate Governance beyond the State." *Global Governance*, 9: 385–399.

Jönsson, Christer. 2010. "Theoretical Approaches to International Organization." In *The International Studies Encyclopedia*, Vol. XI, ed. Richard Denemark, 7028–7045. Oxford: Wiley-Blackwell.

Keck, Margaret E. and Kathryn Sikkink. 1998. *Activists beyond Borders: Advocacy Networks in International Politics.* Ithaca, NY: Cornell University Press.

Keohane, Robert O. 1984. *After Hegemony: Cooperation and Discord in the World Political Economy.* Princeton, NJ: Princeton University Press.

Keohane, Robert O. and Joseph S. Nye, eds. 1972. *Transnational Relations and World Politics.* Cambridge, MA: Harvard University Press.

Keohane, Robert O. and Joseph S. Nye. 1977. *Power and Interdependence: World Politics in Transition.* Princeton, NJ: Princeton University Press.

Keohane, Robert O. and David G. Victor. 2011. "The Regime Complex for Climate Change." *Perspectives on Politics*, 9: 7–23.

Khagram, Sanjeev and Saleem H. Ali. 2008. "Transnational Transformations: From Government-centric Interstate Regimes to Cross-sectoral Multi-level Networks of Global Governance." In *The Crisis of Global Environmental Governance: Towards a New Political Economy of Sustainability*, ed. Jacob Park, Ken Conca, and Matthias Finger, 132–162. Abingdon: Routledge.

Levin, Kelly, Benjamin Cashore, and Jonathan Koppell. 2009. "Can Non-state Certification Systems Bolster State-Centered Efforts to Promote Sustainable Development through the Clean Development Mechanism?" *Wake Forest Law Review*, 44: 777–798.

Lipschutz, Ronnie D. and Judith Mayer. 1996. *Global Civil Society and Global Environmental Governance: The Politics of Nature from Place to Planet.* Albany, NY: State University of New York Press.

Lövbrand, Eva, Teresia Rindefjäll, and Joakim Nordqvist. 2009. "Closing the Legitimacy Gap in Global Environmental Governance? Lessons from the Emerging CDM Market." *Global Environmental Politics*, 9: 74–100.

Merriam, Charles E. 1944. *Public and Private Government.* New Haven, CT: Yale University Press.

Neumann, Iver B. and Ole J. Sending. 2010. *Governing the Global Polity: Practice, Mentality, Rationality.* Ann Arbor: University of Michigan Press.

Newell, Peter. 2005. "Towards a Political Economy of Global Environmental Governance." In *Handbook of Global Environmental Politics*, ed. Peter Dauvergne, 187–201. Cheltenham: Edward Elgar.

Okereke, Chukwumerije, Harriet Bulkeley, and Heike Schroeder. 2009. "Conceptualizing Climate Governance beyond the International Regime." *Global Environmental Politics*, 9: 58–78.

Parthan, Binu, Marianne Osterkorn, Matthew Kennedy *et al.* 2010. "Lessons for Low-Carbon Energy Transition: Experience from the Renewable Energy and Energy Efficiency Partnership (REEEP)." *Energy for Sustainable Development*, 14: 83–93.

Paterson, Matthew. 2000. *Understanding Global Environmental Politics.* Basingstoke: Palgrave Macmillan.

Paterson, Matthew, David Humphreys, and Lloyd Pettiford. 2003. "Conceptualizing Global Environmental Governance: From Interstate Regimes to Counter-Hegemonic Struggles." *Global Environmental Politics*, 3: 1–10.

Pattberg, Philipp and Johannes Stripple. 2008. "Beyond the Public and Private Divide: Remapping Transnational Climate Governance in the Twenty-First Century." *International Environmental Agreements: Politics, Law and Economics*, 8: 367–388.

Pierre, Jon. 2000. "Introduction: Understanding Governance." In *Debating Governance: Authority, Steering, and Democracy*, ed. Jon Pierre, 1–12. Oxford: Oxford University Press.

Ranganathan, Janet. 2011. "GHG Protocol: The Gold Standard for Accounting for Greenhouse Gas Emissions," http://insights.wri.org/news/2011/10/ghg-protocol-gold-standard-accounting-greenhouse-gas-emissions (accessed January 1, 2012).

Rosenau, James N. 1990. *Turbulence in World Politics*. Princeton, NJ: Princeton University Press.

Rosenau, James N. 1992. "Governance, Order, and Change in World Politics." In *Governance without Government: Order and Change in World Politics*, ed. James N. Rosenau and Ernst O. Czempiel, 1–29. Cambridge: Cambridge University Press.

Rosenau, James N. 1997. *Along the Domestic–Foreign Frontier: Exploring Governance in a Turbulent World*. Cambridge: Cambridge University Press.

Rosenau, James N. 2003. *Distant Proximities: Dynamics beyond Globalization*. Princeton, NJ: Princeton University Press.

Rosenau, James N. and Ernst O. Czempiel, eds. 1992. *Governance without Government: Order and Change in World Politics*. Cambridge: Cambridge University Press.

Ruggie, John Gerard. 2004. "Reconstituting the Global Public Domain: Issues, Actors, and Practices." *European Journal of International Relations*, 10: 499–531.

Stokke, Olaf S. 1997. "Regimes as Governance Systems." In *Global Governance: Drawing Insights from the Environmental Experience*, ed. Oran R. Young, 27–63. Cambridge, MA: MIT Press.

Szulecki, Kacper, Philipp Pattberg, and Frank Biermann. 2011. "Explaining Variation in the Effectiveness of Transnational Energy Partnerships." *Governance*, 24: 713–736.

Thompson-Feraru, Anne. 1974. "Transnational Political Interests and the Global Environment." *International Organization*, 28: 31–60.

Toly, Noah J. 2008. "Transnational Municipal Networks in Climate Politics: From Global Governance to Global Politics." *Globalizations*, 5: 341–356.

WCED (World Commission on Environment and Development). 1987. *Our Common Future*. Oxford: Oxford University Press.

Weiss, Thomas G. 2011. *Thinking about Global Governance: Why People and Ideas Matter*. Abingdon: Routledge.

Willets, Peter. 1982. *Pressure Groups in the Global System: The Transnational Relations of Issue-Oriented Non-governmental Organizations*. London: F. Pinter.

Yergin, Daniel. 1977. *Shattered Peace: The Origins of the Cold War and the National Security State*. Boston, MA: Houghton Mifflin.

Young, Oran R., ed. 1997. *Global Governance: Drawing Insights from the Environmental Experience*. Cambridge: Cambridge University Press.

Ziegler, Andreas, Timo Busch, and Volker H. Hoffmann. 2011. "Disclosed Corporate Responses to Climate Change and Stock Performance: An International Empirical Analysis." *Energy Economics*, 33: 1283–1294.

Chapter 10

Global Environmental Security

Simon Dalby

Introduction

Security is a key term in the contemporary political lexicon. Security's multiple meanings and linkages to numerous issue areas make formulating environment in relation to it much more tricky than might at first seem apparent. Both as a conceptual matter and as a guide to policy-making the juxtaposition of security and environment is a fraught endeavor. This has become even more so in recent years as the theme of climate change has been linked to security in a discussion of "climate security." Much of the recent policy debate about climate security either forgets or ignores earlier research work on the links between security, conflict, and environment. Popular accounts frequently reinvent Malthusian fears of scarcities and disruptions as a cause of violence, or naively assume the invocation of security to deal with the problems will necessarily produce sensible and effective policies.

The disruption of the relatively stable climate system in recent decades has been caused by the use of abundant fossil fuels by the huge global economy that has spread into most parts of the planetary biosphere. The scale of human activity in the biosphere and the rapidly growing scientific understanding of these processes have both changed our understanding of humanity's role in the biosphere and made clear the need for policies to deal with our rapidly changing circumstances (Steffen et al. 2011). This is necessary to ensure that we don't endanger the key conditions necessary for human civilization to exist. This is very obviously now a matter of global security broadly construed (Dalby 2009). It is more specifically a matter of who decides such things as the future composition of the planetary atmosphere. These emerging new circumstances have changed how politics, technology, and economy link to security policy, but much of the debate still draws on earlier formulations

The Handbook of Global Climate and Environment Policy, First Edition. Edited by Robert Falkner.
© 2013 John Wiley & Sons, Ltd. Published 2013 by John Wiley & Sons, Ltd.

that usually presuppose that the expansion of the global economy is essential to the provision of security.

As this chapter shows, the changing understanding of humanity's place in the biosphere, and the importance of our growing disruptions of it, have to be worked carefully into the analysis if security is to provide useful policy guidance in coming decades. The chapter first looks back to the history of some key moments in the evolution of a global sensibility that links security to environment. Subsequent sections look at the debate in the 1990s about environmental security and how various formulations shaped the discussion of the links between conflict, war, and environment. The final sections outline how the contradictions in these earlier discussions have become especially pressing in light of the failures of climate mitigation policies to arrest the pace of global change. Climate change now makes environmental factors a crucial consideration for global security, but, as this is a fairly recent innovation, it remains unclear how this will play out in terms of specific policies and their consequences (Webersik 2010). But some important political choices have become unavoidable.

The Many Meanings of Security

All this requires very careful reflection on what security means in particular circumstances, and how it is connected into discussions of war and peace, and the larger stabilities of the global political system that supposedly ought to prevent interstate warfare from recurring. The most important aspect of security is that it is used in political discourse to refer to whatever is the highest priority. Invoking security frequently implies an emergency, which in turn justifies extraordinary powers and sometimes the suspension of aspects of normal civilian life (Buzan *et al.* 1998). But such invocations draw widely on cultural assumptions about danger and safety, and the ability to successfully make a claim that emergency measures are appropriate is a powerful political capability; national security frequently trumps democracy or rights in a crisis situation (Williams 2007). Thus linking environment to security is tied into claims that it is the most important matter needing attention, and a source of danger requiring priority action. When the notion of global security is linked to environment, and now specifically to climate, this clearly is a matter that those who use such language think ought to be a top priority for policy-makers. Activists have sometimes used military analogies with the Second World War to try to suggest that a similar mobilization of society and industry is needed to deal with the threat of climate change (Spratt and Sutton 2008).

However, as Daniel Deudney (1990) pointed out forcefully at the beginning of the major debate on environmental security in the early 1990s, invoking national security, and in particular the role of the military in addressing environmental matters, doesn't usually produce an appropriate policy response. It doesn't because the military is ill-equipped, and certainly not trained to undertake environmental actions; this is a major mismatch between the agency involved and the nature of the problem. If security involves the heavy-handed application of emergency measures it may be completely counterproductive in dealing with rebuilding economies in a sustainable way or expanding citizenship rights and effective participation in the necessary decision-making. There are thus compelling arguments to "de-securitize" many aspects of the environmental security discussion and return it to the normal processes of political deliberation rather than treat matters as emergencies (Floyd 2010). Nonetheless, as climate change interacts with the increasingly artificial circumstances of urban life and maintaining infrastructure becomes a priority, then security is stretched to encompass matters only indirectly related to traditional concerns with political threats and international violence.

While national security, the protection of the political order and territorial integrity of states, is the most prominent version of security in the modern world, used by state elites to maintain domestic control quite as much as to deal with external threats, the international dimensions of security are crucially important insofar as war between states is understood as the primary danger. Between nuclear-armed powers the dangers are very obvious, and preventing such conflict is a priority in maintaining a state of international security, but even the preparation and testing of nuclear weapons is dangerous, and here the links between radioactive fallout and security connected environment to security long before concerns about climate change were raised. There are few obvious scenarios in which climate change or other environmental matters might trigger a major interstate war, both because the costs of going to war obviously outweigh most plausible environmental benefits, and in many places aggrieved parties do not have a military option (German Advisory Council on Global Change 2008). Bangladesh simply does not have a military option to force states to stop greenhouse gas emissions that are leading to the sea level rise that endangers its citizens on much of its low-lying territory.

Where national security and territorial integrity had been the operating principles of the international system, now human security, shifting the priority from states to people, has raised complicated new discussions of sovereignty and international law. This is now linked to the principle of the Responsibility to Protect, wherein governments are obligated to protect the human security of their populations (ICISS 2001), although how this might be applied to the environment is far from clear, beyond obvious concerns with infrastructure planning (Pascal 2010). Making people rather than states the "referent object" of security is especially important when it comes to matters of environmental security and now, most recently, climate security, which is obviously also very much a global concern (Brauch *et al.* 2011). But linking the focus on small-scale local environmental conflict as a development problem with global atmospheric stability is not easy to do. Not least because fossil-fueled development is precisely what is causing large-scale atmospheric disruptions. The conceptual confusion around security needs to be unpacked carefully if useful policy implications are to be extracted from the discussion, but all of this requires careful reflection on how security has been invoked in the past in relation to the atmosphere, and how this is changing in light of new research, if it is to be appropriately contextualized now.

Atmospheric Security: Nuclear Weapons, Ozone, and Climate

The nuclear destruction of Hiroshima and Nagasaki in August 1945 introduced the world to the dangers of radioactive fallout. Deadly rains, spreading the debris from a nuclear explosion in the atmosphere widely and contaminating food, water, and buildings, made the environmental dimensions of security clear in at least one easily understandable image. The unfortunately named Japanese trawler *Lucky Dragon*, badly contaminated by the fallout from an American nuclear weapons test in the Pacific in March 1954, reinforced the popular understanding of the environmental aspects of these devices well before active campaigns on the part of pediatricians to collect baby teeth to document the spread of Strontium 90 in particular brought the issue home to Americans (Miller 1986). Documenting the consequences of radiation on populations continues to be a major source of controversy concerning nuclear power, only most recently revived in the case of unexplained death rate increases following the Fukushima nuclear meltdown in 2011 (Mangano and Sherman 2012). The partial test ban treaty of the early 1960s that effectively moved nuclear testing underground to prevent further atmospheric contamination was more an international environmental agreement than an arms control treaty, but the link between nuclear weapons, environment, and global security was clearly indicated in its formulation (Soroos 1997).

Subsequently as industrialization spread rapidly in the latter part of the twentieth century, other environmental pollutants came to widespread attention in the 1960s, substances such as DDT and other pesticides, as well as smog caused by coal-burning and subsequently automobile emissions (McNeill 2000). Part of the solution to local pollution was to build taller smokestacks, ensuring that the pollution was diluted and spread more widely rather than causing immediate damage and health hazards to local populations. Ironically this simply made a local problem a global one when winds moved pollution across frontiers and acid rain formed as a result of atmospheric chemistry. In particular the destruction of forest and lacustrine environments as a result of the transboundary movement of pollution from the United States into Canada and European pollution over Scandinavia added an important international dimension to this and triggered a series of international efforts to reduce sulfur emissions (Park 1987).

The late 1980s also witnessed dramatic alarms about deforestation in the Amazon, and a noteworthy drought in North America in 1988, severe enough to drastically constrain river transport on the Mississippi, and in the process cause economic disruptions in the United States. The global atmosphere had become a matter of concern in the 1980s, and while initially concerns had been about a new ice age, once the recent global climate data were carefully compiled it became clear that rapidly rising levels of carbon dioxide in particular were likely to warm the global climate noticeably (Schneider 1989). Putting these concerns with acid rain, ozone depletion, and then concerns about climate change together raised awareness and political concern about global environmental matters with enough urgency to get them considered as a matter of global security.

In the 1980s the environmental dimensions of nuclear war also came to prominence in the discussion of what was quickly dubbed "nuclear winter" (Turco *et al.* 1983). Where earlier concerns about radiation and the damage to the stratospheric ozone layer by the use of nuclear weapons had been noted, in the early 1980s scientists asked what might happen if huge dust and smoke clouds were lofted into the upper atmosphere by numerous nuclear explosions. The models of the atmosphere used to investigate these things suggested that the Earth would be shaded by debris and smoke to such an extent that the northern hemisphere in particular would be noticeably cooled. The fear was that they would be cooled to such an extent that crops would probably fail and the direct destruction of cities and industries by the nuclear explosions, fallout, and ozone layer disruptions would be seriously augmented by the collapses of many ecosystems and agriculture. Thus the indirect ecological effects of nuclear war might be even more serious than the immediate destruction (Sagan and Turco 1990). All of this added to the arguments against nuclear war and linked environmental matters once again directly into the discussions of international security. They also emphasized the fragility of the planet's climate system and the potential for humanity to change its basic parameters.

These concerns were emphasized a few years later when it suddenly became clear that 1970s fears of stratospheric ozone depletion were being realized over the South Pole in particular, although with a substantial reduction of stratospheric ozone over the North Pole in northern hemisphere winters too. In part this is because of the pattern of winds in the polar vortex over Antarctica that make a very cold environment in the polar winter night, which facilitates the breakdown of CFCs and the subsequent chemical "scavenging" of ozone in those regions. This matters greatly because in effect stratospheric ozone acts as a shield for life from harmful UVB solar radiation. Hence action needed to be taken rapidly to phase out the production and use of CFCs and related chemicals. The Montreal Protocol and subsequent follow-on agreements have stopped the production of most of these substances, although polar winter ozone holes persist, and will do so for decades to come until atmospheric CFCs diminish to low levels (Benedick 1991).

In this context of alarm about global security and environment, many of the themes that now shape the current international discussion were initially drawn together in Toronto in the June 1988 international conference on "The Changing Atmosphere: Implications for Global Security." While the conference was concerned about ozone depletion as well as acid rain issues, the biggest concern the conference statement identified was "climate warming, rising sea level, altered precipitation patterns and changed frequencies of climatic extremes caused by the 'heat trap' effects of greenhouse gases." All this mattered, the delegates thought, because the consequences of these changes would in the long run be profoundly disruptive for all states. Such major disruptions were understood as a matter of global security because they could lead to conflict in many ways. In the words of the conference closing statement:

> The best predictions available indicate potentially severe economic and social dislocation for present and future generations, which will worsen international tensions and increase the risk of conflict among and within nations. It is imperative to act now.

What was also clear was that international cooperation was going to be needed. In the words of the Toronto conference statement: "No country can tackle this problem in isolation. International cooperation in the management and monitoring of, and research on, this shared resource is essential."

Tentative steps in this direction came from the subsequent Earth Summit in Rio de Janeiro in 1992 at which the United Nations Framework Convention on Climate Change (UNFCCC) was agreed to by many states. The overall purpose of the convention clearly states that it is necessary to prevent humanity causing dangerous levels of climate change. While there are arguments about how much climate change is dangerous, by the mid-1990s there was a widespread agreement that the climate should not warm more than 2 °C. However, this was more a political compromise than a scientific evaluation. Two decades later current projections frequently suggest that we are headed well past this mark unless things change very soon (Anderson and Bows 2011). Major ecological changes will inevitably follow if present atmospheric trends continue. In Rio it was obvious that international cooperation was key to solving this problem and dealing with curbing greenhouse gas emissions. But developing countries were also very clear that those who were creating the problem, the rich developed states that used huge amounts of carbon fuels, were going to have be those who started to solve it, and would need to pay compensation to the poorer states for forgone development opportunities (Kjellen 2008).

Sustainable Development and Environmental Security

Simultaneously with the emerging awareness of atmospheric fragilities, there was a parallel global discussion of resource management, environmental degradation, and development. This lengthy discussion has also led into the global security discussion related to environment, but from the ground up, as it were, in contrast to the top-down global atmosphere discussion. The formulation of sustainable development was effectively institutionalized by the World Commission on Environment and Development (WCED) in its report *Our Common Future* in 1987. Taken for granted in the document is the assumption that violence is likely in imminent struggles for access to scarce resources. And at least implicit in much of the discussion is the argument that renewable resources are a key part of this problem and that such shortages will likely be aggravated by environmental degradation of various sorts. This line of argument fed the initial formulations of what became the discourse of environmental security and has continued to shape many of the discussions since (Dalby 2002).

One theme that emerged in the late 1980s was fears of North–South hostilities as impoverished states took action against the rich North. A variation on that theme suggested that conflict in Southern states, driven by environmental difficulties, might cause spillover effects as migrants caused political difficulties for destination states. All of this might lead to global security issues if the resultant instabilities led to interstate warfare (Homer-Dixon 1991). This discussion was part of a larger re-evaluation of international security, in the United States in particular, that came about due to the collapse of the Soviet Union and with it the end of the Cold War, coincident with the military mobilization of the coalition that reversed Saddam Hussein's invasion of Kuwait in 1990 in the brief Gulf War early in 1991 (Allison and Treverton 1992). Among the many new contenders for priority concern for security were ethnic nationalism, international migration, the drugs trade, nuclear proliferation, emergent diseases, and the global environment (Klare and Chandrani 1998). In the Soviet Union itself matters of environment and the need to think very differently about global security had been raised following the Chernobyl nuclear meltdown in 1986, although these were dismissed by commentators and politicians in the West, not least because the Soviet Union had an appalling record on environmental matters (Dabelko 2008).

These two key questions – first the empirical one concerning if and when environmental changes might cause conflicts and if so, where and how, and second the question of the appropriate framework for security planning in the new geopolitical circumstances – structured the debate through much of the 1990s. These are explicitly linked by implicit assumptions concerning what is to be secured, and whose environments matter for security (Barnett 2001). Subsequently, as later sections of this chapter will show, these themes have returned to influence how, most recently, "climate security" is formulated as a global issue.

The initial premise of the WCED (1987) that environmental degradation would cause conflict was widely accepted among the commentators at the time. What is notable about Thomas Homer-Dixon's (1991) intervention in the early 1990s is that he, for one, did not accept the basic premise and turned from a wide-ranging policy discussion to try some detailed empirical work that would establish the parameters of how environmental change might be a problem, and specifically how it might lead to what he called acute conflict. He subsequently concluded that inadequate political institutions were crucial to explaining where conflict was most likely (Homer-Dixon 1999). How precisely these connect to more traditional matters of security has been under debate in the last couple of decades, but Malthusian fears of population growth and resource shortages have been reinvented frequently despite the robust empirical research that suggests that violence is not a frequent result of environmental change (Kahl 2006) or, most recently, specifically a response to climate-change-induced droughts (Theisen *et al.* 2011).

In the early years of the new millennium this discussion was effectively turned upside-down when the "greed not grievance" arguments suggested that abundance rather than scarcity was related to violence in the "new wars" of the 1990s (de Soysa 2002). Here the suggestion is that resources that are worth fighting for when few other economic options are available are the source of organized violence. One fights to control revenue from natural resources if one has few other options. Thus the discussion of conflict diamonds, the violence surrounding oil resources in many places, and the destruction of tropical forests to support insurgencies suggested a very different set of circumstances relating to conflict, and more directly tied concerns with violence in the peripheries into discussions of consumption in the metropoles of the global economy (Le Billon 2012). This suggests a very different set of ideas about the sources of violence and which security policies might be appropriate.

Critics have suggested that the concept of security is so caught up with the processes of the global economy that it isn't a helpful way of engaging contemporary politics (Neocleous 2008). Although this line of critique refers largely to the militarization of policing as part of the war on terror, it is worth remembering that at least so far much of security has been about the protection and support of the mode of economic development that has set in motion the transformation of most of the Earth's ecosystems, and the atmosphere in particular. Jon Barnett's (2000, 2001) critique of environmental security focused on why in these circumstances so much of the literature in the 1990s in particular focused on the global South and insecurities there, rather than on the larger political economy that was causing the global disruptions.

The critical literature on development has suggested that the quasi-imperial way that carbon sinks are now arranged and offsets calculated is in many ways a reinvention of colonizing practices of the past (Lohmann 2006). Indeed development discourse is increasingly related to matters of instability in peripheral places, where violence and military interventions are part of the process of governing and security merges with development (Duffield 2007). Attempts by richer states to secure supplies of food in the face of likely future disruptions are in places perpetuating the dispossession of marginal and poor peoples as practices of land-grabbing spread (Matondi *et al.* 2011). The political protests around the world in 2011 have, however, now raised questions about global governance in ways that are connecting matters of global economics with larger matters, including climate change.

The overall logic of sustainable development on the one hand and the immense wealth involved in petroleum and gas industries on the other have combined to maintain fossil-fueled economic growth as the priority, despite repeated warnings that more fundamental rethinking and reorientation of the global economy is necessary. Nearly all the discussion thus far has focused on mitigating climate change on the entirely sensible understanding that preventing climate change is the first priority, not least because, as the influential British review of the economics of climate change, the so-called Stern Review (Stern 2007) pointed out forcefully, it is much cheaper to pay for incremental changes now than try to fix the problem after disastrous disruptions are underway. But it has become clear that contemporary policies are not likely to lead to a situation where the climate system remains in the configuration we have taken for granted as a premise for global security until very recently, a situation that now requires a much more comprehensive re-evaluation of security risks (Mabey *et al.* 2011). Clearly the quasi-imperial view of an external Earth to be managed from the metropoles, while maintaining their modes of consumption, is not an adequate formulation of security if sustaining a viable biosphere into the future is the goal of climate security (Steffen *et al.* 2011).

Climate Security

Concerns over atmospheric changes and sustainable development concerns came together at the Earth Summit in Rio de Janeiro in June of 1992, where the UNFCCC was signed. It formally entered into force in March 1994. While the UNFCCC wasn't a document that contained enforceable limits on greenhouse gas emissions, it clearly stated that its purpose was to constrain greenhouse gas concentrations at levels that would prevent "dangerous anthropogenic interference with Earth's climate system." While this isn't phrased in terms of security, clearly it is both a matter of global concern and something that deals with a threat to human civilization at the largest scale. Instead the convention focuses on mitigating climate change dangers by reducing greenhouse gas emissions.

In 2007 once again the environment forcefully entered the security discussion at the largest of scales when the American security establishment finally paid attention to the issue of climate change and released several major reports raising the alarm about the possible violent consequences of climate change (Campbell *et al.* 2007; CNA Corporation 2007; Campbell 2008; Pumphrey 2008). Memories of the destruction caused to New Orleans by the flooding that followed Hurricane Katrina in 2005, which posed the question of vulnerabilities bluntly, were then combined with the attention paid to the fourth assessment report of the Intergovernmental Panel on Climate Change (IPCC) released in 2007. While the IPCC report did not consider the security implications of climate change, the attention paid to both this report and Al Gore's movie *An Inconvenient Truth*, which together won the Nobel Peace Prize in 2007, revived the environmental security agenda by reworking it in terms of climate change. The German advisory committee on security and climate change released its own comprehensive evaluation of numerous security risks in 2008. In 2009 the United Nations Secretary General released a report suggesting that climate change might act as a conflict enhancer and explicitly linked climate to security (United Nations Secretary General 2009).

But as with the 1990s discussion, much of this literature still focused on potential violence in underdeveloped states and the potential for violence if climate disrupted rural subsistence, rather than focusing on the larger ecological transformations in motion or the increasing vulnerabilities of urban populations dependent on complicated infrastructure and lengthy commodity chains for their survival (Dalby 2009). Africa continues to gain most of the attention given the supposed vulnerabilities of its rural populations (Brown *et al.* 2007), despite research that suggests that the adaptive capabilities of even very poor people are largely misunderstood by much of the development literature (Carr 2011). American preoccupations with national security (Busby 2008) or military responses to migration and related matters (Smith 2007) continue to focus on external threats to metropolitan security (Briggs 2010; Moran 2011), rather than engaging in a more fundamental analysis of the forces driving global change. Where this has been the focus, in the lengthy processes of negotiating mitigation measures designed to reduce carbon emissions and enhance the "sink capabilities" of forest ecosystems, it is clear that measures taken so far have failed to substantially reduce the overall rate of increase of greenhouse gases.

As other chapters in this volume make clear (Chapter 1 and Chapter 20) over the last couple of decades mitigation hasn't even slowed the rate of accumulation of greenhouse gases in the atmosphere. Attempts to provide a follow-on to Kyoto and extend Kyoto were a matter of intense political argument in COP meetings, notably in Copenhagen in 2009, and it was only in Durban in December 2011 that states finally agreed to negotiate a binding agreement by 2015 that will come into force in 2020. By then many climate scientists now argue it will be too late to keep climate changes within the range fairly close to that which made civilization possible in the first place; there is an "emissions gap" between the political aspirations in the Copenhagen and Durban COP agreements and the rising CO_2 levels in the atmosphere (United Nations Environment Programme 2010). In these circumstances policies of adaptation, trying to change societies in ways that deal with the consequences of climate change and minimize the vulnerabilities of societies to more extreme weather and unpredictable droughts, storms, and floods, simply have to be addressed.

Adaptation

Rising sea levels, changed storm tracks, hotter summers, more extreme rain events, reduced Arctic ice cover, possible new disease vectors, and numerous other matters are possible consequences of climate change. Adaptation requires the simple recognition that what has been understood to be the normal conditions for particular societies in terms of temperature patterns, water supplies, and crop-growing conditions no longer apply (Pascal 2009). While much of this is specifically local, when disaster strikes numerous international implications may occur, and disaster diplomacy is part of the larger considerations of adaptation that now require global attention (Adger *et al.* 2009). Not least this is because disasters may have implications for international relations quite as much as for the immediate victims of infrastructure failures. The interconnectedness of the global economy and its fuel supplies was made abundantly clear in 2005 when hurricanes disrupted American production and refining facilities in the Gulf of Mexico, driving the international price of oil upward (Yergin 2011).

Over the last few decades humanity has become a species that lives in cities. Long before this happened, urban-based economies dominated rural areas, drawing materials, food, and fuel from hinterlands to make urban life possible. Urban populations are increasingly dependent on the communications infrastructure, on roads, rails, pipelines, sewage systems, electricity, and phone systems, and these are highly vulnerable, to storms in particular (Graham 2010). Coastal cities are especially vulnerable to flooding, as the residents of Bangkok discovered in 2011. In the short run emergency aid, often provided by military forces, is a key mode of adaptation, providing food, water, shelter, medical assistance, and sometimes evacuation to stricken people. International cooperation in these matters is growing and as such global security is enhanced as peoples are assisted in periods of danger. Climate change might indeed precipitate peace (Gartzke 2012). Cooperation in the face of adversity is a much better response than treating victims of disasters as a potential threat. But at best these are stopgap measures, unless they lead, as they sometimes do, to reducing political tensions and encouraging cooperative mitigation and adaptation measures. Coping with such disruptions is now a key part of security policy-making in many places (Brauch *et al.* 2011).

Migration is one of the major issues related to adaptation but has mostly been seen as a problem in the security literature (Smith 2007). The contradiction between national security and a global vision that takes adaptation seriously poses one of the major policy issues for coming decades. In the last couple of generations boundaries and borders have been settled and assumptions that populations are stable and fixed within certain geographical areas has become the norm of a fixed territorial order. Movement is the most basic adaptation measure of natural systems to large-scale environmental changes. Species move, and indeed now are moving wholesale in response to climate changes. If people try to do the same thing across national frontiers, will the states into which they try to move see these people as a threat to political stability (Guild 2009)? The current popularity of boundary-fence construction in many places where there is a notable disparity of wealth across a frontier suggests that political elites are already trying to use territorial means to protect "national security" from migration (Jones 2012). While this is frequently done in terms of anti-terror measures, it is clear that these policies can easily be invoked in the face of migration caused by disasters in the short run or longer-term environmental disruptions.

All of this is made most difficult, because, despite the frequent use of the term environmental refugee, there is no such legal category, nor any international convention that recognizes environmental causes as a legitimate reason for migrants to claim refugee status or international protections (Piguet *et al.* 2011). While climate change may set people in motion due to the indirect effects of agricultural changes or outright failures caused by droughts, floods, or other disruptions, the territorial structure of the present geopolitical order is ill-equipped to deal with the human consequences (Pascal 2010). Technically this may not be a matter of global security understood in the traditional manner of things that might lead to large-scale political disruptions and interstate conflict, but it clearly is a matter of global change induced human insecurity.

Geoengineering

But now just as adaptation measures are beginning to be taken seriously by infrastructure planners, given the failures of mitigation policies over the last couple of decades, the science of climate change is pointing to the increasingly likely occurrence of very dangerous climate change. This is what will happen when the climate system crosses some of the tipping points or thresholds beyond which the climate system will begin to operate in entirely new and largely unpredictable ways (Lenton *et al.* 2008). While a good deal of the argument about how much climate change is dangerous has long assumed that 2 °C average global warming is "safe" in that it will not cause rapid transformation, and the scientific basis of that claim has long been understood to be dubious, if present trends continue, global temperatures will increase more than this, with all sorts of unpredictable results (Mabey *et al.* 2011).

If mitigation strategies, which ought to be both easier and safer given that they are about preventing the problem rather than trying desperate experiments after the fact, have failed, then there may be no alternative to geoengineering, unless, that is, humanity effectively decides that a radically changing climate isn't a bad thing, and that there is no good reason why we should live in a biosphere that has two polar ice caps. The planet after all has had periods in the geological past in which no permanent ice existed at the poles; it may well face this prospect once again, and indeed it seems that this is the future that current human activity is now setting in motion. But it is precisely the potential for massive human suffering and death, not least because of huge agricultural disruptions in the process of so transforming the planet's climate, that makes scientists invoke notions of global security in arguing that such a course with its numerous imponderables is just too risky to seriously contemplate (Schneider *et al.* 2010). In these new circumstances global security now means keeping the planetary climate system within the parameters that we have known for the last few millennia; that is, close to the conditions that gave rise to human civilization in the first place.

Thus serious consideration is increasingly being given to artificially changing the atmosphere in ways that will counteract the enhanced warming effects of carbon dioxide and methane (Royal Society 2009; Bipartisan Policy Center 2011). Under the rubric of "geoengineering" discussions of solar radiation management are taking place around scenarios for such things as injecting aerosols into the atmosphere to reflect sunlight back into space rather than have it heat Earth's surface. In theory this can be done, mimicking the consequences of large volcanic eruptions that put sulfur aerosols in the high atmosphere and shade the planet. Other technical suggestions include creating artificial clouds over the oceans using mobile automatic ships to spray water into the atmosphere. While there is some fascinating speculation about science-fiction-type scenarios involving the construction of huge mirrors in space, they undoubtedly require more lift capacity than space-travel programs can provide.

What is clear is that these ideas are no more than stopgap measures to buy time while more fundamental rethinking is done (Steffen *et al.* 2011).

Geoengineering raises numerous new questions of governance and environmental policy, not least because widespread agreement would seem to be necessary prior to initiating major planetary engineering efforts (Humphreys 2011). To return to classical considerations of security in international relations, the inevitable question becomes what happens if one state decides to take matters into its own hands and starts inserting aerosols into the upper atmosphere to cool the planet, an industrial enterprise now within the capabilities of at least some of the larger countries (Dyer 2008). Once again the intersection of technology and an endangered atmosphere is unavoidable, but now rather than environmental change being an unintended consequence of nuclear warfare or unrestricted CFC production, the atmosphere has become an arena for deliberate manipulation, perhaps even by using technology derived from the missiles and environmental engineering that were earlier seen as part of the threat to humanity.

Conclusion

Having effectively taken our collective fate into our own hands, now the traditional assumption of environment as a relatively benign backdrop for human activities is no longer a valid assumption in thinking about global security (Dalby 2009). There are many gaps in global governance and there is no deliberative body that decides on how many polar ice caps the planet ought to have, or what the average atmospheric temperature considered optimal for human life should be. But these decisions are effectively being taken by the carbon-fueled mode of economic activity that continues to expand rapidly despite 20 years of discussion about greenhouse gas emission levels and the existence of the UNFCCC, explicitly committed to ensuring that dangerous levels of anthropogenic atmospheric change are prevented.

The failures of international diplomacy to solve the problem of rising greenhouse gas concentrations have, however, spawned numerous new experiments in trying to tackle climate change (Hoffman 2011). Innovative business models are being used by a growing economic sector that is mobilizing carbon markets and the possibilities of cap-and-trade systems to re-engineer economies in ways that reduce carbon fuel use (Newell and Paterson 2010). Albeit very late in the day, numerous organizations and institutions are beginning to try to reduce emissions and operate in more sustainable manners, suggesting quite clearly that if sustainable security is to be provided it will likely come from new innovations in technology and energy systems as well as their governance rather than from international treaty initiatives by the major powers (Lilliestam *et al.* 2012). A new geopolitics with new ecological notions of security may now be in the making.

In the face of such considerations the question of who is securing what future is unavoidable. So far the political institutions of the modern nation-states system and the political economy of carbon-fueled industrialism suggest that if present tendencies remain on track, in the face of mounting disruptions, global security will be an extension of what Paul Rogers (2010) calls "keeping the violent peace," where military forces are used to quell insurrections and maintain the existing political and economic arrangements. Political elites may well decide to use the wealth accumulated by carbon-fueled economic growth to build the technologies for geoengineering instead of trying to tackle the more fundamental political questions of inequality and instability that threaten global security.

The alternative is a more radical political orientation that takes seriously the logic of the UNFCCC and recognizes that new modes of energy use and a much more just system that facilitates participation in decision-making by a much larger portion of humanity are necessary for a sustainable economic system that makes a stable climate system the basis for global security. This vision suggests a much less militarized version of the future, one not dependent on technological manipulation of the planetary environment by a self-appointed elite (Klein 2011). Which course is taken in the next couple of decades will not only determine how human societies evolve, it will probably also quite literally determine how many polar ice caps the planet has, and the course of evolution of life itself for many millions of years. Nothing less is involved now in attempts to secure the globe.

References

Adger, Neil, Irene Lorenzoni, and Karen O'Brien, eds. 2009. *Adapting to Climate Change: Thresholds, Values, Governance*. Cambridge: Cambridge University Press.

Allison, G. and G.F. Treverton, eds. 1992. *Rethinking America's Security: Beyond the Cold War to a New World Order*. New York: W.W. Norton.

Anderson, K. and A. Bows. 2011. "Beyond Dangerous Climate Change: Emission Scenarios for a New World." *Philosophical Transactions of the Royal Society A*, 369: 20–44.

Barnett, Jon. 2000. "Destabilizing the Environment-Conflict Thesis." *Review of International Studies*, 26: 271–288.

Barnett, Jon. 2001. *The Meaning of Environmental Security*. London: Zed Books.

Benedick, Richard. 1991. *Ozone Diplomacy*. Cambridge, MA: Harvard University Press.

Bipartisan Policy Center. 2011.*Geoengineering: A National Strategic Plan for Research on the Potential Effectiveness, Feasibility, and Consequences of Climate Remediation Technologies*. Washington, DC: Bipartisan Policy Center.

Brauch, Hans Günter, Ursula Oswald Spring, Patricia Kameri-Mbote *et al.*, eds. 2011. *Coping with Global Environmental Change, Disasters and Security Threats: Challenges, Vulnerabilities and Risks*. Berlin, Heidelberg, and New York: Springer-Verlag.

Briggs, Chad. 2010. "Environmental Change, Strategic Foresight, and Impacts on Military Power." *Parameters*, Autumn: 1–15.

Brown, O., A. Hammill, and R. McLeman, 2007. "Climate Change as the 'New' Security Threat: Implications for Africa." *International Affairs*, 83(6): 1141–1154.

Busby, J.W. 2008. "Who Cares about the Weather? Climate Change and U.S. National Security." *Security Studies*, 17: 468–504.

Buzan, Barry, Ole Waever, and Jaap deWilde. 1998. *Security: A New Framework for Analysis*. Boulder, CO: Lynne Rienner.

Campbell, Kurt M., ed. 2008. *Climatic Cataclysm: The Foreign Policy and National Security Implications of Climate Change*. Washington, DC: Brookings Institution.

Campbell, Kurt M., Jay Gulledge, J.R. McNeill *et al.* 2007. *The Age of Consequences: The Foreign Policy and National Security Implications of Global Climate Change*. Washington, DC: Center for Strategic and International Studies and Center for a New American Security.

Carr, Edward. 2011. *Delivering Development: Globalisation's Shoreline and the Road to a Sustainable Future*. Basingstoke: Palgrave Macmillan.

CNA Corporation. 2007. *National Security and the Threat of Climate Change*. Alexandria, VA: CNA Corporation.

Dabelko, Geoff. 2008. "An Uncommon Peace: Environment, Development and the Global Security Agenda." *Environment*, 50(3): 32–45.

Dalby, Simon. 2002. *Environmental Security*. Minneapolis, MN: University of Minnesota Press.

Dalby, Simon. 2009. *Security and Environmental Change*. Cambridge: Polity.

De Soysa, Indra. 2002. "Ecoviolence: Shrinking Pie or Honey Pot?"*Global Environmental Politics*, 2(4): 1–34.

Deudney, Daniel. 1990. "The Case against Linking Environmental Degradation and National Security." *Millennium*, 19: 461–476.

Duffield, M. 2007. *Development, Security and Unending War: Governing the World of Peoples*. Cambridge: Polity.

Dyer, Gwynne. 2008. *Climate Wars*. Toronto: Random House.

Floyd, Rita. 2010. *Security and the Environment: Securitization Theory and US Environmental Security Policy*. Cambridge: Cambridge University Press.

Gartzke, Erik. 2012. "Could Climate Change Precipitate Peace?" *Journal of Peace Research*, 49(1): 177–192.

German Advisory Council on Global Change. 2008. *Climate Change as a Security Risk*. London: Earthscan.

Graham, Steve, ed. 2010. *Disrupted Cities: When Infrastructure Fails*. Abingdon: Routledge.

Guild, Elspeth. 2009. *Security and Migration in the 21st Century*. Cambridge: Polity.

Hoffman, Matthew J. 2011. *Climate Governance at the Crossroads: Experimenting with a Global Response to Kyoto*. Oxford: Oxford University Press.

Homer-Dixon, Thomas 1991. "On the Threshold: Environmental Changes as Causes of Acute Conflict." *International Security*, 16(1): 76–116.

Homer-Dixon, Thomas 1999. *Environment, Scarcity, and Violence*. Princeton NJ: Princeton University Press.

Humphreys, David 2011. "Smoke and Mirrors: Some Reflections on the Science and Politics of Geoengineering." *Journal of Environment & Development*, 20(2): 99–120.

ICISS (International Commission on Intervention and State Sovereignty). 2001. *The Responsibility to Protect*. Ottawa: International Development Research Centre.

Intergovermental Panel on Climate Change. 2007. *Climate Change 2007: The Fourth Assessment Report*. Cambridge: Cambridge University Press.

Jones, Reece. 2012. *Border Walls: Security and the War on Terror in the United States, India, and Israel*. London: Zed Books.

Kahl, Colin. 2006. *States, Scarcity and Civil Strife in the Developing World*. Princeton, NJ: Princeton University Press.

Kjellen, Bo. 2008. *A New Diplomacy for Sustainable Development: The Challenge of Global Change*. New York: Routledge.

Klare, Michael and Yogesh Chandrani, eds. 1998. *World Security: Challenges for a New Century*. New York: St. Martin's Press.

Klein, Naomi. 2011. "Capitalism vs. the Climate." *The Nation*, November 28, http://www.thenation.com/article/164497/capitalism-vs-climate# (accessed November 2, 2012).

Le Billon, Philippe. 2012. *Wars of Plunder: Conflicts, Profits and the Politics of Resources*. London: Hurst.

Lenton, T.M., H. Held, E. Kriegler *et al.* 2008. "Tipping Elements in the Earth's Climate System." *Proceedings of the National Academy of Sciences*, 105(6): 1786–1793.

Lilliestam, Johan, Antonella Battaglini, Charlotte Finlay *et al.* 2012 "An Alternative to a Global Climate Deal May be Unfolding before Our Eyes." *Climate and Development*, 4(1): 1–5.

Lohmann, Larry. 2006. *Carbon Trading: A Critical Conversation on Climate Change, Privatisation and Power*. Development Dialogue No. 48. Uppsala: Dag Hammarskjöld Centre.

Mabey, Nick, Jay Gulledge, Bernard Finel, and Katherine Silverthorne. 2011. *Degrees of Risk: Defining a Risk Management Framework for Climate Security*. London: E3G.

Mangano, Joseph J. and Janette D. Sherman. 2012. "An Unexpected Mortality Increase in the United States Follows Arrival of the Radioactive Plume from Fukushima: Is There a Correlation?" *International Journal of Health Services*, 42(1): 47–64.

Matondi, Prosper B., Kjell Havnevik, and Atakilte Beyene, eds. 2011. *Biofuels, Land Grabbing and Food Security in Africa*. London: Zed Books.

McNeill, J.R. 2000. *Something New under the Sun: An Environmental History of the Twentieth Century*. New York: W.W. Norton.

Miller, Richard L. 1986. *Under the Cloud: The Decades of Nuclear Testing*. New York: Free Press.

Moran, Daniel, ed. 2011. *Climate Change and National Security: A Country-Level Analysis*. Washington, DC: Georgetown University Press.

Neocleous, Mark. 2008. *Critique of Security*. Montreal: McGill-Queen's University Press.

Newell, Peter and Matthew Paterson. 2010. *Climate Capitalism*. Cambridge: Cambridge University Press.

Park, Christopher. 1987. *Acid Rain: Rhetoric and Reality*. London: Methuen.

Pascal, Cleo. 2009. "From Constants to Variables: How Environmental Change Alters the Geopolitical and Geoeconomic Equation." *International Affairs*, 85(6): 1143–1156.

Pascal, Cleo. 2010. *Global Warring: How Environmental, Economic and Political Crises will Redraw the World Map*. Toronto: Key Porter.

Piguet, Étienne, Antoine Pécoud, and Paul de Guchteneire, eds. 2011. *Migration and Climate Change*. Cambridge: University of Cambridge Press.

Pumphrey, Carolyn, ed. 2008. *Global Climate Change: National Security Implications*. Carlisle, PA: U.S. Army War College, Strategic Studies Institute.

Rogers, Paul. 2010. *Losing Control: Global Security in the 21st Century*. London: Pluto.

Royal Society. 2009. *Geoengineering the Climate: Science, Governance and Uncertainty*. London: The Royal Society.

Sagan, Carl and Richard Turco. 1990. *A Path Where No Man Thought: Nuclear Winter and the End of the Arms Race*. New York: Random House.

Schneider, Stephen. 1989. *Global Warming: Are We Entering the Greenhouse Century?* San Francisco: Sierra Club Books.

Schneider, Stephen, Armin Rosencranz, Michael D. Mastrandrea, and Kristin Kuntz-Duriseti. 2010. *Climate Change Science and Policy*. Washington, DC: Island Press.

Smith, Paul J. 2007. "Climate Change, Mass Migration and the Military Response." *Orbis*, 51(4): 617–633.

Soroos, Marvin. 1997. *The Endangered Atmosphere*. Columbia, SC: University of South Carolina Press.

Spratt, D. and P. Sutton. 2008. *Climate Code Red: The Case for a Sustainability Emergency*. Fitzroy: Australian Friends of the Earth.

Steffen, Will, Asa Persson, Lisa Deutsch *et al.* 2011. "The Anthropocene: From Global Change to Planetary Stewardship." *Ambio*, 40: 739–761.

Stern, N., 2007. *The Economics of Climate Change: The Stern Review*. Cambridge: Cambridge University Press.

Theisen, Ole Magnus, Helge Holtermann, and Halvard Buhaug. 2011. "Climate Wars? Assessing the Claim that Drought Breeds Conflict." *International Security*, 36(3): 79–106.

Turco, R.P., O.B. Toon, T.P. Ackerman *et al.* 1983. "Nuclear Winter: Global Consequences of Multiple Nuclear Explosions." *Science*, 222: 1283–1292.

United Nations Environment Programme. 2010. *The Emissions Gap Report*. Nairobi: United Nations Environment Programme.

United Nations Secretary General. 2009. *Climate Change and its Possible Security Implications*, A/64/350, September. New York: United Nations.

Webersik, Christian. 2010. *Climate Change and Security: A Gathering Storm of Global Challenges*. Santa Barbara: Praeger.

Williams, Michael. 2007. *Culture and Security: Symbolic Power and the Politics of International Security*. Abingdon: Routledge.

World Commission on Environment and Development. 1987. *Our Common Future*. Oxford: Oxford University Press.

Yergin, Daniel. 2011. *The Quest: Energy, Security, and the Remaking of the Modern World*. New York: Penguin Books.

Chapter 11

International Environmental Law

Daniel Bodansky

Introduction

International environmental law (IEL), as a substantive field, consists of the body of norms relevant to environmental problems of an international character – for example, because they implicate more than one state or relate to areas beyond national jurisdiction. More broadly, it includes the processes by which international environmental norms develop and are implemented, as well as the institutions that play a role in these processes.

Although IEL developed as a sub-field of public international law, it encompasses topics not traditionally considered to be part of international law, including the roles of non-state actors and of non-legal ("soft law") norms. Moreover it has developed its own distinctive doctrines, legal processes, and institutions:

- Distinctive doctrines include the precautionary principle and the principle of common but differentiated responsibilities (CBDR).
- Distinctive legal processes include the framework convention/protocol approach, tacit amendment procedures, and non-adversarial, forward-looking approaches to non-compliance.
- Distinctive institutions include the annual conferences of the parties (COPs) established by many multilateral environmental agreements (MEAs).

This chapter focuses on these distinctive features of IEL. The chapter begins by considering the standard-setting process, focusing in particular on the negotiation of international agreements and on the emergence and role of more general doctrines such as the duty to prevent transboundary harm. Next, it considers the implementation and compliance processes. It concludes by surveying the main international

The Handbook of Global Climate and Environment Policy, First Edition. Edited by Robert Falkner.
© 2013 John Wiley & Sons, Ltd. Published 2013 by John Wiley & Sons, Ltd.

institutions relevant to the development and application of international environmental law.

Standard-Setting

Treaties

From the inception of international environmental law, treaties and other international cooperation have been the primary means of achieving international cooperation. Negotiated agreements offer several advantages over more informal mechanisms:

- They enable states to address issues in a purposive, rational manner.
- They promote reciprocity by allowing states to delineate precisely what each is expected to do.
- They provide greater certainty about the applicable norms than non-treaty sources of international law, which lack a written, canonical form.
- Finally, they allow states to tailor a regime's institutional arrangements and mechanisms to fit the particular problem.

Traditionally, treaties were comparatively static arrangements, memorializing the rights and duties of the parties as agreed at a particular point in time. Today, international environmental agreements are usually dynamic arrangements, establishing ongoing regulatory processes (Gehring 1994). The result is that, in most environmental regimes, the treaty text itself represents just the tip of the normative iceberg. The majority of the norms are adopted through more flexible techniques, which allow international environmental law to respond more quickly to the emergence of new problems and new knowledge.

Along with the principle of *pacta sunt servanda* (which says that agreements must be kept), the most fundamental rule of treaty law is that treaties depend on state consent. This is perhaps one reason why treaty norms are often characterized as "commitments" rather than "obligations" – to emphasize the self-binding quality of treaty law. Treaty norms are not obligations imposed on states; rather, they are commitments that a state voluntarily undertakes.

The development of international environmental agreements involves many design choices, including membership rules, substantive scope, legal or non-legal form, choice of regulatory instrument, stringency of commitments (both in general and for particular countries), precision, voting rules, financial incentives, reporting and review procedures, non-compliance institutions, minimum participation requirements, and ease of exit through reservations or withdrawal (Koremenos *et al.* 2001; Raustiala 2005).

One important focus of recent scholarship has been on the interconnections between these design elements (Boockmann and Thurner 2002; Barrett 2003; Gilligan 2004; Raustiala 2005). A legally binding agreement may be stronger than a non-binding instrument, but attract less participation. Inclusion of more stringent substantive commitments may make states less willing to accept strong compliance mechanisms. Ultimately, the effectiveness of an environmental regime is a function not only of the stringency of its commitments, but the degree to which states participate and comply (Barrett 2003). So we must consider different design elements in conjunction with one another, rather than in isolation, both to understand the design choices made in existing regimes and in developing new regimes.

A second insight of recent scholarship has been to understand international environmental regimes as dynamic systems that evolve over time. International environmental law has promoted the evolution of regimes through a variety of mechanisms. These include:

• *Regular scientific assessments to help produce "consensus" knowledge.* International scientific assessments can help overcome arguments against environmental regulation based on scientific uncertainty (Mitchell 2006). For example, in the late 1970s, some European states argued against international regulation of ozone-depleting substances on grounds of scientific uncertainty. A major international scientific assessment jointly undertaken by the US National Aeronautics and Space Administration (NASA), the World Meteorological Organization (WMO), and other international and national bodies in 1986 helped set the stage for the adoption of the Montreal Protocol on Substances that Deplete the Ozone Layer the following year (Parson 2003: 251–252).

• *Soft-law instruments such as codes of conduct and guidelines.* Even when there is some agreement about the need for international standards, states may be reluctant to lock themselves into legally binding commitments and prefer to develop "soft-law" instruments such as codes of conduct and guidelines, which are less costly than treaty commitments to exit or violate (Abbott and Snidal 2000). Starting with soft-law approaches allows states to become comfortable with a regulatory approach before formalizing the approach in a treaty. For example, in developing a prior informed consent (PIC) regime for exports of hazardous wastes and dangerous chemicals, states first elaborated PIC procedures through non-binding guidelines and codes of conduct, before legally mandating the procedures in the 1989 Basel Convention on the Control of Transboundary Movement of Hazardous Wastes and the 1998 Rotterdam Convention on the Prior Informed Consent Procedure for Certain Hazardous Chemicals and Pesticides in International Trade. Similarly, in the North Sea pollution regime, states were unable to adopt ambitious, precise commitments in a legally binding treaty. So they started by adopting such commitments in political declarations (Skjærseth 1998). In essence, non-binding approaches represent a type of risk-management strategy, reducing the risk to states of being bound by norms that they may ultimately deem undesirable. Although they involve a lesser degree of commitment than treaties, they can also be quite effective in changing behavior (Shelton 2004).

• *The framework convention-protocol approach.* Another technique that allows states to proceed in an incremental manner is to start by negotiating a framework agreement establishing the basic system of governance for a given issue area – the core institutions, decision-making procedures, and norms – and then develop more specific regulatory standards in subsequent protocols, after greater cognitive and normative consensus has emerged (Bodansky 1999). The theory is that once a framework convention is adopted, the international law-making process takes on a momentum of its own. States that were initially reluctant to undertake substantive commitments, but that could not object to the seemingly innocuous process established by the framework convention, will feel increasing pressure not to fall out of step as that process gains momentum. (But for a critical view, see Downs *et al.* 2000.) This approach was first used in the European acid rain regime. In the late 1970s, there was insufficient consensus to adopt regulatory requirements limiting emissions of sulfur dioxide or other precursors of acid rain. So, instead, states adopted a framework agreement – the 1979 Long-Range Transboundary Air Pollution Convention – which in essence serves as the "constitution" for the regime. Later, after greater consensus had developed (in part as a result of information produced through the regime's monitoring program), states adopted a series of regulatory protocols addressing sulfur dioxide, nitrogen oxides, volatile organic compounds, heavy metals, and persistent organic pollutants (Levy 1993; Wettestad 2002). Similarly, in the regime to protect the stratospheric ozone layer, states began by adopting the 1985 Vienna Convention on the Protection of the Ozone Layer, which expressed concern about the problem but imposed no regulatory requirements. Two years later, parties to the Vienna Convention negotiated and adopted the Montreal Protocol on Substances that Deplete the Ozone Layer, which imposed quantitative limits on states' consumption and production of ozone-depleting substances (Parson 2003).

- *Tacit amendment procedures.* International environmental regimes need to be able to develop quickly, in response either to changes in a problem itself or to our scientific understanding of a problem. For example, when the Montreal Protocol was first adopted, chlorofluorocarbons (CFCs) and halons were seen as the principal ozone-depleting substances. Now we know that a series of other chemicals – including carbon tetrachloride and methyl bromide – also contribute to the depletion of the ozone layer. In the ozone case, scientific knowledge developed so quickly that "the CFC reduction rates agreed . . . in September 1987 were already obsolete by the time the protocol entered into force" (Sand 1991: 15). Traditionally, treaties were hard to update, since amendments require ratification by states. But most international environmental agreements now put their regulatory requirements in an annex or schedule that can be amended more easily than the main body of the treaty, through a tacit acceptance procedure under which amendments apply automatically to all parties unless a party specifically opts out. For example, the detailed regulations on whaling under the International Whaling Convention are included in a schedule that can be amended by a three-quarters majority vote, and apply to all parties unless a party specifically objects. In essence, these flexible amendment procedures vest environmental treaty regimes with ongoing regulatory authority.

- *Differential standards, to take account of differences between states in historical responsibility, capacity, and national circumstances.* States differ substantially in their historical responsibility for international environmental problems, their capacity to address these problems, their national circumstances, and their regulatory and political cultures. So a one-size-fits-all approach to international environmental problems is unlikely to get widespread support. To promote broader participation, many international environmental regimes include differentiated commitments, which are stronger for some countries and weaker for others (Rajamani 2007). For example, the Montreal Protocol gives developing countries a 10-year grace period in which to comply with its basic regulatory requirements. The Kyoto Protocol negotiations took the principle of differentiation to an extreme, by elaborating emissions limitations requirements on carbon dioxide and other greenhouse gases only for developed countries, while specifically excluding any new commitments for developing countries.

- *Elaboration through decisions of the parties.* Decisions of the parties offer another mechanism for elaborating a regime. Often, treaties specifically direct the Conference of the Parties (COP) to develop rules on complex topics that are too difficult or time-consuming to address in the treaty text itself. The Montreal Protocol, for example, directs the parties to develop a compliance procedure; similarly, the Kyoto Protocol established a number of market mechanisms, including emissions trading and the Clean Development Mechanism, but left the elaboration of the detailed rules for how these mechanisms would work to the COP. Decisions of the parties can also give more determinate content to vague provisions in a treaty, such as the "wise use" requirement for wetlands found in the 1971 Ramsar Convention on Wetlands, or address an issue not dealt with in the treaty itself. The compliance procedure for the Convention on International Trade in Endangered Species, for example, was developed almost entirely through decisions of the parties (Reeve 2002).

Non-treaty Norms

In addition to treaty norms, which bind only parties and usually address only a limited subject area, international environmental law includes a number of more general norms:

- *The duty to prevent significant transboundary harm* was first articulated in the 1941 *Trail Smelter* case and is reflected in various non-binding instruments, including the 1972 Stockholm Declaration on the Human Environment and the 1992 Rio Declaration on Environment and Development. This duty underpins much of the rest of international environmental law, which in essence is an elaboration of this core duty (Handl 2007).
- *The precautionary principle* addresses the issue of scientific uncertainty. Formulations of the precautionary principle vary widely, but provide at a minimum that scientific uncertainty should not be a basis for inaction (Trouwborst 2002).
- *The principle of common but differentiated responsibilities and respective capabilities (CBDRRC)* provides a rationale for differentiating the commitments of developed and developing countries (Rajamani 2007). As noted above, this principle is reflected in a variety of environmental treaties, including the Montreal Protocol and the Kyoto Protocol.

In contrast to treaties, which are the product of a purposive process of negotiation, the source and legal status of these general norms of international environmental law is uncertain. Conventional accounts of the sources of international law identify two sources other than treaties: custom and general principles. In theory, the two sources differ in that customary norms are generated through the regular practice of states, engaged in out of a sense of legal obligation, while general principles are norms that reflect fundamental propositions of law, shared by legal systems around the world. But, in practice, the distinction between the two is often blurred. It is not clear, for example, whether the duty to prevent transboundary harm is a rule of customary law, reflecting the actual practice of states, or a general principle of law. Even the International Court of Justice, in proclaiming the duty to be part of the corpus of general international law, did not identify its legal basis (Bodansky 2010: 200).

In addition to their uncertain legal status, general principles such as the duty to prevent transboundary harm and the precautionary principle are very general, leaving states with significant leeway in deciding what to do. States have a duty to prevent transboundary environmental harm, but what constitutes "significant" harm, and what standard of care must states use to avoid harm? States ought to undertake precautionary action, but in what circumstances and to what degree? Because virtually any behavior that a state might wish to engage in, for self-interested reasons, could be reconciled with these very general standards, states are able to interpret these norms in self-serving ways, with little cost to their reputation. Although courts could potentially give non-treaty norms more determinate content by applying them in particular cases, international environmental law lacks tribunals with general jurisdiction over environmental disputes, so these norms are rarely applied judicially. As a result, norms of non-treaty law operate primarily as meta-rules. They do not determine the result in particular disputes or negotiations; rather, they serve a discursive function, setting the terms of international debates about environmental issues and providing evaluative standards that actors can use either to justify their own proposals or arguments or to criticize those of others.

Implementation

The explosion of international environmental law-making over the past several decades makes it easy to fall prey to the view that the development of international environmental agreements, in itself, represents progress – that texts matter and that stronger texts mean better environmental protection. But words on paper are not enough. Although they represent an important first step, what matters, in the final analysis, is not the number of treaties that have been negotiated or even ratified, but rather their effectiveness in improving the quality of the environment. Accordingly, political scientists and international lawyers have given increasing attention to the issue of effectiveness over the last two decades (Brown Weiss and Jacobson 1998; Victor *et al.* 1998; Miles *et al.* 2002).

Implementation is the process by which policies get translated into action and is integral to effectiveness. It can encompass a wide range of measures, such as elaborating a policy through more specific laws or regulations, educating people about what a rule requires, building a new power plant that emits less pollution, and monitoring and enforcing compliance. In a broad sense, all of these measures can be considered part of the implementation process.

Implementation is a particular challenge for international environmental law, because it typically aims to control not merely state conduct but private conduct. Success depends on a wide variety of factors, including:

- *The depth or stringency of the commitment.* The bigger the required change from the status quo, the more likely it is that implementation will be costly and will conflict with entrenched interests.

- *The type of commitment involved.* Commitments to engage in particular conduct (for example, adopting an oil pollution discharge standard) are more directly under a party's control than commitments to achieve some general result (reducing national emissions by a specified amount, as in the Kyoto Protocol), which depend on a multitude of factors that may be difficult to change.
- *The capacity of the state.* Implementation generally requires resources and expertise to draft laws, monitor behavior, administer a permitting scheme, prepare reports, bring prosecutions, and so forth.
- *The degree to which implementation converges with other domestic policy objectives.* For example, a country is more likely to implement a commitment to reduce carbon dioxide emissions if doing so will also reduce urban air pollution or contribute to energy security.

As in most areas of international law, states serve as the primary transmission belt for putting international environmental rules into effect. International environmental agreements impose obligations on states and rely on states to implement their commitments. For this reason, the success of multilateral environmental agreements depends on the degree to which they are "domesticated" (Hanf and Underdal 1998).

Treaties vary considerably in how much freedom they give states in the choice of implementation methods. At one end of the spectrum, some agreements set forth quite specific obligations of conduct that leave little discretion. For example, the International Convention for the Prevention of Pollution from Ships (MARPOL) requires flag states to prescribe precise rules for the construction and design of oil tankers, and to prohibit and sanction violations of these standards by vessels operating under their authority. Often, however, international law does not specify any particular implementation method, leaving it up to each state to decide how it will fulfill its international obligations in accordance with its own domestic law. A typical formulation on implementation, found in many treaties, simply requires states to take "appropriate" measures. This allows each state to take into account its own legal system, regulatory culture, and other national circumstances in determining what measures are "appropriate." At the far extreme, treaties establishing an obligation to achieve some overall result, such as the national emissions targets in the Kyoto Protocol, give states almost complete flexibility in determining how they will reach the required outcome – whether by means of taxes, product standards, emission limits, voluntary agreements with industry, subsidies, education, and so forth.

A threshold issue in treaty implementation is whether implementation requires legislation. For a variety of reasons, sometimes the answer is no. A treaty may focus on governmental actions such as reporting, which can be performed by the executive branch on its own authority, without any need for legislative approval. Or, under a country's constitution, treaties may have the force of domestic law directly, making additional legislative implementation unnecessary. Or existing legislation might provide the necessary authority to implement a treaty's obligations.

Even when implementing legislation is needed, the adoption of legislation is usually only the first step in the implementation process. Most treaties require various types of administrative implementation, such as further rule-making to give greater specificity to general legislative mandates, monitoring and assessment, preparation of reports, issuance of permits, and the investigation and prosecution of alleged violations. Consequently, as a recent study of implementation observed, "[o]ne cannot simply read domestic legislation to determine whether countries are complying... [Compliance] involves assessing the extent to which governments follow through on the steps that they have taken" (Jacobson and Brown Weiss 1998: 2, 4).

National courts have played a relatively modest role thus far in enforcing international environmental law (Bodansky and Brunnée 2002). In a few cases, national courts have used international environmental law to review governmental action or to interpret national law. For example, in *Minors Oposa*, the Philippine Supreme Court applied the principle of intergenerational equity to allow a group of children to challenge licenses to harvest timber. But, generally, national courts have become involved in the implementation process only indirectly, through their role in applying a state's domestic implementing legislation.

Compliance

Most states may comply with most of their international commitments most of the time, as Louis Henkin famously proclaimed (1979: 47). But violations remain a problem. Even comparatively easy, procedural commitments, such as the obligation to file reports, often go unfulfilled by states. As a result, we cannot rely on states to implement their international environmental commitments. International measures are also sometimes needed to make international environmental law effective.

International environmental regimes have developed a wide variety of institutions and mechanisms to address the problem of compliance (UNEP 2007). Some of these are specified in the treaty text itself, others have been elaborated through decisions of the parties, and still others have developed more informally through practice over time.

Two Models of Compliance

Scholars have developed two models of compliance, which reflect different assumptions about state behavior, the causes of non-compliance, and the role of the international system in responding. The enforcement model views states as unitary, rational actors that will violate an agreement when it suits their interests, and concludes that sanctions are needed to induce states to comply (Downs *et al.* 1996). In contrast, the managerial model of compliance sees states as complex organizations that have a propensity to comply with treaties unless strong countervailing circumstances are present, and explains most non-compliance as the result of mistakes, changes in circumstances, or lack of capacity, rather than of a deliberate decision to violate (Chayes and Chayes 1995). On this view, the function of a compliance system is not to punish non-compliance, but rather to encourage and facilitate compliance – for example, by providing financial and technical assistance to states, thereby lowering the costs of compliance; clarifying the content of international obligations; or requiring states to file reports and prepare national implementation plans, which help mobilize and empower domestic constituencies

Generally, the managerial approach to compliance predominates in international environmental law (Breitmeier *et al.* 2006). The response to Russia's non-compliance with the Montreal Protocol in the mid-1990s provides an illustration. Rather than recommend sanctions, the other parties (through the Protocol's Implementation

Committee) in essence negotiated a phase-out plan with Russia, involving subsidies from the World Bank to close the Russian facilities that produced CFCs. As a result, Russia closed its last production facility in 2002, thereby coming into compliance with the Protocol (Yoshida 1999: 135–139).

Sources of Information

Regardless of which model of compliance one adopts, obtaining accurate information is a critical first step. States that deliberately violate an agreement will be deterred by sanctions only to the extent that they fear discovery. The efficacy of enforcement measures is thus a function not only of the magnitude of the sanctions but also of the likelihood of detection.

Generally, national reporting is the primary source of information concerning implementation and effectiveness (Raustiala 2001). In addition, NGOs are an important, independent source of information. Greenpeace, for example, monitors whaling activities and trade in hazardous wastes, whereas TRAFFIC gathers information on illegal trade in wildlife products.

Comparatively few international environmental agreements have formal procedures for the review of national reports, but many have more informal arrangements either to review the accuracy of the information provided in national reports (a process usually referred to as verification) or to evaluate performance (Raustiala 2001). The UN climate change regime establishes perhaps the most detailed review process to date, involving review of individual developed country reports by expert review teams.

Promoting Compliance

Multilateral environmental agreements generally take a proactive approach: they do not merely respond to non-compliance *ex post*, but actively seek to promote compliance *ex ante* through the provision of various types of financial and technical assistance (Sand 1999). Virtually all multilateral environmental agreements provide some implementation assistance. In some cases, MEAs provide only quite limited support, for example, to prepare reports or provide training. In other cases, they provide much more significant assistance to implement substantive requirements designed to reduce pollution or conserve resources. Beginning with the 1973 World Heritage Convention, multilateral environmental agreements have often established special funds to assist with implementation. The World Heritage Fund is quite small, with an annual budget of about only US$4 million to help countries identify and propose sites for inclusion on the World Heritage List, prepare management plans, and train personnel. In contrast, the Montreal Protocol's Multilateral Fund provides more than US$150 million per year (and more than US$2.8 billion since its inception in 1990) to support specific projects to phase out the use of ozone-depleting substances, including through technology transfer.

Responding to Non-compliance

Historically, international law sought to address issues of non-compliance through dispute settlement initiated by the injured against the culpable state. But although the last decade has witnessed a modest rise in environmental litigation, traditional dispute settlement still plays a small role in the implementation of international environmental law. The international law of state responsibility is geared primarily to bilateral enforcement by the "injured state" against the non-compliant state, not to global commons problems such as ozone depletion or climate change, where the harms are widely distributed and where, as a result, no individual state is likely to have a sufficient incentive to undertake enforcement actions.

Rather than rely on traditional dispute settlement to address the problem of non-compliance, many multilateral environmental regimes have developed flexible, political approaches that aim to identify the sources of non-compliance in a particular case and find appropriate responses (Wolfrum 1998). In contrast to traditional dispute settlement, these new treaty-based compliance regimes are:

- *Political and pragmatic, not legalistic.* They view compliance and non-compliance as part of a continuum, not in all-or-nothing terms. On this continuum, the difference between a small and a big violation, or between bare compliance and overcompliance, may be more significant than the difference between compliance and breach.
- *Forward- not backward-looking.* Their goal is to manage environmental problems in order to achieve a reasonable level of compliance in the future, not to establish legal rights and duties or to rectify past breaches. Accordingly, one of the principal responses to non-compliance is to provide assistance – an approach that seems bizarre from the perspective of traditional dispute settlement because it arguably rewards a state for its internationally wrongful act.
- *Non-adversarial rather than contentious in nature.* The procedures are collective rather than bilateral in nature. Any state may initiate a case, with no need to show injury. In many cases thus far, the non-compliant state itself has initiated proceedings.

The Montreal Protocol's Non-compliance Procedure exemplifies this more flexible approach and has served as the model for several other agreements. Today, most multilateral environmental agreements have either already adopted a non-compliance procedure or are considering doing so.

Few multilateral compliance procedures rely significantly on sanctions (UNEP 2007: 117–118). Typically, the most significant "sanction" imposed by international environmental regimes is exposure. Although exposure may seem to be a modest penalty, it can result in significant costs. It subjects a state to adverse publicity both at home and abroad, it makes future treaty negotiations more difficult, and it can "infect other aspects of the relationship between the parties" (Chayes and Chayes 1995: 152) and even a state's status as a member in good standing of the international community (Young 1992: 176–177). In addition to exposure, some international environmental regimes require delinquent states to develop compliance action plans that detail how they will bring themselves back into compliance, on the assumption that non-compliance is usually the result of poor planning and lack of capacity.

What additional sanctions might be possible? Trade measures offer a potential lever, and the Montreal Protocol provides for trade restrictions as a response to both non-participation and non-compliance. The use of trade measures to promote participation and compliance has proven highly controversial, however, and other international environmental agreements have thus far not followed the Montreal Protocol's lead.

Financial penalties are also sometimes suggested as a sanction, but they have proven to be politically unacceptable. In any event, they would not solve the enforcement problem because they themselves require enforcement. (If a state violates an environmental commitment, what reason is there to think that it will comply with an obligation to pay a financial penalty?) At most, non-compliance may result in a loss of eligibility for existing funding, rather than the imposition of penalties. But even that penalty is unusual because states fear that cutting assistance will exacerbate rather than solve non-compliance. So, in practice, non-compliance more often leads to greater rather than less financial assistance – exactly the opposite of an enforcement model.

A final possibility would be to impose sanctions directly on the individuals responsible for violating international environmental law, rather than on the state. Whether criminal punishment would be appropriate is debatable, however, since most environmental problems result from everyday activities rather than from "bad" actors. Even with respect to deliberate, widespread environmental damage, which might merit criminal punishment, proposals to designate "ecocide" as an international crime (Gray 1996) have attracted little support. Individual criminal responsibility for environmental offenses is rare even in domestic law and seems unlikely anytime soon at the international level.

Institutions

International environmental law has no international institution with general governance functions – it has no World Environmental Organization to match the World Trade Organization. Instead, a patchwork of international institutions address environmental issues, leading to concerns about overlap, duplication of effort, lack of coordination, and even conflict. Some institutions are global, others regional or bilateral. Some relate to a particular issue area such as whaling or forestry, others have a broader environmental mandate, and still others, a mandate encompassing non-environmental as well as environmental issues. Some are scientific in orientation, others focus on capacity building or have a more policy-oriented role. (See generally DeSombre 2006.)

The international institution with the broadest competence over environmental issues is the United Nations Environment Programme (UNEP), established in the wake of the 1972 Stockholm Conference. In contrast to UN specialized agencies, UNEP does not have a separate treaty basis. Instead, like the UN Development Programme and the Commission on Sustainable Development, it derives its authority from the UN General Assembly, which created it (and which, in turn, derives its authority from the UN Charter). UNEP is small, with only a few hundred professional staff and a budget of under US$220 million per year, and it lacks significant decision-making authority. Instead, it has played a largely informational and catalytic role, helping to spur the negotiation of treaties such as the regional seas agreements in the 1970s and 1980s, the 1987 Montreal Protocol, the 1989 Basel Convention on hazardous wastes, and the 1992 Biodiversity Convention, as well as the development of various soft-law instruments.

Perhaps the most distinctive types of international environmental institutions are those established by individual multilateral environmental agreements (MEAs) (Churchill and Ulfstein 2000). Virtually every MEA now establishes a COP, which meets on a regular basis (usually annually), is open to all treaty parties, and serves as the supreme decision-making body for its constitutive agreement. These meetings go by different names in different treaty regimes. In the whaling regime, for example, the annual meeting of the parties is styled the International Whaling Commission (IWC), and the state representatives are referred to as "commissioners." In contrast, the meeting of the parties to the Long-Range Transboundary Air Pollution Convention is called the Executive Body, even though it is open to all of the treaty parties. Powers of the COP may include negotiating and adopting new protocols or annexes, amending existing agreements, and making decisions to elaborate or interpret the existing treaty rules. The decision-making authority and procedures of COPs vary from agreement to agreement. Some have limited authority (usually by a two-thirds or three-fourths majority vote) to adopt new environmental rules that bind all of the treaty parties, except those that file a specific objection. Other powers may include establishing subsidiary bodies, reviewing implementation, and monitoring compliance.

In addition to a regular meeting of the parties, most international environmental regimes have recognized the utility of a permanent secretariat. Even the Antarctic Treaty system, which for years had declined to establish a secretariat, recently decided to do so. Treaty secretariats perform largely administrative functions, such as organizing meetings, gathering and transmitting information, and administering training and capacity-building programs. But they may also play more substantive roles such as commissioning studies, setting agendas, compiling and analyzing data, providing technical expertise, mediating between states, making compromise proposals, monitoring compliance, and providing financial and technical assistance (Biermann and Siebenhüner 2009).

Why do states create international environmental institutions such as these? To what extent are these institutions merely creatures of the states that created them, as opposed to actors in their own right? How influential and effective are they in addressing environmental problems?

According to functionalist theories of international organizations, states establish international institutions to perform functions that states have difficulty performing individually. Among these functions are collecting information, monitoring compliance, and, in general, addressing collective action problems and providing public goods. The most basic rationale for international institutions is efficiency: international governance can be provided more easily and efficiently through a permanent institution than on a purely ad hoc, decentralized basis. Imagine the difficulties of addressing ozone depletion if every time states wanted to do something collectively, they had to organize a diplomatic conference – choosing a time and place, designating a secretariat, deciding on rules of procedure, and agreeing on relevant sources of information. International institutions, like business firms, reduce transaction costs by eliminating the need to define procedures and roles on a constantly recurring basis, and by allowing decisions to be made in a centralized, coordinated manner. This not only promotes efficiency, but also creates greater predictability and makes commitments by states to address a particular problem through international cooperation more credible.

According to this functionalist, statist approach to international organizations, international organizations are essentially agents of states, which exercise delegated authority. As agency theory teaches, however, agents have their own interests and do not necessarily act exactly as their principals might have wished. The same is true of international environmental institutions. Although they are created by states, they are usually not merely vessels for the transmission of state preferences. Rather, they are actors in their own right, with their own functions, decision-making rules, and organizational cultures, and often their own personnel (who serve as international civil servants rather than as state representatives) (Vaubel 2006).

In analyzing international institutions, we can array them along a spectrum, based on their degree of autonomy from states. At one extreme, an international institution such as the G8 or the Antarctic Treaty Consultative Meeting serves merely as an intergovernmental forum; at the other, the European Court of Human Rights operates as an autonomous actor in deciding cases under the European Convention on Human Rights, with a stable budget and independent judges. International law uses the concept of "legal personality" to denote the point along this spectrum at which an international institution is considered sufficiently autonomous to have a separate legal existence and to be able to act in its own right for certain legal purposes – asserting claims, entering into treaties, and exercising other implied powers that are necessary for it to fulfill its functions.

Most international institutions lie somewhere in between the two extremes of intergovernmental creature and autonomous actor. They have a dual or hybrid character, usually with different components reflecting their intergovernmental as opposed to their more autonomous/independent elements. The United Nations, for example, consists of the General Assembly and Security Council on the one hand, composed of states, and the secretariat on the other, composed of international officials. Similarly, the World Bank consists of a Board of Governors, representing the member-states, as well as a permanent staff headed by a president and Board of Directors. In referring to the United Nations or the World Bank, it is important to be clear which component one means. When commentators criticize the UN for failing to stop the genocide in Darfur, for example, do they mean the secretariat, or the member-states, or some combination of the two? Or when analysts write that the World Bank has the authority to develop operational policies relating to the environment, do they mean that the Board of Directors and permanent staff can do so on their own, or with the approval of the Board of Governors?

Conferences of the parties lie toward the intergovernmental rather than the supranational end of the spectrum. Even if meetings of the parties are only forums for states to meet and interact, however, they play a crucial role in keeping attention focused on an issue. The annual meetings of the International Whaling Commission, for example, help ensure that whaling remains on the international policy agenda, just as meetings of the parties to the Convention on International Trade in Endangered Species (CITES) provide a focal point for efforts to limit trade in elephant ivory, rhino horn, or sturgeon. In contrast, the 1940 Western Hemisphere Convention, which failed to provide for any institutional follow-up, is largely forgotten, with little if any effect on state behavior, despite strong substantive provisions. Regular meetings serve to enmesh states in an international process that takes on a life of its own. Attendance at regular meetings helps to socialize state representatives; they begin to develop a collective culture that tends to make them act differently, as a group, than they would act individually as agents of their states. In this manner, a COP can develop into something more than simply a vehicle for the transmission of state preferences and lead to different results than if states acted on their own.

To the extent that international institutions allow voting (rather than simply unanimous or consensus decision-making) or include only a subset of the treaty parties, they assume an even more clearly corporate character. By participating in an institution that allows decisions to be made by a qualified majority vote, or that establishes bodies with limited membership (such as the UNEP Governing Council, the Global Environmental Facility (GEF) Council, or the CITES Standing Committee), a state accepts a process that can result in decisions that it opposes. To be sure, most multilateral environmental agreements give objecting states the right to opt out of decisions with which they disagree. But exercising this right can be difficult, particularly for weaker states, which fear alienating other treaty parties. As a result, states may end up acquiescing to decisions that they dislike. For example, southern African countries such as Botswana and South Africa ultimately accepted the ban on trade in elephant ivory adopted by CITES in 1990, even though they had argued strongly that the ban should not apply to them because they had successfully controlled poaching.

Some fear that the autonomy of international institutions can create pathologies – perhaps most importantly, lack of accountability (Barnett and Finnemore 1999). This concern, though valid, needs to be kept in perspective. As with any organization, international institutions can produce agency costs. At the same time, even the strongest environmental institutions are still comparatively weak. They lack independent resources and are dependent on states for funding. They even lack general authority to adopt binding rules or decisions. In short, international institutions do not replace anarchy with hierarchy but, rather, with looser forms of governance. They depend for their influence not on material power, but on their perceived neutrality, expertise, and ability to provide benefits to states, all of which contribute to a belief, more generally, in the legitimacy of multilateral governance.

Evaluating the New Directions in International Environmental Law

In its brief history, international environmental law has failed to solve many pressing problems (most notably, climate change and loss of biodiversity), but it has also had some notable successes (including protection of the ozone layer and prevention of oil pollution). In achieving these, it has displayed impressive ingenuity, developing a wide range of mechanisms to set standards and promote implementation.

On the standard-setting side, international environmental law has developed distinctive approaches, including the framework convention/protocol approach and tacit amendment procedures. Similarly, on the compliance side, the picture also looks quite different from the standard approach of international law, which focuses on the concepts of breach, state responsibility, invocation of responsibility by the injured state, dispute settlement, and remedies such as restitution and compensation. In contrast to this traditional model, international environmental regimes have developed their own *sui generis* arrangements, aimed not so much at determining state responsibility and imposing remedies as at making the regime more effective in the future.

Finally, in terms of institutions, the central international environmental institution – the COP – represents a new form of international cooperation. From the perspective of general international law, it is neither an intergovernmental conference nor a traditional international organization but a combination of the two.

Together, these changes have transformed international environmental law into a distinct field, with its own characteristic methodologies and techniques. In the process, they have blurred not only the (already fuzzy) line between international law and politics, but also the lines between public and private, and international and domestic. In international environmental law, the private sector engages in the quintessential public task of general standard-setting through regimes such as the International Organization for Standardization and the Forest Stewardship Council. And in MARPOL, private-sector actors play a key role in the compliance process through the inspection and certification of oil tankers.

Some express concern about these developments, fearing that they erode the fundamental distinctiveness of law as a social instrument (Koskenniemi 1993). However, the emergence of new approaches to standard-setting and compliance represents an understandable and appropriate response to the distinctive characteristics of international environmental problems:

- These problems are physical as well as legal and political and involve a great deal of technical complexity.
- They result primarily from private rather than governmental conduct.
- They are highly uncertain and rapidly changing.

In order to address international environmental problems, we therefore need to develop dynamic regulatory regimes that can respond flexibly to new knowledge and problems, and that take a pragmatic and forward-looking approach to issues of compliance and effectiveness.

This facilitative approach to international environmental law is less ambitious but more realistic than the common goal of developing international laws "with teeth" (i.e. coercive powers). It views international environmental law as a process to encourage and enable, rather than require, international cooperation. Instead of pushing for the development of supranational institutions, it accepts state sovereignty as a given. It attempts to help states achieve mutually beneficial outcomes, for example, by building scientific and normative consensus and by addressing barriers to compliance, such as mistrust between states and lack of domestic capacity.

This is a comparatively modest agenda. Over time, however, it can contribute to greater international cooperation and thereby to the solution of environmental problems such as climate change. To be effective, international environmental law must understand not only its role but its limits. It must focus on those aspects of a problem where it can make a difference, recognizing that it is part – but only part – of the solution.

Acknowledgments

This chapter incorporates materials from chapters 1, 6, 8, 10, and 11 of Bodansky (2010). These materials are reprinted with permission of Harvard University Press.

References

Abbott, Kenneth W. and Duncan Snidal. 2000. "Hard and Soft Law in International Governance." *International Organization*, 54(3): 421–456.

Barnett, Michael N. and Martha Finnemore. 1999. "The Politics, Power, and Pathologies of International Organizations." *International Organization*, 53: 699–732.

Barrett, Scott. 2003. *Environment and Statecraft: The Strategy of Environmental Treaty-Making*. Oxford: Oxford University Press.

Biermann, Frank and Bernd Siebenhüner, eds. 2009. *Managers of Global Change: The Influence of International Environmental Bureaucracies*. Cambridge, MA: MIT Press.

Bodansky, Daniel. 1999. *The Framework Convention/Protocol Approach*. Framework Convention on Tobacco Control Technical Briefing Series, Doc. WHO/NCD/TFI/99.1. Geneva: World Health Organization.

Bodansky, Daniel. 2010. *The Art and Craft of International Environmental Law*. Cambridge, MA: Harvard University Press.

Bodansky, Daniel and Jutta Brunnée. 2002. "Introduction: The Role of National Courts in the Field of International Environmental Law." In *International Environmental Law in National Courts*, ed. Michael Anderson and Paolo Galizzi, 1–22. London: British Institute of International and Comparative Law.

Boockmann, Bernhard and Paul W. Thurner. 2002. *Flexibility Provisions in Multilateral Environmental Treaties*. Centre for European Economic Research (ZEW) Discussion Paper No. 02–44. Mannheim: Centre for European Economic Research.

Breitmeier, Helmut, Oran R. Young, and Michael Zürn. 2006. *Analyzing International Environmental Regimes: From Case Study to Database*. Cambridge, MA: MIT Press.

Brown Weiss, Edith and Harold Jacobson, eds. 1998. *Engaging Countries: Strengthening Compliance with International Environmental Accords*. Cambridge, MA: MIT Press.

Chayes, Abram and Antonia Handler Chayes. 1995. *The New Sovereignty: Compliance with International Regulatory Agreements*. Cambridge, MA: Harvard University Press.

Churchill, Robin R. and Geir Ulfstein. 2000. "Autonomous Institutional Arrangements in Multilateral Environmental Agreements: A Little-Noticed Phenomenon in International Law." *American Journal of International Law*, 94: 623–659.

DeSombre, Elizabeth R. 2006. *Global Environmental Institutions*. Abingdon: Routledge.

Downs, George W., Kyle W. Danish, and Peter N. Barsoom. 2000. "The Transformational Model of International Regime Design: Triumph of Hope or Experience?" *Columbia Journal of Transnational Law*, 38: 465–514.

Downs, George W., David M. Rocke, and Peter N. Barsoom. 1996. "Is the Good News about Compliance Good News about Cooperation?" *International Organization*, 50: 379–406.

Gehring, Thomas. 1994. *Dynamic International Regimes: Institutions for International Environmental Governance*. Frankfurt: Peter Lang.

Gilligan, Michael J. 2004. "Is There a Broader-Deeper Trade-off in International Multilateral Agreements?" *International Organization*, 58: 459–484.

Gray, Mark Allan. 1996. "The International Crime of Ecocide." *California Western International Law Journal*, 26: 215–271.

Handl, Gunther. 2007. "Transboundary Impact." In *The Oxford Handbook of International Environmental Law*, ed. Daniel Bodansky, Jutta Brunnée, and Ellen Hey, 531–550. Oxford: Oxford University Press.

Hanf, Kenneth and Arild Underdal. 1998. "Domesticating International Commitments: Linking National and International Decision-Making." In *The Politics of International Environmental Management*, ed. Arild Underdal, 149–170. Dordrecht, the Netherlands: Kluwer.

Henkin, Louis. 1979. *How Nations Behave: Law and Foreign Policy*, 2nd edn. New York:
Columbia University Press.

Jacobson, Harold K. and Edith Brown Weiss. 1998. "A Framework for Analysis." In *Engaging
Countries: Strengthening Compliance with International Environmental Accords*, ed. Edith
Brown Weiss and Harold K. Jacobson, 1–18. Cambridge, MA: MIT Press.

Koremenos, Barbara, Charles Lipson, and Duncan Snidal. 2001. "The Rational Design of
International Institutions." *International Organization*, 55: 761–799.

Koskenniemi, Martti. 1993. "Breach of Treaty or Non-compliance? Reflections on the
Enforcement of the Montreal Protocol." *Yearbook of International Environmental Law*,
3: 123–162.

Levy, Marc A. 1993. "European Acid Rain: The Power of Tote-Board Diplomacy." In *Institutions for the Earth: Sources of Effective International Environmental Protection*, ed. Peter
M. Haas, Robert O. Keohane, and Marc A. Levy, 75–132. Cambridge, MA: MIT Press.

Miles, Edward L., Aril Underdal, Steinar Andresen *et al*. 2002. *Environmental Regime Effectiveness: Confronting Theory with Evidence*. Cambridge, MA: MIT Press.

Mitchell, Ronald B. 2006. *Global Environmental Assessments: Information and Influence*.
Cambridge, MA: MIT Press.

Parson, Edward A. 2003. *Protecting the Ozone Layer: Science and Strategy*. Oxford: Oxford
University Press.

Rajamani, Lavanya. 2007. *Differential Treatment in International Environmental Law*.
Oxford: Oxford University Press.

Raustiala, Kal. 2001. *Reporting and Review Institutions in 10 Multilateral Environmental
Agreements*. Nairobi: UNEP.

Raustiala, Kal. 2005. "Form and Substance in International Agreements." *American Journal
of International Law*, 99: 581–614.

Reeve, Rosalind. 2002. *Policing International Trade in Endangered Species: The CITES Treaty
and Compliance*. London: Royal Institute of International Affairs.

Sand, Peter. 1991. *Lessons Learned in Global Environmental Governance*. Washington, DC:
World Resources Institute.

Sand, Peter. 1999. "Carrots without Sticks? New Financial Mechanisms for Global Environmental Agreements." *Max Planck Yearbook of United Nations Law*, 3: 363–388.

Shelton, Dinah. 2004. *Commitment and Compliance: The Role of Non-binding Norms in the
International Legal System*. Oxford: Oxford University Press.

Skjærseth, Jan Birger. 1998. "The Making and Implementation of North Sea Commitments:
The Politics of Environmental Participation." In *The Implementation and Effectiveness of
International Environmental Commitments: Theory and Practice*, ed. David Victor, Kal
Raustiala, and Eugene R. Skolnikoff, 327–380. Cambridge, MA: MIT Press.

Trouwborst, Arie. 2002. *Evolution and Status of the Precautionary Principle in International
Law*. The Hague: Kluwer Law International.

UNEP (United Nations Environment Programme). 2007. *Compliance Mechanisms under
Selected Multilateral Agreements*. Nairobi: UNEP.

Vaubel, Roland. 2006. "Principal-Agent Problems in International Organizations." *Review
of International Organizations*, 1: 125–138.

Victor, David, Kal Raustiala, and Eugene R. Skolnikoff, eds. 1998. *The Implementation and
Effectiveness of International Environmental Commitments: Theory and Practice*. Cambridge, MA: MIT Press.

Wettestad, Jørgen. 2002. *Clearing the Air: European Advances in Tackling Acid Rain and
Atmospheric Pollution*. Aldershot: Ashgate.

Wolfrum, Rüdiger. 1998. "Means of Ensuring Compliance and Enforcement of International
Environmental Law." *Recueil des Cours*, 272: 9–154.

Yoshida, O. 1999. "Soft Enforcement of Treaties: The Montreal Protocol's Non-compliance Procedure and the Functions of International Institutions." *Colorado Journal of International Environmental Law*, 10: 95–141.

Young, Oran R. 1992. "The Effectiveness of International Institutions: Hard Cases and Critical Variables." In *Governance without Government: Change and Order in World Politics*, ed. James N. Rosenau and Ernst-Otto Czempiel, 160–194. New York: Cambridge University Press.

Chapter 12

Green Growth

Michael Jacobs

Introduction

Over recent years the concept of "green growth" has burst onto the international policy scene. A term rarely heard before 2008, it now occupies a prominent position in the policy discourse of international economic and development institutions. The World Bank, along with five other multilateral development banks, has committed itself to this goal (World Bank 2012a, 2012b). The OECD has adopted a "green growth strategy" of research and publications (OECD 2012a). A new international body, the Global Green Growth Institute (GGGI), supported by a number of governments, has been created to advise countries on its implementation. Using its own preferred label of "the green economy," the United Nations Environment Programme (UNEP) has published a 600-page report (UNEP 2011). These four institutions have jointly established a "Green Growth Knowledge Platform" to provide a locus for research and knowledge about the field (World Bank 2012c). A number of high-level meetings and networks have been established.[1] Several countries have adopted green growth as an explicit policy objective (OECD 2012a), while at the G20 Summits in France and Mexico in 2011 and 2012, the largest economies in the world committed themselves to its promotion (Government of France 2011; Government of Mexico 2012). The "green economy" was a major focus of the Rio+20 United Nations Summit in June 2012 (UNCSD 2012).

The core meaning of the concept of green growth can be simply stated. It is economic growth (growth of gross domestic product or GDP) which also achieves significant environmental protection. The "significant" matters. Few doubt the compatibility of growth and some kinds of environmental improvement: this would not require a special term.[2] But *how* significant, the concept leaves open. In early uses of the term the focus was entirely on the mitigation of climate change (Huberty *et al.*

2011), but it now more normally covers a wider range of environmental resources (soil, water, fish stocks, habitats, and so on). Some definitions leave the precise degree of environmental protection undetermined: thus to the World Bank (2012b), green growth is

> growth that is efficient in its use of natural resources, clean in that it minimizes pollution and environmental impacts, and resilient in that it accounts for natural hazards and the role of environmental management and natural capital in preventing physical disasters.

But others apply a more stringent "sustainability" standard. For the OECD (2011), "green growth means fostering economic growth and development, while ensuring that natural assets continue to provide the resources and environmental services on which our well-being relies."

What these definitions have in common, however, is made clear in the analysis which follows them: it is a level of environmental protection which is not being met by current or "business-as-usual" patterns of growth. It is this in turn which gives the concept its political traction.

It is not the only one occupying this terrain, however. A range of sister concepts are now also in frequent use, most of them seeking to widen the idea of economic growth to become the more socially equitable "development." Such development may be "low carbon," "low emissions," "climate-compatible" and/or "green" (Climate and Development Knowledge Network 2012). UNEP's definition of a "green economy" captures these ideas: it is one that "results in improved human well-being and social equity, while significantly reducing environmental risks and ecological scarcities" (UNEP 2011). While these terms do not have exactly the same meaning as "green growth," they should nevertheless be seen as variants of the same concept, both because they all embrace the same core idea of growth compatible with environmental protection, and because the networks and institutions in which they are being discussed and supported are largely the same (Green Economy Coalition 2012; World Bank 2012c).

This chapter first seeks to place the concept of green growth within the history of recent discourses of environmental protection. It will then distinguish between a "standard" version of green growth and a "strong" interpretation which seeks to present a much bolder argument to policy-makers. Three different forms of this argument will be identified, and the evidence for them surveyed. Finally, the chapter asks whether the idea of green growth is likely to be "successful": will its arguments prove sufficiently convincing, and the interests gathered around it sufficiently strong, to change the priorities of economic policy-making?

From Sustainable Development to Green Growth: The Role of Environmental Discourses

The concept of economic growth which also meets environmental objectives is not new. Indeed it lay at the heart of the discourse of "sustainable development," first popularized by the Brundtland Report (World Commission on Environment and Development 1987) and subsequently institutionalized by the Rio Earth Summit in 1992 (Dresner 2008). Sustainable development remains the core principle of international environmental policy-making, and of national environmental planning in many countries. Indeed, the official institutions now promoting green growth insist that it is not a substitute for sustainable development but a way of achieving it (OECD 2011; UNEP 2011; World Bank 2012b). But why then invent a whole new discourse around it?[3]

The answer (though not officially acknowledged) is that the concept of sustainable development has had decreasing traction on economic policy-making over recent years. In the period immediately following the 1992 Earth Summit the sustainable development goal was widely adopted by governments and others, and in many countries had a tangible impact on the priority given to environmental objectives. The 1990s saw a clear upsurge in environmental legislation and policy and, in the business sector, environmental management. Yet by the early years of the new century momentum had significantly slowed. Moreover, it became clear that countries' apparent commitment to sustainable development had not been sufficient to reverse the historic decline in the health of the global environment that had led to its invention: almost all significant global indicators have continued to worsen. The evidence of dangerous human-made climate change, in particular, demonstrated that something much more profound had to be done. An existing concept, already universally supported, could not help here: sustainable development was too much part of the furniture of government commitments to motivate more radical change.

Yet at the same time policy-makers were highly conscious that an environmental discourse focused on costs and limits and the need to constrain growth to address them would be unlikely to attract political support in a world where GDP growth (and the employment it generates) remains the core interest of voters and businesses and the overriding policy objective of governments. This is especially true in the field of climate change, where the dominant discourse has centered on the economic cost of mitigation and international negotiations have been concerned with how the global "burden" should be distributed (Stern 2007).

The purpose of the discourse of green growth has therefore been to shift from this negative and politically unattractive framing to something more positive. Like sustainable development, it seeks to show that environmental protection need not come at the expense of prosperity. Unlike sustainable development, however, it faces the issue of growth head-on. Sustainable development was a deliberate exercise in holding together a wide coalition of political support by sidestepping the question of the fundamental compatibility of growth and environmental protection and reframing the economic objective as "development." Green growth not only insists on that compatibility, but claims that protecting the environment can actually yield *better* growth. In this it reflects its different provenance: whereas the concept of sustainable development came out of the environmental movement, where ideological argument about the "limits to growth" was widespread, green growth has emerged from the more mainstream and pragmatic community of environmental-economic policy-makers.

This also makes green growth a much more focused concept. As frequently observed, sustainable development was a baggy idea, incorporating a variety of often ill-defined objectives. Its meaning was contested, interpreted in more conservative or more radical ways by different interests (Jacobs 1999). By contrast green growth is more or less self-explanatory: it might attract fewer adherents (and already has some "green" opponents), but it's fairly clear what it means.

In this sense green growth is indeed something new. It is a child of sustainable development. But it is a response to its inadequacies, and to the particular focus on both climate change and economic growth which have dominated mainstream policy debate in recent years.

The Standard Argument for Green Growth

Embedded in the concept of green growth is an assertion. This is that economic growth can occur even while environmental impacts are significantly reduced. In this sense – rather more than was true of sustainable development – green growth is not just a normative ideal, but carries within it a strong economic claim, both theoretical and empirical.

Since the historical record of industrialization in every country is that economic growth is associated with a wide range of forms of environmental damage – from resource depletion to climate change – the claim that this relationship is contingent, not necessary, clearly needs some theoretical basis. At a simple level that was provided in the early 1970s by John Holdren and Paul Ehrlich (1974), who proposed the explanatory identity $I = PCT$, in which I = environmental impact, P = population, C = consumption, and T = technology (or more specifically, the productivity of technology in relation to environmental use). The equation showed that with a rising population and (given economic growth) rising consumption, environmental impact would inevitably increase *unless* the rate of technological improvement was sufficient to overcome it. The maths was straightforward but daunting. If over a given period of, say, 50 years, population doubled and consumption quadrupled (which would happen with an annual compound growth rate of no more than 3%), the "environmental productivity" of technology would have to improve eightfold to prevent worsening environmental damage – and by much more if impact was to be reduced to "sustainable" levels. But subsequent analysis showed that such improvements were not in principle impossible: through a whole variety of methods, including use of renewable resources and new materials, industrial and consumer productivity improvements, recycling of wastes, sustainable harvesting practices, and changes to the structure of the economy (especially by making it more "intellectually" than "materially" based), it was possible to conceive of very dramatic technological and social changes which could at least in theory allow growth to occur even while resources were sustained and environmental quality improved (Jacobs 1991; Ekins 2000).

These arguments provide the theoretical foundation for the claim that growth can be green. The modern "green growth" concept, however, rests on a more empirical basis. Two claims are made. First, that the costs of tackling environmental damage are not so great that they reduce the natural growth rate of a well-performing economy to zero. And second, that if such damage is *not* tackled, the costs to growth of a worsening environment will be greater.

These claims were most famously argued for in the Stern Review of the economics of climate change (Stern 2007). Building on similar work done by others, the report modeled the cost of stabilizing greenhouse gas emissions (using well-designed policies, implemented early) as between 4% and −2% of GDP, with a central estimate of only 1%. By contrast, if the world failed to reduce emissions, the economic costs of the climate change that would then follow would be much larger: the equivalent of between 5% and 20% of GDP every year. So the core message of the Stern Review was straightforward: the costs of acting on global warming were significant but manageable, they were compatible with the continuation of economic growth, and they were much less than the costs of not acting.

Though the Stern Review did not itself use the term, this is what we might call the "standard" argument for green growth. It is rooted in the economics of climate change (rather than consideration of a wider set of environmental costs), and relies on a relatively simple cost–benefit analysis of alternative paths of economic development.

It did not go uncontested, however. Critics of the Stern Review focused in particular on the estimates of the costs of allowing climate change to occur. Many argued that these were so far in the future as to be incommensurable with the present costs of preventing it: in the future human societies would be richer (because of economic growth) and would develop the technologies to adapt to or otherwise prevent warming (Nordhaus 2007). Though they did not convince everyone, these criticisms nevertheless politically weakened the claims of "green growth." While in the long term controlling environmental damage might be beneficial, in the short term it was not obvious that the costs to GDP were worth the almost entirely future benefits that would result. When the economic climate worsened after the financial crash of 2008, a strategy whose first effect was to slow growth down did not look very attractive to policy-makers.

The "Strong" Versions of Green Growth

So it was at this point that a much stronger argument for green growth began to be made. From 2008, as the term itself came to be used, its proponents made a much more immediate claim. Environmental protection was not just compatible with continued economic growth: it could positively promote it. So far from slowing the economy down, policies to make it greener could be a driver of higher output and rising living standards. And they could do so in the relatively short term, not merely in the long.

Over the last few years three different kinds of argument and evidence have been used to justify and explain this claim, each using a different framework of economic theory. They are not mutually exclusive – many of those writing about green growth use many or all of them. Some are applicable only in some circumstances and some kinds of countries, others at least in principle to all. Each leads to a different set of policy conclusions. (For helpful surveys, see Bowen and Fankhauser 2011 and Huberty *et al.* 2011.)

Green Keynesianism: Environmental Stimulus in Recession

The original case for green growth made in the wake of the 2008 financial crisis was primarily a short-term one. Economies which had experienced a huge recessionary shock could be stimulated back into growth – particularly employment growth – partly through measures aimed at improving the environment (Pollin *et al.* 2008). Fields such as energy efficiency, renewable energy, water quality improvement, agricultural and landscape management, public transport, and pollution control provided opportunities to get people into work and to increase demand for a wide range of goods and services. Almost all countries which introduced fiscal stimulus packages in 2008/2009 included within them significant "green" programmes of these kinds. South Korea's environmental spending was generally acknowledged to be the largest, estimated at around 79% (US$59 billion) of its total stimulus package. But others were also significant: China identified a third of its stimulus package, a total of US$219 billion, as "green'; the USA around 12% (US$118 billion). While individual European states' figures varied, as much as 60% (US$23 billion) of the European Union's collective stimulus package was environmental in content (Barbier 2010b; Robins *et al.* 2010).

The core argument used to justify these sums was the simple Keynesian one that in a slump, governments should sustain aggregate demand in the economy by replacing lost private-sector demand with public expenditure. This in turn creates a multiplier effect that generates further income and employment growth. Such spending does not have to be green, but given the extent of the environmental opportunities available, and the various additional amenity and health benefits they offer, a green stimulus package offers particular advantages. A particular rationale can be given for spending in those areas where green investments are in due course going to be required anyway, such as to replace aging power stations or upgrade transmission lines. In these cases, the Keynesian stimulus would merely bring forward investment from the future to the present, where it could both have a larger stimulatory effect and benefit from the cheaper labor, materials, and financing costs available in a recession (Bowen *et al.* 2009). Governments do not even necessarily have to spend or borrow themselves to achieve a green multiplier: regulatory or tax policies which force or incentivize firms to invest in environmental improvements can have the same impact without increasing public deficits (Zenghelis 2012).

These kinds of general Keynesian arguments can in principle be applied as much to "brown" or non-environmental spending as to "green." But some proponents of environmental spending go further, arguing that green measures in a recession are *better* for short-term growth. They point out in particular that many environmental measures are labor-intensive, and so give greater employment growth per dollar spent than non-green measures (Green New Deal Group 2008; Engel and Kammen 2009). Making buildings more energy-efficient, for example, can employ large numbers of relatively unskilled workers, distributed widely in terms of geography. Much environmental spending is for various kinds of construction and resource management activities (wind turbines, solar panels, agricultural and water management) which, because they are location-specific, are not susceptible to "offshoring" in the manner of much manufacturing. At the same time, it is argued, improvements in energy efficiency (and other forms of resource efficiency) are cost-saving to the economy, releasing resources for firms and households to spend elsewhere, and thus generating their own stimulus effect (Roland-Holst 2008).

Estimates of the impact of the green stimulus measures taken in 2008–2010 give some credence to these arguments. Around half a million net jobs were estimated to have been created by the environmental elements of the US stimulus package

(Barbier 2010a), with as many as 960 000 by the similar measures taken in South Korea (OECD 2010). Indeed there is some evidence that in terms of job creation the environmental stimulus measures may have out-performed (as their proponents predicted) the non-green elements: one estimate for the USA suggests 20% more jobs were created by green measures than by traditional infrastructure spending (Houser *et al.* 2009).

Both the theoretical and empirical claims, however, are disputed. Indeed, unsur-prisingly, the case for a specifically green form of Keynesianism has become mired in long-standing economic disputes about the effectiveness of Keynesian policy in gen-eral. Critics argue that stimulus spending – whether green or brown – is ultimately ineffective in creating employment or growth. Public expenditure simply crowds out private spending by forcing interest rates upwards; as governments spend more, rational firms and consumers save more, since they know that eventually taxes will have to rise to pay for public borrowing (Moore 2011). Keynesians counter that in a slump there is no private spending going on which can be crowded out: what happens rather is simply a new recessionary equilibrium, with low investment and low demand. Firms and consumers do not act with collective rationality: on the contrary, a "paradox of thrift" operates, in which lack of prospective demand leads private actors to save rather than spend, which in turn ensures the lack of demand. Only government action can reverse this vicious circle (Zenghelis 2012).

Both sides of this argument can point to analysis of the economic impacts of the 2008/2009 stimulus packages to justify their claims. While official estimates of the multiplier effect of the US package, for example, are strongly positive, others are negative (US Congressional Budget Office 2009; Mulligan 2010). (A negative multiplier would mean that the stimulus package actually destroyed output and jobs.) Much of the disagreement revolves around the time frame in which effects are judged. Critics of Keynesian policy, green and otherwise, argue that any observed job gains are temporary, and will be ultimately offset by the jobs lost as a result of the higher interest rates demanded by the budget deficits which finance them. Proponents argue that Keynesian stimulus measures are only intended to be short-term, to be applied when the economy is in recession, and in those conditions perform a critical role in raising the economy's equilibrium output and employment levels. In fact many green Keynesians have extended this argument into a stronger claim that environmental measures can drive economic growth in the medium and long term as well (Spencer *et al.* 2012). In doing so, however, they have gone beyond purely Keynesian arguments to wider theories of growth.

Growth Theory: Correcting Market Failures

The second (and central) case that environmental improvement can positively con-tribute to economic growth is based on the core framework of economic theory which explains why and how growth occurs (Hallegatte *et al.* 2011). Economic out-put results from the bringing together of factors of production or capital: labor, physical capital, and technology and human capital. Growth in output occurs when these factors increase either in absolute size or productivity. The different forms of capital depreciate over time but can be increased in size and productivity by invest-ing in them a proportion of output – such as in improved technology and better education and health of the workforce. If the rate and forms of investment are sufficient, economic history shows that the outcome will be economic growth.

Green growth theory then starts from the simple observation that the natural environment is also a factor of production, but one which both classical growth theory and historic patterns of economic growth in practice have largely ignored (Nordhaus 1974; Solow 1974; Smulders 1999; Brock and Taylor 2005). The environment acts as a form of capital in three ways: it provides resources, it assimilates wastes, and it performs various "environmental services" which sustain life, including climatic regulation and ecosystem health. This "natural capital" has been undervalued both in economic theory and practice because it has been largely unpriced, provided as an apparently free gift of nature. Many of the environment's functions occur as common or collective goods without the property rights which attach to other factors of production, and without therefore the private incentive to value them properly in economic terms (Jacobs 1991).

The standard economic concept to describe this is that of "market failure." Markets "fail" when they do not take into account the full value of the activities within them. The production and consumption decisions which economic actors take are therefore distorted relative to those they would take if the environment were properly valued, in a whole series of ways. Natural resources tend to be overexploited: soil eroded, fisheries depleted, water overabstracted. Ecosystems which provide valuable services, such as wetlands and forests, are allowed to be degraded or destroyed. Resources such as energy and materials are used inefficiently, with an excessive generation of waste (and therefore pollution). And the amenity, health, and cultural value of natural environments are underappreciated.

In all these ways, green growth theory argues that current patterns of economic growth are *prima facie* sub-optimal. They misallocate resources between the different factors of production. They underinvest in natural capital, and overinvest in activities which cause its degradation. If these systematic market failures were corrected, growth might be higher. Indeed, the situation is worse than this, because in many countries the environmental costs of using natural resources are not just unpriced, but their exploitation is actually subsidized. Subsidies for extracting and using fossil fuels, and for other forms of resource extraction and agriculture, are estimated at around US$1.1 trillion per annum (Dobbs *et al.* 2011). Such subsidies further distort production and consumption decisions away from their optimal path. From these premises, advocates of green growth argue that a range of different environmental measures and policies can be growth-generating.

In developing countries, much of the emphasis has been on the conservation and enhancement of natural capital, such as soil quality, fisheries, forests, and habitats such as mangrove swamps and coral reefs. Arguing that in economies dependent on these resources, the net depreciation of natural capital is a retardant of growth in the same way that the net depreciation of physical capital would be, the UNEP has gathered considerable evidence on the positive growth impact available from the conservation and sustainable management of natural resources (UNEP 2011). In some cases this arises from higher productivity in production of the resource; in others from the development of secondary, value-adding, products that conservation of the resource allows; in some from the development of related industries, such as tourism. The UNEP report points out that many of these resources are controlled by the poor, and so strategies to conserve them and enhance their productivity are poverty-reducing as well as growth-enhancing.

Some of these growth benefits clearly show up in higher incomes, so are captured by the conventional growth indicator of GDP (gross domestic product). But others are unmeasured: it is difficult to capture the value which preservation of a mangrove swamp has for coastal defense, for example, or a forest for water supply. For this reason some of the focus of environmental policy has been on the creation of systems of payment which enable monetary value to reflect ecological value: payments for forest conservation (e.g. under Reducing Emissions from Deforestation and Forest Degradation (REDD) schemes) provide the prime example (UNEP 2011). If international financial assistance can enable those living around tropical forests to generate value out of forest preservation, its contribution to economic growth can become real and not just "intangible."

These arguments for green growth are not universally accepted. It is quite possible to argue that, though sustainable management of natural resources can lead to growth in these ways, much higher rates of growth can be achieved through their unsustainable exploitation. This was, after all, how most developed countries grew during their own industrialization. They exploited resources to the full – creating severe environmental degradation and human health costs in the process – in order to build a foundation of wealth and productive knowledge which then enabled the creation of a different kind of advanced technology and service economy altogether. Effectively, natural capital was substituted by physical (human-made) capital, leading to a higher total capital stock, and therefore higher growth rates. Advocates of "brown" growth in developing countries argue that this is the right strategy for them too: the rapid exploitation of natural resources can generate much faster growth in the short term than conservation, providing a base from which industrialization can then be achieved. Such exploitation cannot last forever, but it can certainly continue for now.

Who is right? To a considerable extent the question hangs on the time period under consideration. The now-industrialized countries exploited resources at a time when nature was abundant, and where the costs of resource depletion were therefore in the far future. Today nature is much scarcer, and resource depletion is already upon us – trends in fish stocks, water supplies, soil quality, and other resources reveal this clearly (Rockström et al. 2009). The wider ecological costs of forest degradation, loss of wetlands, coral damage, and so on are evident, and already occurring. So the green growth argument is that brown growth *was* possible, but is no longer: we have now reached the moment when overexploitation has to stop. If it does not, it will undermine itself: the net loss of natural capital will not be compensated by the creation of physical capital. Environmental damage will cause more cost to output and welfare than it will generate benefit. By contrast, advocates of brown growth insist that that time has not yet come. Depending on the length of the period over which growth rates are projected, and the resources and countries in question, empirical evidence can be adduced on both sides.

Whereas in developing countries the focus of green growth arguments has been the conservation and enhancement of natural capital, in developed and emerging economies it has been the way in which environmental policies, as well as tackling environmental costs, can address *other*, non-environmental market failures which inhibit growth. Growth theory acknowledges that current patterns of economic activity are far from optimal. At least four kinds of market failure have relevance to environmental policy.

First, energy and other resources are not used efficiently. For a variety of behavioral and structural reasons, firms and households fail to use myriad means of improving energy efficiency which would be of net benefit to them – that is, which would save them more money than they would cost (Gillingham *et al.* 2009). At the same time, energy use has externalities – particularly the role of fossil-fuel emissions in causing climate change – which are not properly captured in energy prices. So in correcting the environmental market failure, environmental policy can also correct the behavioral and structural ones. Such a policy includes taxes and emissions trading schemes that put a price on carbon; regulations which require minimum energy efficiency standards for buildings, vehicles and appliances; and public spending to promote innovation. When such policies are introduced, firms and consumers respond to the higher cost of energy use by raising their energy efficiency, and they innovate in doing so. Emissions fall, the economy saves costs, and the productivity of an important factor of production rises. The result is greener growth. The same effects can occur if environmental policy is applied to other resources, including water, material commodities, and wastes (Dobbs *et al.* 2011).

Second, markets left to themselves underinvest in key productivity-improving activities such as research and development (R&D) and the creation of economic networks (relationships between firms and activities which enhance productivity and innovation). This is because these activities have "spillover" benefits which cannot be captured exclusively by those who invest in them (Aghion and Howitt 1992). Since much environmental policy aims at promoting innovation, including specific efforts to support R&D and to create new networks of firms or infrastructure (such as industrial clusters or distributed energy systems), it may stimulate innovation and network benefits more widely in the economy, with positive impacts on growth in general (Porter and van der Linde 1995; Ambec *et al.* 2011).

Third, environmental policy can have a range of co-benefits addressing other externalities. Improving energy efficiency and using non-fossil fuels both help to reduce air pollution, which can have a major impact on health costs and labor productivity (Graff Zivin and Neidell 2011). They can also improve energy security, which reduces the costs caused by the volatility of energy prices (Rozenborg 2010). Cutting transport emissions can simultaneously cut urban transport congestion, with significant benefits for overall productivity (OECD 2012b).

Fourth, environmental policy can improve the economic efficiency of the taxation system. Where taxes are used to achieve environmental goals, policy-makers have the option of reducing other taxes to compensate. Since income and labor taxes, considered in isolation, penalize economic activity, their reduction in favor of environmental taxes may be growth-generating (Bosquet 2000).

In all these ways, growth theory allows a positive role for environmental policy to move the economy closer to an optimal growth path. Whether it will do so in practice, of course, depends on the scale of the costs which environmental policies impose on the economy. There is certainly evidence from a variety of countries that energy efficiency measures can generate growth and employment, releasing cost savings which are then spent in other areas of economic activity. (This "rebound" effect of higher energy efficiency on demand means that the overall emissions reduction is smaller than anticipated: that is, growth is generated, but it is less green than advocates may believe (Jenkins *et al.* 2011).) But it is not clear that the same is true of investments in low-carbon energy sources or other forms of environmentally motivated resource substitution. In general renewable energy has been considerably more expensive than fossil fuels, and therefore required substantial public subsidy. Though these costs to the economy have fallen as the scale of generation has increased and technological innovation has occurred, the overall effect on growth in the past has (in most countries) probably been negative. On the other hand, if at some point in the future – as is already beginning to occur in some countries – the cost of renewables reaches parity with that of fossil fuels, these past costs may be seen as an investment which has generated future growth. Again, assessment of the trade-off between the costs and benefits to growth of environmental policy depends to a large extent on the time period over which it is considered.

Comparative Advantage and Technological Revolution: Innovation and Industrial Policy

The third kind of argument for green growth relates to the commonly made claim that environmental policy creates significant numbers of new jobs in environmental industries. To meet higher environmental standards, firms and households require new products and services (energy-efficient equipment, pollution control technologies, engineering services, and so on). The global environmental sector is now estimated to be worth over US$5 trillion per annum, and to be growing at over 3% a year (Department for Business, Innovation and Skills 2012).

The creation of "green jobs" dependent on environmental policy does not of course mean that such policy is driving economic growth. More jobs might be being displaced in "brown" sectors and across the economy as a whole by the higher costs imposed. So two kinds of argument are used to justify the claim that environmental policy can be an engine of growth in this way.

The first is that countries which introduce stringent environmental policies give their domestic environmental sector firms a head start over those in other countries. Forced to innovate, they develop goods and services in response which enable them to win not only domestic business but export markets (Porter and van der Linde 1995; Lanoie *et al.* 2008). Such a "first mover advantage" is held to account for the success of the Danish and Spanish wind turbine sectors, the German solar industry, and a range of Japanese and American environmental technology firms, each of which first developed in response to domestic environmental policies, and then grew to become world-leading businesses (Ambec *et al.* 2011). Their contribution to domestic growth, it is argued, outweighs any economic costs which the original policies may have imposed.

The development of comparative advantage in international trade through environmental policy is obviously not something which can be replicated by all countries. In this sense it cannot be a means to generalized global green growth. But that does not mean that it is not a viable growth path for individual countries. A number of governments around the world (among them Denmark, South Korea, South Africa, and Ethiopia) are now pursuing industrial strategies with this object in mind (Global

Green Growth Institute 2012; OECD 2012a). China is perhaps the most notable example: of the seven "strategic industries" on which investment will be focused in its Twelfth Five Year Plan (2011–2015), three – alternative energy, alternative-fuel cars, and energy saving and environmental protection – are green. China is already the world's leading exporter of both wind and solar technologies (China Greentech Initiative 2012).

The second argument for regarding "green jobs" as the harbinger of wider green growth involves a larger and more general claim. This is that low-carbon energy systems and other environmental technologies are on the brink of creating a "new industrial revolution" (Stern and Rydge 2012). Their pervasive impact, it is argued, will unleash a wave of innovation in production methods, products, and lifestyles that will transform the economy in the same way as previous technologies such as the steam engine, the railways, the internal combustion engine, and the microprocessor. Those making this argument observe that economic growth has tended to occur in "long waves" of around 50–60 years, driven by technological change but then encompassing, and restructuring, whole systems of production, distribution, and consumption. The next long wave of growth, it is claimed, will be driven primarily by information technologies, but if combined with various forms of low-carbon and "smart" energy systems, new agricultural and production technologies, new materials, and new systems for recycling, it has the potential to create an economy with dramatically lower environmental impact (Perez 2010).

This is the most radical form of the green growth argument. Here, environmental improvement is seen not just as a possible motor of economic growth but as one of its primary engines. In some versions this appears to be expressed almost as an inevitability – environmental policy is needed initially to set the new industrial revolution in train but it will then take off on its own. But in others a much stronger or "entrepreneurial" role for the state is envisaged, in which the necessary innovation is guided and funded by public policy and new infrastructures of production and consumption are developed through public spending, regulation, and planning (Mazzucato 2011). An impressive array of evidence can be cited for the role which the state has played in previous technological revolutions. For critics, however, this is precisely what undermines the case. It is not just markets which fail: the idea that the state should seek to "pick winners" among competing technologies and direct the patterns of growth is a recipe for *government* failure (Winston 2006). If this is what green growth depends on, such critics argue, it is not likely to succeed.

Conclusion: The Political Economy of Green Growth

As these different forms of the green growth argument have been articulated and debated in the period since 2008, each has found supporters and critics among governments and international institutions, and within the wider academic and policy communities. It is clear that in most countries the dominant economic view remains that the case for green growth is unproven: strong environmental policy continues to be seen largely as a drag on growth rather than a driver of it, particularly in the difficult economic conditions which have prevailed in most of the developed world since the financial crash. But amongst the disagreements, two clear conclusions do emerge.

One is that the theory of green growth (on whichever body of economic thought it is based) cannot determine the question of whether any particular green growth strategy or path will achieve the claims made for it. That will be an empirical matter. It is quite plausible that *some* environmental policies will be growth-enhancing, but others will act as a constraint. The difference could arise from the nature of the environmental problem being addressed, the stringency of the objective, or the efficiency of the policy instruments being used. So there isn't a general conclusion that green growth is or is not possible. It will depend on what kind of green growth is in question.

At the same time, it is clear that the case for green growth is stronger the further ahead one's frame of reference looks. Seen over a very short period, the costs of environmental policy loom large, and the output benefits uncertain. But over a longer time frame, the potential for technological innovation to reduce costs and drive growth becomes greater, while the economic costs of failing to protect the environment – as resource depletion, pollution, and ecosystem loss reach critical levels – become larger. Few would now dispute the "standard" green growth argument that in the long run protecting the environment will cost less than not doing so. The question posed by the claims of "strong" green growth is how close that long run now is.

It was to cast light on this question that UNEP conducted a new form of economic modeling for its Green Economy report (UNEP 2011). Most modeling of environmental policy has sought to measure its costs against a "reference scenario" in which such policy is absent – but where there are also no costs of environmental damage. UNEP sought to model an arguably more realistic reference scenario in which, because of the lack of environmental policy, considerable environmental costs are projected to occur. The results are instructive. The "green growth" path, in which 2% of global GDP goes into environmental investments, starts off lower than the reference scenario (in which no such specific investments are made). But as the costs of environmental damage begin to constrain growth in the latter, the benefits of environmental policy in the green scenario begin to emerge, and the two growth trajectories cross over. Within seven years from the 2010 start date, the "green growth" scenario has a higher rate of growth than the base case, and real GDP is greater by just after 2020.

No modeling exercise can in itself prove much: too many of the assumptions and equations are open to question. But the UNEP results clarify the nature of the green growth argument. It is that protecting the environment does have costs in the short term. But these should really be understood as the investments needed to generate growth in the medium to long term. There should be no surprise about this: growth theory tells us that growth results from investment, which inevitably subtracts from consumption now. There is a particular urgency about environmental investment, because in every year in which it is not made, environmentally damaging and high-carbon capital will be laid down in its place, locking in high emissions and resource depletion for years to come. The case for green growth can in this sense be redefined: it is the case for a growth path which can be sustained over more than just the next few years.

None of this means that green growth is about to become universal economic orthodoxy. But there are nevertheless grounds for believing that it might prove a more effective discourse than sustainable development in marshaling momentum for strong environmental policy.

The first is that environmental degradation is far more evident now than it was 20 years ago. Climate change, water scarcity, food insecurity, and rising commodity prices have made the environmental consequences of growth much more immediate to mainstream policy-makers (Haas 2012). Second, the focus of the green growth discourse on economic growth gives it much greater purchase on mainstream economic policy-making. Indeed, it is striking how much of the advisory literature on green growth is about the importance of creating the right conditions for growth *per se* – in labor markets, fiscal policy, effective government institutions, and so on – and not simply about well-designed environmental policy (OECD 2011; World Bank 2012b).

Third, and perhaps most importantly, the green growth discourse has a much stronger base of support among economic interests than did sustainable development. The huge growth over the last two decades of environmental industry sectors has meant that there are now many more, and more powerful, businesses with a direct commercial interest in strengthening environmental policy. For any discourse to succeed in shifting the goals of policy it must gather around it a "discourse coalition" of actors with a strong interest in its success (Hajer 1995). In the low-carbon energy and environmental sectors – allied to important elements of the environmental movement – green growth is developing just such a supporter base (see e.g. Aldersgate Group 2012).

Yet at the same time there are perhaps even more powerful interests which are threatened by it. For the fossil-fuel, extractive, and resource-intensive industries, stronger environmental policy means higher costs and greater operational restrictions. Even if such policy can generate growth overall, it does not mean that every sector will benefit from it. On the contrary, a green-growth path in practice means a transformation in the structure of national economies which is bound to create losers as well as winners.

For this reason, the political battles over green growth will not take place simply at the level of discourse. It is clear that high-carbon and resource-intensive industries will seek to ensure that the concept of green growth does not make intellectual or political headway. But even more vociferously they will oppose the particular environmental policies which are put forward to stimulate it. In practice it will therefore often be in these specific disputes that the more general battle will be manifest. What is not yet clear is whether articulation of the green growth discourse will help shift the balance of such policy debates towards the environmental side, providing an economic counterweight to the familiar claims of jobs, competitiveness, and growth on which the "anti-environmental" case will rest. That is its purpose. But whether it succeeds will ultimately be a consequence of the balance of economic and political forces which are marshaled on either side. The impact of the green-growth discourse, that is, will rest less on its economic theory than on its political economy.

This will not be a straightforward struggle. For the concept of green growth is not only opposed by high-carbon and resource-intensive industries. A number of civil society organizations have also come out against it. Some development groups see green growth simply as a rubric under which existing patterns of capitalist

development are maintained with a green veneer, a means by which developed countries prevent developing ones from escaping poverty (Hoffman 2011; Lander 2011).
Meanwhile many environmentalists continue to reject the claim that continuous
exponential economic growth can ever be compatible with long-run environmental
sustainability (Jackson 2009).

In doing so they offer a sharp reminder that in the end the validity of green
growth as a concept will depend not on whether, by creating demand for environmental products, correcting market failures, and stimulating innovation, economic
growth can be greener over the short to medium term. Ultimately the question will
be whether such improvements in the productivity of environmental use can be
sustained indefinitely. For 9 billion people aspiring to developed-world living standards, facing already severe pressures on planetary resources, is there any rate of
global economic growth *in practice* which will nevertheless allow the environment
to be properly sustained over the long term? For how long can global economic output be progressively "dematerialized" at the required rate (Jackson 2009; Hepburn
and Bowen 2013)? With respect to this question, green growth could yet prove to be
a savior, or an illusion.

Notes

1   See, e.g., www.globalgreengrowthforum.com and http://greengrowthleaders.org.
2   There has been considerable debate about whether some kinds of environmental improvement, par-
    ticularly in local pollutants, are a "normal" outcome of economic growth. For a survey of this
    "environmental Kuznets curve" literature, see Ekins 2000.
3   It is not the only rival discourse. In the academic literature the concept of "ecological modernization"
    was developed in the 1990s to provide a more theoretical framework for the same basic set of ideas
    as applied to developed economies. For a survey see Mol and Spargaaren 2000.

References

Aghion, Philippe and Peter Howitt. 1992. "A Model of Growth through Creative Destruc-
    tion." *Econometrica*, 60(2): 323–351.
Aldersgate Group. 2012. "Green Growth," http://www.aldersgategroup.org.uk/themes/green-
    growth (accessed October 6, 2012).
Ambec, Stefan, Mark A. Cohen, Stuart Elgie, and Paul Lanoie. 2011. *The Porter Hypothesis at
    20: Can Environmental Regulation Enhance Innovation and Competitiveness?* Washington,
    DC: Resources for the Future.
Barbier, Edward B. 2010a. *A Global Green New Deal: Rethinking the Economic Recovery.*
    Cambridge: Cambridge University Press.
Barbier, Edward B. 2010b. "Green Stimulus, Green Recovery and Global Imbalances." *World
    Economics*, 11(2): 1–27.
Bosquet, Benoit. 2000. "Environmental Tax Reform: Does It Work? A Survey of the Empirical
    Evidence." *Ecological Economics*, 34: 19–32.
Bowen, Alex and Samuel Fankhauser. 2011. "The Green Growth Narrative: Paradigm Shift
    or Just Spin?" *Global Environmental Change*, 21: 1157–1159.
Bowen, Alex, Samuel Fankhauser, Nicholas Stern, and Dimitri Zenghelis. 2009. "An Outline
    of the Case for a 'Green' Stimulus." Policy Brief. London: Grantham Research Institute on
    Climate Change and the Environment, London School of Economics.

Brock, William A. and M. Scott Taylor. 2005. "Economic Growth and the Environment: A Review of Theory and Empirics." In *Handbook of Economic Growth*, vol. 1A, ed. Philippe Aghion and Stephen N. Durlauf, 1749–1821. Amsterdam: Elsevier.

China Greentech Initiative. 2012. "The National People's Congress Approved the 12th Five-Year Plan," March 14, http://www.china-greentech.com/node/1666 (accessed October 6, 2012).

Climate and Development Knowledge Network. 2012. "Green Growth: Debates and Resources," http://cdkn.org/2011/07/greengrowth/ (accessed October 6, 2012).

Department for Business, Innovation and Skills. 2012. *Low Carbon Environmental Goods and Services (LCEGS) Report for 2010/11*. London: Department for Business, Innovation and Skills.

Dobbs, Richard, Jeremy Oppenheim, Fraser Thompson *et al.* 2011. *Resource Revolution: Meeting the World's Energy, Materials, Food and Water Needs*. London: McKinsey Global Institute.

Dresner, Simon. 2008. *The Principles of Sustainability*, 2nd edn. London: Earthscan.

Ekins, Paul. 2000. *Economic Growth and Environmental Sustainability*. London: Routledge.

Engel, Ditlev and Daniel M. Kammen. 2009. "Green Jobs and the Clean Energy Economy." Copenhagen: Copenhagen Climate Council.

Gillingham, K., Richard G. Newell, and Karen Palmer. 2009. "Energy Efficiency Economics and Policy." *Annual Review of Resource Economics*, 1: 597–619.

Global Green Growth Institute. 2012. *Annual Report 2011*. Seoul, Korea: Global Green Growth Institute, http://www.gggi.org/sites/www.gggi.org/files/attachment/20120625/2011_GGGI_Annual_Report_2.pdf (accessed October 6, 2012).

Government of France. 2011. "Cannes Summit Final Declaration," http://www.g20-g8.com/g8-g20/g20/english/for-the-press/news-releases/cannes-summit-final-declaration.1557.html (accessed October 6, 2012).

Government of Mexico. 2012. "G20 Leaders' Declaration," http://g20.org/images/stories/docs/g20/conclu/G20_Leaders_Declaration_2012_1.pdf (accessed October 6, 2012).

Graff Zivin, Joshua S. and Matthew J. Neidell. 2011. "The Impact of Pollution on Worker Productivity." Working Paper 17004. Cambridge, MA: National Bureau of Economic Research.

Green Economy Coalition. 2012. "Background Paper: Surveying the Green Economy Landscape Post-Rio+20," http://www.greeneconomycoalition.org/updates/background-paper-surveying-green-economy-landscape-post-rio20 (accessed October 6, 2012).

Green New Deal Group. 2008. *A Green New Deal*. London: New Economics Foundation.

Haas, Peter M. 2012. "The Political Economy of Ecology: Prospects for Transforming the World Economy at Rio+20." *Global Policy*, 3(1): 94–101.

Hajer, Maarten A. 1995. *The Politics of Environmental Discourse: Ecological Modernization and the Policy Process*. Oxford and New York: Oxford University Press.

Hallegatte, Stéphane, Geoffrey Heal, Marianne Fay, and David Treguer. 2011. "From Growth to Green Growth: A Framework." Policy Research Working Paper 5872. Washington, DC: World Bank.

Hepburn, Cameron and Alex Bowen. 2013. "Prosperity with Growth: Economic Growth, Climate Change and Environmental Limits." In *Handbook on Energy and Climate Change*, ed. R. Fouqet. Cheltenham: Edward Elgar.

Hoffman, Ulrich. 2011. "Some Reflections on Climate Change, Green Growth Illusions and Development Space." Discussion Paper 205. Geneva: United Nations Conference on Trade and Development.

Holdren, John and Paul Ehrlich. 1974. "Human Population and the Global Environment." *American Scientist*, 62 (May–June): 282–292.

Houser, Trevor, Shashank Mohan, and Robert Heilmayr. 2009. "A Green Global Recovery? Assessing US Economic Stimulus and the Prospects for International Coordination." Policy Brief 09-3. Washington, DC: Peterson Institute for International Economics.

Huberty, Mark, Huan Gao, and Juliana Mandell. 2011. *Shaping the Green Growth Economy*. Berkeley, CA: Green Growth Leaders, http://www.uncsd2012.org/content/docu ments/Shaping-the-Green-Growth-Economy_report.pdf (accessed October 6 2012).

Jackson, Tim. 2009. *Prosperity without Growth: Economics for a Finite Planet*. London: Earthscan.

Jacobs, Michael. 1991. *The Green Economy: Environment, Sustainable Development and the Politics of the Future*. London: Pluto Press.

Jacobs, Michael. 1999. "Sustainable Development as a Contested Concept." In *Fairness and Futurity: Essays on Environmental Sustainability and Social Justice*, ed. Andrew Dobson, 21–45. Oxford: Oxford University Press.

Jenkins, Jesse, Michael Shellenberger, and Ted Nordhaus. 2011. *Energy Emergence, Rebound and Backfire as Emergent Phenomena*. Oakland, CA: Breakthrough Institute.

Lander, Edgardo. 2011. *The Green Economy: The Wolf in Sheep's Clothing*. Amsterdam: Transnational Institute.

Lanoie, Paul, Michel Patry, and Richard Lajeunesse. 2008. "Environmental Regulation and Productivity: Testing the Porter Hypothesis." *Journal of Productivity Analysis*, 30(2): 121–128.

Mazzucato, Mariana. 2011. *The Entrepreneurial State*. London: Demos.

Mol, Arthur P.J. and Gert Spargaaren. 2000. "Ecological Modernization Theory in Debate: A Review." *Environmental Politics*, 9(1): 17–49.

Moore, Stephen. 2011. "Obama's 'Investment' Charade." *Wall Street Journal*, January 27.

Mulligan, Casey. 2010. "Simple Analytics and Empirics of the Government Spending Multi- plier and other 'Keynesian' Paradoxes." Working Paper 15800. Washington, DC: National Bureau of Economic Research.

Nordhaus, William. 1974. "Resources as a Constraint on Growth." *American Economic Review*, 64: 22–26.

Nordhaus, William. 2007. "A Review of the Stern Review on the Economics of Climate Change." *Journal of Economic Literature*, 45(3): 686–702.

OECD (Organisation for Economic Co-operation and Development). 2010. *Interim Report of the Green Growth Strategy: Implementing our Commitment for a Sustainable Future*. Paris: OECD.

OECD (Organisation for Economic Co-operation and Development). 2011. *Towards Green Growth*. Paris: OECD.

OECD (Organisation for Economic Co-operation and Development). 2012a. "OECD Work on Green Growth," http://www.oecd.org/greengrowth/oecdworkongreengrowth.htm (accessed October 6, 2012).

OECD (Organisation for Economic Co-operation and Development). 2012b. *Transport Out- look 2012: Seamless Transport for Greener Growth*. Paris: OECD.

Perez, Carlota. 2010. "Technological Revolutions and Techno-Economic Paradigms." *Cam- bridge Journal of Economics*, 34: 185–202.

Pollin, R., Heidi Garrett-Peltier, James Heintz, and Helen Scharber. 2008. *Green Recovery: A Program to Create Good Jobs and Start Building a Low Carbon Economy*. Washington, DC: Center for American Progress; and Amherst, MA: PERI, University of Massachusetts.

Porter, Michael and Claas van der Linde. 1995. "Toward a New Conception of the Environment–Competitiveness Relationship." *Journal of Economics Perspectives*, 9(4): 97–118.

Robins, Nick, Robert Clover, and D. Saravanan. 2010. *Delivering the Green Stimulus*. London: HSBC Global Research.

Rockström, J., W. Steffen, K. Noone *et al.* 2009. "A Safe Operating Space for Humanity." *Nature*, 461(24): 472–475.

Roland-Holst, David. 2008. "Energy Efficiency, Innovation and Job Creation in California." Research Paper. Berkeley, CA: Center for Energy, Resources and Environmental Sustain- ability.

Rozenborg, J., Stéphane Hallegatte, Adrien Vogt-Schilb *et al.* 2010. "Climate Policies as a Hedge against the Uncertainty on Future Oil Supply." *Climate Change Letters*, 101(3): 663–669.

Smulders, Sjak. 1999. "Endogenous Growth Theory and the Environment." In *Handbook of Environmental and Resource Economics*, ed. J.C.J.M. van den Berg, 610–621. Cheltenham: Edward Elgar.

Solow, R.M. 1974. "The Economics of Resources or the Resources of Economics." *American Economic Review*, 64: 1–14.

Spencer, Thomas, Lucas Chancel, and Emmanuel Guérin. 2012. "Exiting the Crisis in the Right Direction: A Sustainable and Shared Prosperity Plan for Europe." Paris: IDDRI (Institute for Sustainable Development and International Relations).

Stern, Nicholas. 2007. *The Economics of Climate Change: The Stern Review*. Cambridge: Cambridge University Press.

Stern, Nicholas and James Rydge. 2012. "The New Energy-Industrial Revolution and an International Agreement on Climate Change." *Economics of Energy and Environmental Policy*, 1: 1–19.

UNCSD (United Nations Conference on Sustainable Development). 2012. "Rio+20: United Nations Conference on Sustainable Development," http://www.uncsd2012.org/rio20/ (accessed October 6, 2012).

UNEP (United Nations Environment Programme). 2011. *Towards a Green Economy: Pathways to Sustainable Development and Poverty Eradication*. Nairobi: UNEP.

US Congressional Budget Office. 2009. "Estimated Impact of the American Recovery and Reinvestment Act on Employment and Economic Output as of September 2009." Washington, DC: US Congressional Budget Office.

Winston, Clifford. 2006. *Government Failure versus Market Failure: Microeconomics Policy Research and Government Performance*. Washington, DC: Brookings Institution Press.

World Bank. 2012a. "MDBs: Delivering on the Promise of Sustainable Development," http : // web.worldbank.org / WBSITE / EXTERNAL / TOPICS / EXTSDNET / 0,,menuPK: 64885113∼pagePK:7278667∼piPK:64911824∼theSitePK:5929282∼contentMDK: 23222131,00.html (accessed October 6, 2012).

World Bank. 2012b. *Inclusive Green Growth: The Pathway to Sustainable Development*. Washington, DC: World Bank.

World Bank. 2012c. "Green Growth Knowledge Platform," http://www.greengrowth knowledge.org/Pages/GGKPHome.aspx (accessed October 6, 2012)

World Commission on Environment and Development. 1987. *Our Common Future*. Oxford: Oxford University Press.

Zenghelis, Dimitri. 2012. "A Strategy for Restoring Confidence and Economic Growth through Green Investment and Innovation." Policy Brief. London: Grantham Research Institute on Climate Change and the Environment, London School of Economics.

Chapter 13

Sustainable Consumption

Doris Fuchs

Introduction

A sustainable consumption perspective on global climate and environmental policy is comprehensive, multifaceted, and, as this chapter will argue, fundamentally challenging for science and politics as well as for society in general. It induces us to critically reflect on core principles, on which our life is based. Indeed, a sustainable consumption perspective, more clearly than any other sustainability approach, forces us to ask whether we are willing and able to fundamentally change the politico-economic pillars of our societies in order to be able to reach a form of development that is sustainable.

Common to all sustainable consumption perspectives is their consideration of social and environmental burdens and necessary political reforms and interventions through the lens of the associated consumption activities and practices. Importantly, consumption activities and practices, in this context, refer not only to consumer choices, but also to the economic, political, social, and cultural contexts of these consumer choices. In other words, scholars and politicians applying a sustainable consumption perspective do not attribute all blame and responsibility for unsustainable consumption practices to the consumer.

Two principal approaches can be distinguished: "weak sustainable consumption" (WSC) and "strong sustainable consumption" (SSC) (Fuchs and Lorek 2005; Lorek and Fuchs 2013). The former supports the notion that sustainable consumption can be achieved via an increase in the efficiency of products, production processes, services, and the provision of these services alone. The latter focuses on the pursuit of fundamental shifts in consumption patterns and reductions[1] in consumption levels (mostly in industrialized countries).[2] It draws its core insights from research on the limited capacity of the Earth's ecosystems and the empirical evidence of rising

The Handbook of Global Climate and Environment Policy, First Edition. Edited by Robert Falkner.
© 2013 John Wiley & Sons, Ltd. Published 2013 by John Wiley & Sons, Ltd.

consumption levels outpacing any gains in efficiency achieved, over the past decades. Although all research inevitably contains a normative basis, the SSC perspective in particular is explicitly normative in orientation. Scholars working in this field link their research to questions of the "good life," of social justice and democracy, as well as "overconsumption" and "misconsumption" (Princen 2002). The present chapter will make the argument that such a normative focus on SSC in governance and research is first among the pivotal contributions a sustainable consumption perspective can make to global climate and environmental policy.

The chapter proceeds as follows. The next section provides an overview of sustainable consumption governance and research to date, discussing in particular the questions of moving beyond the efficiency focus as well as potential normative ambitions of sustainable consumption governance and research. Applying an SSC approach, the following section develops new insights on global climate policy, identifying opportunities as well as barriers to improvements in its reach and effectiveness. Finally, the concluding section summarizes the chapter with an outlook on research and governance needs, in particular those arising from an SSC approach to climate policy.

Global Sustainable Consumption Governance to Date

On the international political level, the Oslo Roundtable defined sustainable consumption as

> the use of services and related products, which respond to basic needs and bring a better quality of life while minimizing the use of natural resources and toxic materials as well as the emissions of waste and pollutants over the life cycle of the service or product so as not to jeopardize the needs of further generations (Ministry of Environment, Norway 1994).

However, sustainable consumption has come to be associated with all kinds of meanings in governance and research activities. It can be linked to consumer health, consumer safety, quality of life, resource efficiency, waste reduction, or life-cycle thinking (Mont and Plepys 2008). Such a wide variation in use is familiar from many other concepts, including sustainable development. The variation can at least partly be explained by discursive, political contests over the meaning of the term and its implications and by the range of disciplines contributing to sustainable consumption research. Thus, it is necessary to always take a close look at the implied meaning of sustainable consumption when it shows up in debates or publications.

The above definition of sustainable consumption was one of the first results of intergovernmental and research processes, building on the arrival and establishment of sustainable consumption as a topic on the global political agenda.[3] Subsequent efforts to frame the term in political processes led to a narrowing of the definition to questions of efficiency improvements and technological innovation, that is, to WSC (Fuchs and Lorek 2005). Consumption levels and patterns were taken as given, while the aim became satisfying them with fewer resources. In addition to limiting the focus on resource efficiency, the political debate emphasized consumer rights and sovereignty. Politically, this is an attractive strategy. After all, the notion of consumer sovereignty implies the existence of consumer ability to make purchasing decisions free from structural constraints. Consumer rights and sovereignty, however, can also be used to imply consumer responsibility. Thus, the concept is often used to argue that it is consumers who should shift the market towards sustainability and that interventions should focus on urging consumers to improve the sustainability characteristics of their consumption choices as well as on enabling consumers to do so.

In this vein, countless activities by international governmental organizations (IGOs) such as the Commission for Sustainable Development (CSD), the Organisation for Economic Co-operation and Development (OECD), the European Union (EU), and the United Nations Environmental Programme (UNEP), as well as individual governments, have aimed to foster research and reforms on improvements in the efficiency of consumption, innumerable reports on sustainable consumption have been sponsored and published by IGOs, and numerous meetings have been held (Fuchs and Lorek 2005; Berg 2011). The work commissioned and carried out by these actors contributed to the increasing availability of information on consumption trends, indicators, and policies. Moreover, it raised awareness for the relevance of sustainable consumption on the governance agenda. Yet, the core focus of the activities was limited to resource efficiency and technological solutions to environmental problems caused by consumption, with a particular interest in innovations for business (Fuchs and Lorek 2005). With the exception of *Consumption Opportunities*[4] published by UNEP in 2001, there was no questioning of the larger societal contexts and implications of Western consumption levels and patterns. On the contrary, the head of UNEP DTIE (Division of Technology, Industry and Economics) stressed that "sustainable consumption is not about consuming less, it is about consuming differently, consuming efficiently, and having an improved quality of life" (UNEP/CDG 2000: 12). The fundamental, underlying notion was the need for improvements in the sustainability of consumption in parallel to continuing economic growth. Moreover, the emphasis was on consumer sovereignty, with a UNEP DTIE report stressing the "rights of free consumers," for instance (Bentley and de Leeuw 2000). Here, the official political agenda closely mirrored business perspectives. A report issued by the World Business Council for Sustainable Development (WBSCD 2002) for the World Summit on Sustainable Development (WSSD) explicitly attributed the key role in shaping markets to consumers.[5]

At the WSSD in Johannesburg in 2002, no significant progress on global sustainable consumption governance was made. Negotiators agreed on a call for governments to "encourage and promote the development of a 10-year framework of programmes in support of regional and national initiatives to accelerate the shift towards sustainable consumption and production" (United Nations 2002: 7). Thus, the political aim was only broadly defined, without any specificity or binding elements. Potential conflicts between the sustainability of consumption and the continued pursuit of consumption-driven growth were not acknowledged. In the subsequent Marrakesh process,[6] the major actors stressed again the importance of de-coupling economic growth and environmental degradation through improvements in the efficiency of resource use. In addition, a series of regional meetings were to serve as platforms for the exchange of knowledge. For this, seven task forces, representing voluntary initiatives of groups of interested countries, were created.[7] However, by the year of the

Rio+20 summit, little tangible progress towards a redirection of global sustainable consumption governance had been achieved (Stakeholder Forum 2012).

At the same time, sustainable governance innovations are developing bottom-up, with increasing numbers of local currencies, community exchanges, food cooperatives, or public gardening initiatives developing (Eberle *et al.* 2006; J. Barber 2007; Lebel and Lorek 2008; Seyfang 2009). While such local initiatives may be encouraging, the crucial challenge is to move their ideas from the micro- to the meso-level to allow them to have a bigger impact.

Sustainable Consumption Research to Date

The following section can only provide a glimpse of the large amount of sustainable consumption research existing today. In this endeavor, it aims to identify core themes. The discussion starts with a brief overview on research on impacts and determinants of consumption as well as intervention strategies. It then points out one of the core controversies in the field, which relates to underlying conceptions of the consumer. Finally, the discussion delineates core challenges to strong sustainable consumption that can be identified on the basis of a critical reading of the history of consumption governance, today.

Impacts, Determinants, and Interventions

Researchers have tried to identify priority areas for sustainable consumption research and governance (Mortensen 2006; Kaenzig and Jolliet 2007). Some of these inquiries have looked at consumption clusters, identifying especially food, mobility, and housing as relevant (Lorek and Spangenberg 2001; EEA 2010). Taking such an approach one step further, other analyses identified, for example, agricultural processes and in particular meat products as particularly relevant in the food sector (Tukker *et al.* 2006), and heating as well as cooling (both in terms of air conditioning, where relevant, and in terms of refrigerators and freezers in the kitchen) as pivotal areas in the housing sector (Bürger 2011). Others have looked at key intervention points with both a high ecological impact and a substantial potential for the steering of consumption (Bilharz 2008). Moreover, scholars have tried to identify trends and model corresponding scenarios. Such analyses are important, because tomorrow's relevant impacts may still be "below the environmental radar" today (Røpke 2011).

Already some of the early work on sustainable consumption focused on the question of its determinants (Røpke 1999). Today, we know that socio-demographic and socio-economic characteristics such as age, income, gender, and education, norms and values, as well as psychological aspects such as perceptions of self-control and constructions of identity have an impact on the willingness to buy energy-efficient products, for instance (Vermeir and Verbeke 2006; Krömker and Dehmel 2011; Luchs and Mooradian 2012). Based on these various factors, scholars have identified consumer groups or lifestyles representing differences in (at least self-identified – see below) consumption levels and patterns (Baiocchi *et al.* 2010). Criticizing the individual-focused approaches, other scholars have shown the consumer environment to be an important factor (Shove 2003; Gram-Hanssen 2010). Besides obvious aspects such as the availability of environmentally or socially superior products in

markets and the provision of relevant information, changes in communication technologies, global finance and trade, and demographics (and the interactions between
them), which induce shifts in job situations, gender roles, and time constraints, in
turn, exert an influence on the sustainability of consumption long before the consumer ever makes a choice (Fuchs and Lorek 2001; Røpke and Godskesen 2007).
Here, the embeddedness of seemingly individual characteristics such as worldviews
in societal structures and practices becomes clear (Seyfang 2007).

With respect to interventions, scholars have investigated the effectiveness of instruments as well as potential roles of various actors. Thus, they have inquired into
the use of command-and-control regulation such as standards and prohibitions,
market-based/economic instruments such as environmental taxes or emission trading schemes, or information-based instruments such as eco-labels or information
campaigns (Daugbjerg and Sønderskov 2011; Wolff and Schönherr 2011). With
respect to the relevant actors, studies have evaluated the role of governmental actors
in shaping the sustainability of consumption via the exercise of public authority,
of business actors via self-regulation, public–private or private–private partnerships,
and of civil society actors via information campaigns or the mobilization of consumers, for instance. Importantly, recent studies have shown regulatory approaches
entailing enforcement and sanctioning mechanisms to be more effective than informational ones (Rehfeld *et al.* 2007). This finding runs counter to the political inclination
to rely on informational instruments due to their lower political costs. Secondly and
related to that point, the effectiveness of private governance approaches is highly controversial. Business-led private standards, in particular, can frequently be shown to
perhaps improve some sustainability deficits of products and productions processes,
but not to address broader sustainability challenges (Fuchs and Boll 2012).

Controversy: The More or Less Sovereign Consumer

One of the most fundamental controversies in the sustainable consumption debate is
related to the understanding of the consumer as such. While some researchers see the
consumer as a *homo economicus*, that is, a rational individual making decisions on
the basis of cost–benefit calculations, others see the consumer as a *homo sociologicus*,
that is, as a norm-driven individual basing consumption choices on social influences
or personal values. Both groups, however, would consider consumers to be able
to make relatively autonomous and flexible consumption choices. A fundamentally
different perspective on consumers sees them instead as locked into consumption
practices due to their habits and routines as well as the structural constraints resulting from their technological, socio-economic, political, and cultural environments
(Røpke 1999; Sanne 2002).

This controversy extends to the question of consumer sovereignty. Scholars
emphasizing the influence of structural constraints on consumers see little room for
consumer sovereignty. In this context, the question of the "distancing" and "shading" of the effects of consumption decisions represents something bigger than merely
the question of information availability (Princen 2002, 2010). Likewise, power asym-
metries between global retail corporations and their political and media presence,
on the one side, and individual consumers, on the other, serve to highlight challenges to the sovereignty of consumers in shaping markets (Fuchs and Lorek 2001;

Fuchs 2007). Moreover, the enormous role (interdependent) practices play in shaping everyday consumption reduces the space for intentional, well-thought-out consumption decisions, which the notion of consumer sovereignty presupposes (Shove 2003). Knowledge–action, attitude–behavior, or behavior–impact gaps show that consumers may choose environmentally or socially inferior products or practices, for instance, despite better knowledge due to the lock-in of consumption decisions or conflicting messages (Lebel *et al.* 2006; Markkula and Moisander 2012; Moraes *et al.* 2012). Even consumers making conscious sustainability choices frequently fail to significantly reduce their overall environmental footprint, because of the large share of ordinary, that is, everyday, habitual, hardly noticed, consumption (Csutora 2012). In sum, critical scholars have long challenged the politically attractive notion of "consumer sovereignty" and the associated "individualization of responsibility" on the grounds of their failure to consider the economic, political, and societal structural constraints within which consumption "decisions" occur (Maniates 2002).

The implications of the above controversies are even clearer when it comes to the recommendations for political interventions that are studied or derived. The rational, sovereign consumer may be convinced by economic incentives to make sustainably superior consumption choices, as well as perhaps through the provision of information appealing to extrinsic values (such as the cost savings to be achieved with an energy-efficient appliance). The norm-driven sovereign consumer may be persuaded through the provision of necessary information appealing to intrinsic values, inducing value change or the creation and promotion of enlightened role models (Munasinghe 2010). Interestingly, less rather than more information may be a promising strategy here, for instance to protect consumers from too many labels with little meaning (Möller 2004). For the locked-in consumer, regulation generating or at least facilitating changes in the living and working environment would appear to be necessary, accompanied perhaps by information or economic incentives inducing a "rethinking" of what would otherwise have been ordinary or routine consumption. Today, we know that all these intervention types have a role to play (Heiskanen and Lovio 2010; Warde 2011). Moreover, for the rational and locked-in consumers, and given the difficulty and long-term nature of value change even for the norm-driven consumer, choice editing as well as restrictions on advertising have to be considered as important political strategies, as long as so many signals drive consumption patterns and levels in an unsustainable direction (Brohmann and Eberle 2006; Yates 2008; Alexander *et al.* 2011; Dhar and Baylis 2011). While this clearly is an interventionist strategy for democratic societies, such interventions have proven acceptable in the past, when actions hurt others (consider restrictions on advertising for cigarettes). The question thus has to be whether the overconsumption of the world's natural resources and over-pollution of its sinks by a small share of the global population does not constitute a similar imposition of burden on others.

The Challenges of Strong Sustainable Consumption

Another related controversy in the sustainable consumption literature and debate is the nature of change required for an effective pursuit of sustainable consumption and the ease or difficulty with which such change can be achieved. Early on, some scholars argued that evolution has equipped us with a disposition based on instinctive and long-learned behavior patterns, which may well include an inclination for accumulation (McDougall 1923). Other scholars would reject such a view. They probably would be less critical, however, of an argument that humans as social beings have an inclination to position themselves in social networks and hierarchies. Such a positioning can be pursued with a range of signifiers, of course. As a long list of thinkers have pointed out, however, we have increasingly learned to use material goods as signifiers in our consumer cultures. Thus, some scholars have called for turning environmentally and socially superior products into signifiers, that is, make the hybrid Smart sexy rather than the SUV. However, this strategy may run into difficulties, if ordinary consumption really determines the major share of our environmental footprint. Moreover, it is potentially of limited effectiveness in an economic system based on mass consumption, that is, if it is not only the individual consumption choice but also the quantity and frequency of consumption choices contributing to social position.

The extent to which the social need for positioning or symbols shapes the sustainability of consumption also has an impact on the potential to persuade consumers to consume less. For decades now, scholars have pointed out that an increasing number of individuals supposedly are intentionally and explicitly choosing "downshifting" (Schor 1998), voluntary simplicity (Elgin 1993), Lifestyles of Health and Sustainability (LOHAS) (Ray 2000), or sacrifice[8] (Maniates and Meyer 2010). Such choices correspond to the critiques of consumer culture as a source of unhappiness, depression, loneliness, and stress rather than happiness, fulfillment, and lasting contentment, which in turn appear to be supported by data showing that increases in income and belongings after a certain level do not lead to similar increases in happiness. But if consumption is a major part of social positioning (Bourdieu 1984; Veblen 1994 [1899]; Howarth 1996; Baudrillard 1998), then strategies of dropping out of the game of "competitive upscaling" (Hirsch 1977) may carry a social cost (Douglas and Isherwood 1996). Needs may be universal (Maslow 1954; Max-Neef 1991), but satisfiers are culturally specific and in the case of today's Western societies frequently of a materialist nature, as pointed out above. In consequence, there may be a challenge to the supposed "double dividend" of downshifting in terms of a happier and at the same time more sustainable life, or rather its existence would require a societal renegotiation of markers of status (Jackson 2005).[9] After all, the empirical data also show that happiness positively correlates with income *within* countries. Similarly, studies have shown the difficulty of "locking in" green consumption patterns given evolved cognitive dispositions (Buenstorf and Cordes 2007). Others have found that even ethical consumers are influenced by the role of "pleasure" in their shopping experiences (Johnston and Szabo 2011). Thus, highly optimistic assumptions that the alternative lifestyles described above can easily be transformed into sufficiently broad movements in a society pervasively characterized by hyperconsumption (B. Barber 2007) and a predisposition to look for a "good deal" (Ruppel Shell 2009) should be treated with some caution. Some individuals may well choose a life with less material consumption, more time, deeper social relationships, and be happier, healthier, and fulfilled. But the jury is still out on the question of the ease with which one may persuade a larger share of the population to not make use of a share of their consumption opportunities.

This leads us to the questions of the nature of change and the depth of interventions needed. Considerable improvements in the resource efficiency of consumption clearly could be achieved with a stringent steering of producers and consumers towards the most efficient available technologies and products as well as investments in further promising technological innovation. As pointed out above, however, there are strong reasons to believe that improvements in efficiency, whether by a factor of 4, 10, or any other number, are not going to suffice. In order to move towards SSC, however, radical changes would be needed. In fact, one has to ask to what extent SSC is possible in capitalist systems. Clearly, it would appear impossible in capitalist systems endlessly pursuing growth, in which increases in well-being are understood and measured in terms of levels of material consumption, and in which growth is understood to depend on mass consumption and efficiency rather than sufficiency (Princen 2005; Jackson 2009; Seyfang 2009; Lorek and Fuchs 2013).

Sustainable Consumption and Global Climate Policy

Applying a sustainable consumption perspective, the following section develops new insights on global climate and environmental policy, identifying opportunities as well as barriers to improvements in its reach and effectiveness. Due to space constraints, the discussion will concentrate only on the consumption clusters food and housing (in terms of household energy use).

Food is a major contributor to global greenhouse gas (GHG) emissions related to consumption. In the UK, for instance, food is responsible for 20% of greenhouse gas emissions (Audsley *et al.* 2009). In Finland, the food chain has been found to contribute 14% of the country's GHG emissions and agricultural processes 69% of that (three-quarters of that from raising livestock, methane emissions from beef and dairy cattle) (Virtanen *et al.* 2011). Among the food-related activities, therefore, agricultural production processes contribute by far the largest share to GHG emissions, with meat production being responsible for a large share of this. The indirect effects from food production and processing (both in the consuming region as well as in other parts of the world) are especially important, and include emissions from livestock, agriculture, and industry on water, soil, and air, waste, transport, and the overuse of fish resources (Mortensen 2006).

In terms of policy intervention, the food sector arrived relatively late on the scene as a target for measures aimed at improving the sustainability of consumption (beyond the issue of food safety) and especially GHG emissions (EUPOPP 2011). Earlier, a focus on reducing transport necessities had dominated the debate both in science and politics and especially in civil society due to the greater awareness of the carbon emissions associated with transport (Wiedmann and Minx 2008; Hillier *et al.* 2009). Thus, NGOs encouraged consumers to buy local or regional foods. From a research perspective, the findings on this aspect were more ambivalent, as local greenhouse production may be much more energy-intensive than organic production in more distant places (this does not include the transport of food by plane). Thus, while the reduction in overall transport needs clearly is necessary, the decision for the appropriate sourcing location of individual products needs to consider the specific context. Moreover, scholars have increasingly revealed the many different aspects contributing to the carbon footprint of food products, which go far beyond the question of transport. Accordingly, the carbon labeling of groceries has come to be discussed as a potential policy tool (Vanclay *et al.* 2011). The difficulties and complexities associated with determining the carbon footprint of a food product should not be underestimated (Mózner Vetöné 2011). Nevertheless, this strategy appears promising as a means to better inform consumers about the climate change impacts of their consumption choices.

At the same time, a number of food retailers have started addressing their own carbon footprint. The activities involved range from improving the fuel efficiency of the transport fleet, to increasing the energy efficiency of store lighting as well as refrigeration and cooling in stores, to experiments with carbon footprinting of selected products in a few very advanced cases. Generally, these activities show some awareness but fall far short of addressing the most important dimensions of the climate impact of food (Fuchs and Boll 2012). Improvements in the fuel efficiency of the transport fleet are necessary, but so is the reduction in miles traveled overall, especially when it comes to miles traveled by plane. This is not necessarily in the interest of a highly concentrated food retail industry with supply chains reaching all over the globe.

Another important strategy would be the reduction of all forms of food waste, with organic waste ranging from products rejected because they lack "standard" size, shape, or color, to agricultural produce rotting before reaching markets, produce not sold in stores and groceries not used in households, as well as the additional packaging waste (Stuart 2009). Fortunately, food waste has increasingly received public and political attention. Strategies have not moved beyond awareness-raising, however, and more effective political intervention has yet to be designed and adopted.

From an SSC perspective, moreover, a reduction in the consumed quantities of certain foods would appear unavoidable. Current levels of meat as well as dairy product consumption in Western industrialized societies clearly are not sustainable and constitute major contributors to climate change (Zhu *et al.* 2006; EUPOPP 2011). This is not a change that can be easily achieved or be a politically popular objective. While the carbon labeling of food products may contribute to a reduction in meat consumption among some consumers, it would appear that more interventionist measures ranging from economic disincentives to choice editing would be needed to obtain real change (Danish Ministry of the Environment 2012). Here, public catering can play an initiating role (Wahlen *et al.* 2012). Public canteens in some cities have started to have a veggie day, that is, not offer the choice of a meat dish one day a week. Unfortunately, such a strategy can only be a start, from a climate change perspective (EUPOPP 2011).

Household energy use contributes 25% to overall GHG emissions in the EU (EEA 2011). More than half of this is indirect emissions result from heating and electricity use (EEA 2011). Globally, the impact of heating (or cooling) is highly contextual and depends on building and construction characteristics, temperatures, and energy sources. Cooling devices are major sources of energy used in the house and the primary contributors to climate change among household appliances (Bürger 2011). Communications technology currently contributes only a comparatively small share to household energy use, albeit with a strongly increasing trend (Bürger 2011). The latter is due to the ever-larger quantity of relevant appliances in households, but more importantly also to the ever-larger screen sizes of televisions, in particular.

Potential intervention strategies reach from mandatory performance standards and subsidies for energy efficient buildings, to the provision of information on optimal heating strategies, in the case of heating, for example (EUPOPP 2011). Similarly, reductions in energy used by household appliances may be achieved through subsidies for the most efficient appliances, especially when combined with relevant information, such as effective labels, and information on the energy use of existing appliances in households and on the availability and potential cost savings of more efficient appliances (Deutsch 2010; Brohmann *et al.* 2011; EUPOPP 2011; Heinzle 2012). Moreover, shifting the times of energy use may help to reduce the overall capacity needed in the system (Gölz 2011; Mert 2011).

Many instruments potentially applicable to the question of household energy consumption address the different uses of energy in general. Thus, households may be induced to reduce their energy consumption through progressive tariffs and/or energy taxes, for instance. Moreover, raising the share of renewable energy sourced in the system clearly contributes to reducing the carbon footprint of the relevant households, which can, in turn, be fostered by subsidies, green quotas, and investments in associated necessary technological innovations such as energy storage systems. Moreover, one could think about not only offering progressive tariffs in terms of the quantity of energy consumed but also in terms of the overall capacity provided at any one point. In Italy, for instance, households traditionally have 3 kW contracts, which are cheaper, but also mean that they cannot use two energy-intensive household appliances simultaneously. This may seem a rather dramatic intervention, from the perspective of other countries, in particular American households. But it does lead to real savings in terms of the energy capacity a system needs to provide at any one point in time. For the future, engineers envision "smart homes" with "smart appliances" that are supposed to help individuals reduce their energy consumption (Gölz 2011; Wahlen *et al.* 2012).

From an SSC perspective, all these may be helpful steps. However, an SSC perspective would argue that the real issues lie with ever-increasing house sizes (in which ever-fewer people live, due to demographic and cultural changes), as well as ever-growing refrigerator and screen sizes, for example. These create new energy needs, which easily eat up efficiency gains achieved via the above strategies. From an SSC perspective, therefore, it would be desirable to at least start a public debate about how much heated (or cooled) space an individual may claim as his or hers. Similarly, an SSC perspective would inquire into the societal desirability of equipping houses with private pools. Finally, an SSC perspective would argue that we have to leave the majority of fossil-fuel resources in the ground rather than think of more efficient ways to use them, as the atmosphere is the limit we face, rather than the fossil fuels themselves (Edenhofer 2011; Princen 2011). Again, these are politically far from popular questions. Yet, they do highlight the real challenges an SSC perspective on energy use in households and climate change reveals.

In terms of broader and politically even less popular questions, an SSC perspective would question the stock-market-driven nature of many food and energy corporations and the divergence of the resulting objectives from public sustainability objectives. In this context, the enormous degree of capital concentration in both sectors would also raise questions. Clearly, from the perspective of our current economic system it is hard to see why it should be in the interest of food retailers to sell fewer meat products or in the interest of large energy companies to dramatically reduce the energy consumption of their customers. Changing the system to achieve a greater degree of convergence between public sustainability objectives and private investor/owner interests is a real challenge here.

Conclusion

This chapter has developed an SSC perspective on global climate and environmental policy. In pursuit of this objective, it has made the argument that such a perspective is particularly valuable in pointing out the fundamental challenges to our lifestyles and politico-economic system any serious attempts at achieving sustainable development will have to involve. An SSC perspective forces us to recognize the insufficiency of attempts to improve the resource efficiency of current Western consumption patterns. After decades of improvements in resource efficiency and the associated improvements in consumer information and options, we have strong evidence that resource savings achieved tend to be overcompensated by rising consumption levels.[10] Thus, improving the fuel efficiency of cars is of little use if ever more people drive ever-longer distances. Similarly, improving the heating characteristics of one's home does not help if we build larger and larger homes or use the money saved on the heating bill to fly to the Maldives for vacation. WSC governance may be helpful in contributing to improvements in the sustainability of consumption, but only if it is accompanied by SSC governance. Accordingly, addressing consumption levels and their drivers needs to be the core objective of sustainable consumption governance and research.

The resulting challenges start with the need for interventions in the "rights" of "sovereign consumers." While mandatory, more relevant and transparent information may help improve the sustainability of consumption choices (which in some contexts may well mean less information), information by itself is not going to be sufficient. As pointed out above, research has shown that effective intervention requires instruments with sanctioning and enforcement potential. In addition, economic incentives will have to be readjusted to the pursuit of the public good, and, in some cases, politically unpopular measures such as choice editing will have to be taken.

Challenges associated with a pursuit of SSC reach to the need to restrict "private" economic activities, which are being used to further consumption and which have come to be "natural" accompaniments of our daily lives, such as marketing. Such steps would allow reducing the "discursive polyphony" about consumption and sustainability (Markkula and Moisander 2012), and especially the continuous, contradictory messages of "buy more" and "use less" addressed at consumers.

More fundamentally, the challenges will involve a rethinking of current methods of measuring development and well-being as well as the definition of growth as a political and societal necessity. In this context, a rethinking of and better balance between the influences of different ideas and interests in democratic politics would appear crucial. More fundamentally, classic debates on what constitutes the good life and what contributes to human prosperity will have to move back into the focus of societal and political debate (Ehrenfeld 2008; Di Giulio et al. 2011). In terms of politico-economic structures, SSC will require the development of alternative systems that foster socially just well-being.

None of these challenges will be easily overcome. As the past decades have shown, strong barriers to change exist. Unwillingness to pursue such fundamental reforms dominates. Political actors shy away from addressing politically costly issues; business actors have developed their business models on the basis of current incentive structures; and even NGOs depend on support from consumers. Still, alliances between NGOs and research may be able to propel societal debate and reforms forward (Fuchs and Lorek 2005; Cohen 2006). They will have to ask in what kind of societies, that is, within which politico-economic frameworks, sustainable consumption can be achieved and inquire into strategies to allow consumption to better contribute to sustainability and well-being. On the basis of answers to these questions, we will have to develop new models of sustainable societies.

Notes

1 For a substantial share of the global population increasing consumption is necessary to reach a level that can be called sustainable. However, a fast-growing "global consumer class" exhibiting increasingly Westernized consumption patterns exists, especially in the BRICS countries (World Watch Institute 2012).
2 For an elaboration see Lorek 2010.
3 Agenda 21 called for the adoption of sustainable consumption patterns (United Nations 1993).
4 Significantly, UNEP failed to effectively pursue insights from this report in the international debate (Fuchs and Lorek 2005).
5 Similarly, the advertising sector report (European Association of Communications Agencies 2002) did not acknowledge any potential problems resulting from advertising's influence on consumption.
6 Named after the site of the first relevant international meeting after the Johannesburg summit.
7 Sustainable Lifestyles, Sustainable Products, Sustainable Procurement, Sustainable Building and Construction, Sustainable Education, Sustainable Tourism, and Cooperation with Africa.
8 See Hall (2010) for an excellent discussion of unintended, false, and hard sacrifices.
9 Findings that even green consumers relate their shopping to the experience of pleasure and insights gained by neurologists into the influence of consumption on brain activity may further provide indications of this challenge.
10 The so-called rebound effect partially explains this dynamic, but overall sources are more complex. Note, however, that the rebound effect is also more complicated than often assumed (Hertwich 2005).

References

Alexander, Jon, Tom Crompton, and Guy Shrubsole. 2011. *Think of Me as Evil? Opening the Ethical Debates in Advertising.* London: Public Interest Research Centre and WWT-UK.
Audsley, Eric, Matthew Brander, Julia Chatterton *et al.* 2009. "How Low Can We Go?" Unpublished manuscript, http://assets.wwf.org.uk/downloads/how_low_report_1.pdf (accessed March 19, 2012).
Baiocchi, Giovanni, Jan Minx, and Klaus Hubacek. 2010. "The Impact of Social Factors and Consumer Behavior on Carbon Dioxide Emissions in the United Kingdom." *Journal of Industrial Ecology*, 14(1): 50–72.
Barber, Benjamin R. 2007. *Consumed: How Markets Corrupt Children, Infantilize Adults, and Swallow Citizens Whole.* New York: W.W. Norton.
Barber, Jeffrey. 2007. "Mapping the Movement to Achieve Sustainable Production and Consumption in North America." *Journal of Cleaner Production*, 15(6): 499–512.

Baudrillard, Jean. 1998. *The Consumer Society*. London: Sage.

Bentley, Matthew and Bas de Leeuw. 2000. "Sustainable Consumption Indicators." Report for UNEP DTIE. Paris: United Nations.

Berg, Annukka. 2011. "Not Roadmaps but Toolboxes." *Journal of Consumer Policy*, 34(1): 9–23.

Bilharz, Michael. 2008. *"Key Points" nachhaltigen Konsums*. Marburg: Metropolis.

Bourdieu, Pierre. 1984. *Distinction: A Social Critique of the Judgement of Taste*. Cambridge, MA: Harvard University Press.

Brohmann, Bettina, Christian Dehmel, Doris Fuchs *et al.* 2011. "Prämienprogramme und progressive Stromtarife als Instrumente zur Förderung des nachhaltigen Elektrizitätskonsums privater Haushalte." In *Wesen und Wege nachhaltigen Konsums*, ed. Rico Defila, Antonietta Di Giulio, and Ruth Kaufmann-Hayoz, 443–453. Munich: oekom.

Brohmann, Bettina and Ulrike Eberle. 2006. "Nachhaltiger Konsum braucht gemeinsame Visionen und übergreifende Strategien." *Umweltpsychologie*, 10(1): 210–16.

Buenstorf, Guido and Christian Cordes. 2007. *Can Sustainable Consumption Be Learned?* Papers on Economics and Evolution 0706. Jena: Max Planck Institute of Economics.

Bürger, Veit. 2011. "Quantifizierung und Systematisierung der technischen und verhaltensbedingten Stromeinsparpotenziale der deutschen Privathaushalte." In *Die politische Förderung des Stromsparens in Privathaushalten*, ed. Doris Fuchs, 17–43. Berlin: Logos.

Cohen, Maurie. 2006. "Sustainable Consumption Research as Democratic Expertise." *Journal of Consumer Policy*, 29(1): 67–77.

Csutora, Maria. 2012. "One More Awareness Gap?" *Journal of Consumer Policy*, 35(1): 145–163.

Danish Ministry of the Environment. 2012. "What Can Policy Makers Do?" http://www.mst.dk/English/Sustainability/scp/green_nordic_retail/WhatCanPolicymakersDo/what_can_policymakers_do.htm#Overview_of_SCO:Instruments (accessed February 29, 2012).

Daugbjerg, Carsten and Kim Sønderskov. 2011. "Environmental Policy Performance Revisited." *Political Studies*, 60(2): 399–419, http://onlinelibrary.wiley.com/doi/10.1111/j.1467-9248.2011.00910.x/pdf (accessed October 20, 2012).

Deutsch, Matthias. 2010. "Life Cycle Cost Disclosure, Consumer Behavior, and Business Implications." *Journal of Industrial Ecology*, 14(1): 103–120.

Dhar, Tirtha and Kathy Baylis. 2011. "Fast-Food Consumption and the Ban on Advertising Targeting Children." *Journal of Marketing Studies*, 48(5): 799–813.

Di Giulio, Antonietta, Bettina Brohmann, Jens Clausen *et al.* 2011. "Bedürfnisse und Konsum – Ein Begriffsystem und dessen Bedeutung im Kontext von Nachhaltigkeit." In *Wesen und Wege nachhaltigen Konsums*, ed. Rico Defila, Antonietta Di Giulio, and Ruth Kaufmann-Hayoz, 47–71. Munich: oekom.

Douglas, Mary and Baron Isherwood. 1996. *The World of Goods: Towards an Anthropology of Consumption*, rev. edn. London: Routledge.

Eberle, Ulrike, Doris Hayn, Regine Rehaag, and Ulla Simshäuser, eds. 2006. *Ernährungswende: Eine Herausforderung für Politik, Unternehmen und Gesellschaft*. Munich: oekom.

Edenhofer, Ottmar. 2011. "Keys to Curbing Climate Change." In *Our Common Future Conference: Summary Report*, ed. B. Reinhold, F. Streiter, and M. Klasen, 84–87. Hanover: Volkswagen Stiftung.

EEA (European Environment Agency). 2010. "Consumption and the Environment." Unpublished manuscript, http://www.cca.europa.eu/soer/europe/consumption-and-environment (accessed March 5, 2012).

EEA (European Environment Agency). 2011. *End-User GHG Emissions from Energy: Relocation of Emissions from Energy Industries to End Users 2005–2009*. EEA Technical Report 19/2011. Copenhagen: European Environment Agency.

Ehrenfeld, John. 2008. *Sustainability by Design*. New Haven, CT: Yale University Press.

Elgin, Duane. 1993. *Voluntary Simplicity*. New York: William Morrow.

EUPOPP. 2011. "Policies to Promote Sustainable Consumption Patterns in Europe," http://www.eupopp.net (accessed March 5, 2012).

European Association of Communications Agencies. 2002. "Industry as a Partner for Sustainable Development." Brussels: EACA and WFA.

Fuchs, Doris. 2007. *Business Power in Global Governance*. Boulder, CO: Lynne Rienner.

Fuchs, Doris and Frederike Boll. 2012. "Emerging Private Voluntary Programs and Climate Change." In *Private Voluntary Programs in Global Climate Policy*, ed. Karsten Ronit, 143–178. New York: United Nations University Press.

Fuchs, Doris and Sylvia Lorek. 2001. *An Inquiry into the Impact of Globalization on the Potential for "Sustainable Consumption" in Households*. ProSus Report 2/01. Oslo: ProSus.

Fuchs, Doris and Sylvia Lorek. 2005. "Sustainable Consumption Governance. A History of Promises and Failures." *Journal of Cleaner Production*, 28(3): 261–288.

Gölz, Sebastian. 2011. "Smart Metering und Feedbacksysteme." In *Die politische Förderung des Stromsparens in Privathaushalten*, ed. Doris Fuchs, 192–213. Berlin: Logos.

Gram-Hanssen, Kirsten. 2010. "Standby Consumption in Households Analyzed with a Practice Theory Approach." *Journal of Industrial Ecology*, 14(1): 150–165.

Hall, Cheryl. 2010. "Freedom, Values, and Sacrifice." In *The Environmental Politics of Sacrifice*, ed. Michael Maniates and John Meyer, 61–86. Cambridge, MA: MIT Press.

Heinzle, Stefanie. 2012. "Disclosure of Energy Operating Cost Information." *Journal of Consumer Policy*, 35(1): 43–64.

Heiskanen, Eva and Raimo Lovio. 2010. "User–Producer Interaction in Housing Energy Innovations." *Journal of Industrial Ecology*, 14(1): 91–102.

Hertwich, Edgar. 2005. "Consumption and the Rebound Effect." *Journal of Industrial Ecology*, 9(1–2): 85–98.

Hillier, Jonathan, Cathy Hawes, Geoff Squire *et al.* 2009. "The Carbon Footprints of Food Crop Production." *International Journal of Agricultural Sustainability*, 7(2): 107–118.

Hirsch, Fred. 1977. *Social Limits to Growth*. Cambridge, MA: Harvard University Press.

Howarth, Richard. 1996. "Status Effects and Environmental Externalities." *Ecological Economics*, 16(1): 25–34.

Jackson, Tim. 2005. "Live Better by Consuming Less?" *Journal of Industrial Ecology*, 9(1–2): 19–36.

Jackson, Tim. 2009. *Prosperity without Growth*: London: Earthscan.

Johnston, Josée and Michelle Szabo. 2011. "Reflexivity and the Whole Foods Market Consumer." *Agriculture and Human Values*, 28(3): 303–319.

Kaenzig, Josef and Oliver Jolliet. 2007. "Prioritising Sustainable Consumption Patterns." *International Journal of Innovation and Sustainable Development*, 2(2): 140–154.

Krömker, Dörthe and Christian Dehmel. 2011. "Plug and Pull: Energy Saving in Private Households." In *Die politische Förderung des Stromsparens in Privathaushalten*, ed. Doris Fuchs, 138–175. Berlin: Logos.

Lebel, Louis, Doris Fuchs, Po Garden, and Dao Giap. 2006. *Linking Knowledge and Action for Sustainable Production and Consumption Systems*. USER Working Paper, WP-2006-09. Unit for Social and Environmental Research (USER), Chiang Mai University, Thailand.

Lebel, Louis and Sylvia Lorek. 2008. "Enabling Sustainable Production-Consumption Systems." *Annual Review of Environment and Resources*, 33(1): 241–275.

Lorek, Sylvia. 2010. *Towards Strong Sustainable Consumption Governance*. Saarbrücken: LAP Publishing.

Lorek, Sylvia and Doris Fuchs. 2013. "Strong Sustainable Consumption Governance: Precondition for a Degrowth Path?" *Journal of Cleaner Production*, 38(1): 36–43.

Lorek, Sylvia and Joachim Spangenberg. 2001. "Indicators for Environmentally Sustainable Household Consumption." *International Journal of Sustainable Development*, 4(1): 101–120.

Luchs, Michael and Todd Mooradian. 2012. "Sex, Personality, and Sustainable Consumer Behaviour." *Journal of Consumer Policy*, 35(1): 127–144.

Maniates, Michael. 2002. "In Search of Consumptive Resistance." In *Confronting Consumption*, ed. Thomas Princen, Michael Maniates, and Ken Conca, 199–236. Cambridge, MA: MIT Press.

Maniates, Michael and John Meyer, eds. 2010. *The Environmental Politics of Sacrifice*. Cambridge, MA: MIT Press.

Markkula, Annu and Johanna Moisander. 2012. "Discursive Confusion over Sustainable Consumption." *Journal of Consumer Policy*, 35(1): 105–125.

Maslow, Abraham H. 1954. *Motivation and Personality*. New York: Harper and Row.

Max-Neef, Manfred. 1991. *Human Scale Development*. London: Zed Books.

McDougall, William. 1923. *An Introduction to Social Psychology*. London: Methuen.

Mert, Wilma. 2011. "Waschen, wenn der Wind weht." In *Die politische Förderung des Stromsparens in Privathaushalten*, ed. Doris Fuchs, 176–191. Berlin: Logos.

Ministry of Environment, Norway. 1994. "Report of the Sustainable Consumption Symposium." Oslo: Ministry of Environment.

Möller, Sabine. 2004. "Die Vermeidung von Consumer Confusion." *Thexis*, 22(4): 27–30.

Mont, Oksana and Andrius Plepys. 2008. "Sustainable Consumption Progress." *Journal of Cleaner Production*, 16(4): 531–537.

Moraes, Caroline, Marylyn Carrigan, and Isabelle Szmigin. 2012. "The Coherence of Inconsistencies." *Journal of Marketing Management*, 28(1–2): 103–128.

Mortensen, Lars F. 2006. "Sustainable Household Consumption in Europe?" *Consumer Policy Review*, 16(4): 141–147.

Mózner Vetöné, Zsófia. 2011. "Applying Consumer Responsibility Principle in Evaluating Environmental Load of Carbon Emissions." *Society and Economy*, 33(1): 131–144.

Munasinghe, Mohan. 2010. "Can Sustainable Consumers and Producers Save the Planet?" *Journal of Industrial Ecology*, 14(1): 4–6.

Princen, Thomas. 2002. "Consumption and Its Externalities." In *Confronting Consumption*, ed. Thomas Princen, Michael Maniates, and Ken Conca, 23–42. Cambridge, MA: MIT Press.

Princen, Thomas. 2005. *The Logic of Sufficiency*. Cambridge, MA: MIT Press.

Princen, Thomas. 2010. "Consumer Sovereignty, Heroic Sacrifice." In *The Environmental Politics of Sacrifice*, ed. Michael Maniates and John Meyer, 145–164. Cambridge, MA: MIT Press.

Princen, Thomas. 2011. "Leave-it-in-the Ground: The Politics of Stopping Fossil Fuels at Their Source." Presentation at ISA Annual Convention 2011, Montreal, March 16.

Ray, Paul. 2000. *The Cultural Creatives*. New York: Harmony Books.

Rehfeld, Katharina-Maria, Klaus Rennings, and Andreas Ziegler. 2007. "Integrated Product Policy and Environmental Product Innovations." *Ecological Economics*, 61(1): 91–100.

Røpke, Inge. 1999. "The Dynamics of Willingness to Consume." *Ecological Economics*, 28(3): 399–420.

Røpke, Inge. 2011. "Ecological Macroeconomics: Challenges for Consumer Studies." Presentation at Sustainable Consumption Conference, Hamburg, November 6–8.

Røpke, Inge and Mirjam Godskesen. 2007. "Leisure Activities, Time and Environment." *International Journal of Innovation and Sustainable Development*, 2(2): 155–174.

Ruppel Shell, Ellen. 2009. *Cheap: The High Cost of Discount Culture*. New York: Penguin Press.

Sanne, Christer. 2002. "Willing Consumers or Locked in?" *Ecological Economics*, 42(1–2): 273–287.

Schor, Juliet. 1998. *The Overspent American*. New York: Basic Books.

Seyfang, Gill. 2007. "Cultivating Carrots and Community." *Environmental Values*, 16(1): 105–123.

Seyfang, Gill. 2009. *The New Economics of Sustainable Consumption*. New York: Palgrave Macmillan.

Shove, Elizabeth. 2003. *Comfort, Cleanliness and Convenience*. London: Berg.

Stakeholder Forum. 2012. "SCP Governance, Sustainable Consumption and Production Governance: A Guide towards Rio+20." Report for the UNCSD 2012, http://www.stakeholderforum.org/fileadmin/files/SDG%20SCP%20Uchita.pdf (accessed March 5, 2012).

Stuart, Tristram. 2009. *Waste: Uncovering the Global Food Scandal*. New York: W.W. Norton.

Tukker, Arnold, Gjalt Huppes, Jeroen Guinée *et al.* 2006. *"Environmental Impact of Products (EIPRO)."* Seville: European Science and Technology Observatory and Institute for Prospective Technological Studies.

UNEP/CDG. 2000. *Sustainable Consumption and Production: Creating Opportunities in a Changing World*. Berlin: Carl Duisberg Gesellschaft.

United Nations. 1993. "Earth Summit: Agenda 21: The United Nations Programme of Action from Rio." New York: United Nations.

United Nations. 2002. "Plan of Implementation of the World Summit on Sustainable Development." Johannesburg: United Nations.

United Nations Environmental Programme. 2001. *Consumption Opportunities: Strategies for Change*. Geneva: UNEP.

Vanclay, Jerome, John Shortiss, Scott Aulsebrook *et al.* 2011. "Customer Response to Carbon Labelling of Groceries." *Journal of Consumer Policy*, 34(1): 153–160.

Veblen, Thorstein. 1994 [1899]. *The Theory of the Leisure Class*. Boston: Houghton Mifflin.

Vermeir, Iris and Wim Verbeke. 2006. "Sustainable Food Consumption: Exploring the Consumer 'Attitude–Behavioral Intention' Gap." *Journal of Agricultural and Environmental Ethics*, 19(2): 169–194.

Virtanen, Yrjö, Sirpa Kurppa, Merja Saarinen *et al.* 2011. "Carbon Footprint of Food: Approaches from National Input–Output Statistics and a LCA of a Food Portion." *Journal of Cleaner Production*, 19(16): 1849–1856.

Wahlen, Stefan, Eva Heiskanen, and Kristiina Aalto. 2012. "Endorsing Sustainable Food Consumption: Prospects from Public Catering." *Journal of Consumer Policy*, 35(1): 7–21.

Warde, Alan. 2011. "Climate Change, Behavior Change and Sustainable Consumption." Presentation at Sustainable Consumption: Towards Action and Impact, International Scientific Conference, Hamburg, November 6–8.

WBSCD. 2002. "The Business Case for Sustainable Development." Geneva: World Business Council for Sustainable Development.

Wiedmann, Thomas and Jan Minx. 2008. "A Definition of 'Carbon Footprint'." In *Ecological Economics Research Trends*, ed. Carolyn Pertsova, 1–11. New York: Nova Science Publishers.

Wolff, Franziska and Norma Schönherr. 2011. "The Impact Evaluation of Sustainable Consumption Policy Instruments." *Journal of Consumer Policy*, 34(1): 43–66.

World Watch Institute. 2012. *State of the World 2012: Creating Sustainable Prosperity*. New York: W.W. Norton.

Yates, Lucy. 2008. "Sustainable Consumption: The Consumer Perspective." *Consumer Policy Review*, 18(4): 96–99.

Zhu, Xueqin, Lia Wesenbeeck, and Ekko Ierland. 2006. "Impacts of Novel Protein Foods on Sustainable Food Production and Consumption: Lifestyle Change and Environmental Policy." *Environmental and Resource Economics*, 35(1): 59–87.

Chapter 14

Climate Change Justice

Edward Page

Introduction

It is now known beyond all reasonable doubt that the human consequences of climate change will be substantial, on balance adverse, and will rise markedly with higher levels of global warming and sea-level rises (Parry *et al.* 2007: 65–68; Stern 2007: 65–103). Those bearing the greatest disadvantages will be populations residing in the developing world (due to geographical vulnerability, limited adaptive capacity, and the reliance of developing state economies on ecosystem services) and vulnerable social groups located in all regions (due to the way the impacts of climate change compound existing social and economic inequalities) (Parry *et al.* 2007: 69; Adger 2010: 282–283). Within this context of vulnerability and risk, policy-makers and normative theorists have become increasingly preoccupied with the concept of "climate change justice," which, for the purposes of the chapter, is defined as the equitable distribution of benefits and burdens arising from global climate change and policies for its management.

Three key challenges arise for any plausible theory of climate change justice. First, to determine the share of the capacity of the Earth's atmosphere to assimilate carbon dioxide (CO_2) and other greenhouse gases that morally relevant agents should be able to exploit as a matter of distributive justice. According to the standard way of approaching this "justice in emissions" problem, the task is to find the correct principle(s) of justice that should regulate the total amount of greenhouse gas that states and agents operating within their territories should be permitted to emit each year over the next century (Shue 1993: 48–50; Caney 2009: 125–126). The international legal background of this task is the "ultimate objective" of the United Nations Framework Convention on Climate Change (UNFCCC) of 1992 to achieve "stabilization of greenhouse gas concentrations in the atmosphere at a level that would

The Handbook of Global Climate and Environment Policy, First Edition. Edited by Robert Falkner. © 2013 John Wiley & Sons, Ltd. Published 2013 by John Wiley & Sons, Ltd.

prevent dangerous anthropogenic interference with the climate system" (UNFCCC 1992: Article 2). Second, the burdens associated with managing climate change and its adverse effects should be equitably allocated amongst the relevant agents. The idea here is that an account of justice in emissions would be theoretically incomplete, as well as practically useless, without an accompanying account of "justice in burdens" that specifies the way in which agential and institutional burdens associated with effective policies of climate mitigation and adaptation should be distributed within and between generations (Page 2008; Caney 2010: 751–752). Third, the duties and entitlements of climate change justice, if they are to be of genuine relevance for policy-makers, must be incorporated into the process whereby national, regional, and global climate policies are selected. A further aspect of this "justice in governance" problem is that, in absence of the integration of normative theory and climate policy-making, attempts to manage climate change through international cooperation have the potential to undermine established norms of global poverty reduction and political legitimacy.

In this chapter, my aim is to give a sense of the progress that normative theorists have made in developing, and applying practically, the concept of climate change justice through an examination of all three problems. For the purposes of simplicity, I assume throughout that the primary agents to whom enforceable duties and entitlements of climate justice can be allocated are *states* rather the individual citizens or corporations operating within, and between, their territories. One reason for this "statist" starting point is that, given the intergenerational and international character of the climate problem, individual citizens lack the properties required to undertake or supervise successful mitigation and adaptation measures. States, by contrast, are the political units at the heart of existing domestic and international environmental law on climate change; states possess many of the political and economic resources necessary to manage climate change that sub-state actors lack; and, as signatories and ratifiers of treaties and conventions, states actively claim legitimacy in areas of policy required to respond to climate change.

Distributing Greenhouse Gas Emissions Justly

Avoiding "dangerous anthropogenic interference" in the climate system over the next century will require a coordinated international response in terms of reducing global flows, and later stocks, of atmospheric greenhouse. The challenge here is for the international community to impose distributive order on the hitherto unregulated use of the atmosphere in a way that cannot be reasonably viewed as unjust by any state (Gardiner 2010: 52). This "justice in emissions" problem has two key dimensions (Caney 2009: 125). The first is the task of establishing and enforcing a global emissions budget over the coming decades consistent with the early peaking of (and subsequent significant rate of reduction in) greenhouse gas emissions that will predictably deliver a high probability of avoiding dangerous anthropogenic interference. The second task is to allocate a set of greenhouse gas emissions entitlements amongst states over this period that can be viewed by all as equitable.

While there exists no academic or policy-maker consensus as to how the goal of dangerous anthropogenic interference in the climate system should be specified to be of genuine policy relevance, an increasing number of analysts and environmental organizations, and the governments of over 190 states, have now endorsed a solution to the emissions trajectory problem that has the objective of limiting global warming to no more than 2 °C over its pre-industrial value (see European Commission 2011: 3; Garnaut 2011: 36–38; UNFCCC 2011: 3). Since the 2 °C objective is merely one of several possible methods of concretizing the goal of dangerous anthropogenic interference, it can usefully be reframed as the objective of "avoiding dangerous climate change." Recent research indicates that achieving a 29–70% chance of avoiding dangerous climate change by meeting the 2 °C objective would require policies being adopted that result in no more than 2000 billion tonnes of carbon dioxide (CO_2)-equivalent being emitted in the 2000–2050 period, of which roughly 400 billion tonnes have already been emitted (Meinshausen *et al.* 2009: 1161). The problem of allocation, assuming this analysis is both correct and endorsed by the international community as the basis of post-Kyoto climate politics, amounts to specifying how the remaining 1600 billon tonnes of CO_2-equivalent should be shared amongst the world's 200 states between 2012 and 2050.

Four substantive allocation principles currently dominate the literature on the just allocation of emissions entitlements (see Vanderheiden 2008a: 221–257; Caney 2009: 127–137; Gardiner 2010: 56–60).

(1) *Emissions grandfathering.* According to "emissions grandfathering," each qualifying state should reduce its emissions by a uniform (or close to uniform) amount in percentage terms relative to some pre-specified base year such as 1990 or 2005. This approach remains the primary method of allocating mitigation responsibilities amongst the developed states listed in Annex B of the Kyoto Protocol to the UNFCCC, which imposed a legally binding average cut in six greenhouse gases of 5% relative to 1990 levels (UNFCCC 1997: Annex B). The normative idea behind emissions grandfathering is essentially that the mitigation commitments required of each state as part of its duty to participate in the collective climate mitigation effort should reflect the fact that national emissions prior to the negotiation of global climate agreements were essentially unregulated and should therefore be treated as defining the baseline for the mitigation effort that can reasonably be requested of each state.

Notwithstanding its success in becoming a major pillar of international environmental law and the domestic environmental policies of many developed states, the obvious problem with emissions grandfathering as an expression of climate change justice is that it assigns an implausible weight to the normative relevance of historic usage of the capacity of the atmosphere to assimilate greenhouse gas. As such, the approach has no real response to the objection that anchoring the emissions entitlements of states to their past emissions profiles would be unfair to states responsible for modest accumulations of atmospheric greenhouse gas since 1750 (Vanderheiden 2008a: 226; Gardiner 2011: 425). Emissions grandfathering could easily result, for example, in the per capita emissions of the developed world continuing to exceed those of the developing world for many decades even if the latter were exempt from the relevant emissions reductions.

(2) *Equal mitigation sacrifices.* According to this approach to justice in emissions, states that pass a simple prosperity test should undertake mitigation activities that impose on each a roughly similar cost in terms of forgone national income or well-being over the 2000–2050 period (Traxler 2002; Miller 2009: 146–151). If the

global climate response requires a level of mitigation activity to avoid dangerous climate change that would impose a 2% loss in the combined incomes of middle- and high-income states, relative to what they would have been in the 2000–2050 period in the absence of the necessity for a coordinated international climate response, the equal sacrifice approach requires that this burden be borne so that the average citizen in each state faces a similar loss in future income expectations (Miller 2009: 147). In this sense, seeking to equalize the burden that each state faces in reducing its greenhouse emissions as part of the fight against dangerous climate change does not seek to establish an idealized pattern of international greenhouse emissions, or the rights authorizing these emissions, over the 2012–2050 period. There is, that is, no individual or collective "right" to emit a certain amount of the greenhouse gas that might be emitted over the next century without triggering dangerous climate change. Rather, the actual pattern of greenhouse emissions over this period is permitted to vary so long as loss borne by the average citizen of each state caused by changes in lifestyle and consumption is equalized.

One fairly obvious normative problem with the equal sacrifice approach is that it seems unjust, given significant disparities in living standards amongst high- and middle-income states, to require that the poorer states in this group (however wealthy they might be compared to low-income states) bear identical economic burdens in percentage terms to their richer counterparts merely because an effective solution to climate change requires a widespread mitigation effort. On Miller's (2009) derivation of the approach, for example, all high-income members of the Organisation for Economic Co-operation and Development (OECD) would be required to forgo the same proportional gains in national income per capita despite the greater than 4 : 1 ratio between their current national incomes. Some versions of the equal sacrifice approach seek to finesse this problem by abandoning national *income* as the metric of sacrifice in favor of national *welfare* in order to capture the non-monetary role that many greenhouse gas emitting activities play in the lives of the average citizen of all states (Traxler 2002). While this change of metric introduces a certain degree of sensitivity to variations in lifestyle, geographical location, and income amongst those states attributed duties of mitigation, the problem remains that equalizing mitigation burdens amongst states whose citizens experience very different average qualities of life means the approach seems both unfair and unlikely to motivate compliance amongst many qualifying states.

(3) *Emissions egalitarianism.* Critics of emissions grandfathering may be attracted to the view that, in the absence of a good reason to the contrary, a principle of equality should dictate the spatial and temporal distribution of emissions rights. That is to say, emissions rights should be allocated amongst states so that each will enjoy a similar level of access to the capacity of the atmosphere to assimilate greenhouse gases without triggering "dangerous anthropogenic interference." Due to the obvious problems posed by variations in state population size for egalitarian approaches, emissions rights egalitarians almost invariably adopt the view that it is the *per capita emissions*, rather than the *absolute emissions*, of states that should be equalized over some pre-specified historical period (Meyer 2000: 56; Baer and Athanasiou 2002: 76–97). The idea here is that global climate change justice will be achieved only if developed and developing states converge on a roughly equal level of

annual greenhouse gas emissions per person. Garnaut (2011: 42–45), for example, suggests that this would optimally involve the average citizen of all states converging in the middle of the century at the greenhouse emissions level of the current average Indian citizen: roughly 2 tonnes of CO_2-equivalent per annum.

Aside from the negative justification of "emissions rights equality" noted above, there are a number of positive arguments that have been adduced in the literature for the equalization of the per capita greenhouse gas emissions of each state. It might be held, for example, that the assimilative capacity of the atmosphere is the common property of mankind and therefore its value must be distributed equally within and between generations, like any other commons. Here, the *common ownership* rights of access of citizens belonging to all states render redundant territorial and historical claims of appropriation and transfer (Singer 2002: 35). Another possibility, however, is that apportioning emissions rights equally amongst states might be the only way to express equal concern and respect for the *vital interests* of citizens of all states who share the capacity of the atmosphere to assimilate greenhouse gases.

Though a popular view among policy analysts, emissions egalitarianism faces a fierce combination of normative objections leveled at both its egalitarian ethos and per capita operationalization. Thus, it has been argued that equalizing the per capita level of each state's greenhouse emissions would reduce neither global inequality nor the number of humans who fail to have their basic needs met due to climate change impacts (Caney 2009: 127ff.). Other critics have argued that neither of the main justifications is cogent: atmospheric commons arguments are at best a weak and inconclusive defense of emissions rights equality since they do not uniquely support an equal right of citizens or states to emit greenhouse gases (Starkey 2011: 116–122) and respecting equal concern implies an equal right to emit no more than it does equal income since people will derive vastly different amounts of satisfaction from different bundles of either (Bell 2011: 36–37). Finally, the "forward-lookingness" of vital interest versions of the approach (this brand of emissions egalitarianism is little, if at all, concerned with past inequalities in per capita emissions) could be criticized for entailing that historically low, and histori-cally high, emitting states should converge on roughly the same per capita emissions levels by the middle of the century. Converging on 2 tonnes of CO_2-equivalent per person, as recommended by the Garnaut Review, for example, will be rejected by states that ascribe to the popular view in the developing world that developed states should make a greater mitigation sacrifice than merely reducing their per capita emissions to the average global emissions level required to prevent dangerous climate change.

(4) *Emissions sufficientarianism.* In response to some of the problems facing emis-sions egalitarianism, some theorists have argued that emissions rights should be distributed so that citizens of every state have access to enough of the atmosphere's capacity to meet their basic needs (such as nutrition, shelter, and basic health care) but not non-basic needs (such as access to air travel and many types of consumer good). As long as each person can meet their basic needs, and global emissions remain on the selected safe emissions trajectory, policy-makers need not aim to bring about an egalitarian pattern of international greenhouse emissions (Shue 1993: 55–56).

Vanderheiden, for example, posits an interconnected set of "basic rights" including a "basic right to climate stability," a right to a "minimum per capita level of emissions," and a "right to develop" in defending such a view (Vanderheiden 2008b: 64). Putting these climate change entitlements together, he argues, entails that "persons have a basic right to their survival emissions but they have lesser rights, if at all, to their luxury emissions" (Vanderheiden 2008a: 243). Such rights, which are posited as inalienable and therefore non-tradable, are possessed by all equally but their fulfillment rests on no present or future person being denied access to a "sufficient" amount of the assimilative capacity of the atmosphere.

The "basic rights" derivation of emissions sufficientarianism raises many of the usual problems associated with accounts of distributive justice that appeal to basic needs or decent lives, namely, the implausibility for many of the claim that there are no distributionally relevant inequalities in emissions, or any other type of scarce resource, above the point where all citizens have their basic needs met (Page 2006: 92–95). Another problem facing the approach is that of distinguishing, in a manner that would make the approach philosophically coherent and operationalizable in practical terms, between "luxury" and "subsistence" emissions (Gardiner 2011: 424–425). Perhaps the strongest objection to the approach, however, emerges from the identification it makes between "emissions sufficiency" and "personal sufficiency." The problem here is that no particular distribution of emissions rights, or the economic value of these rights, would predictably maximize the number of people leading lives of a decent quality. This is because the quality of any particular human life is shaped by a broad range of subjective and objective factors and many of these factors are only loosely connected to the way in which the atmospheric commons is shared within or between generations.

I hope to have shown in this brief discussion that the "justice in emissions" problem has no technical solution. That is to say, *none* of the most commonly involved principles has yet been developed to deliver a normatively persuasive pattern of emissions when considered in isolation; and, though I have not the space to explore this thought further, solutions that appeal to more than one principle could be expected to generate no less intractable problems of justification and operationalization. Consequently, the practical contribution of the philosophical literature surveyed above does not seem to lie in its ability to supply a single view suitable for direct implementation by the institutions of the UNFCCC, but rather the clarification of some key normative issues prior to free and fair negotiations amongst state representatives seeking a principled, yet context-sensitive, distribution of emissions rights with which each state has reason to conform.

Which Burdens, Whose Responsibility?

Suppose an international agreement could be reached as to how much greenhouse gas each state, and thus the average citizen within that state, could justly exploit each year over the coming decades. The problem would remain of distributing amongst states the financial and non-financial burden associated with coordinating the structural, technological and attitudinal changes required to realize any preferred greenhouse emissions profile. The costs of effective adaptation, moreover, would also remain. Three principles have dominated the debate concerning how burdens of mitigation and adaptation should be distributed amongst states and I survey the normative justification and distributive consequences of each below (Caney 2006, 2010; Page 2008, 2011; Dellink *et al.* 2009).

(1) According to the Contribution-to-Problem Principle (CPP), states should bear the costs of managing climate change and its adverse effects in proportion to their share of cumulative global greenhouse gas emissions. According to the standard operationalization of the CPP, each state would be allocated a share of the total cost of the global adaptation and mitigation response in relation to its overall contribution to anthropogenic climate change as measured in terms of the greenhouse gas emissions emanating from within its borders since 1750. At first glance, the CPP seems to provide a powerful interpretation of the principle of "common but differentiated responsibilities" adopted by the UNFCCC (1992: Article 3) as a basis for international burden sharing. This is because it links the distribution of climate burdens to the varying contributions of each state to the physical processes that drive climate change (Neumayer 2000: 187).

An immediate practical problem arising with the CPP is the problem of linking the *causes* of climate change (accumulations of atmospheric greenhouse gas of all states) with the disadvantageous human *effects* brought about by these accumulations (such as increases in morbidity and mortality in other states arising from extreme weather events) (Miguez 2002: 25–32; Gardiner 2011: 24–41). The national historical responsibility for climate change invoked by the CPP presupposes (i) the existence of identifiable harmful acts or events which (ii) befall identifiable agents such that (iii) they can be traced back to the climate-altering behavior of specific states. The problem is that these conditions appear at odds with certain integral features of the climate problem. Greenhouse gases, for example, are "well mixed" in that they become evenly distributed throughout the atmosphere shortly after being emitted irrespective of the nature of the activity involved or its geographical location. Since 1 tonne of CO_2-equivalent emitted *anywhere* results in the same amount of climate changing potential being exerted *everywhere*, robust and reliable protocols that trace particular climatic events to any particular state's accumulated emissions currently lie beyond the grasp of climate modelers. Neither is it currently possible to identify any extreme weather events that would not have occurred "but for" anthropogenic climate change. Such considerations complicate greatly the attempt to specify the amount of adaptation or mitigation for which each state might be held accountable by the CPP (Allen and Lord 2004: 552).

On the assumption that the causation problem could be solved, perhaps through a probabilistic rather than a deterministic approach to climatic harm, additional normative objections could be raised against the CPP. First, should national responsibility for climate change be calculated in terms of the accumulated greenhouse gas released within the territorial borders of each state (the production method of apportionment) or in terms of the accumulated emissions released worldwide for which each state can be held accountable as the end-users of the associated activities (the consumption method of apportionment)? While the former method remains a central pillar of the UNFCCC architecture, the selection of apportionment method is a hotly contested issue on both normative and political grounds, in part because it would lead to very large variations in attributions of climatic responsibility in combination with the CPP (Garnaut 2011: 38).

Second, in grounding climate burden attributions on the causal processes whereby a state's cumulative greenhouse emissions contribute to adverse future outcomes, the CPP is open to the objection that states cannot reasonably be held responsible for much of the past greenhouse gas emitted by actors located within the borders of ancestral political units. The problem here is that there have been many instances when the continuity of state identity has broken down during the time frame in which the CPP must operate. The CPP, as a "backward-looking approach," does not cope well with the numerous changes evident in state boundaries due to wars, secessions, and internal political events that have transformed their institutional character since 1750. This is because these political changes seem to undermine the normative conditions for holding an existing state responsible for the cumulative emissions of ancestral geopolitical units as if they were one and the same entity (Caney 2006: 469–470; Miller 2009: 151ff.).

A third normative problem is that, although developed states may be causally responsible for the bulk of the CO_2 emitted since 1750, it is not at all clear that even constitutionally stable states should be held *morally* responsible for the cost of policy measures designed to manage the adverse effects of historical activities for which they could be held *causally* responsible. This is because attributions of moral responsibility, whether applied at the level of citizens or states, are generally thought to require more demanding standards of agential capability than those of causal responsibility. Thus, to be morally responsible for redressing a disadvantage an agent has contributed to causally, is in the process of causing, or is expected with great confidence to cause in the future an agent should have possessed the ability "to choose and to control his conduct in accordance with his choice" (Honoré 1999: 32). The problem in the climate context is that states seem to lack this ability as regards the majority of greenhouse emissions for which they can be held causally responsible since 1750 due to widespread ignorance of the climate problem amongst the policy-makers and general citizenry of most states until the late twentieth century. The implication of this "excusable ignorance" problem (Caney 2010: 208; Page 2011: 416–417) is that the CPP could apply only subsequent to a moment in history after which policy-makers in each state could not reasonably claim ignorance of, and hence the ability to control, the climate-changing activities for which they are causally responsible. Needless to say, the identification of a non-arbitrary time period over which the CPP would apply poses serious theoretical challenges not least because the moment at which policy-makers can no longer reasonably invoke the excusable ignorance defense will vary across states.

(2) According to the Ability-to-Pay Principle (or APP), states should bear climatic burdens in proportion to their relative capacities to bear such burdens: the more they are able to remedy climatic disadvantage in terms of implementing effective policies of mitigation and adaptation, that is, the more they should do so (Shue 1999: 537). All things being equal, the APP implies that the developed states should shoulder the burden of climate justice because they are uniquely able to harness their "mitigative capacity" and "adaptative capacity" both domestically and internationally. This superior mitigative and adaptive capacity comes as a result of various privileged features of high development including a greater ability to deploy viable low-emissions technologies; superior social and human capital in the relevant areas of social policy connected to successful adaptation; and a greater capacity to bear the

developmental sacrifices associated with emitting less greenhouse gas in general terms. In this way, the APP is a "forward-looking" rather than "backward-looking" justification of differential climatic burdens: the burdens associated with dealing with the negative externalities associated with climate change should be borne by the most able, irrespective of past behaviors that can be traced to subsequent modifications of the composition of the atmosphere.

At first glance, the APP has much to recommend it. As a forward-looking approach to climatic burden sharing it is not subject to the problems associated with the national excusable ignorance or national boundary problems. Moreover, it appeals to the common-sense conviction that, where issues of moral responsibility and liability are unclear, apportioning the greatest remedial burdens to states that have the capacity to solve a problem, and a significant interest in the problem being solved, amounts to a progressive solution to the climate problem. Nevertheless, the APP is questionable as a single-principle solution of the "justice in burdens" problem for at least three reasons.

First, a burden differentiation problem arises between "responsibly rich states" and "irresponsibly rich states." Suppose two states enjoy a roughly equivalent capacity to respond to climate change but whereas one developed using highly efficient (climatically less damaging) technology, the other developed using far less efficient (climatically far more damaging) technology. Although the first state will have contributed far less to climate change, the APP applied in isolation implies that the climate burdens each should bear will be similar if not identical. This appears a rather implausible approach to the distribution of climatic burdens amongst rich states.

Second, a similar set of worries arises when developing world states are disaggregated into those that have begun to develop using low-carbon technologies and practices and those states that have begun to develop using more climatically damaging technologies and practices. Although it might be thought that *all* states must bear at least some climatic burdens to achieve a sustainable solution to the climate problem, it seems unfair to developing states which developed in a cleaner manner that they bear the same burden as developing states that did not. This thought experiment is not an exercise in science fiction since significant differences in cumulative emissions records exist amongst developing states with similar resources in terms of the income and wealth at their disposal (Boden *et al.* 2011).

Third, and perhaps most decisively, the APP leaves unanswered the deeper, normative, question of *why* those who have the most resources should bear the greatest climate burdens, other than because they *can* (Page 2008: 561–562). It may be the case that climatic burden allocation should be filtered through a weak "ought implies can" principle that excludes burdens being foisted on states that could only take on these burdens if they neglect the basic needs of their citizens. But a stronger "ought implies can" principle requiring that climatic burdens are attributed to states in proportion to a combination of their mitigative and adaptive capability irrespective of the origins of this capacity seems implausible in the absence of a background theory of global justice that makes sense of such attributions. Even armed with such a theory, moreover, the APP would not be a likely basis of a stable agreement amongst representatives of states seeking a climate agreement that none can reasonably reject as biased towards the interests of any of their number.

(3) According to the Beneficiary Pays Principle (or BPP), states should bear climatic burdens in proportion to how much they have benefited from the economic and social activities associated with the rise in concentrations of greenhouse gas since 1750. For a state not to pay their fair share of the cost of the climate response as determined by the BPP would be unjust as it would amount to profiting from environmental damaging activities originating within and between its territorial boundaries (Gosseries 2004: 43–46; Page 2011: 420–422). It could be maintained that the economies of all states have benefited from agricultural and industrial activities that have released greenhouse gas into the atmosphere since 1750. Nevertheless, developed states are picked out by the BPP as having a peculiarly strong responsibility to bear climate burdens because their high development can be traced fairly directly to past and present activities, such as access to abundant energy supplies sourced from fossil-fuel combustion, that drive climate change.

As described, the BPP has certain theoretical advantages over the APP and CPP. The BPP is in one respect superior to the CPP since it avoids the problem of holding present-day states morally responsible for the behavior of ancestral political units. The BPP is also not prone to the production/consumption apportionment problem since it focuses on the ultimate destination of the benefits produced by greenhouse gas-emitting activities and not the jurisdictional origins of global greenhouse gas emissions. The BPP also seems superior to the APP in the important respect that it offers an explanation of the special responsibility of developed states that is not reducible to the mere fact of their superior wealth. Despite its advantages, however, the BPP is a controversial, and relatively undeveloped, basis for climatic burden attribution. One reason for this is that it has struck some theorists as unfair to require later generations to surrender benefits in order to sponsor policies of mitigation and adaptation when earlier generations have enjoyed similar benefits and not surrendered any benefit (Caney 2006: 473). This "chronological unfairness" objection is especially troublesome in contexts where some states have consumed or despoiled the benefits they inherited from industrialization, since the most obvious interpretation of the BPP is to restrict "burden disgorgement" to states that have retained climate-change-linked benefits and, moreover, can surrender these benefits without thereby subjecting their own citizens to great hardship.

A second problem for the BPP is that it is immensely difficult to separate the part of the present wealth of developed states that arose from activities that caused climate change from the part that can be attributed to other factors. The BPP, then, seems incapable of being fully operationalized. One dimension of this problem is that distinguishing between the two types of wealth presupposes a distinction between benefits that were "caused" by fossil-fuel-driven industrialization ("climatic benefits") and benefits that merely "correlated" with fossil-fuel-driven industrialization in that they owe their origins for the most part to historical contingency or entrepreneurial flair ("non-climatic benefits"). If no clear distinction can be discerned between climatic and non-climatic benefits then it will not be possible to distinguish between the benefits an agent should be prepared to sacrifice to combat climate change and the benefits that an agent should be permitted to retain as the expression of activities independent of climate change. Since the CPP and APP do not trace climate burdens to the receipt of climate benefits, they do not face the problem of benefit identification or disaggregation.

I hope to have shown in this brief discussion that none of the three common justifications of differential national climatic responsibility has yet to be formulated in a way that would deliver, in isolation, a plausible solution to the "justice in burdens" problem. Though I do not have the space to demonstrate it here, it is at best unclear whether hybrid solutions will prove any more successful in this task (Page 2011: 425–426). As with the "justice in emissions" problem, then, there appears to be no technical solution to the "justice in burdens" problem: parties to the UNFCCC must negotiate a solution that balances the strengths and weaknesses of alternative principles through a process of negotiation that furnishes the resulting agreement with a measure of normative legitimacy that is independent of any philosophical account of the selected burden-sharing rules.

Climate Change Justice and the Global Policy Mix

How might our best solutions to the "justice in emissions" and "justice in burdens" problems be translated into an effective set of climatic institutions, policies, and mechanisms? Let us focus, for the purposes of simplicity, on climate mitigation policy. The full range and content of mitigation policies defies analysis in the space available, but the main options can usefully be summarized as the following (Gupta and Tirpak 2007: 56–59; Stern 2007: 349–354):

1. *Direct governmental regulation* (e.g. limits on industrial or household greenhouse gas emissions enforced by legal rules and orders).
2. *Government expenditure* (e.g. subsidies designed to encourage renewable energy use in developed states or research and development assistance for developing states seeking to implement low-carbon technologies).
3. *Market-based mechanisms* (e.g. carbon tax or emissions trading schemes designed to make participating firms or states internalize the full social costs of their exploitation of the atmospheric commons).
4. *Voluntary measures* (e.g. energy efficiency agreements reached between regulators and associations representing particular economic sectors).
5. *Informational measures* (e.g. environmental labeling of consumer products and services).

An extensive literature has now emerged seeking to evaluate climate policies such as those listed above, as well as the domestic and international policy frameworks that systematize these policies, according to a range of normative considerations (see, for example, Bodansky 2004: 19–62). Such evaluations are generally framed in terms of two types of normative desiderata. "Consequentialist" (or "teleological") reasoning evaluates acts and social policies according to the way in which their expected outcomes fit with some desired set of outcomes such as the equalization of welfare or fulfillment of basic needs. Here, the justness of a climate policy is specified by the desirability of its expected outcomes. "Non-consequentialist" (or "deontological") reasoning, by contrast, evaluates acts or social policies in terms of the legitimacy or fairness of their origins. Here, the justness of a climate policy is specified independently of the desirability of its expected outcomes. The distinction between outcomes and procedures is a useful starting point for the normative analysis of alternative climate policies and policy frameworks, and both types of principle play a vital public motivation role in securing support from citizens and policy-makers for an effective international climate response. Here, due to space constraints, I discuss briefly three normative desiderata that can be applied to policy solutions to the "justice in emissions" and "justice in burdens" problems.

Economic Efficiency

Economic efficiency, in the climate policy context, is a consequentialist desideratum, meaning quite simply that the international community should seek to prevent dangerous climate change using the most affordable possible means available. Thus, according to the text of the UNFCCC (1992: Article 2), "policies and measures to deal with climate change should be cost-effective so as to ensure global benefits at the lowest possible cost." The demand for economic efficiency in climate policy-making draws on the common-sense normative idea that no public policy should impose more costs than are required to secure the environmental quality benefits it is designed to bring about. To implement a cost-inefficient policy response to climate change would, according to this line of argument, violate the consequentialist norm that environmental policies should not be adopted that bring about a lower quality of life for existing and future generations than could have been secured through an alternative policy response.

If economic efficiency was the only consideration of relevance to climate policy beyond the primary desideratum of environmental effectiveness, climate policy would simply seek to equalize the marginal mitigation and adaptation costs of all atmospheric users with the selection of policy mechanisms being wholly derivative to this aim. Consider climate mitigation policy. The idea here is that the global policy response should bring about a situation where atmospheric users located in each state will only emit an additional unit of CO_2-equivalent if no other atmospheric user located in this, or another state, could produce more benefit by doing so. However, minimizing marginal mitigation costs in the early stages of the global climate response could result in the achievement of a lower probability of avoiding dangerous climate change since there may be economically inefficient policy combinations that lead to a more speedy stabilization in global concentrations of greenhouse gases. In this sense, climate mitigation policy is inherently a matter of balancing "the costs of action and the perils of inaction" and thus involves profound issues of balance between lower/higher mitigation and lower/higher net economic costs of mitigation policies (Nordhaus 2007: 687).

Distributive Equity

Distributive equity concerns the just distribution of resources and opportunities across agents possessing competing needs, interests, and entitlements. As we have seen, climate change justice cannot truly be said to obtain until a just distribution of climatic entitlements, and burdens associated with enforcing those entitlements, exists amongst states. However, an additional problem arises from the need to avoid a situation whereby the global response to climate change introduces new inequalities in the distribution of resources in other spheres of human life or exacerbates previously existing inequalities in these spheres. The problem is that it is far from obvious which principles of justice should be used to settle distributive conflicts between the alternative climate policies described above. There is a need to incorporate norms of fair process and environmental responsibility into our evaluation of rival policies, and the impact of policies of mitigation and adaptation on established goals of global development must also be taken into consideration. We might call this the "global equity" problem. The global equity problem arises not merely because different combinations of policies, as well as different approaches to the reconciliation of the various normative desiderata of climate policy, will affect the interests of different populations in different ways but also because these populations will *value* these effects according to different desiderata.

The complexity of the global equity problem can be illustrated through an example from the literature on climate mitigation policy-making. A broad range of economists and policy-makers favor the imposition of a transparent, and increasing, price being placed on greenhouse emissions worldwide as an effective means of achieving economically efficient and environmentally effective climate mitigation (Nordhaus 2007; Stern 2007; Garnaut 2011). The normative problem that global carbon pricing raises is that it expresses inequality of opportunity. Rich and poor states could not possibly participate on fair and equal terms since the former could draw on their superior financial resources to emit far more greenhouse gas than the latter. The danger here is that developing states that participate in such international emissions trading schemes would be encouraged to reduce their current emissions at the cost of their development goals as a consequence of the core objective of the scheme to discourage emissions amongst all but the most efficient converters of emissions into economic benefits. In this way, the concern to maximize the economic efficiency of the global climate response could endanger the achievement of established goals of global equity and development such as those entrenched into the UN Millennium Development Goals.

Political Legitimacy

Following Buchanan and Keohane (2006: 411), to claim that an institution is legitimate means that it is "morally justified in making rules and attempting to secure compliance with them" and that agents "subject to those rules have moral content-independent reasons to follow them and/or to not interfere with others' compliance with them." A climate governance institution or policy will be legitimate in this *normative* sense so long as the agents whose conduct it constrains have moral reasons to conform to, and support publicly, the rules propagated that do not turn on the specific content of these rules (Bodansky 1999: 601–602). Climate governance institutions and policies will be, in addition, *sociologically* legitimate if there is a widespread belief amongst relevant affected agents that the political authority involved has been wielded justifiably (Eckersley 2007: 307–308). There are two key qualities separating legitimate and illegitimate climate institutions. "Participatory legitimacy" requires that agents whose behavior or condition is modified by climate policy play an active role in its construction and implementation such that they enjoy a genuine sense of ownership over the rules and norms propagated by the institution (Paavola 2005:

314). "Accountability legitimacy," by contrast, requires that agents whose behavior is constrained by climate institutions and policies are adequately informed of the aims, objectives, and mechanisms associated with the climate response and, in addition, that they possess the capacity to sanction the associated institutions in the event of abuse of power or other violations of duty (Hale 2008: 75–76).

As has been shown by many studies of transnational and global environmental policy, both properties are vulnerable in the climatic context, and tensions swiftly arise between these qualities and norms of economic efficiency and environmental effectiveness. Thus, popular market-based policies, such as emission trading or carbon tax schemes, tend to purchase economic efficiency benefits at the cost of removing democratic controls on the spatial location of desired mitigation activity, which atmospheric users must perform these actions, or what sort of technology should be used (Baldwin 2008). Direct government regulation of emissions installations and subsidies for low-carbon energy technologies, moreover, despite shifting control over climate policy one step nearer the level of citizens and other non-state actors, do so at the apparent cost of forgoing the potential of market-based mechanisms to promote environmental quality more cost-effectively (Stern 2007: 351–367; Garnaut 2011: 77–88). In this way, climate policies, and the architectures that systematize them, are subject to a permanent legitimation crisis arising from attempts to reconcile fair processes with effective and affordable environmental outcomes (Eckersley 2007: 307–308).

In the above, I briefly explored the normative desiderata of climate policy that shape the background for the "justice in emissions" and "justice in burdens" problems. For a third time, we find that there is no obvious technical solution to the practical application, or reconciliation, of the various desiderata. First, how should we characterize the relationship between the various desiderata? It seems fairly clear that environmental effectiveness is of preeminent importance and should be allocated some priority over subsequent desiderata. But how do cost efficiency, distributive equity, and political legitimacy relate to one another? Second, how might the various desiderata be applied across political, social, and cultural boundaries? It is tempting to believe that the normative approach selected for one jurisdiction will also be suitable for extension to other jurisdictions. But this itself threatens the acknowledgment of the values contained in some of the desiderata themselves, in particular distributive equity and political legitimacy.

Conclusion

In this chapter, I have aimed to provide the reader with a sense of the progress that normative theorists have made in clarifying the concept of climate change justice and suggested how it might contribute to the construction of global climate policy. We have found that neither normative theorists nor policy analysts have been able to define a uniquely plausible, and practically useful, solution to any of the three problems of climate justice that could be adopted in an action-guiding manner prior to negotiations amongst parties to the UNFCCC. This is not necessarily a cause for alarm, either normatively or practically. While normative theorizing can help clarify rival burden-sharing principles for policy-makers and negotiators – as well as explore the fit between these principles and established norms of justice – the selection of policies of mitigation and adaptation is best seen as a matter of deliberation amongst states seeking agreement on a climate solution that none could reasonably reject. The approach suggested here rejects both the pessimistic vision of "climate change realists" (who tend to view global climate policy-making as a mere matter of politics and power) and the reductionist theorizing of many "climate change idealists" (who tend to reduce the problem of international climate negotiations to a matter of imposing a favored normative approach to the three problems as if they were soluble to a technical or "moral mathematical" solution). Normative theory, by contrast, is a more subtle weapon in the arsenal of the global climate community if it is conceived as a mechanism whereby principles can be articulated, developed, and interpreted before being injected into a process of free and fair negotiation that has no predetermined conclusion.

References

Adger, Neil. 2010. "Climate Change, Human Well-Being and Insecurity." *New Political Economy*, 15(2): 275–292.

Allen, Myles and Richard Lord. 2004. "The Blame Game: Who Will Pay for the Damaging Consequences of Climate Change?" *Nature*, 432, December 2: 551–552.

Baer, Paul and Tom Athanasiou. 2002. *Dead Heat: Global Justice and Global Warming*. New York: Seven Stories Press.

Baldwin, Robert. 2008. "Regulation Lite: The Rise of Emissions Trading." *Regulation and Governance*, 2(2): 193–215.

Bell, Derek. 2011. "Climate Duties, Human Rights and Historical Emissions." In *China's Responsibility for Climate Change*, ed. Paul Harris, 25–46. Bristol: Policy Press.

Bodansky, Daniel. 1999. "The Legitimacy of International Governance: A Coming Challenge for International Environmental Law?" *American Journal of International Law*, 93(3): 596–624.

Bodansky, Daniel. 2004. *International Climate Efforts beyond 2012: A Survey of Approaches*. Washington, DC: Pew Center on Global Climate Change.

Boden, Tom, Gregg Marland, and Bob Andres. 2011. *Global, Regional, and National Fossil-Fuel CO_2 Emissions*. Oak Ridge, TN: Carbon Dioxide Information Analysis Center, Oak Ridge National Laboratory, US Department of Energy. doi:10.3334/CDIAC/00001_V2011.

Buchanan, Allen and Keohane, Robert. 2006. "The Legitimacy of Global Governance Institutions." *Ethics and International Affairs*, 20(4):405–437.

Caney, Simon. 2006. "Environmental Degradation, Reparations, and the Moral Significance of History." *Journal of Social Philosophy*, 37(3): 464–482.

Caney, Simon. 2009. "Justice and the Distribution of Greenhouse Gas Emissions." *Journal of Global Ethics*, 5(2): 125–146.

Caney, Simon. 2010. "Climate Change and the Duties of the Advantaged." *Critical Review of International Social and Political Philosophy*, 13(1): 203–228.

Dellink, Rob, Michel Den Elzen, Harry Aiking *et al.* 2009. "Sharing the Burden of Financing Adaptation to Climate Change." *Global Environmental Change*, 19: 411–421.

Eckersley, Robin. 2007. "Ambushed: The Kyoto Protocol, the Bush Administration's Climate Policy and the Erosion of Legitimacy." *International Politics*, 44: 306–324.

European Commission. 2011. *A Roadmap for Moving to a Competitive Low Carbon Economy*. Brussels: European Commission, http://eur-lex.europa.eu/LexUriServ/LexUriServ.do?uri=COM:2011:0112:FIN:EN:PDF (accessed March 11, 2012).

Gardiner, Stephen. 2010. "Ethics and Climate Change: An Introduction." *Wiley Interdisciplinary Reviews: Climate Change*, 1: 54–66.

Gardiner, Stephen. 2011. *A Perfect Moral Storm: The Ethical Tragedy of Climate Change.* Oxford: Oxford University Press.

Garnaut, Ross. 2011. *The Garnaut Review 2011: Australia in the Global Response to Climate Change.* Cambridge: Cambridge University Press.

Gosseries, Axel. 2004. "Historical Emissions and Free-Riding." *Ethical Perspectives*, 11(1): 36–60.

Gupta, Sujata and Dennis A. Tirpak. 2007. "Policies, Instruments and Co-operative Arrangements." In *Climate Change 2007: Mitigation*, ed. Bert Metz, 745–807. Cambridge: Cambridge University Press.

Hale, Thomas N. 2008. "Transparency, Accountability and Global Governance." *Global Governance*, 14(1): 73–94.

Honoré, Tony. 1999. *Responsibility and Fault.* Oxford: Hart Publishing.

Meinshausen, Malte, Nicolai Meinshasen, William Hare *et al.* 2009 "Greenhouse-Gas Emissions Targets for Limiting Global Warming to 2 °C." *Nature*, 458, April 30: 1158–1162.

Meyer, Aubrey. 2000. *Contraction and Convergence: The Global Solution to Climate Change.* Bristol: Green Books.

Miguez, José. 2002. "Equity, Responsibility and Climate Change." In *Ethics, Equity and International Negotiations on Climate Change*, ed. Luiz Pinguelli-Rosa and Mohan Munasinghe, 7–35. Cheltenham: Edward Elgar.

Miller, David. 2009. "Global Justice and Climate Change." In *The Tanner Lectures on Human Values.* Salt Lake City: University of Utah: 119–156.

Neumayer, Eric. 2000. "In Defence of Historical Responsibility for Greenhouse Gas Emissions." *Ecological Economics*, 33: 185–192.

Nordhaus, William D. 2007. "A Review of the Stern Review on the Economics of Climate Change." *Journal of Economic Literature*, 45(3): 686–702.

Paavola, Jouni. 2005. "Seeking Justice: International Environmental Governance and Climate Change." *Globalizations*, 2(3): 309–322.

Page, Edward A. 2006. *Climate Change, Justice and Future Generations.* Cheltenham: Edward Elgar.

Page, Edward A. 2008 "Distributing the Burdens of Climate Change." *Environmental Politics*, 17(4): 556–575.

Page, Edward A. 2011. "Cashing In on Climate Change: Political Theory and Global Emissions Trading." *Critical Review of International Social and Political Philosophy*, 14(2): 59–79.

Parry, Martin L., Osvaldo F. Canziani, and Jean Palutikof. 2007. "Technical Summary." In *Climate Change 2007: Impacts, Adaptation and Vulnerability*, ed. Martin Parry, Osvaldo F. Canziani, and Jean Palutikof *et al.*, 23–78. Cambridge: Cambridge University Press.

Shue, Henry. 1993. "Subsistence Emissions and Luxury Emissions." *Law & Policy*, 15(1): 39–60.

Shue, Henry. 1999. "Global Environment and International Inequality." *International Affairs*, 75(3): 531–545.

Singer, Peter. 2002. *One World: The Ethics of Globalization.* London: Yale University Press.

Starkey, Richard. 2011. "Assessing Common(s) Arguments for an Equal Per Capita Allocation." *Geographical Journal*, 177(2): 112–126.

Stern, Nicholas. 2007. *The Economics of Climate Change: The Stern Review.* Cambridge: Cambridge University Press.

Traxler, Martino. 2002. "Fair Chore Division for Climate Change." *Social Theory and Practice*, 28(1): 101–134.

United Nations Framework on Climate Change (UNFCCC). 1992. "United Nations Framework Convention on Climate Change," http://unfccc.int/resource/docs/convkp/conveng.pdf (accessed October 20, 2012).

United Nations Framework on Climate Change (UNFCCC) 1997. "The Kyoto Protocol to the United Nations Framework Convention on Climate Change," http://unfccc.int/resource/docs/convkp/kpeng.pdf (accessed October 20, 2012).

United Nations Framework on Climate Change (UNFCCC) 2011. "Report of the Conference of the Parties on its Sixteenth Session. Part Two: Action Taken by the Conference of the Parties at its Sixteenth Session," http://unfccc.int/resource/docs/2010/cop16/eng/07a01.pdf (accessed October 20, 2012).

Vanderheiden, Steve. 2008a. *Atmospheric Justice: A Political Theory of Climate Change.* Oxford: Oxford University Press.

Vanderheiden, Steve. 2008b. "Climate Change, Environmental Rights, and Emissions Shares." In *Political Theory and Climate Change*, ed. S. Vanderheiden, 43–66. Cambridge, MA: MIT Press.

Part III   Global Actors, Institutions, and Processes

Chapter 15

The Nation-State, International Society, and the Global Environment

Robert Falkner

Introduction

Global environmental protection has become a well-established and to some extent routine aspect of foreign policy. Hardly a day passes without an international gathering of environmental experts and diplomats debating issues from species loss to air pollution or global warming. But it was not always thus. A century ago, most states considered environmental concerns to be part of domestic, not international, politics. The League of Nations was not given an environmental mandate, and even the United Nations did not initially have a separate body dealing with environmental matters. It was only from the early 1970s onwards that international society began to take a more systematic interest in matters relating to regional and global environmental protection. What explains this relatively recent surge of interest in green diplomacy? How committed are states, and the great powers in particular, to the emerging norm of global environmental responsibility? Does the rise of international environmental politics signify a lasting process of normative change in international relations, that is, a greening of international society and the nation-state?

This chapter reviews recent scholarship on the role that global environmental protection plays in states' foreign policy and the changes in international society that have promoted a green dimension in international diplomacy. It opens with a discussion of the relationship between the nation-state and the global environment, and between a territorially defined international system and the interdependencies of global ecosystems. Environmentalists have traditionally considered this relationship to be deeply problematic, although some have more recently speculated on the emergence of the "green state." The chapter then considers how domestic forces shape foreign environmental policy, before reviewing the international factors that have promoted a greater engagement by states with global environmental concerns.

The Handbook of Global Climate and Environment Policy, First Edition. Edited by Robert Falkner.
© 2013 John Wiley & Sons, Ltd. Published 2013 by John Wiley & Sons, Ltd.

The Nation-State and the Ecological Challenge

Environmentalists have long considered the nation-state to be a dysfunctional form of political organization when it comes to addressing global environmental problems. There are several reasons for the deep-seated anti-statism in environmental thinking and activism (Paterson 2009). For one, the division of the world into sovereign nation-states stands in the way of the collective action that is needed to address global environmental threats. In the absence of a central authority such as a world government, individual states are driven to pursue a narrowly defined, short-term, national interest that ignores the universal and long-term concerns of the global environment. A fundamental mismatch exists between the political borders of international society and the physical boundaries of the ecological systems that span the world. In a fragmented and decentralized international system, the authority to deal with various local, regional, and global environmental challenges rests with political entities that are "widely seen to be both too big and too small" (Hurrell 2007: 216) to provide effective solutions.

Furthermore, despite enjoying formal sovereignty within their own borders, many nation-states do not exercise effective control over the environmental destruction that goes on within, or emanates from, their territory. This is evidently the case with so-called weak or failing states that are unable to provide even a modicum of domestic order and governance. Impoverished and unstable countries such as Haiti and Somalia come to mind in this context. Environmental destruction is rampant in such conditions, as local state officials either ignore environmentally damaging activities or are complicit in them (Hurrell 2007: 221). But a certain form of "state failure" can also be found in the so-called strong states of the industrialized world. As various authors have pointed out, some of the most advanced liberal democracies suffer from a control deficit when it comes to addressing newly emerging but potentially catastrophic ecological risks (Jänicke 1990; Beck 1995). As ecological problems take on ever more global dimensions, the ability of *all* types of states to direct social and economic dynamics towards greater sustainability is being called into question. The global environmental crisis has thus unearthed a much more profound crisis of sovereign statehood, with globalization and a shift of power from public to private actors eroding the regulatory power of the nation-state (Mathews 1997; Strange 1999).

Set against this ecological critique of the nation-state, which operates at both an empirical and normative level, is the assertion by other scholars that state-based political institutions remain central to the search for global environmental solutions (Meadowcroft 2005; Hurrell 2007). In this view, the nation-state and the international system have turned out to be more resilient to the corrosive effects of increased economic globalization and ecological interdependence than critics have suggested. Moreover, the growing awareness of the global dimensions of the environmental crisis has if anything increased the demands for state intervention in the global economy. Even if individual states on their own are unable to provide effective solutions to environmental problems, their central role in establishing the international environmental agenda, creating international environmental institutions, and negotiating environmental treaties has served to strengthen the legitimacy of state-centric forms of global governance.

Indeed, as the international states system is unlikely to be replaced with a different form of global political organization, a growing number of environmental scholars argue that any global environmental rescue needs to involve the political authority of the nation-state. States may be myopic in their pursuit of the national interest, but they remain central to any attempt at organizing collective environmental action, whether at the domestic or global level. Only states possess the authority and steering capacity to direct powerful global economic actors towards greater environmental sustainability (Barry and Eckersley 2005). Where different national interests can be aligned to tackle global problems, as was the case with international cooperation to combat ozone layer depletion, the established institutions of international diplomacy can form the basis for effective remedial action at the global level (Benedick 1991). From this perspective, the challenge is thus to work out the conditions for successful interstate cooperation and institution-building, and to identify the political leadership that can promote green values and concerns in foreign policy (Eckersley 2004: 253–254).

More recent contributions to this debate have focused on the question of the nation-state's changing character, with some pointing to the beginning of a transformation in the state's central purpose and "the possible genesis of an ecological state, a state that places ecological considerations at the core of its activity" (Meadowcroft 2005: 3). At the heart of this argument is the idea that the state is not an immutable entity but a historically contingent political construct, and that its socially defined purpose has shifted over time, embracing industrialism and liberal democracy in the nineteenth century, welfare provision and social democracy in the twentieth century, and now environmentalism in the late twentieth and early twenty-first century. The basis for such a transformation can be found in the emergence of what Eckersley and others refer to as "ecological democracy," an evolution of liberal democracy to a state in which

> all those potentially affected by a risk should have some meaningful opportunity to participate or otherwise be represented in the making of the policies or decisions that generate the risk (Eckersley 2004: 111).

Others point to the gradual constitutionalization of environmental rights that advances both substantive and procedural environmental principles in policy-making (Hayward 2005). While most discussions of the evolution of the green state focus on the domestic sources of change (Dryzek et al. 2003), some authors highlight the inevitably international dimension of this process, as the aims of the ecological transformation of statehood can only be secured through international collective action (Eckersley 2004; Meadowcroft 2005: 12–13).

Whether such a transformation towards an ecological state is already underway and is likely to succeed remains a matter of debate and contention (Reus-Smit 1996). Some of the contours of this gradual and difficult transformation are already discernible, and recent scholarship has fleshed out both the normative foundations of what might be called "ecologically responsible statehood" (Eckersley 2004: 2) and its empirical manifestations. Others, however, warn that the intensification of environmental problems in countries with limited state capacity may promote a different kind of green state, one that combines authoritarian rule rather than democracy with a developmental model that seeks to balance the underlying growth imperative with emerging social and environmental pressures (Beeson 2010).

Much of the empirical debate about the international dimensions of the greening of the state has focused on Europe and the European Union's global environmental leadership. The notion of the EU as a unique political entity that reflects post-national values, including global environmental protection, has played a significant role in the wider debate about Europe's role in international affairs. Reflecting its unique supranational character and reliance on civilian rather than military power (Whitman 1998), some scholars have identified the EU as a "normative power" that defines itself in part out of a concern for environmental sustainability (Manners 2002). Other political values such as human rights and peaceful conflict resolution are usually cited as the core elements of Europe's distinctive normative identity. However, since acquiring an environmental policy competence in the late 1980s, the EU has increasingly made sustainable development and environmental protection one of its core principles and has also inscribed them into various constitutional treaties (McCormick 2001; Baker 2006). As a consequence, sustainability has risen in importance as a guiding principle of European foreign policy, and European leaders now routinely claim an environmental leadership role in international politics (Vogler 2005).

The notion of the EU as a green normative power has raised a number of questions and objections. Some scholars question the depth of the EU's environmental commitment and point to serious contradictions between its progressive stance on issues such as climate change and biosafety and its relatively weaker role in areas such as fisheries and agriculture. Others highlight major contradictions between the EU's commitment to sustainability and its other normative principles, such as economic freedom and trade liberalization (Zito 2005). Scholars working within a political economy perspective also point to the close links between domestic interest constellations and European foreign environmental policy, which determine the degree to which the EU assumes a global leadership role (Falkner 2007). In this view, green normative power is better seen as a strategy of regulatory export, with the drive towards the global adoption of European environmental regulations serving as much an economic as an environmental interest.

Inside-out: Domestic Sources of Foreign Environmental Policy

Why do states pursue environmental objectives as part of their foreign policy? The answers to this question fall into two broad categories: *inside-out* explanations that focus on the role of domestic factors in shaping foreign policy; and *outside-in* explanations that reverse the domestic logic and trace a state's stance in global environmental politics back to its position within the structure of the international system (for an overview of theories of foreign environmental policy, see Barkdull and Harris 2002). This section deals with the former, while the next section discusses the latter perspective.

Inside-out explanations can be subdivided into two broad strands: societal explanations that focus on the role of public opinion and competition between domestic interest groups; and statist explanations that identify the sources of green foreign policy within the institutional structures of the state. Their main difference concerns the extent to which the state, and specific actors within state institutions, can be assumed to be autonomous in deciding a state's foreign environmental policy.

Societal Interests: Environmental NGOs and Business

The societal perspective adopts a bottom-up logic to foreign policy-making in which the state is assumed to be a largely neutral actor that mostly responds to the demands and pressures arising from domestic politics.

This perspective is intuitively convincing if we consider long-term historical trends, which suggest that a state's conduct in international environmental politics broadly reflects domestic societal values and preferences. That societies that place greater emphasis on environmental protection tend to pursue a proactive foreign environmental policy can be seen from the creation of the international environmental agenda in the 1970s. The countries behind the first UN environment conference in 1972 – mainly the Scandinavian countries and the United States – were the first to experience a dramatic rise in domestic environmental awareness and green political campaigning. Environmental concerns still play a prominent role in Scandinavian societies today, and countries such as Norway, Sweden, and Denmark are noted for their green diplomacy. In contrast, a declining societal interest in environmental issues in America has coincided with a retreat of the USA from an agenda-setting role in global environmental affairs and a transfer of global environmental leadership from the USA to the EU (Kelemen and Vogel 2010; Vogel 2012). Even if the causal link between public opinion and foreign policy is not a straightforward one, long-term trends in public opinion provide at least a partial explanation of broader shifts in foreign environmental policy.

One of the factors that help explain the transmission of public opinion into governmental policy is the level of political mobilization and organization of environmental interests. The existence of a strong and highly organized environmental movement is of relevance here as it can directly play into electoral politics, either by shaping the electoral preferences of swing voters or by giving rise to the formation of green parties that compete directly in parliamentary elections. During the 1980s, for example, West Germany's Green Party gained in political prominence. As it won a growing number of seats in national and regional parliaments, other parties were forced to take environmental concerns more seriously. Ever since the arrival of the Greens in national politics, Germany has been in a leading international position on issues such as climate change, and particularly so after the Green Party was able to form a coalition government with the Social Democrats in 1998 (Hatch 2007).

To be sure, the strength of the environmental movement alone is no reliable indicator of a country's likely stance on specific environmental issues. For example, America is home to well-organized and experienced environmental campaign groups, but the USA has turned its back on environmental multilateralism and has failed to ratify most of the international environmental treaties negotiated since the 1992 Rio Earth Summit (Brunnée 2004). Environmental interests compete with other powerful domestic interests that may oppose ambitious environmental policies. It is thus in the interplay between pro- and anti-regulatory interests that we can find the strongest explanation of how domestic forces drive a country's foreign environmental policy. Much of the research literature has therefore focused on the competition between environmental campaigners and business groups as the key domestic constituencies of global environmental politics.

That environmental campaigners should call for their government to take a lead on global environmental issues is unsurprising. But that environmentalists could find allies in the business community in their push for international environmental regulation is a phenomenon that requires some explanation. Although business actors have traditionally resisted calls for stricter regulation, whether at national or international level, the field of business lobbying has changed dramatically since the 1980s, when global business groups first came to accept a greater responsibility for environmental protection (Falkner 2008: 5–7). Faced with strong regulatory pressure at home and international competition from countries with low environmental standards, some business groups have opted for a strategy of regulatory export to create a global level playing field or gain a first mover advantage.

It is the competitive dynamic of an increasingly global marketplace that has led certain business groups to join forces with environmental groups in pushing for international treaties such as the Montreal Protocol (Benedick 1991; Falkner 2008: chapter 3) or resisting international agreements such as the North American Free Trade Agreement (NAFTA) (Gallagher 2004). As DeSombre (2000) argues in her study of US foreign environmental policy, domestic support from environmentalists and industry groups has been a critical push factor behind various attempts by US administrations to internationalize domestic regulation. DeSombre's analysis supports the broader conclusion that once an environmentally progressive country has introduced new environmental regulations within its own jurisdiction, environmental campaigners will usually support the extension of those standards to the global level. Once industry groups start calling for regulatory export to extend domestic regulation to their competitors, a so-called "Baptists-and-bootleggers-coalition" becomes the key driving force behind the government's foreign environmental policy. Similar patterns of domestically driven attempts to internationalize national environmental regulations have been observed in a number of contexts, most notably in the EU (Pollack 1997; Darst and Dawson 2008; Kelemen 2010).

As the role of business actors has received more attention in environmental scholarship (see Chapter 17 in this volume), the divisions within the business community have come under closer scrutiny (Falkner 2008). Business is rarely united in its stance on international environmental regulation, and divisions between different sectors or companies provide opportunities for environmental campaigners and state actors to create pro-regulatory alliances with progressive business forces. The political space that so-called business conflict creates has been recognized more widely in the literature. Students of social movements, for example, have identified various "industry opportunity structures" (Schurman 2004) that empower activists in their effort to build political support for specific regulatory approaches. Similarly, state actors themselves may seek to mobilize supportive industry interests to bolster their case for international environmental policies. Interaction between domestic interest groups and governments is thus not a one-way street but involves state actors themselves seeking to shape the domestic basis of foreign policy.

Statist Approaches

In contrast to societal models of foreign environmental policy which give domestic political actors causal priority, statist approaches place greater emphasis on the (semi-)autonomous role played by state actors themselves. Societal explanations may dominate the literature on environmental politics, with but a small number of statist scholars have demonstrated how an exclusive focus on pluralist interest competition tends to underplay the significant degree of policy choice that actors within the state have. Statist explanations take on different forms, ranging from a focus on presidential leadership and executive–legislative relations to the study of intra-bureaucratic power struggles and the role of policy ideas and ideologies (see Barkdull and Harris 2002: 79–84).

Hopgood's (1998) study of America's engagement with global environmental affairs is one of the defining examples of a statist explanation of foreign environmental policy. Surveying the period from the 1972 Stockholm Conference to the 1992 Rio Summit, Hopgood shows how key members of successive US administrations played a central role in determining US foreign policy objectives. Although exposed to persistent lobbying by environmentalists and other interest groups, state officials enjoyed a considerable degree of autonomy in developing diplomatic strategies. Indeed, international environmental politics provided "an enhanced opportunity for officials to turn their preferences into policy" (Hopgood 1998: 222). Rather than merely responding to the policy input from domestic interest groups, these senior executive branch officials used societal actors as a power resource and support base in their political struggles within the core state institutions. Key individuals such as Russell Train thus acted as pivotal policy entrepreneurs, exerting influence over official policy that is independent from the strength of environmental lobbying in American politics (Barkdull 2001).

To some extent, the statist perspective challenges the view that global environmental politics is increasingly shaped by non-state actors. Environmental campaign groups and scientists may be influential in creating awareness and concern for environmental problems, but state actors remain central to the formulation of environmental policy and the creation and implementation of multilateral environmental agreements (Economy and Schreurs 1997: 3). Yet, it would be too simplistic to portray this debate as a zero-sum game in which state and non-state actors compete for control over international policy-making. Instead, more recent scholarship suggests a focus on emerging linkages – and potential synergies – between societal and state actors in shaping international environmental cooperation.

One such strand of research has focused on the growing internationalization of environmental politics, with domestic and international processes more closely intertwined and the state operating in a complex field of overlapping networks of actors. As Economy and Schreurs (1997: 2) argue, "[t]his internationalization of environmental politics is transforming the relationship among actors within and among states . . . Agenda setting, policy formulation, and implementation are becoming increasingly internationalized." States may remain in a powerful position to shape the international agenda, but international actors and institutions "reach down into the state to set domestic policy agendas and influence policy formation and implementation processes" (Economy and Schreurs 1997: 6). Increasingly dense vertical linkages between domestic environmental politics, state institutions, and the field of global environmental governance are thus reshaping established patterns of policy-making within the state and beyond (Busch *et al.* 2012).

Another strand has focused on the growth in horizontal environmental policy networks that connect state actors across national boundaries. In *A New World Order* (2004), Slaughter points to the growth in transgovernmental networks, comprising environmental regulators from different jurisdictions that form "clubs" of experts with the purpose of enhancing the flow of policy-relevant information, assisting with the enforcement of national laws, and promoting the international harmonization of laws and regulations. In this way, state actors have responded to the challenges of an increasingly interdependent world in which states can no longer come up with effective policy responses on their own. Such networks increasingly include non-state actors too, for example in standard-setting organizations with a mixed membership such as the International Organization for Standardization (ISO) and public–private partnerships that promote sustainable development objectives (Prakash and Potoski 2006; Pattberg *et al.* 2012).

Outside-in: The Greening of the International System

An alternative view of the relationship between states and environmental politics places greater emphasis on the international system and its impact on state identity and behavior. In this outside-in perspective, foreign environmental policy is explained with the help of international power structures or the evolution of international norms. It also considers the role that policy transfer, diffusion, and learning play in the spread of green practices thoughout the international system.

Power and Hegemony

Power-based explanations of state behavior focus on the distribution of power in the international system. Although they have played a fairly marginal role in the study of international environmental politics (Barkdull and Harris 2002: 70), the contours of a structural approach can be detected in discussions of the role of hegemonic states in advancing the international environmental agenda (Falkner 2005). To employ the concept of hegemony in a meaningful way in this context, one has to move beyond narrow conceptions of military power and consider its economic and political dimensions, but also broader questions of the social construction of legitimate hegemonic power (Clark 2011: chapter 9). Borrowing ideas from hegemonic stability theory, some scholars have pointed to the leadership provided by the United States in the creation of international environmental institutions and agreements, particularly in the 1970s and 1980s. Hegemons may not be able to impose environmental accords, but their leadership can help foster international consensus on certain regulatory solutions (Young 1989: 88).

Environmental leadership can take many forms, such as injecting policy entrepreneurship into negotiations, developing regulatory blueprints that facilitate international policy diffusion, and using economic incentives and sanctions to encourage international cooperation (Young 1991; Vogel 1997; Ovodenko and Keohane 2012). Hegemony is neither a necessary nor sufficient condition for such leadership, but powerful states are usually in a better position to succeed in shaping the international environmental agenda in this way. Environmental leadership has thus been attributed to the United States until at least the 1992 Rio Earth Summit (DeSombre 2000: 5), while its absence in American foreign policy has been noted widely since (Paarlberg 1999; Falkner 2005). More recently, the EU has increasingly been credited with exercising global environment leadership. Its economic might as the world's largest import market has given it the clout to push for higher environmental standards internationally, whether through multilateral negotiation or *de facto* standard-setting based on the "trading-up" mechanism (Vogel 1997; Falkner 2007; Kelemen 2010). But questions persist about the EU's ability to shape the global agenda. Apart from questions about its coherence and capability as an international actor (Bretherton and Vogler 2005), observers of multilateral processes such as the climate negotiations have highlighted the EU's lack of clout in situations where multiple great powers resist more ambitious environmental objectives and issue complexity makes it difficult to find mutually agreeable solutions (Haug and Berkhout 2010).

The experience with American and European leadership in international environmental affairs raises broader questions about the explanatory value of power-based theories, and particularly hegemonic stability theory. The first question concerns the correlation between a state's hegemonic position and its foreign environmental policy stance. As the American case shows, the fact of preponderant power does not in itself determine whether a hegemonic state is likely to promote or hinder international environmental governance. Throughout the last four decades, the US has pursued multilateral *and* unilateral strategies and has supported *and* blocked international environmental treaties. Hegemonic stability theory cannot explain the particular policy choices that a hegemon makes, a key limitation of structural approaches that has been noted in other areas too (Lake 1993: 477).

Furthermore, structural theory provides only a first-cut explanation of the likely outcome of international environmental bargaining. In the environmental field, in particular, hegemonic power is of only limited use in producing successful outcomes in a multilateral setting (Mitchell 2003). Even when hegemons seek to advance global environmental protection, it is far from clear how they could "force" other states to agree to more ambitious environmental policies and implement them domestically. The sheer scale and complexity of global environmental destruction makes hegemonic power a blunt and mostly ineffective tool. Moreover, where hegemons act as laggards or veto powers, they may not always succeed in preventing other groups of states from developing international environmental agreements. Several instances can be cited that suggest that environmental regime-building is possible without, and even against the interests of, the hegemon: for example, the 1982 Convention on the Law of the Sea (Young 1994: 117), the 1997 Kyoto Protocol on climate change, and the 2000 Cartagena Protocol on Biosafety (Falkner 2007).

The analytical shortcomings of hegemony-based explanations notwithstanding, power theory remains of vital importance to our understanding of the international politics of the environment. Imbalances of power between developed and developing countries have shaped the way in which international society has addressed global environmental problems. They have ensured that Northern environmental interests usually end up being prioritized on the international agenda. More recently, the rise of emerging powers and new coalitions such as the BRIC or BASIC countries have left their mark on international negotiations. New veto players (e.g. China, India) have emerged, as has become evident in the climate change negotiations. At the same time, established powers such as the USA have been able to challenge some of the key elements of the existing climate regime, leading to a reinterpretation of the main equity norms ("common but differentiated responsibilities") and a corresponding shift in the international bargaining dynamic (Hurrell and Sengupta 2012).

Green Norms in International Society

A different outside-in perspective on foreign environmental policy is provided by those that focus on the emergence of environmental norms in international society. This perspective draws broadly on social theories of international relations, particularly constructivism, historical sociology, and English School theory. It sees environmentalism as a potentially transformative force that affects the normative structure of international relations. According to this logic, environmental norms become embedded in the social structure of the international system and contribute to the ongoing redefinition of state interests and identities towards greater inclusion of environmental sustainability concerns (Falkner 2012).

The notion that the international system is undergoing a process of greening has been put forward by several authors. Sociologists associated with the Stanford School have advanced a world society perspective on the rise of global environmentalism and argue that a world culture based on scientific rationality, combined with environmental mobilization and organization across boundaries, has led to the emergence of a "world environmental regime" (Meyer *et al.* 1997). In similar fashion, Deudney (1993) provides a functionalist account of how growing awareness of the global environmental crisis is ushering in a new era of international cooperation, based on international institution building and the growth of post-national cosmopolitan values. In both these accounts, environmentalism is seen as a transformative force that ends up transcending the states system. States increasingly assume the responsibility of promoting global environmental objectives, but it is the transnational forces of scientific rationality and cosmopolitanism that are at the root of the global political transformation.

In contrast, writers within the English School tradition locate the political change that the rise of environmental values brings about within the society of states. In *The Global Covenant* (2000), for example, Jackson identifies the new environmental ethic that environmentalists promote as the source of a new guardianship norm in international society that is explicitly addressed to states and their key representatives. State leaders have become "chief trustees or stewards of the planet . . . because they have the authority and power to address the problem" (Jackson 2000: 176). In similar vein, Buzan (2004) speaks of environmental stewardship as a new primary institution of international society, alongside more established institutions such as sovereignty, territoriality, diplomacy, and the market.

How well established is this new international environmental norm, or primary institution in English School parlance? As Hurrell reminds us, while

> the ecological challenge has indeed been one of the most important factors contributing to the changes that have taken place in the changing normative structure of international society . . . there is a real danger that transformationist claims overstate the scale of the changes that have actually taken place (Hurrell 2007: 236).

Even though environmentalism pushes global politics beyond the limits of state-centrism, "there is little chance of escaping from the centrality of the state" (2007: 235). It is therefore critical to understand the barriers to deep-seated change in the normative structure of international society. To have a lasting effect, environmental values need to permeate what Reus-Smit calls the constitutional order of international relations. They need to change the very purpose of the state, its *raison d'être*, which until recently has been defined by a focus on industrial growth and the concomitant exploitation of natural resources. Modern environmentalism as it arose in the late twentieth century started to challenge this industrial purpose but, as Reus-Smit warns, the results of this "ideological reevaluation ... remain unclear" (Reus-Smit 1996: 119).

What is becoming clear, however, is that the greening of international society has left distinctive traces in the behavior of states. At a minimum, states have come to accept a basic commitment to environmental multilateralism as a procedural norm. Environmental protection has become a widely accepted concern in foreign policy, and all major powers now participate in international environmental conferences and negotiations as a matter of routine. In what could be considered an "environmental citizenship" norm, "states are now expected to participate in the ever-expanding scope of environmental standard-setting and treaty-making" (Falkner 2012: 517). It is also becoming clear that the emerging norm of environmental responsibility has left its mark on existing international norms, or primary institutions, even though the extent of their greening remains a matter of debate. National sovereignty is being reinterpreted and the purpose of the nation-state is beginning to shift "from environmental exploiter and territorial defender to that of environmental protector, trustee, or public custodian of the planetary commons" (Eckersley 2004: 209). International law has undergone a gradual evolution to include more far-reaching and innovative legal concepts and approaches, such as the precautionary principle and the harm prevention norm. And the market principle is being redefined to include corrective state interventions that seek to internalize the often hidden environmental costs of market transactions. Yet, the transformation of the foundational principles of international society is far from complete and involves ongoing processes of normative challenge and accommodation that often leave environmental responsibility as the weaker norm (Falkner 2012).

Policy Diffusion

A third strand of research on the greening of international relations focuses on the spread of specific policies and regulatory models throughout the international system. This perspective investigates horizontal and vertical processes of environmental policy diffusion between states and/or international organizations. It starts from the observation that states increasingly adopt policies and instruments developed elsewhere, whether or not they have agreed to be bound by international environmental treaties. Policy diffusion, learning, and transfer can thus be identified as mechanisms that help to bring about a convergence in state behavior even in the absence of a strong international environmental regime.

Recent scholarship has identified a number of cases of innovative environmental policy approaches being diffused from one country to another. For example, a majority of governments have adopted strategic approaches to long-term environmental planning and sustainable development over the last three decades (Jörgens 2004); eco-labels were first adopted in Germany in 1978 and spread to Scandinavia, the United States, and Japan in the 1980s, and later throughout the rest of Europe and to New Zealand and Australia (Kern *et al.* 2001); and energy taxation became widely used in European countries despite the failure to create an EU-wide carbon tax in the early 1990s (Busch *et al.* 2005: 159).

The reasons behind the growth in policy diffusion are varied and complex, and scholarly debate continues on how best to explain this phenomenon. Some scholars point to a predominantly functionalist logic that is based on mimicry. In this perspective, governments that face similar environmental policy challenges copy the behavior of other governments that have developed innovative policy solutions and have shown some degree of success in implementing them. Under conditions of uncertainty over the consequences of policy choices, governments find it in their interest to adopt innovative policy models in order to reduce uncertainty and transaction costs (Ovodenko and Keohane 2012).

Others highlight the role that power plays in directing policy diffusion. Leading states that have invested political energy and incurred economic costs when establishing domestic environmental regulations may wish to see them adopted in other countries too, not least to ensure that their competitors operate under similar regulatory constraints and costs. The United States and the EU in particular have actively sought to spread their regulatory models worldwide, setting off processes of regulatory competition and polarization (Bernauer 2003; Drezner 2007). Yet other explanations focus on the ideational environment within which new policy ideas emerge, are adopted in one country, and then spread to other countries. Thus, the growing use of emissions trading schemes and self-regulatory agreements with industry reflects the popularity of a market-based approach to regulation since the 1980s (Bernstein 2001). Finally, domestic interest groups shape the pattern of policy diffusion by creating supportive environments for the adoption of new policies or vetoing their transfer if they oppose them (Falkner 2008; Meckling 2011).

In most cases, a combination of the above factors will be at play in international diffusion processes. To take the example of carbon emissions trading, mimicry and path dependence were involved in the US administration's initial support for emissions trading during the Kyoto Protocol negotiations, following the successful implementation of emissions trading as part of the US sulfur dioxide reduction program. That the USA succeeded in embedding emissions trading in the Kyoto Protocol's flexibility instruments, against European and developing countries' initial resistance, is more a reflection of the strength of American power than any form of international policy learning. Thereafter, the EU developed its own carbon emissions system as part of its commitment under the Kyoto Protocol, building in part on the institutional experiences of BP's and the UK's pilot trading schemes. Mimicry has again played an important role in the diffusion of emissions trading from Europe to other parts of the world that are seeking to learn from European experiences in designing a cost-efficient regulatory approach to carbon emission reductions (Damro and Méndez 2003; Meckling 2011).

Conclusion

This chapter has provided an overview of recent scholarship on the role of the state in international environmental politics. It is clear from the above discussion that, despite the many challenges that the global ecological crisis poses to the capacity and legitimacy of state-based environmental governance, global environmental policy will need to continue to focus on how states and international society shape global environmental strategies. States are not going to disappear any time soon, nor are non-state actors likely to replace the state and its institutions with alternative global governance mechanisms. Understanding the potential for innovative state-centric policy approaches thus remains of critical importance to any discussion of how to organize a global response to the environmental crisis.

Recent scholarship has identified several ways in which domestic actors seek to influence states' foreign policy on environmental issue. A domestic politics-based perspective that focuses on the role of public opinion and the interplay between environmental and business interest groups offers a powerful explanation of foreign environmental policy. Environmentally progressive states often pursue a strategy of regulatory export, with key domestic constituencies pushing for the international-ization of domestic regulation. Building grassroots support for a proactive green diplomacy is thus an important element in any strategy to build support for ambi-tious international environmental policy. Especially in democracies but increasingly also in countries with autocratic regimes, the roots of strong environmental policy abroad are to be found in domestic politics.

However, despite the important influence of societal and business actors, the state cannot simply be seen as a neutral arbiter of domestic interests. Instead, the insti-tutional structures of the state and the interests and ideas of key actors within core state institutions can often determine the direction of foreign environmental policy. This statist perspective plays a somewhat marginal role in the research literature but deserves closer attention. Proactive and charismatic state leaders have helped to shape environmental agendas and promote progressive policies at the domestic and international level. There is room for political leadership by individuals in the greening of state policies, even if domestic and international constraints restrict the scope for such forms of political agency.

The constraints on foreign environmental policy are most clearly felt when we consider the international political environment within which states operate. The absence of strong and centralized international institutions that could enforce envi-ronmental rules and compel individual states to comply with them must count as one of the chief obstacles for greater global environmental sustainability. In an anarchic international system, power differences and shifts in the international power balance inevitably constrain state behavior. But unlike in other areas of global policy-making, power theory provides only a limited explanation of the emergence and direction of international environmental politics. Great powers (and hegemons in particular) can be influential in putting environmental concerns on the international agenda, but power alone has rarely achieved successful outcomes in international environmental negotiations.

Social theories of international relations add a different perspective on the inter-action between states and the international system. In this perspective, the rise of global environmentalism contributes to the ongoing transformation in the norma-tive structure that underpins international society. We can thus identify a gradual "greening" of international relations that manifests itself in the growing acceptance of key environmental norms and of the need to engage in international processes of environmental regime-building. States now recognize both a substantive commit-ment to environmental sustainability and a procedural commitment to environmental multilateralism. They learn from one another and diffuse innovative policies through international and transgovernmental networks. In this sense, international society has embarked on a profound process of normative expansion and transformation, even if this process is ongoing and incomplete. Further research is needed to develop a better understanding of the forces that can promote this process and of the conflict that arises between the emerging norm of environmental responsibility and existing norms of international society.

References

Baker, S. 2006. "Environmental Values and Climate Change Policy: Contrasting the European Union and the United States." In *Values and Principles in European Union Foreign Policy*, ed. S. Lucarelli and I. Manners, 77–96. Abingdon: Routledge.

Barkdull, John. 2001. "U.S. Foreign Policy and the Ocean Environment: A Case of Executive Branch Dominance." In *The Environment, International Relations, and U.S. Foreign Policy*, ed. Paul G. Harris, 134–156. Washington, DC: Georgetown University Press.

Barkdull, John and Paul G. Harris. 2002. "Environmental Change and Foreign Policy: A Survey of Theory." *Global Environmental Politics*, 2(2): 63–91.

Barry, John and Robyn Eckersley. 2005. "W(h)ither the Green State?" In *The State and the Global Ecological Crisis*, ed. John Barry and Robyn Eckersley, 255–272. Cambridge, MA: MIT Press.

Beck, Ulrich. 1995. *Ecological Politics in an Age of Risk*. Cambridge: Polity.

Beeson, Mark. 2010. "The Coming of Environmental Authoritarianism." *Environmental Politics*, 19(2): 276–294.

Benedick, Richard E. 1991. *Ozone Diplomacy: New Directions in Safeguarding the Planet*. Cambridge, MA: Harvard University Press.

Bernauer, Thomas. 2003. *Genes, Trade, and Regulation: The Seeds of Conflict in Food Biotechnology*. Princeton, NJ: Princeton University Press.

Bernstein, Steven. 2001. *The Compromise of Liberal Environmentalism*. New York: Columbia University Press.

Bretherton, Charlotte and John Vogler. 2005. *The European Union as a Global Actor*, 2nd edn. Abingdon: Routledge.

Brunnée, Jutta. 2004. "The United States and International Environmental Law: Living with an Elephant." *European Journal of International Law*, 15(4): 617–649.

Busch, Per-Olof, Aarti Gupta, and Robert Falkner. 2012. "International–Domestic Linkages." In *Global Environmental Governance Reconsidered. New Actors, Mechanisms, and Interlinkages*, ed. Frank Biermann and Philipp Pattberg, 199–218. Cambridge, MA: MIT Press.

Busch, Per-Olof, Helge Jörgens, and Kerstin Tews. 2005. "The Global Diffusion of Regulatory Instruments: The Making of a New International Environmental Regime." *ANNALS of the American Academy of Political and Social Science*, 598(1): 146–167.

Buzan, Barry. 2004. *From International to World Society? English School Theory and the Social Structure of Globalisation*. Cambridge: Cambridge University Press.

Clark, Ian. 2011. *Hegemony in International Society*. Oxford: Oxford University Press.

Damro, Chad and Pilar Luaces Méndez. 2003. "Emissions Trading at Kyoto: From EU Resistance to Union Innovation." *Environmental Politics*, 12(2): 71–94.

Darst, Robert and Jane I. Dawson. 2008. "'Baptists and Bootleggers, Once Removed': The Politics of Radioactive Waste Internalization in the European Union." *Global Environmental Politics*, 8(2): 17–38.

DeSombre, Elizabeth R. 2000. *Domestic Sources of International Environmental Policy: Industry, Environmentalists, and U.S. Power*. Cambridge, MA: MIT Press.

Deudney, Daniel. 1993. "Global Environmental Rescue and the Emergence of World Domestic Politics." In *The State and Social Power in Global Environmental Politics*, ed. Ronnie D. Lipschutz and Ken Conca, 280–305. New York: Columbia University Press.

Drezner, Daniel W. 2007. *All Politics is Global: Explaining International Regulatory Regimes*. Princeton, NJ: Princeton University Press.

Dryzek, John S., David Downes, Christian Hunold, David Schlosberg, with Hans-Kristian Hernes. 2003. *Green States and Social Movements: Environmentalism in the United States, United Kingdom, Germany, and Norway*. Oxford: Oxford University Press.

Eckersley, Robyn. 2004. *The Green State: Rethinking Democracy and Sovereignty*. Cambridge, MA: MIT Press.

Economy, Elizabeth and Miranda A. Schreurs. 1997. "Domestic and International Linkages in Environmental Politics." In *The Internationalization of Environmental Protection*, ed. Miranda A. Schreurs and Elizabeth Economy, 1–18. Cambridge: Cambridge University Press.

Falkner, Robert. 2005. "American Hegemony and the Global Environment." *International Studies Review*, 7(4): 585–599.

Falkner, Robert. 2007. "The Political Economy of 'Normative Power' Europe: EU Environmental Leadership in International Biotechnology Regulation." *Journal of European Public Policy*, 14(4): 507–526.

Falkner, Robert. 2008. *Business Power and Conflict in International Environmental Politics*. Basingstoke: Palgrave Macmillan.

Falkner, Robert. 2012. "Global Environmentalism and the Greening of International Society." *International Affairs*, 88(3): 503–522.

Gallagher, Kevin P. 2004. *Free Trade and the Environment: Mexico, NAFTA, and Beyond*. Palo Alto, CA: Stanford University Press.

Hatch, Michael T. 2007. "The Politics of Climate Change in Germany: Domestic Sources of Environmental Foreign Policy." In *Europe and Global Climate Change: Politics, Foreign Policy and Regional Cooperation*, ed. Paul G. Harris, 41–62. Cheltenham: Edward Elgar.

Haug, Constanze and Frans Berkhout. 2010. "Learning the Hard Way? European Climate Policy after Copenhagen." *Environment: Science and Policy for Sustainable Development*, 52(3): 20–27.

Hayward, Tim. 2005. "Greening the Constitutional State: Environmental Rights in the European Union." In *The State and the Global Ecological Crisis*, ed. John Barry and Robyn Eckersley, 139–158. Cambridge, MA: MIT Press.

Hopgood, Stephen. 1998. *American Foreign Environmental Policy and the Power of the State*. Oxford: Oxford University Press.

Hurrell, Andrew. 2007. *On Global Order: Power, Values, and the Constitution of International Society*. Oxford: Oxford University Press.

Hurrell, Andrew and Sandeep Sengupta. 2012. "Emerging Powers, North–South Relations and Global Climate Politics." *International Affairs*, 88(3): 463–484.

Jackson, Robert. 2000. *The Global Covenant: Human Conduct in a World of States*. Oxford: Oxford University Press.

Jänicke, Martin. 1990. *State Failure: The Impotence of Politics in Industrial Society*. Cambridge: Polity Press.

Jörgens, Helge. 2004. "Governance by Diffusion: Implementing Global Norms through Cross-National Imitation and Learning." In *Governance for Sustainable Development. The Challenge of Adapting Form to Function*, ed. William M. Lafferty, 246–283. Cheltenham: Edward Elgar.

Kelemen, R. Daniel. 2010. "Globalizing European Union Environmental Policy." *Journal of European Public Policy*, 17(3): 335–349.

Kelemen, R. Daniel and David Vogel. 2010. "Trading Places: The Role of the United States and the European Union in International Environmental Politics." *Comparative Political Studies*, 43(4): 427–456.

Kern, Kristine, Helge Jörgens, and Martin Jänicke. 2001. *The Diffusion of Environmental Policy Innovations. A Contribution towards Globalising Environmental Policy.* Discussion Paper FS II 01-302. Berlin: Wissenschaftszentrum Berlin für Sozialforschung.

Lake, David A. 1993. "Leadership, Hegemony, and the International Economy: Naked Emperor or Tattered Monarch with Potential?" *International Studies Quarterly*, 37(4): 459–489.

Manners, Ian. 2002. "Normative Power Europe: A Contradiction in Terms?" *Journal of Common Market Studies*, 40(2): 235–258.

Mathews, Jessica T. 1997. "Power Shift." *Foreign Affairs*, 76(1): 50–66.

McCormick, John. 2001. *Environmental Policy in the European Union.* Basingstoke: Palgrave Macmillan.

Meadowcroft, James. 2005. "From Welfare State to Ecostate." In *The State and the Global Ecological Crisis*, ed. John Barry and Robyn Eckersley, 3–23. Cambridge, MA: MIT Press.

Meckling, Jonas. 2011. *Carbon Coalitions: Business, Climate Politics, and the Rise of Emissions Trading.* Cambridge, MA: MIT Press.

Meyer, John W., David John Frank, Ann Hironaka *et al.* 1997. "The Structuring of a World Environmental Regime, 1870–1990." *International Organization*, 51(4): 623–651.

Mitchell, Ronald B. 2003. "International Environment." In *Handbook of International Relations*, ed. Walter Carlsnaes, Thomas Risse, and Beth A. Simmons, 500–516. London: Sage.

Ovodenko, Alexander and Robert O. Keohane. 2012. "Institutional Diffusion in International Environmental Affairs." *International Affairs*, 88(3): 523–541.

Paarlberg, Robert L. 1999. "Lapsed Leadership: U.S. International Environmental Policy Since Rio." In *The Global Environment: Institutions, Law, and Policy*, ed. Norman J. Vig and Regina S. Axelrod, 236–255. London: Earthscan.

Paterson, Matthew. 2009. "Green Politics." In *Theories of International Relations*, ed. Scott Burchill, Andrew Linklater, Richard Devetak *et al.*, 260–283. Basingstoke: Palgrave Macmillan.

Pattberg, Philipp, Frank Biermann, Sander Chan and Aysem Meert, eds. 2012. *Public–Private Partnerships for Sustainable Development: Emergence, Influence and Legitimacy.* Cheltenham: Edward Elgar.

Pollack, Mark A. 1997. "Representing Diffuse Interests in EC Policy-Making." *Journal of European Public Policy*, 4(4): 572–590.

Prakash, Aseem and Matthew Potoski. 2006. *The Voluntary Environmentalists: Green Clubs, ISO 14001 and Voluntary Environmental Regulations.* Cambridge: Cambridge University Press.

Reus-Smit, Christian. 1996. "The Normative Structure of International Society." In *Earthly Goods: Environmental Change and Social Justice*, ed. Fen Osler Hampson and Judith Reppy, 96–121. Ithaca, NY: Cornell University Press.

Schurman, Rachel. 2004. "Fighting 'Frankenfoods': Industry Opportunity Structures and the Efficacy of the Anti-biotech Movement in Western Europe." *Social Problems*, 51(2): 243–268.

Slaughter, Anne-Marie. 2004. *A New World Order.* Princeton, NJ: Princeton University Press.

Strange, Susan. 1999. "The Westfailure System." *Review of International Studies*, 25(3): 345–354.

Vogel, David. 1997. "Trading Up and Governing Across: Transnational Governance and Environmental Protection." *Journal of European Public Policy*, 4(4): 556–571.

Vogel, David. 2012. *The Politics of Precaution: Regulating Health, Safety, and Environmental Risks in Europe and the United States.* Princeton, NJ: Princeton University Press.

Vogler, John. 2005. "The European Contribution to Global Environmental Governance." *International Affairs*, 81(4): 835–850.

Whitman, Richard G. 1998. *From Civilian Power to Superpower? The International Identity of the European Union.* London: Macmillan.

Young, Oran R. 1989. *International Cooperation: Building Regimes for Natural Resources and the Environment.* Ithaca, NY: Cornell University Press.

Young, Oran R. 1991. "Political Leadership and Regime Formation: On the Development of Institutions in International Society." *International Organization*, 45: 281–309.

Young, Oran R. 1994. *International Governance. Protecting the Environment in a Stateless Society.* Ithaca, NJ: Cornell University Press.

Zito, Anthony R. 2005. "The European Union as an Environmental Leader in a Global Environment." *Globalizations*, 2(3): 363–375.

Chapter 16

Transnational Environmental Activism

Susan Park

Introduction

While environmental politics was once considered "low politics" in international relations, much of the work of raising its importance has come not just from increasing scientific knowledge about environmental problems but overwhelmingly from the ongoing activities of concerned individuals, groups, networks, and movements. This chapter identifies the actors, aims, and agency that constitute environmental activism in global environmental politics. It does so in five sections: first, the variety of actors engaged in environmental activism is identified; second, the aims and activities of environmental activists are examined; third, the agency or ability of environmental activists to bring about change in global environmental politics is investigated; fourth, a review of environmental activism is undertaken in order to further improve environmental outcomes particularly in relation to climate change; and finally, the chapter reflects on the global policy implications of current environmental activist efforts.

Defining Environmental Activists

While individuals engage in environmental activism, the benefits of cooperation and coordination have been realized through the creation of non-government organizations (NGOs) to further environmental goals. There has been a dramatic increase in the number of NGOs from the beginning of the twentieth century (Princen and Finger 1994: 1–26 (Introduction); Sikkink and Smith 2002: 30). Scholars estimated that 6000 international NGOs were operating by 1990, increasing to over 50 000 by 2005 (Clapp and Dauvergne 2011: 80; UIA 2011).[1] The United Nations Economic and Social Council (ECOSOC) formalized the definition of NGOs and their role in

The Handbook of Global Climate and Environment Policy, First Edition. Edited by Robert Falkner.
© 2013 John Wiley & Sons, Ltd. Published 2013 by John Wiley & Sons, Ltd.

world politics during the United Nation's inception (Gotz 2008: 238; Betsill and Corell 2008: 4). It currently recognizes 3500 NGOs, attributing them consultative status, with 2432 NGOs working on sustainable development (ECOSOC 2011). Relying on this data can be problematic, because there is no generally accepted definition of the term "NGO" and conceptual confusion continues (Gotz 2008). Our understanding of Environmental NGOs (ENGOs) is further hampered by the blurring of what constitutes the "environment" when problems cross a range of social, economic, and political areas (Conca 1996: 104).

At the most basic level NGOs include any organization that does not represent government. This may include trade unions, business councils, criminal organizations such as the Mafia, and religious orders such as the Catholic Church. While this definition of NGOs is widely accepted, others further distinguish NGOs from other non-state actors based on their purpose: between organizations whose aim is to maximize the material wealth or power of their members and those that aim to further altruistic public policy goals like conserving the environment for the common good. Accordingly, Chasek *et al.* define an NGO as an "independent, non-profit organization not beholden to a government or profit-making organization" (2006: 73). This means that the term "'NGO' is established as describing a broad range of private organizations serving public purposes" (Gotz 2008: 235).

Environmental NGOs

Environmental NGOs (ENGOs) promote the conservation of the environment. Chasek *et al.* (2006: 74) argue that ENGOs have international influence for three reasons. First, because they possess expert knowledge about their issue and can be innovative in thinking through how to respond to the problem; second, they are dedicated to the environment beyond national or sectoral interests compared to other actors; and finally, because NGOs represent citizens who can be mobilized to support environmental outcomes in traditional domestic political processes.

ENGOs vary in terms of their objectives and ideology, affiliation, structure, and funding. ENGOs have specific environmental objectives and a strategy based on the political philosophy which determines how they meet their goals. These may be divided according to NGOs that: undertake environmental projects and programs (otherwise known as operational NGOs, see Willets 2011); conduct environmental research such as collecting data and monitoring trends; provide education and raise awareness; undertake advocacy through political lobbying and networking; provide policy recommendations and draft treaty texts; or a combination of these. ENGOs may be research institutes or think tanks that conduct and publicize policy positions on environmental issues such as the World Resources Institute (WRI), the World Conservation Union (IUCN), the International Institute for Environment and Development (IIED), and the Centre for Science and the Environment (CSE) (Chasek *et al.* 2006: 74).

Other ENGOs target specific environmental issues. For example, Greenpeace International has seven issue areas: climate change, forests, oceans, agriculture, toxic pollution, nuclear safety, and peace and disarmament (Greenpeace International 2011). Ideology also informs how ENGOs establish strategies to meet their objectives (Gulbrandsen and Andresen 2004: 56; Alcock 2008; McCormick 2011: 102).

Greenpeace began by opposing nuclear issues through a strategy of physical protest and publicity stunts to raise awareness of environmental harm underpinned by an ideology of bearing witness through non-violent opposition (Wapner 1995: 320). For example, Greenpeace raises awareness and advocacy via direct action including chasing and harassing whaling boats; monitoring and reporting on environmental conditions such as tracking the trade in toxic chemicals; and lobbying governments on climate change.

Other large international ENGOs operate differently (Alcock 2008). The World Wide Fund for Nature (WWF; known as the World Wildlife Fund in the United States) focuses on biodiversity conservation at the local and international levels through undertaking biodiversity projects all over the world and lobbying governments. Ideology influences which environmental issues NGOs will target for their campaigns and how they will do so: Earthfirst!, an American NGO, is engaged in eco-sabotage to further its biocentric radical beliefs, whereas the Centre for International Environmental Law (CIEL) lobbies government and intergovernmental organizations (IOs) on the legality of their practices in international and domestic law. ENGOs may work together on the same issue despite these ideological differences or their efforts might work against one another (Princen 1994; Jordan and van Tuijl 2000).

Of course ENGOs vary in size, structure, and funding as well as the level at which they operate: local, national, regional, or international. ENGOs include small-scale grassroots organizations that operate at the community level. They may be national and focused on environmental problems domestically, with or without a focus on the international or global dimension of the problem, such as the Royal Society for the Protection of Birds in Britain, the Bund fur Vogelschutz in Germany (McCormick 2011: 103), and the Australian Conservation Foundation in Australia (ACF 2011). Well-known domestic American ENGOs include the Sierra Club, the National Audubon Society, the National Wildlife Federation (NWF), the Natural Resources Defense Council (NRDC), and Environmental Defense (ED) (Chasek et al. 2006: 75). ENGOs may also operate at the international level, or be affiliated with international NGOs. WWF is based in Switzerland, operates in over 100 countries, and has 5 million supporters. Based in the Netherlands, Friends of the Earth International (FoEI) is a confederation of 76 independent national member groups composed of over 2 million members. With headquarters in the Netherlands also, Greenpeace International operates in 41 countries and has nearly 3 million supporters.

Interactions between NGOs and governments vary depending on the political institutions in place and whether a state is open or closed to influence from groups that might be considered "enemies of the state." In Asia and Eastern Europe ENGOs have been one of the means to channel opposition to the state (Lee and So 1999), and in the case of the former Soviet Union were able to help change the political system (Fisher 1992). ENGOs' relationships with the state varies but it also influences how they are funded, which in turn shapes their operations. Some states will only accept state-sanctioned NGOs. Some "quasi-NGOs" are wholly or partially funded by official donors, states, or even transnational corporations (TNCs) (which may in fact advocate continuing environmental degradation; see Rowell 1996). As a result, ENGOs may engage in strategies of confrontation, collaboration, or complimentary activities with the state in which they operate to achieve their objectives (Najam 1996). The degree of state openness can also affect the efforts of transnational networks to bring about global environmental change (Newell 2000: 134).

Environmental Movements and Networks

While the focus thus far has been on ENGOs, there are of course a wide variety of other looser associations where activists have come together to fight for the environment. Many groups do not constitute NGOs in a formal way, with a charter or constitution, formal legal standing as recognized by the state, an office, logo, and budget. Many local grassroots organizations are informal, and many larger movements do not have formal members (see Pradyumna and Suganuma 2008). There is fluidity between the distinctions between ENGOs and environmental movements, where environmentalists have come together to protect the environment. Examples of environmental movements include the Green Belt movement in Kenya, the Chipko movement in India, and the Rubber Tappers movement in Brazil, all of which have pressed for the sustainable use of natural resources for community survival and began as informal associations (Chasek *et al.* 2006: 77). Informal grassroots organizations may become more formal and hierarchical over time, just as domestic NGOs may grow to become, or join, large international ones.

Above and beyond these classifications larger umbrella organizations may be established to pool resources. Third World Network is an example of an international umbrella NGO comprising individuals and organizations that aim to "contribute to policy changes in pursuit of just, equitable and ecologically sustainable development" for developing countries (TWN 2011). Owing to the dramatic increase in telecommunications (Wapner 1995: 317) such as the mobile phone, the internet, and social media, environmental movements and ENGOs do not need to scale up to achieve international campaign support.

Beyond intermediary NGOs that facilitate and amplify the operations of national and sub-national organizations and associations, informal NGO *networks* have also emerged around particular issues in order to discuss and legitimize the aims and strategies of environmentalists. A prime example is the Climate Action Network (CAN), which formed in 1989 and is dominated by Greenpeace International, the WWF, ED, and Friends of the Earth (Newell 2000: 126). It is a network of over 700 NGOs from over 90 states that "promotes government and individual action to limit human-induced climate change to ecologically sustainable levels" (Alcock 2008: 82; Betsill 2008; CAN 2011; Gulbrandsen and Andresen 2004: 61). CAN plays an "indispensable role in the coordination of strategy and campaigning activity, by orchestrating common positions among NGOs and keeping them informed of the latest developments in climate policy debates" (Newell 2000: 126–127).

The experience of CAN is comparable to the emergence of the environmental movement Climate Justice Now! (CJN!), which was launched on the final day of the Conference of the Parties 13 (COP 13) of the United Nations Framework Convention on Climate Change (UNFCCC) in Bali in 2007. Comprising groups and individuals that "were dissatisfied with the positions and processes of CAN," CJN! is critical of market solutions to climate change, while raising justice as a major concern in climate negotiations. In preparation for Copenhagen in 2009 (COP 15), other European and

Southern radical groups formed another association, the Climate Justice Action network (Fisher 2010: 15; CJA 2011; CJN! 2011; Guerrero 2011: 121). CAN has different priorities in climate negotiations on sustainability, efficiency, and equity (Alcock 2008: 82).

NGOs, individuals, local social movements, the media, foundations, churches, trade unions, consumer organizations, intellectuals, professionals, and parts of governments and IOs may also come together on particular issues to form transnational advocacy networks or TANs (Keck and Sikkink 1998: 9). Environmental TANs may form around issues such as deforestation, anti-oil and anti-dam campaigns (McAteer and Pulver 2009; Park 2010; Rodrigues 2004), as well as changing the practices, policies, and identities of IOs such as the World Bank Group (Park 2010). The emergence of TANs is based upon the coming together of a range of concerned actors around a particular issue, particularly where the international aspect of the network can bring pressure to bear on governments where domestic ENGOs have no leverage. This is known as the boomerang pattern (Keck and Sinkkink 1998: 12). TANs use four methods to achieve their goals: symbolism; generating and spreading information; holding actors accountable; and using leverage to assist less powerful actors (Keck and Sikkink 1998). However TANs may fail to form despite ripe conditions (Botetzagias *et al.* 2010) and inequality and tensions may arise in TANs, particularly where Southern NGOs become dependent on Northern NGOs or where the environmental problem is viewed differently along North–South lines (Jordan and van Tuijl 2000; Newell 2000: 126; Rohrschneider and Dalton 2002; Alcock 2008).

Global Civil Society

Overlaps exist between what some scholars call social movements (O'Brien *et al.* 2000; Khagram *et al.* 2002) and Keck and Sikkink's TANs. There is a lot of conceptual confusion over these categories. The main difference between social movements and NGOs, Sidney Tarrow argues, is behavioral, such that social movements are engaged in

> sustained contentious interaction with states, multinational actors or international institutions, whereas INGOs are engage in routine transactions with the same kinds of actors and provide services to citizens of other states (Tarrow 2001: 12).

Further, TANs may contain NGOs and social movements that work towards a common goal, where TANs are "informal and shifting structures" through which these and other actors interact (Tarrow 2001: 13).

Others go beyond identifying movements and networks to speak of a global civil society (Wapner 1995; Lipschutz 1996). Wapner, for example, argues that transnational activist efforts are evidence of a "world civic politics" where the activities of environmentalists go beyond pressure group politics and advocacy, through, for example, Northern NGOs funding conservation projects in developing countries (1995: 312). Global civil society captures these activities that are above the individual but below the state and take place across state borders (1995: 312–313). Lipschutz similarly locates activists within global civil society in order to counter the dominance of thinking about environmental politics through the Westphalian sovereign state system (1996). Others argue that using the concept global civil society does not give enough room for the role of agency (Keck and Sikkink 1998: 33), while Tarrow goes so far as to say that the global civil society thesis is "unspecified, deterministic and undifferentiated" (2001: 14; see also Rohrschneider and Dalton 2002). Bearing in mind these classificatory pitfalls, we can readily identify how environmental actors change world politics.

Aims and Activities of Environmental Activism

Environmentalists have been concerned with not only conservation, but, beginning in the 1960s and 1970s, actively challenging everyday economic, political, and social practices that contribute to environmental degradation. Environmentalists have had a number of high-profile successes in challenging the activities of TNCs, IOs, and states to take account of their environmental impacts.

Changing Corporate Practices

Much of the literature highlights the increasing role of TNCs as actors in global environmental politics based on their global reach and influence over industry activities (Cashore *et al.* 2004; Levy and Newell 2005; Falkner 2008; see also Chapter 17 in this volume). Yet there has been an "activist discovery and manipulation of economic means of power" (Wapner 1995: 330). This section highlights six targets of successful environmentalist campaigns against: (1) specific types of industries (the extractive industries); (2) specific companies; (3) products; (4) environmentally harmful international trade; and in favor of proscribing and prescribing TNC activity through (5) shareholder and investor activism; and (6) the promotion of voluntary environmental codes of conduct.

First, environmentalists have attempted to halt the environmentally degrading extraction, production, and trade of goods. Transnational activist campaigns have been undertaken against the extractive industries of oil, gas, and mining (Jordan and van Tuijl 2000; McAteer and Pulver 2009; Park 2010). The extractive industries provide energy for industry and individual consumption but are dirty industries in terms of extraction, transportation, waste, and their impact as fossil fuels on the global climate system. Measures that environmentalists have used include: direct protest at the site of extraction and production and corporate headquarters; social media and online as well as traditional media campaigns for spreading information, raising awareness, and advocating for change; boycotting the product and the company; filing law suits against corporate practices that break domestic law; linking harmful practices to international treaties; promoting domestic and international regulation over industry practices such as minimum pollution levels and safe waste disposal. In the case of anti-oil campaigns in Ecuador, TANs used investor activism to change TNC policy and limit some of their dirtiest activities, with mixed success (Jordan and van Tuijl 2000; McAteer and Pulver 2009).

Second, environmentalists have challenged industry-wide practices through targeting a *specific company*. The largest company in an industry may be targeted to change industry-wide operations. For example, in the late 1980s activists were successful in changing McDonald's, the world's largest fast-food provider at the time, use of ozone-depleting Styrofoam packaging of its burgers. Activists included the Citizens' Clearing House for Hazardous Wastes, the Earth Action Network, and Kids Against Pollution. They organized a "send-back" campaign in which people mailed McDonald's packaging to the national headquarters. In 1991 McDonald's bowed to activist pressure by stopping the use of its traditional Styrofoam hamburger boxes, despite not seeing Styrofoam packaging as an ecological problem (Wapner 1995: 327). This changed McDonald's practices in its 11 000 restaurants around the world as well as changing the packaging practices of its competitors.

Third, environmental campaigns have also targeted *specific products*. An international campaign began in 1985 to stop the accidental killing of dolphins while catching tuna. The Earth Island Institute, Greenpeace, and FoE among others campaigned to stop the use of drift-net and purse-seine fishing by tuna fleets in the Eastern Tropical Pacific Ocean that entangle dolphins. They advocated boycotting all canned tuna, demonstrated at shareholders' meetings, and held rallies against the Tuna Boat Association. The Earth Island Institute then assisted in producing a documentary, *Where Have All the Dolphins Gone?*, which was shown throughout the USA and across the world. The documentary promoted the idea of dolphin-safe tuna labeling to market environmentally sensitive brands. Environmental activism was crucial in stopping the slaughter of dolphins by tuna companies. This action has contributed to a sea-change in the tuna industry, stopping fishing practices that might accidentally catch dolphins and contributing to protecting dolphin populations around the world. Of course this success was undermined by the increasing strength of the global trade regime, where environmentalists' efforts to protect the environment ran up against the rules of the World Trade Organization (see also Chapter 24 in this volume).

Fourth, environmentalists have also aimed to *prohibit international trade*. The emergence of the IUCN and later the WWF were critical for the creation of the Convention on the Trade of Endangered Species (CITES). Much of the work on conservation has been led by ENGOs. ENGOs were key in promoting specific bans on engendered species products such as ivory through CITES and mobilizing public opinion, which helped contribute to a collapse of consumer demand (Princen 1994: 143). Activists were also instrumental in establishing the 1989 Basel Convention Controlling Transboundary Movement of Hazardous Wastes and their Disposal (Ford 2005: 323). Greenpeace played a vital role in shifting states' position in favor of ratifying Basel by monitoring and reporting on the trade in toxic waste (Chasek *et al.* 2006: 131), while the International Toxic Waste Action Network and the Basel Action Network continue to campaign for states to sign the Basel Convention and to oppose the trade of toxics between developed and developing states (Clapp 2001; Ford 2005: 232). Lastly, environmentalists such as the Pesticide Action Network (PAN) comprising over 600 organizations and individuals from over 60 countries, and the International POPs Elimination Network made up of over 700 public interest NGOs (IPEN 2011), have played an ongoing role in lobbying for the phasing out of chemicals such as Persistent Organic Pollutants (POPs) while mobilizing grassroots support for the 2001 Stockholm Convention (Chasek *et al.* 2006: 80).

Fifth, environmentalists have attempted to stop TNCs from getting financing for their operations. They have done so through engaging in shareholder activism, investor activism, and investor screening to limit environmentally and socially damaging corporate behavior. TANs target institutional shareholders such as pension funds and socially responsible investment firms that hold shares in companies, to hold the company to account for any decision they make that may negatively impact on the environment. Shareholder activism is on the increase, with 359 resolutions being filed against publicly listed US corporations by shareholders on socially responsible topics in 1997 (McAteer and Pulver 2009: 4). There are currently 67 investor advocacy networks, most of which have emerged over the last decade. Investor activism tends to be single-issue driven, for example the Carbon Disclosure Project, which aims to promote companies that reduce carbon emissions in order to tackle climate change. Further, investor screening also occurs where investors exclude from their portfolio corporations that engage in negative practices such as knowingly damaging the environment (MacLeod and Park 2011: 54).

Finally, activists have also challenged corporations to establish market-based voluntary environmental codes of conduct (Alcock 2008). Corporations have responded with elective public reporting (Gleckman 2004), which is often done on a firm-by-firm basis, and varies considerably in terms of what information is provided on Corporate Social Responsibility (CSR) activities and environmental and social governance (ESG) measures. TNCs have also created the International Chamber of Commerce's (ICC) Business Charter for Sustainable Development, and the World Sustainable Development Business Council. Environmentalists such as Greenpeace and FoEI pushed unsuccessfully for an international environmental accountability treaty to govern TNCs in the run-up to the World Summit on Sustainable Development (WSSD) in 2002 (Clapp 2005: 293–294). As a result, efforts to mitigate the environmental impact of TNCs remain voluntary, although they do accord with international guidelines for corporations such as the OECD Guidelines for Multinational Enterprises, the UN Global Compact, and the ISO 14000 environmental management standards.

The best-known voluntary code of conduct is the Coalition for Environmentally Responsible Economies (CERES). Previously known as the Valdez Principles after the *Exxon Valdez* oil spill in 1989 (Wapner 1995), CERES is a non-profit coalition of 130 member organizations that aims to "help business transition to a sustainable economy." Its members include NGOs, Fortune 500 companies, and institutional investors. CERES launched the Global Reporting Initiative (GRI), which is considered to be the most successful voluntary corporate environmental code and is used by over 1800 corporations for reporting on environmental, social, and economic performance (Levy *et al.* 2009; CERES 2011). The CERES Principles, the GRI, and the Carbon Disclosure Project (CDP) are general codes of conduct for corporations.

Voluntary environmental corporate codes of conduct tended to be based on an industry or product (Vogel 2008: 269). Opposition from environmentalists contributed to corporations creating self-defined implementation standards (Gleckman 2004). For example, protests against banks like Citigroup contributed to change in the finance industry, leading big private banks to create the Equator Principles for project finance. These are voluntary environmental and social guidelines for private sector financiers of major infrastructure and industrial projects. Environmentalists have pushed corporations to create self-financed certification standards such as certificates for forest and marine products, as well as cocoa, coffee, and flowers (Cashore *et al.* 2004; Vogel 2008; Bostrom and Hallstrom 2010).

Businesses and NGOs have together created sector-specific codes and labels. For example, the international Forest Stewardship Council (FSC) forest certification system was created by environmental groups such as WWF, in conjunction with industry and landowners (Cashore *et al.* 2004), and WWF was instrumental in establishing the Marine Stewardship Council for certifying marine products with Unilever. While standards such as the FSC aim to create stability between landowners and environmentalists, it remains difficult to identify the extent to which corporations follow standards like the Equator Principles for project finance (Wight 2012).

Changing International Organizations

In terms of IOs, environmentalists have been most successful in challenging their business-as-usual approach to incorporate environmental considerations in relation to the World Bank Group (Rich 1994; Wade 1997; Park 2010) but less so for the World Trade Organization (WTO) and the International Monetary Fund (IMF) (O'Brien *et al.* 2000). Placing the greening process in context, large-scale bureaucracies driven by sovereign states have difficulty implementing environmental ideas into their activities even without pressure from environmentalists (on the UN, see Conca 1996).

While pressure from environmental activists has been crucial in bringing about policy change, the overall discussion has been focused on whether IOs can become green or whether they have merely green-washed their operations. Evaluating the greening of IOs incorporates two factors: the push for IOs to address environmental issues and the extent to which IOs have responded to environmental pressure. Overwhelmingly, debates on why IOs have become green have focused on the World Bank. The World Bank initially adopted environmental concerns in the 1970s for a mixture of economic, political, and intellectual reasons, yet the "push" for a comprehensive re-evaluation of the Bank's environmental concerns came from mass environmental campaigns in the 1980s.

Yet scholars surmised that the Bank had not internalized environmental concerns. For example, Wade (1997) argued that while the Bank had shifted from "environment versus growth" to "environmentally sustainable development" it had not changed its internal incentive system, thus undermining environmental rigor (which some called "green neoliberalism": Goldman 2005). Environmental activist Bruce Rich (1994) agreed that the Bank had only green-washed its operations because its environmental criteria had not been implemented properly and that the Bank's loan approval culture confounded attempts to further environmentally sustainable development.

Alternatively, Haas and Haas (1995) argued that the World Bank had analyzed how environmental concerns fit within its organizational aims through a re-evaluation of its beliefs about cause and effect, resulting in a change of the organization's goals to employ new environmental criteria rather than superficial operational changes. They argued that only the United Nations Environment Programme (UNEP) and the Bank were capable of integrating environmental considerations into their traditional responsibilities. While recognizing that the Bank has substantially incorporated environmental ideas, the distinction between adaptation and learning tends to separate an organization's tactical responses to states and non-state actors on the one hand, and complex learning on the other. Yet tactical concessions are often seen, in longer-term analysis, to be the first step in a process of norm adherence (Park 2010).

Gutner (2002) also found that the World Bank was a greener bank, because it "finances projects with primary environmental goals and attempts to integrate environmental thinking into the broader set of strategic goals it develops." Others agreed that the World Bank had become greener, as evinced by an increase in environmental projects and loans, and an increase in environmental staff and environmental monitoring. The cause of the greening was the result of increased oversight by the Bank's member-states and targeted action by Bank management that aligned with the culture and incentive structure for staff within the Bank (Nielson *et al.* 2006). Yet this approach overlooked significant input from environmentalists that determined that the World Bank should change, including how it should do so, both through direct interactions with the World Bank and indirectly through powerful member-states. As a result of concerted environmentalist pressure, these standards would be diffused to the private sector through the project finance and political risk insurer arms of the World Bank Group, be taken up by the OECD for its political risk insurers, and by private banks through the Equator Principles (Park 2010).

Influencing Global Summits and Multilateral Environmental Agreements

Environmentalists have not just targeted TNCs and IOs; they have taken an active role in global environmental summits and international treaty negotiations. Clark *et al.* (2005) argue that the number of NGOs has increased in relation to global summits: 250 accredited NGOs attended the UN Conference on the Human Environment in 1972, while 1400 attended the United Nations Conference on Environment and Development (UNCED) in 1992. At the WSSD in 2002, 3200 were accredited (Betsill and Corell 2008: viii). While their involvement in the official proceedings is determined by states, parallel NGO forums have been operating since 1972 – 18 000 NGOs attended the parallel forum at UNCED in 1992 and over 20 000 attended the WSSD (Clark *et al.* 2005: 298, 297).

Since 1992 ENGOs have been involved in national and regional preparatory processes in the lead-up to environmental summits, although as with specific environmental treaty negotiations, the preparatory committee meetings became more exclusionary the closer they were to the summit (Clark *et al.* 2005: 302). Newell argues that the "UNCED System" privileges the better-resourced, primarily Northern ENGOs while encouraging the "the formulation of common positions by groups of interests," which has led to the reduction of "very disparate demands to the status of a lowest-common-denominator set of diluted policy suggestions." The end result has been, he argues, that governments find it easier to reject these proposals or to respond "via tokenistic, incremental policy changes" rather than substantive policy change (Newell 2000: 139). While there is little agreement on what influence ENGOs have had at global environmental summits, the WSSD signified a shift in the activities of ENGOs, states, and IOs with the move towards working together in public–private partnerships for achieving sustainable development (see Andonova 2010).

In relation to specific multilateral environmental agreements (MEAs), much of the environmental lobbying is done at the domestic level in preparations for negotiations (Newell 2000: 128; Betsill and Correll 2008; McCormick 2011). Environmental NGOs have been able to influence states' positions and have been active in negotiations on transboundary air pollution, regional stocks, climate change, ozone depletion, and biodiversity (Wapner 1995; Alcock 2008: 78; Betsill and Corell 2008; McCormick 2011). While states retain the ability to sign and implement MEAs, since the mid- to late 1980s ENGOs have focused on incorporating environmental concerns into the outcome of interstate negotiations, often through shaping public opinion (Newell 2000: 128, 136).

For example, ENGOs were able to use public pressure to press the US government to be part of the UNFCCC (Newell 2000: 130). ENGOs such as Greenpeace also played an important role in establishing the 1982 moratorium on whaling, thus helping to shift the whaling regime from pro- to anti-whaling. Greenpeace has continued to play an influential role in interstate whaling politics, arguably as a result of paying state membership dues and being part of states' delegations to the International Whaling Commission (Skodvin and Andresen 2003: 80). Typically, ENGOs may have "access to the conference venue, presence during meetings, interventions during debate, face-to-face lobbying of delegations, and [receive the] distribution of documents" (Gulbrandsen and Andresen 2004: 59). Often they are given observer status, which may or may not entail the opportunity to interject in debates. Some ENGOs have been included in official state delegations, giving them a greater opportunity to influence proceedings (Newell 2000: 137).

Betsill and Corell (2008) have attempted to identify the extent of ENGO influence in MEAs. Through participation ENGOs can frame the issue, set the agenda, and influence the positions of key states. Yet ENGOs have had varying levels of influence over the outcome of environmental treaties (Betsill and Correll 2008: 11). They argue that ENGOs may have low influence in international negotiations if there is no observable effect of their activity on the negotiation process or outcome despite active participation. ENGOs may have moderate influence if they can shape the negotiation process but did not affect the outcome, as was the case with negotiations over the Kyoto Protocol of the UNFCCC. ENGOs have a high degree of influence if they are able to shape the negotiation process and the outcome, as witnessed in the UN Convention to Combat Desertification (1993–1994) (Betsill and Correll 2008; McCormick 2011: 108).

Global Policy Dimensions: How Do Environmentalists Impact Climate Change?

While many everyday practices continue to have harmful effects on the natural environment, evidence provided above demonstrates that environmentalists can bring about positive change by challenging the practices of TNCs, IOs, and states in MEAs. Activists can do so because they are able to create new categories of meaning and action, such as establishing a consensus in favor of a moratorium on whaling and establishing projects and programs to protect the biosphere with state, corporate, and IO support. Environmentalists have soft power, which enables them to create these new categories and to spread environmental ideas and actions to other actors through framing issues, setting agendas, engaging in persuasion, symbolism, cognitive and social influence, and legitimacy and accountability politics (Princen and Finger 1994; Wapner 1995; Keck and Sikkink 1998; Newell 2000: 124, 129; Bostrom and Hallstrom 2010; Park 2010).

It is not surprising to think of environmentalists as diffusers of ideas, helping to create and shape international conventions on whaling, biodiversity, POPs, the trade in toxics, and CITES. They are able to influence states' positions in international negotiations and as members of IOs, and as domestic regulators for TNCs. Yet they also have material power using their financial resources from membership and donations in order to undertake technical specialist research, monitoring, and advocacy for achieving change in order to protect the environment. In many cases, ENGOs technical expertise has been able to determine (developing) states' policy in areas like climate change (Newell 2000: 132, 142). Some even allege that ENGOs use their financial position to back states in international negotiations such as in the IWC (Skodvin and Andresen 2003).

How are we to evaluate the success of environmentalists in world politics? We can examine the extent to which they have contributed to a change in government policy; strengthened a regime through the inclusion of environmental provisions in treaty texts; advocated stricter positions in international negotiations; and monitored and shamed state, TNC, and IO activity in relation to international treaties, standards, and commitments. We can assess whether their efforts at the international level have led to an improved environmental outcome overall (such as a reduction of carbon emissions), or for a period of time (such as the moratorium on whaling, the ban on ivory, and the recovery of species to sustainable levels on the CITES endangered list). Parsing out the influence of ENGOs is crucial, however, in order to evaluate the extent to which environmentalists have played a catalytic factor in improving environmental conditions.

Scholars have done just this in relation to climate change negotiations. Peter Newell highlights how ENGOs played an important role in setting the climate agenda in the late 1980s by organizing international workshops, leading governments to respond with the establishment of the Intergovernmental Panel on Climate Change (Newell 2000: 131). ENGOs were excluded from meetings where key decisions were made by states in the lead-up to the UNFCCC in 1992, and later at COP 2 in Geneva, although "climate negotiations were regarded as being at the forefront of attempts to open up international negotiations to NGO participation" (Newell 2000: 137). Yet ENGOs have played an important mediating role between states, persuading countries like Brazil and India to negotiate, leading to a developing country position that enabled the Berlin Mandate in 1995 for a binding protocol to the UNFCCC (Arts 1998; Newell 2000: 144).

Michele Betsill (2008) argues that ENGO influence was moderate in negotiations to create the Kyoto Protocol to the UNFCCC between 1995 and 1997. The Climate Action Network was able to mobilize the support of India and China for a Protocol (Newell 2000: 144). CAN, spearheaded by Greenpeace, FOEI, and WWF, "served as the voice of the environmental community during Kyoto negotiations" (Alcock 2008: 82; Betsill 2008: 46), through lobbying governments, producing a daily newsletter, and establishing a "fossil of the day" award for states that had been the most obstructive in negotiations to shame them. They also provided technical information and detailed knowledge on the draft negotiating texts for states, despite being increasingly excluded as Protocol negotiations progressed (Betsill 2008: 47–49).

CAN had four objectives throughout the negotiations: to push for strong targets and timetables for emissions reductions; CAN was split on favoring or opposing emissions trading; they opposed including sinks in the negotiations; and they wanted strong monitoring and compliance for the Protocol, although the latter was not on the agenda (Betsill 2008: 50). ENGO influence on Kyoto was moderate, because CAN's "positions are not reflected in the Protocol's texts." However, CAN was able to shape the negotiating process by "catalyzing debate over emissions trading and sinks" and by influencing the "position of key states on the issue of targets and timetables." Counterfactually, the absence of ENGOs may have made states' positions on the Protocol weaker (Arts 1998: 110; Betsill 2008: 44–58).

Gulbrandsen and Andresen (2004) take up where Betsill left off in examining whether ENGOs were able to influence the implementation measures for the Kyoto Protocol in relation to compliance, flexibility mechanisms, and sinks. These were negotiated from the COP 6 and COP 7 leading to the Marrakesh Accords in 2001. They argue that ENGOs were able to see some of their ideas on compliance adopted, such as being able to submit technical and factual information to the enforcement branch of the Kyoto compliance system, but their participation in compliance negotiations was restricted, and the compliance system was less participatory than they had advocated (2004: 64). In relation to the flexibility mechanism to supplement domestic emissions reductions, contra other CAN members, Environmental Defense (ED) was able to help influence the US position on market mechanisms for emissions reductions, which ultimately led the EU to agree to introduce emissions trading (2004: 65). While the majority of ENGOs in CAN did not get their preferences for compliance on Kyoto adopted, ED was aligned with the US position and therefore did, although this was not because of the persuasive efforts of ED (2004: 66). In short, ENGO activities in the Kyoto negotiations demonstrate that environmental activists have limited influence on outcomes but are able to maintain pressure on decision-makers to achieve the objectives of the Kyoto Protocol.

Can Institutions and Processes Be Made to Work Better?

There continues to be discussions as to whether there are better ways of managing environmental issues at all levels (locally, nationally, regionally, and internationally). Two trends shape these discussions: the shift away from the state towards both market mechanisms and the power of TNCs (Newell 2000: 125), and towards global governance (Alcock 2008). As detailed in the earlier section on "Global Policy Dimensions: How Do Environmentalists impact Climate Change?" environmental activists have played important roles in challenging the operations of TNCs, IOs, and states to mitigate the worst environmentally damaging activities that result from globalization and to make global environmental governing processes more effective. Environmentalists have therefore been able to have some influence in world politics, despite not being able to change the overall trend towards greater environmental degradation globally.

While arresting the degree to which globalization trumps environmental concerns remains key to much transnational environmental agency, scholars have also examined how the structure of global environmental governance could be made more effective. While there is a large literature on the democratic deficit of global governance in international relations, environmentalists have been playing an increasing role in governing environmental problems, particularly in market-driven governance mechanisms such as the FSC (Jordan and van Tuijl 2000; Cashore *et al.* 2004; Bostrom and Hallstrom 2010). However, environmentalist participation tends to reflect rather than overcome the unequal structural power in the exchange (Bostrom and Hallstrom 2010: 57). Some MEAs have had strong ENGO influence owing to their access and capacity to drive solutions to the specific environmental problem, but these have often been in areas of low importance to powerful states (Betsill and Corell 2008). While environmentalist influence in climate negotiations has been moderate (Betsill 2008), scholars have been looking at how global governance and MEAs can be made more democratic in order to be both more effective and more legitimate (Dryzek *et al.* 2011).

Deliberative democratic processes involving individuals from across the globe, as opposed to states representing their citizens in international negotiations, could overcome institutional barriers that stem from the construction of the climate regime, for example. The dramatic increase in the number of environmentalists attending the climate negotiations at COP 15 in Copenhagen in 2009 revealed how unwieldy international negotiations could become. Over 12 048 NGOs were registered at COP 15 in 2009, compared to 979 at COP 1 in 1995 (Cabre 2011: 11). At most negotiations, half of those registered are NGOs, at Copenhagen two-thirds were. Fisher argues that the "the massive expansion of civil society participation at Copenhagen was not only accompanied by civil society disenfranchisement, it actually contributed to it" (2010: 11, 12). Precisely because of the sheer volume of observers, access was limited for many environmentalists, thus preventing their involvement in negotiation.

Opportunities for environmentalists to deliberate over the future of global environmental problems such as climate change may shift the dynamics away from the short-term interests of states and TNCs. In short, having ENGOs at the negotiating table is not enough as states predetermine their "political goals that are not amenable to significant modification though international bargaining, and hence the degree of influence NGOs can exercise remains restricted" (Newell 2000: 137). Making international negotiations on climate change more democratic would also render "lobbying, bargaining, threats, and inducements" that are part of the current system obsolete (Dryzek *et al.* 2011: 40). Recasting the nature of decision-making could therefore improve global environmental governance and change the balance of power between states, IOs, TNCs, and environmentalists.

Conclusions

This chapter has discussed how environmentalists expend a great deal of energy to bring environmental concerns to international attention. It first identified the variety of actors engaged in environmental activism; second, it examined the aims and activities of environmental activists in relation to changing the activities of TNCs, IOs, and states through global summits and environmental treaties; third, it investigated the agency of environmental activists to bring about global environmental change in relation to industries, specific corporations, products, and codes of conduct; fourth, it reviewed how environmental activism is undertaken to improve environmental outcomes; and finally, the chapter reflected on the global policy implications of current environmental activist efforts in relation to climate change.

The chapter demonstrated that environmentalists are on occasion able to stop TNCs, states, and IOs from engaging in environmentally harmful activities but that their successes are outweighed by the overall structure of the international economic and political system. ENGOs have been ingenious in devising tactics and strategies to prevent environmental harm, ranging from investor screening, to YouTube clips, to traditional protests and boomerang politics. These actions sit alongside environmentalist efforts to influence multilateral negotiations on issues like climate change. New ways of making global environmental policy-making more legitimate and effective were identified through introducing alternative institutions such as deliberative democracy. Transnational environmental activism continues to evolve through combining traditional protest politics with new methods of online social campaigns and investor activism. While there is no magic bullet, environmentalists are using combined strategies to harness soft power in relation to environmental problems like climate change.

Note

1   On the difficulties of collecting data on international NGOs see Sikkink and Smith (2002: 26); McCormick (2011: 101).

References

ACF (Australian Conservation Foundation). 2011. "About ACF," http://acfonline.org.au/default.asp?section_id=231 (accessed December 7, 2011).

Alcock, Frank. 2008. "Conflicts and Coalitions within and across the ENGO Community." *Global Environmental Politics*, 8(4): 66–91.

Andonova, Liliana. 2010. "Public–Private Partnerships for the Earth: Politics and Patterns of Hybrid Authority in the Multilateral System." *Global Environmental Politics*, 10(2): 25–53.

Arts, Bas. 1998. *The Political Influence of Global NGOs. Case Studies on the Climate and Biodiversity Conventions*. Utrecht: International Books.

Betsill, Michele. 2008. "Environmental NGOs and the Kyoto Protocol Negotiations: 1995–1997." In *NGO Diplomacy: The Influence of Nongovernment Organizations in International Environmental Negotiations*, ed. Michele Betsill and Elisabeth Correll, 43–67. Cambridge, MA: MIT Press.

Betsill, Michele and Elisabeth Correll. 2008. *NGO Diplomacy: The Influence of Nongovernment Organizations in International Environmental Negotiations*. Cambridge, MA: MIT Press.

Bostrom, Magnus and Kristina Hallstrom. 2010. "NGO Power in Global Social and Environmental Standard-Setting." *Global Environmental Politics*, 10(4): 36–59.

Botetzagias, Iosif, Prue Robinson, and Lily Venizelos. 2010. "Accounting for Difficulties Faced in Materializing a Transnational ENGO Conservation Network: A Case-Study from the Mediterranean." *Global Environmental Politics*, 10(1): 115–151.

Cabre, Miquel Muñoz. 2011. "Issue-Linkages to Climate Change Measured through NGO Participation in the UNFCCC." *Global Environmental Politics*, 11(3): 10–22.

CAN (Climate Action Network). 2011. "About Climate Action Network," http://www.climatenetwork.org/about/about-can (accessed December 1, 2011).

Cashore, Ben, Graeme Auld, and Deanna Newsom. 2004. *Governing through Markets: Forest Certification and the Emergence of Non-state Authority*. Cambridge, MA: MIT Press.

CERES. 2011. "History and Impact," www.ceres.org (accessed December 9, 2011).

Chasek, Pamela, David L. Downie, and Janet Welsh Brown. 2006. *Global Environmental Politics*, 4th edn. Boulder, CO: Westview Press.

CJA (Climate Justice Action). 2011. "About," http://www.climate-justice-action.org/about/organizations/ (accessed December 20, 2011).

CJN! (Climate Justice Now!). 2011. "About Us," http://www.climate-justice-now.org/about-cjn/history/ (accessed December 20, 2011).

Clapp, Jennifer. 2001. *Toxic Exports: The Transfer of Hazardous Wastes from Rich to Poor Countries*. Ithaca, NY: Cornell University Press.

Clapp, Jennifer. 2005. "Transnational Corporations and Global Environmental Governance." In *Handbook of Global Environmental Politics*, ed. Peter Dauvergne, 284–297. Cheltenham: Edward Elgar.

Clapp, Jennifer and Peter Dauvergne. 2011. *Paths to a Green World: The Political Economy of the Global Environment*, 2nd edn. Cambridge, MA: MIT Press.

Clark, Ann Marie, Elisabeth Friedman, and Kathryn Hochstetler. 2005. "The Sovereign Limits of Global Civil Society: A Comparison of NGO Participation in UN World Conferences on the Environment, Human Rights and Women." In *The Global Governance Reader*, ed. Rorden Wilkinson, 292–321. Abingdon: Routledge.

Conca, Ken. 1996. "Greening the UN: Environmental Organisations and the UN System." In *NGOs, the UN, and Global Governance*, ed. Thomas G. Weiss and Leon Gordenker, 103–120. Boulder, CO: Lynne Rienner.

Dryzek, John, Andre Bachtiger, and Karolina Milewicz. 2011. "Towards a Deliberative Democracy Citizen's Assembly." *Global Policy*, 2(1): 33–42.

ECOSOC. 2011. "Basic Facts about ECOSOC Status." Department of Economic and Social Affairs, the United Nations, http://csonet.org/index.php?menu=17 (accessed December 1, 2011).

Falkner, Robert. 2008. *Business Power and Conflict in International Environmental Politics*. Basingstoke: Palgrave Macmillan.

Fisher, Dana. 2010. "COP-15 in Copenhagen: How the Merging of Movements Left Civil Society Out in the Cold." *Global Environmental Politics*, 10(2): 11–17.

Fisher, Duncan. 1992. "The Emergence of the Environmental Movement in Eastern Europe and its Role in the Revolutions of 1989." In *Green Plant Blues: Environmental Politics from Stockholm to Rio*, ed. Ken Conca, Michael Alberty, and Geoffrey Dabelko, 107–115. Boulder, CO: Westview Press.

Ford, Lucy. 2005. "Challenging the Global Environmental Governance of Toxics: Social Movement Agency and Global Civil Society." In *The Business of Global Environmental Governance*, ed. David Levy and Peter Newell, 305–328. Cambridge, MA: MIT Press.

Gleckman, Harris. 2004. "Balancing TNCs, the States, and the International System in Global Environmental Governance: A Critical Perspective." In *Emerging Forces in Environmental Governance*, ed. Norichi Kanie and Peter Haas, 203–215. Hong Kong: United Nations University Press.

Goldman, Michael. 2005. *Imperial Nature: The World Bank and Struggles for Social Justice in the Age of Globalization*. New Haven, CT: Yale University Press.

Gotz, Timothy. 2008. "Reframing NGOs: The Identity of an International Relations Non-starter." *European Journal of International Relations*, 14(2): 231–258.

Greenpeace International. 2011. "What We Do," http://www.greenpeace.org/international/en/ (accessed November 30, 2011).

Guerrero, Dorothy. 2011. "The Global Climate Justice Movement." In *Global Civil Society 2011: Globality and the Absence of Justice*, ed. Martin Albrow and Haken Seckinelgin, 120–127. Basingstoke: Palgrave Macmillan.

Gulbrandsen, Lars and Steinar Andresen. 2004. "NGO Influence in the Implementation of the Kyoto Protocol: Compliance, Flexibility Mechanisms, and Sinks." *Global Environmental Politics*, 4(4): 54–75.

Gutner, Tamar. 2002. *Banking on the Environment: Multilateral Development Banks and Their Environmental Performance in Central and Eastern Europe.* Cambridge, MA: MIT Press.

Haas, Peter and Ernst Haas. 1995. "Learning to Learn: Improving International Governance." *Global Governance*, 1: 255–285.

IPEN (International POPs Elimination Network). 2011. "About IPEN," http://www.ipen.org/ ipenweb/firstlevel/about.html (accessed December 20, 2011).

Jordan, Lisa and Peter van Tuijl. 2000. "Political Responsibility in Transnational NGO Advocacy." *World Development*, 28(12): 2051–2065.

Keck, Margaret and Kathryn Sikkink. 1998. *Activists beyond Borders: Advocacy Networks in International Politics.* Ithaca, NY: Cornell University Press.

Khagram, Sanjeev, James V. Riker, and Kathryn Sikkink, eds. 2002. *Restructuring World Politics: Transnational Social Movements, Networks, and Norms.* Minneapolis, MN: University of Minnesota Press.

Lee, Yok-Shiu and Alvin Y. So, eds. 1999. *Asia's Environmental Movements: Comparative Perspectives.* Armonk, NY: M.E. Sharpe.

Levy, David, Halina Szejnwald Brown, and Martin de Jong. 2009. "The Contested Politics of Corporate Governance: The Case of the Global Reporting Initiative." *Business and Society*, 20(10): 1–27.

Levy, David and Peter Newell. 2005. *The Business of Global Environmental Governance.* Cambridge, MA: MIT Press.

Lipschutz, Ronnie. 1996. *Global Civil Society and Global Environmental Governance,* Albany, NY: State University of New York Press.

MacLeod, Michael and Jacob Park. 2011. "Financial Activism and Global Climate Change: The Rise of Investor-Driven Governance Networks." *Global Environmental Politics*, 11(2): 54–74.

McAteer, Emily and Simone Pulver. 2009. "The Corporate Boomerang: Shareholder Transnational Advocacy Networks Targeting Oil Companies in the Ecuadorian Amazon." *Global Environmental Politics*, 9(1): 1–30.

McCormick, John. 2011. "The Role of Environmental NGOs in International Regimes." In *The Global Environment: Institutions, Law, and Policy*, ed. Regina Axelrod, Stacy VanDeveer, and David Downie, 92–109. Washington, DC: CQ Press.

Najam, Adil. 1996. Nongovernmental Organizations as Policy Entrepreneurs: In Pursuit of Sustainable Development. PONPO Working Paper 231. Yale University, Program on Non-Profit Organizations.

Newell, Peter. 2000. *Climate for Change: Non-state Actors and the Politics of Climate Change.* Cambridge: Cambridge University Press.

Nielson, Daniel, Michael Tierney, and Catherine Weaver. 2006. "Bridging the Rationalist–Constructivist Divide: Re-engineering the Culture at the World Bank." *Journal of International Relations and Development*, 9: 107–139.

O'Brien, Robert, Anne Marie Goetz, Jan Aart Scholte, and Marc Williams. 2000. *Contesting Global Governance: Multilateral Institutions and Global Social Movements.* Cambridge: Cambridge University Press.

Park, Susan. 2010. *World Bank Group Interactions with Environmentalists: Changing International Organisation Identities.* Manchester: Manchester University Press.

Pradyumna, P. Karan and Unryu Suganuma, eds. 2008. *Local Environmental Movements: A Comparative Study of the United States and Japan*. Lexington, KY: University of Kentucky Press.

Princen, Thomas. 1994. "The Ivory Trade Ban: NGOs and International Conservation." In *Environmental NGOs in World Politics: Linking the Local and the Global*, ed. Thomas Princen and Mathias Finger, 121–159. London: Routledge.

Princen, Thomas and Mathias Finger. 1994. *Environmental NGOs in World Politics: Linking the Local and the Global*. London: Routledge.

Rich, Bruce. 1994. *Mortgaging the Earth: The World Bank, Environmental Impoverishment, and the Crisis of Development*. Boston: Beacon Press.

Rodrigues, Maria Guadalupe Moog. 2004. *Transnational Advocacy Networks in Brazil, Ecuador, and India*. Albany, NY: State University of New York Press.

Rohrschneider, Robert and Russell Dalton. 2002. "A Global Network? Transnational Cooperation among Environmental Groups." *Journal of Politics*, 64(2): 510–533.

Rowell, Andy. 1996. *Green Backlash: Global Subversion of the Environmental Movement*. London: Routledge.

Sikkink, Kathryn and Jackie Smith. 2002. "Infrastructures for Change: Transnational Organizations, 1953–93." In *Restructuring World Politics: Transnational Social Movements, Networks and Norms*, ed. Sanjeev Khagram, James V. Riker, and Kathryn Sikkink, 24–46. Minneapolis, MN: University of Minnesota Press.

Skodvin, Tora and Steinar Andresen. 2003. "Nonstate Influence in the International Whaling Commission, 1970–1990." *Global Environmental Politics*, 3(4): 61–86.

Tarrow, Sidney. 2001. "Transnational Politics: Contention and Institutions in International Politics." *Annual Review of Political Science*, 4: 1–20.

TWN (Third World Network). 2011. "Introduction to the Third World Network," http://www.twnside.org.sg/twnintro.htm (accessed December 1, 2011).

UIA (Union of International Associations). 2011. "Appendix 3: Table 1. Number of International Organizations in this Edition by Type (2005/2006)." In *Yearbook of International Organizations Statistics 2005/2006*. Brussels: Union of International Associations, http://www.uia.org/statistics/organizations/types-2004.pdf (accessed December 8, 2011).

Vogel, David. 2008. "Private Global Business Regulation." *Annual Review of Political Science*, 11: 261–282.

Wade, Robert. 1997. "Greening the Bank: The Struggle over the Environment 1970–1995." In *The World Bank: Its First Half Century*, ed. Davesh Kapur, John Lewis, and Richard C. Webb, 611–734. Washington, DC: Brookings Institution Press.

Wapner, Paul. 1995. "Politics beyond the State: Environmental Activism and World Civic Politics." *World Politics*, 47: 311–340.

Wight, C. 2012. "Global Banks, the Environment, and Human Rights: The Impact of the Equator Principles on Lending Policies and Practices." *Global Environmental Politics*, 12(1): 56–77.

Willetts, Peter. 2011. "What is a Non-governmental Organization?" In *UNESCO Encyclopedia of Life Support Systems*, Section 1: Institutional and Infrastructure Resource Issues, Article 1.44.3.7, Non-governmental Organizations, http://www.staff.city.ac.uk/p.willetts/CS-NTWKS/NGO-ART.HTM#Part10 (accessed December 1, 2011).

Chapter 17

Business as a Global Actor

Jennifer Clapp and Jonas Meckling

Introduction

Business actors play an important role in global environmental policy-making and governance. Transnational corporations in particular are responsible for a significant portion of global economic activity, which in turn has enormous implications for both climate change and broader environmental conditions. And because policy frameworks attempt to shape the behavior of business in order to mitigate their impacts on the environment, these actors naturally want to be actively involved in trying to influence policy outcomes. Being part of the process does ensure their buy-in, and thus it is important to have them involved at least to some degree. But the involvement of business actors in environment and climate policy is not always transparent. These actors often operate behind the scenes or in subtle ways that influence the broader policy-making context whether or not they are involved in official policy-making processes. As such, business actors can carry their influence into the policy process through numerous channels.

This chapter seeks to unpack the multiple avenues of involvement of business actors, transnational corporations in particular, in the global environment and climate policy-making process. It shows that through these different avenues of involvement, business actors possess and make use of different kinds of power to influence the process. Their key avenues of influence include lobby activity, market influence, rule-setting participation, and issue-framing exercises. Business influence is exercised in overlapping and complex ways, through formal and informal interventions, public and private arenas, and global through to local scales. At the same time, business actors, while having access to numerous means by which to influence policy, do not always get their way with respect to policy outcomes. There are intervening factors that affect the extent to which business actors can wield power over policy and

The Handbook of Global Climate and Environment Policy, First Edition. Edited by Robert Falkner.
© 2013 John Wiley & Sons, Ltd. Published 2013 by John Wiley & Sons, Ltd.

governance processes. These include, for example, NGO campaigns, business conflict, and business coalitions, each of which can push global environmental and climate policy in different directions – sometimes enhancing, and sometimes tempering the influence of business actors.

The Significance of Business Actors

Global business actors encompass a range of entities that include but are not limited to private firms. Large transnational corporations (TNCs) are indeed significant actors. These firms operate in more than one country, and many have a truly global scope to their operations. Business lobby groups and business-oriented nongovernmental organizations also fall under the heading of business actors, as their activities are often geared specifically to advancing the views of the industries that they represent and/or work with directly.

Collectively, TNCs are important players in the global economy, and their weight has grown significantly since the 1970s. In 2010 the number of TNC parent firms stood at 103 346, over 70% of which were located in developed countries. This compares with just 7000 TNCs that were in operation in 1970. The number of TNC affiliate firms in operation in 2010 was 886 143, over half of which were located in developing countries. Foreign direct investment (FDI) flows are another indicator of the significance of TNCs in the global economy. FDI flows have increased from US$9.2 billion in 1970 to US$1.24 trillion in 2010 (UNCTAD 2008; UNCTAD 2011). This phenomenal grown in TNCs and FDI has been reflected in greater production, sales, employment, and assets of these firms. Indeed, in 2010, the economic activity of TNCs accounted for around one quarter of world GDP. The economic weight of TNCs is also highly concentrated. The ETC group points out, for example, that in 2007 just 147 firms collectively controlled 40% of the value of all TNCs (ETC Group 2011).

The significant economic weight of TNCs in the global economy has important implications for the environment (Newell 2008). TNCs tend to dominate in sectors that have a high environmental impact, including manufacturing, mining, oil extraction, industrial agriculture, forestry, and chemicals. TNC have historically been associated with environmental damage in these sectors, especially in the developing world (e.g. Dauvergne 2001; Leighton et al. 2002). Foreign direct investment by TNCs has been especially strong in the primary sector, including mining and oil extraction, since the 1990s (UNCTAD 2006: 7). TNCs are also significant contributors to climate change, which is not surprising given that sectors in which they are most active are associated with high levels of greenhouse gas emissions (Morgera 2004: 215). Corporate concentration characterizes some of these high environmental impact sectors.

Most environmental regulation is geared toward influencing economic behavior and as such directly affects the operations of business actors. Indeed, business actors are often direct targets of regulation and policy, specifically because of the impact that business can potentially have on the environment. In the 1960s and 1970s, the approach was largely top-down. States imposed environmental regulations in a command-and-control-type fashion in which firms had to meet certain environmental standards. This kind of regulatory approach, however, was not often workable at an international scale within treaties on environmental issues that began to emerge and proliferate in the 1980s–1990s. Different states had different regulatory capacities, making the negotiation of uniform global regulatory environmental standards impossible. In addition, industrialized countries were seen by many to have had the bulk of responsibility for key environmental problems such as climate change and ozone depletion, with the result that global agreements incorporated stricter targets for those countries compared with developing countries.

The approach of negotiating different targets for different countries raised questions about the environmental responsibilities of TNCs that operate in multiple locations around the world. In the 1970s–1980s, the UN Centre for Transnational Corporations floated the idea of a global code of conduct for TNCs, with the aim of holding this specific type of corporate actor accountable for any negative impacts of their activities, including environmental damages, regardless of where they operated (Clapp 2005). This initiative, however, did not develop further, as the UNCTC was disbanded just prior to the Rio Earth Summit in 1992. The Earth Summit took a decidedly different approach to TNCs and global environmental problems. Instead of seeking to regulate them in any kind of uniform fashion, it sought instead to embrace business actors as partners in the quest for sustainable development. Voluntary measures and public–private partnerships were proposed at Rio and have since become an increasingly important governance approach for global environmental policy-making.

In this new governance environment at the international level, firms, lobby groups, and business NGOs have taken on increasingly significant roles in shaping global environmental and climate policy. Corporate lobby groups – long a fixture at the national level – have become important actors on the global stage as well. Business NGOs have also grown in number and influence, particularly as actors in voluntary and certification-type schemes and public–private partnerships for environmental protection.

Channels of Influence

The power of business actors to influence processes of international environmental policy-making has been significant and on the rise in recent decades. Several studies have documented the growth in their political activity in global environmental governance (Falkner 2003; Clapp 2005; Levy and Newell 2005). This work builds on earlier studies in the field of international political economy that examined the structural power of corporate actors (Strange 1988; Gill and Law 1989). With a focus on the authority and legitimacy of business as an actor in global governance, this more recent work has focused not just on the influence firms have as a result of their economic weight in the broader economy, but also on their growing role in shaping environmental policy in a variety of ways (Clapp 1998; Cutler *et al.* 1999; Fuchs 2005a, 2005b). Specifically, this work has shed light on how corporate actors can influence the contours of the very policies that seek to shape their environmental behavior, thus enabling them to ensure that policies are established and implemented in ways that do not threaten their economic viability or profitability.

Channels of business influence in environmental policy-making include more traditional means, such as business lobbying, as well as less transparent forms of influence such as market or structural power, private rule-setting roles, and the sway they are able to hold on the development of public debate around the severity of certain environmental issues and views on the most appropriate policy tools to address those issues. The extent of their influence over policy outcomes has raised important questions about how firms have gained authority to engage in global policy-making, and whether governance processes in which they participate include a sufficiently wide set of voices to render them legitimate in the eyes of the broader public. The various channels in which corporate actors engage with global environmental governance are examined in more depth below.

Lobbying

Business actors have traditionally attempted to influence policy-making processes in a direct fashion through the practice of lobbying. By hiring consultants to meet regularly with government policy-makers, business actors can have their concerns voiced directly to those who matter in the process. Lobbying is seen as a form of "instrumental" or "compulsory" power, which refers to instances where one actor can wield direct influence over the behavior of another (Fuchs 2005a). Individual firms often join business associations that represent the shared interests of a group of firms or an entire sector of the economy. A single lobbyist representing a wide group of private economic actors can have significant influence on the policy-making process, often greater than a firm could achieve if it acted individually.

Business actor lobbying has long taken place at the domestic level to influence the kinds of environmental policies and regulations adopted by individual governments. But business actors have increasingly lobbied multiple governments in international contexts, and at multiple scales of government, as a means by which to influence the negotiation of international environmental agreements. As more international environmental agreements were negotiated in the 1980s and 90s, industry groups began to attend negotiation meetings as well as larger environmental conferences. As observers to these official government meetings, they are able to keep close tabs on how rules are formed and the likely areas for regulatory change. Business actors also use these meetings to voice their concerns and proposals in side-events and in the corridors during breaks.

Business actors are now regular participants in most international environmental gatherings – from broad environmental conferences to specific treaty negotiations regarding particular environmental issues. Their now ubiquitous engagement is a marked change from the 1972 Stockholm Conference on the Human Environment, where business actors made only a 15-minute intervention in the proceedings. At the Rio Earth Summit in 1992, business actors were brought into the conference not only as participants, but were integrated directly into conference themes articulated in Agenda 21 that included an emphasis on corporate voluntary initiatives. A strong presence by industry actors at the 1992 Earth Summit was facilitated by the conference secretary-general, Maurice Strong, who actively encouraged participation by groups such as the World Business Council for Sustainable Development (WBCSD) (Clapp and Dauvergne 2011: 185–186).

Business participation in major global environmental meetings continued with the 2002 Johannesburg World Summit on Sustainable Development, a 10-year follow-up to the Rio Conference. A coalition of business groups that represented the combined forces of the International Chamber of Commerce (ICC) and WBCSD formed a new business lobby group, Business Action on Sustainable Development (BASD, representing 161 TNCs), in 2001, with the explicit purpose of lobbying at the WSSD. The group again formed in the run-up to the 2012 Rio+20 Conference, under the name BASD 2012, specifically to lobby on behalf of its members at this international gathering. The group is convened by the ICC, WBCSD, and the UN Global Compact. Eleven international industry associations that represent a wide range of TNCs are among its members, including for example the World Steel Association, the Global Oil and Gas Industry Association, the International Council on Mining and Metals, and the International Council of Chemical Associations. One of the key aims of the group is to "Express global business positions on key sustainability issues to assist governments in making decisions that allow for a sustainable global business engagement" (BASD 2012).

International business lobby groups also engage with specific international environmental treaty negotiations in an attempt to shape rules that might affect their business. During the negotiation of the Cartagena Protocol on Biosafety, for example, business actors had a strong presence at the meetings. Individual firms sent representatives to observe the meetings and to engage in lobbying, including agricultural chemical and biotech giants Monsanto, Dupont, and Syngenta. Industry associations also had a strong presence that grew in size throughout the negotiations – from 8 groups engaged in the process in 1996 to some 20 industry groups in 1999. These included the ICC as well as biotechnology-specific industry associations such as the Biotech Industry Organization (BIO), which represents a number of firms in the biotechnology industry. By 1999, the industry groups also collaborated through a joint group that referred to itself as the Global Industry Coalition, which claimed at the time to represent some 2200 firms in 130 countries. The aim of the group was to try to influence the treaty outcomes with "one voice." In particular, it sought to lobby against the idea of incorporating the precautionary principle into the agreement (Clapp 2003).

Business actors use lobbying in different ways, in order to play to their strengths. Levy and Egan (1998, 2003) demonstrate how US firms tried to keep any regulatory efforts on climate change at the national level and away from the international level because they could have more influence on a single key player, given that over 140 countries were involved in the international negotiations. However, the degree of influence at the domestic level is determined by the political system. Pluralist and neo-corporatist arrangements grant firms different forms of access to policy-makers. The pluralist political system of the USA gives many different actors access to decision-makers but also increases competition among interests groups, thus encouraging a short-term adversarial culture. The neo-corporatist political systems of many countries in continental Europe, instead, have developed co-operative long-term relationships between a few key social actors. This reduces the number of players inside the system, but makes access for groups outside the system more difficult. In the European Union, the regional level has become the most important level for business lobbying on environmental issues, as the EU has enacted over 700 environmental regulations (McCormick 2001). Coen (2005) argues that the lobbying style in Brussels has become more American to the extent that the European institutions have become more accessible to lobbyists, and more competitive interest politics have emerged.

Market Influence

Firms also influence international environmental policy-making and governance in more diffuse ways that are shaped by their significance within both individual domestic economies and the global economy more broadly. The ability to move their operations across borders to set up shop in other locations, in particular, gives large TNCs a considerable amount of influence that can affect the style and extent of environmental regulations imposed by governments. Governments may be concerned that the imposition of more stringent regulations will push firms to relocate their activities, taking investment dollars and jobs with them, and as a result may fail to strengthen regulations facing particular sectors that are important to their economy. This can happen if firms threaten to relocate when faced with the prospect of more stringent environmental regulations, but it can also happen even if the firms do not threaten to leave openly. This phenomenon, sometimes referred to as the "regulatory chill," is difficult to prove, but anecdotally is widely seen to be in operation in practice (Neumayer 2001).

The ability of firms to influence regulatory outcomes based primarily on their material position within an economy gives them a degree of what is referred to in the literature as structural or market power. This kind of power enables firms, through their dominant position in the global economy, not only to influence the formation and functioning of environmental policy, but also to shape mainstream ideology, which in turn affects the nature of the policy choices contemplated by governments. Attention to structural power has been especially prominent within the historical materialist school of International Relations that focuses on the role that a "bloc" of actors – including TNCs and powerful states – can play in shaping policy more broadly due to their material position within the global economy. Gill and Law, for example, argue that the distribution of power and the interests of actors are determined by the structure of global capitalism (Gill and Law 1989). Other conceptions of structural power draw out the role of different structural factors that lend power, such knowledge, and different ways in which market influence of business actors plays out in different sectors (e.g. Strange 1988; Falkner 2008).

The market influence of business actors is also the product of the key role that they play in the development and use of technology. Business actors take a leading role in deciding which technologies they will invest in and adopt which in turn is fundamentally important for determining shifts in technological usage that can have profound environmental impacts. Not only is the largest share of total research and development in OECD countries in private hands, but it is also concentrated in a small number of firms (Mytelka 2000). Firms that control research and development for new technologies, particularly those that can lead to environmental improvements, can have significant leverage to shape policy and regulatory developments, especially when they have property rights over the knowledge from which those technologies are developed. In such cases, firms can use their leverage to encourage policies that enhance their own market position by requiring use of those technologies.

Technological knowledge, in other words, shapes the regulatory options available (Falkner 2005; Tjernshaugen 2012).

Business decisions regarding investment in technology also play into the effectiveness of the implementation of multilateral environmental agreements. The case of ozone politics is a classic example of the role of corporate research and development investment in regime evolution. DuPont had successfully explored substitutes for chlorofluorocarbons (CFCs), which caused the firm to change its strategy from fierce opposition to support of the Montreal Protocol in order to secure first mover advantages. Once technological investment became relevant in the political process as technological knowledge, the firm emerged as a key actor in the international political dialogue around rules to reduce ozone depletion (Falkner 2008: chapter 3).

Rule-setting

The establishment of privately set standards and rules is a further avenue by which business actors influence international environmental policy and governance (see also Chapter 23 in this volume). With the globalization of the world economy, in particular the development of truly global commodity chains, business actors have begun to establish voluntary rules and standards that govern their operations, and those of others, within those chains. Their participation in such activities has been part of their wider strategy of pursuing corporate social responsibility (CSR). The proliferation and acceptance of private environmental governance schemes has enabled corporate actors to directly influence governance decisions. In some cases those rules have not just supplemented state-based regulations, but have replaced them. In this way, adherence to private standards and rules can become *de facto* obligatory for other market participants, and may also have distributional consequences for less powerful actors within those markets (Fuchs 2005a). According to Fuchs, the rise of private rule-setting activities gives firms a form of structural power. Firms' role in shaping these kinds of standards, even if they are largely voluntary on the part of firms, gives them in effect the ability to set the rules of the game by which others are forced to play.

There is a range of private environmental governance initiatives featuring business actors, including non-binding rules such as principles and codes of conduct, reporting and disclosure schemes, voluntary environmental management standards, and hybrid forms of co-regulation (Utting and Clapp 2008; Clapp and Thistlethwaite 2012). Codes of conduct and principles to which firms sign on are typically set from within industry itself, and adherence to these rules tends not to be monitored or enforced by outside parties. By the late 1990s, most large TNCs had signed on to one or more of these kinds of initiatives. The Global Compact, for example, is a set of principles that firms sign onto that include environmental practices. Similarly, Responsible Care in the chemical industry is a set of best-practice rules that was established by the chemical industry without much external input, and which are self-monitored (Prakash 2000). Reporting and disclosure schemes include measures such as the Global Reporting Initiative and the Carbon Disclosure Project, which encourage firms to report on their environmental performance (Kolk *et al.* 2008; Brown *et al.* 2009).

Firms have also been active participants in the development of environmental management systems (EMS) and market certification schemes that have more external oversight than codes, principles, and disclosure schemes. Environmental management standards, such as the International Organization for Standardization's ISO 14000 series, are in effect a quasi-private, quasi-public regime of standards that are certified by third-party auditors. Industry players took a key role not only in setting the agenda for these standards, but also in setting out the specific rules (Clapp 1998). Concern has been expressed that the ISO 14000 standards, which were not developed fully in the public sphere, have been given public sector legitimacy, for example, by being named by the World Trade Organization (WTO) as international standards for EMS standards. There is also concern when states choose to ease up on monitoring of firms that adhere to these standards (Krut and Gleckman 1998).

Market-based rules that focus on certification of particular products to environmental and social standards are designed to create market incentives for firms to correct or improve environmental practices at crucial points along the product's supply chain (Cashore 2002). Market certification schemes are sometimes controlled entirely within an industry, but they can also be undertaken with the involvement of non-state actors such as NGOs and other independent oversight providers such as third-party auditors. When undertaken with multiple stakeholders, these types of certification schemes are seen to be forms of "co-regulation" (Pattberg 2005). These kinds of market-based regulations include initiatives such as sustainable forestry certification under the Forest Stewardship Council and Fairtrade certification overseen by Fairtrade Labelling Organizations International (Raynolds 2000; Cashore 2002). Supermarket firms have also increasingly begun to certify produce to "good agricultural practices" through the Global GAP program – a scheme that industry players took a lead role in developing (Fuchs *et al.* 2009). These various certification schemes have enormous influence on other market participants, who, if they wish to sell into global commodity chains, must adhere to the rules, which also means taking on the risks and costs of certification.

Through their strong participation in these various private regimes, firms are able to influence the shape of these kinds of rules. Moreover, their position in global economic networks enables them to have a strong influence on which private standards and labels become widely adopted. Industry actors may prefer to adopt standards and labels that have more industry input over NGO-created standards and labels. At the same time, firms that sign on to these kinds of initiatives can and often do use them to advertise their environmental credentials and to make the case that external regulations beyond these voluntary measures are not needed to achieve positive environmental change in firms' behavior.

Issue-framing

A further way in which business actors have engaged in environmental policy-making is through their attempts to shape public discourses surrounding conceptions of environmental sustainability and the role corporations play in the promotion of environmental goals. By making strategic use of ideas, firms can shape the broader public's understanding of the nature of environmental problems and the possible solutions to those problems (Sell and Prakash 2004). For example, business actors often tend to frame environmental problems as being the product not of industry activity, but rather individual consumption choices or forces out of the control of society. By framing ideas in ways that favor the interests of industry, business actors hold what is referred to as "discursive power." When those ideas spread widely it translates into indirect influence over the policy-making process (Fuchs 2005b).

One strategy used by industry to frame environmental problems is to interpret scientific evidence for a broad audience. Business actors have challenged the science behind environmental problems, as was the case early on with respect to climate change. Scientific knowledge is not simply imposed on the political process by the scientific community but is essentially contested in discursive battles (Litfin 1995). Tactics employed by firms include public campaigns or the creation of so-called "Astroturf" organizations. The latter are set up as grassroots organizations in order to give the impression of civil society activity, but are actually run by individual firms or business associations. Such strategies have been used in climate change politics (Levy and Egan 2003). This discursive power based on technological knowledge plays out in debates on substitutes for harmful substances (e.g. ozone) or technologies (e.g. energy technologies) and in the evaluation of the environmental risks and benefits of particular technologies (e.g. biotechnology).

Business actors also frame environmental problems and solutions to make environmental policy more compatible with business interests. While popular discourse often depicts firms as the cause of global environmental problems, the discourse of ecological modernization – where environmental improvements are seen to occur alongside economic growth – is promoted by business actors because it puts firms center stage in the solution to environmental problems (Young 2000). Key to this view is the idea of "win-win" solutions, which benefit both the economy and the environment. Regarding environmental policy, the discourse on ecological modernization suggests market-based measures, such as emissions trading schemes and voluntary agreements. Neo-Gramscian scholars in particular point to the role industry plays in marketing only "market-friendly" solutions to environmental problems that fit within a neoliberal economic framework (Levy and Egan 2003). In climate politics, for instance, business actors increasingly attempt to depict climate change as a business opportunity for low-carbon technologies and for productivity improvement through increases in energy efficiency. Through their engagement in international gatherings on the theme of sustainable development, business actors have portrayed the voluntary corporate initiatives and industry-set standards, discussed above, as more efficient and effective than government command-and-control regulation (Finger and Kilcoyne 1997; Clapp 2005).

A notable example of issue-framing is clear in the case of business actors and their portrayal of agricultural biotechnology. Monsanto, for example, has actively engaged in public debate about the issue. It has made the case that genetically modified foods serve the poor, and that they are vitally important to solve world hunger and promote environmentally sustainable agricultural production. This message has been reiterated through its annual reports, its web site, and through public statements and advertisements. Industry associations to which major agricultural biotechnology firms belong, such as the Biotechnology Industry Organization and CropLife, have also put forward a similar framing of agricultural biotechnology in an attempt to promote this particular view of agricultural biotechnology, and to argue that these types of crops should not be held back by what they consider to be overly stringent regulation (Williams 2009).

Intervening Factors

Firms possess a range of power resources that allow them to influence global environmental policy-making through the four channels we discussed above. Yet power capabilities do not equal actual political influence. While power refers to an actor's capabilities to effect political change, influence refers to the actual effect of an actor's behavior on political outcomes (cf. Corell and Betsill 2008: 24). In fact, the translation of power resources into actual political influence is mitigated by a number of intervening factors. The literature on non-state actors in global politics has pointed to a series of factors that play into the level of influence of non-state actors. These include, for instance, the rules of access, the political stakes of an issue, the existence of policy crises, and state allies as political opportunity structures (Betsill 2008; Meckling 2011b). Other authors emphasize the role of strategy and in particular the framing of a cause as a source of advocacy effectiveness (Sell and Prakash 2004).

In this chapter, we focus on three select intervening variables, which relate primarily to the role of transnational political competition and cooperation in mitigating corporate influence. More recently, work in international political economy (IPE) on transnational pluralism has emphasized competition and cooperation among state and non-state actors as pivotal forces in world politics (Mattli and Woods 2009; Avant et al. 2010; Cerny 2010). The central assumption is that individual and collective actors "engage in processes of conflict, competition, and coalition-building in order to pursue those interests" (Cerny 2010: 4). Building on this assumption, we focus on three key variables that mitigate or enhance corporate influence in global environmental politics: (1) competition from NGOs, (2) conflict within the business community, and (3) coalition-building among business actors.

NGO Campaigns

Environmental groups are key non-state actors in global environmental politics, assuming the role of advocate for environmental action (see also Chapter 16 in this volume). While sometimes NGOs and firms cooperate, often enough NGOs act as a countervailing force to business interests. Other non-state actors that may act as countervailing forces are consumers and labor (Ronit 2007). The relative strength of environmental NGOs vis-à-vis business groups thus affects the likelihood of the lobbying success of firms. Research on transnational advocacy networks has demonstrated how NGOs organize transnational campaigns (Keck and Sikkink 1998; Khagram et al. 2001). NGOs can attack individual firms or entire industries through naming-and-blaming campaigns, thus undermining in particular their legitimacy and discursive clout. Alternatively, NGO campaigns may be directed at states in an attempt to influence political agendas and outcomes.

NGO campaigns targeted directly at firms aim to change corporate behavior. Often the goal is to disclose green-wash activities by holding firms accountable. For example, in the late 1980s the Earth Island Institute and other environmental groups launched a campaign against tuna fishing methods that led to bycatch of dolphins

(Wapner 1995). Their campaign included a boycott of canned tuna, demonstrations at shareholder meetings, and public rallies. In response, the three largest tuna companies stopped using those fishing strategies. In a number of instances, naming-and-blaming campaigns against firms or industries have led to self-regulation through codes of conduct (Haufler 2001). The heightened rule-setting activity of corporations is thus partially a result of the power of civil society organizations and their ability to scrutinize corporate conduct (Falkner 2003).

NGO campaigns directed at states aim to change state behavior or shape rules and laws. The more influential the NGO campaign is, the more limited its business influence. For example, Schurman (2004) demonstrates how the anti-biotech movement in Europe successfully fought the agricultural biotechnology industry over rules on biosafety. Its advocacy for a precautionary approach ultimately shaped environmental regulation. The success of the NGO campaign was facilitated by the existence of industry vulnerabilities, so-called "industry opportunity structures." Different economic and cultural features of firms and industries may render them more or less vulnerable to NGO campaigns.

As in the case of corporate lobbying, the efficacy of NGO campaigns depends on the level of coalition-building or conflict in the NGO community. Alcock (2008) suggests that conflict or cooperation between environmental groups depends on their values and the governance approaches they advocate. In climate politics, for instance, conflict erupted between the Climate Action Network and Environmental Defense over the use of markets in designing a governance response to climate change. NGOs, thus, face similar challenges as firms in global environmental politics.

Business Conflict

Though we often speak of a business as a monolithic interest group, different business actors often hold diverging policy preferences. Political struggles extend into the business community itself. Divisions between different firms and/or different industry sectors may diminish the clout of the overall business community (Falkner 2008). This is the central argument of the "business conflict" model in IPE (Cox 1996; Skidmore-Hess 1996). Firms may have different policy preferences for mainly two reasons: first, environmental policy has differential distributional effects; and, second, firms operate in different political and institutional environments.

The distributional effects of environmental policy and regulatory policy in general lead sectors and firms to different policy preferences, which can result in inter-industry or intra-industry conflict over policy preferences and political strategies. After all, "political competition follows in the wake of economic competition" (Epstein 1969: 142). Such distributional effects exist when environmental regulation causes lower aggregate costs to an industry as a whole compared to other industries; when environmental regulation generates rents for some industries or firms while it erects barriers to other industries and firms; and when environmental regulation causes differential costs across firms in the same industry (Keohane *et al.* 1998). Conflict over political strategy may not only erupt due to different economic interests but also due to being embedded in different social and political institutions (DiMaggio and Powell 1991). These institutions exist at different levels, such as the national level (Vogel 1986) or at the level of organizational fields (Hoffman 2001; Dingwerth and Pattberg 2009). Some authors have emphasized the role of home country-level institutions in leading to divergent business strategies (Skjærseth and Skodvin 2001). Others suggest that economic interests have a stronger effect on corporate strategies than different institutional environments (Levy and Newell 2000; Clapp 2003).

Business conflict has played a role in mitigating business influence in a number of cases of global environmental policy-making (Falkner 2008). In the early phase of climate politics, fossil-fuel industries were capable of organizing an influential anti-regulatory coalition, which started to disintegrate with the emergence of business conflict in the run-up to the Kyoto negotiations. The line of conflict ran in particular between American and European oil companies (Pulver 2005). While US firms forcefully opposed targets and timetables, European firms, but also some US companies, accepted mandatory targets and timetables as long as market mechanisms were available for implementation (Meckling 2008, Meckling 2011a). Against this backdrop the Kyoto Protocol was created. Thereafter, the business division between anti-regulatory and pro-regulatory forces deepened. The split in the business community was widely seen to have weakened the opposition toward mandatory climate policy. While business conflict has played an important role in some cases of global environmental politics, Falkner (2008) stresses that the analytical possibility of business conflict does not imply its empirical ubiquity. In fact, as we discuss below, business is often able to organize unity.

Business Coalitions

Given external and internal political competition, firms face the challenge of organizing collective action to achieve political clout. As Cerny argues, political outcomes "are determined ... by how coalitions and networks are built in real-time conditions among a plurality of actors" (Cerny 2003: 156). Alliance-building helps to leverage and pool sources of power such as funding or legitimacy, thus being a strategic form of power (cf. Levy and Scully 2007). Coalition-building requires a certain level of coordination among actors. They are often embedded in broader policy networks of actors that share interests and norms but that do not necessarily coordinate their strategies (Mahoney 2007).

With regard to business, we have witnessed the emergence of "complex multi-level and institutional advocacy coalitions" in Europe and the United States and at the international level (Coen 2005). Such coalitions emerged in virtually all major environmental policy fields. Coalitions that oppose environmental regulation often consist of firms only. Coalitions that promote a particular policy proposal may include environmental organizations as allies, as "firms have learned to *mix and match* their political alliances with various environmental and business interests groups to create flexible advocacy coalitions" (Coen 2005: 216). Alliances between business groups and green groups have been referred to as "Baptist-and-bootlegger" coalitions (Yandle 1983; DeSombre 2000). The term connotes the cooperation of unlike interest groups in advocacy for a common cause.

In the realm of environmental politics, we could observe the emergence of transnational business coalitions in a number of cases, with climate change, ozone, and biosafety being high-profile cases (Newell and Levy 2006; Meckling 2011b). Here, we discuss the latter two. In an effort to prevent further ozone regulations in the

USA, the chlorofluorocarbon (CFC) industry created the Alliance for Responsible CFC Policy in 1980 (Haas 1992; Falkner 2008). Representing the interests of US firms, the Alliance first opposed the regulation of CFCs in the USA. Due to increasing scientific certainty about the ozone problem and shifting political dynamics, the coalition shifted its strategy to support international caps on CFCs. This political shift was led by DuPont in 1986. Henceforth, DuPont and the Alliance for Responsible CFC Policy became the hub of a larger transatlantic coalition aiming to shape the international rules of ozone regulation. On the other side of the Atlantic, it was in particular the German chemical company Hoechst that pushed the new transnational industry agenda (Falkner 2008). The new pro-regulatory industry stance strengthened the position of the US Department of State, which was in favor of an international agreement.

In biosafety politics, the Global Industry Coalition was the unified voice of biotechnology companies in Europe and the United States (Newell 2003). In addition, it served as a forum for information exchange. Created in 1998, the coalition represented mostly seed companies, pharmaceutical companies, commodity traders, and food manufacturers (Reifschneider 2002). The coalition was led by association leaders from Canada, Europe, and the United States. It claimed to represent more than 2200 companies from over 130 countries (Clapp 2003). The transnational coalition allowed industry to speak with one voice on fundamental issues while governments – in this case the EU and the USA – were divided.

Conclusion

Business actors – in particular multinational corporations – have a preponderance of power resources and shape global environmental policy and governance through a number of different channels. These avenues of influence include direct lobbying at both the national and international levels, market influence due to their ability to move capital across borders and control technology research and development, participation in rule-setting schemes of various types, and framing issues that shape broader understandings of environmental problems and their solutions. Business actors often employ more than one of these strategies, often simultaneously, to effect influence over the outcome of environmental policy and governance.

At the same time, corporate influence over environmental policy and governance varies according to the circumstances under which their activities aim at shaping the process take place. The extent to which they influence policy outcomes, in other words, is mediated through processes of transnational political contention and cooperation. Political competition – from both within and outside the business community – is a force that limits the influence of business in global environmental policy-making. Successful coalition-building, instead, may strengthen the influence of firms. Analyses of business actors in global environmental policy and governance must take these complexities into account.

What are the lessons for the future? The history of corporate involvement in global environmental politics and governance suggests a trend from business opposing environmental regulation to business shaping the regulatory options, style, and rules of governance arrangements. Business not only shaped understandings of environmental problems, it also shaped possible solutions, and supported in particular the emergence of market-based forms of environmental governance such as private governance and environmental markets. Analysts should therefore continue to follow closely the role corporate influence plays in the choice of governance approaches and in the specific design of global environmental governance mechanisms.

References

Alcock, Frank. 2008. "Conflicts and Coalitions within and across the ENGO Community." *Global Environmental Politics*, 8(4): 66–91. doi:10.1162/glep.2008.8.4.66.

Avant, Deborah D., Martha Finnemore, and Susan K. Sell, eds. 2010. *Who Governs the Globe?* Cambridge: Cambridge University Press.

BASD 2012. 2011. "Business Calls on Governments to Set Policies that Accelerate Progress toward a Green Economy," http://basd2012.org/483/business-calls-on-governments-to-set-policies-that-accelerate-progress-toward-a-green-economy/ (accessed April 12, 2012).

Betsill, Michele M. 2008. "Reflections on the Analytical Framework and NGO Diplomacy." In *NGO Diplomacy: The Influence of Nongovernmental Organizations in International Environmental Organizations*, ed. Michele M. Betsill and Elisabeth Corell, 177–206. Cambridge, MA: MIT Press.

Brown, Halina Szejnwald, Martin de Jong, and Teodorina Lessidrenska. 2009. "The Rise of the Global Reporting Initiative: A Case of Institutional Entrepreneurship." *Environmental Politics*, 18(2): 182–200.

Cashore, Benjamin. 2002. "Legitimacy and the Privatization of Environmental Governance: How Non-state Market-Driven (NDSM) Governance Systems Gain Rule-Making Authority." *Governance*, 15(4): 503–529. doi:10.1111/1468-0491.00199.

Cerny, Philip G. 2003. "The Uneven Pluralization of World Politics." In *Globalization in the 21st Century. Convergence or Divergence?*, ed. Axel Hülsemeyer, 153–175. Basingstoke: Palgrave Macmillan.

Cerny, Philip G. 2010. *Rethinking World Politics: A Theory of Transnational Neopluralism*. New York: Oxford University Press.

Clapp, Jennifer. 1998. "The Privatization of Global Environmental Governance: ISO 14000 and the Developing World." *Global Governance*, 4(3): 295–316.

Clapp, Jennifer. 2003. "Transnational Corporate Interests and Global Environmental Governance: Negotiating Rules for Agricultural Biotechnology and Chemicals." *Environmental Politics*, 12(4): 1–23. doi:10.1080/09644010412331308354.

Clapp, Jennifer. 2005. "Global Environmental Governance for Corporate Responsibility and Accountability." *Global Environmental Politics*, 5(3): 23–34. doi:10.1162/152638 0054794916.

Clapp, Jennifer and Peter Dauvergne. 2011. *Paths to a Green World: The Political Economy of the Global Environment*, 2nd edn. Cambridge, MA: MIT Press.

Clapp, Jennifer and Jason Thistlethwaite. 2012. "Private Voluntary Programs in Environmental Governance: Climate Change and the Financial Sector." In *Private Voluntary Programs in Global Climate Policy: Pitfalls and Potentials*, ed. Karstin Ronit, 43–76. Tokyo: United Nations University Press.

Coen, David. 2005. "Environmental and Business Lobbying Alliances in Europe: Learning from Washington?" In *The Business of Global Environmental Governance*, ed. David L. Levy and Peter J. Newell, 197–222. Cambridge, MA: MIT Press.

Corell, Elisabeth and Michele M. Betsill. 2008. "Analytical Framework: Assessing the Influence of NGO Diplomats." In *NGO Diplomacy: The Influence of Nongovernmental Organizations in International Environmental Negotiations*, ed. Michele M. Betsill and Elisabeth Corell, 19–42. Cambridge, MA: MIT Press.

Cox, Ronald W. 1996. "Introduction: Bringing Business Back In: The Business Conflict Theory of International Relations." In *Business and the State in International Relations*, ed. Ronald W. Cox, 1–7. Boulder, CO: Westview Press.

Cutler, C.A., Virginia Haufler, and Tony Porter. 1999. *Private Authority and International Affairs*. Albany, NY: State University of New York Press.

Dauvergne, Peter. 2001. *Loggers and Degradation in the Asia-Pacific: Corporations and Environmental Management*. New York: Cambridge University Press.

DeSombre, Elizabeth R. 2000. *Domestic Sources of International Environmental Policy: Industry, Environmentalists, and U.S. Power*. Cambridge, MA: MIT Press.

DiMaggio, Paul J. and Walter D. Powell. 1991. "The Iron Cage Revisited: Institutional Isomorphism and Collective Rationality in Organizational Fields." In *The New Institutionalism in Organizational Analysis*, ed. Walter D. Powell and Paul J. DiMaggio, 41–62. Chicago: University of Chicago Press.

Dingwerth, Klaus and Philipp H. Pattberg. 2009. "World Politics and Organizational Fields: The Case of Sustainability Governance." *European Journal of International Relations*, 15(4): 707–743. doi:10.1177/1354066109345056.

Epstein, Edwin. 1969. *The Corporation in American Politics*. Englewood Cliffs, NJ: Prentice-Hall.

ETC Group. 2011. "Who Will Control the Green Economy?" http://www.etcgroup.org/sites/www.etcgroup.org/files/publication/pdf_file/ETC_wwctge_4web_Dec2011.pdf (accessed April 18, 2012).

Falkner, Robert. 2003. "Private Environmental Governance and International Relations: Exploring the Links." *Global Environmental Politics*, 3(2): 72–87. doi:10.1162/152638003322068227.

Falkner, Robert. 2005. "The Business of Ozone Layer Protection." In *The Business of Global Environmental Governance*, ed. David L. Levy and Peter J. Newell, 105–134. Cambridge, MA: MIT Press.

Falkner, Robert. 2008. *Business Power and Conflict in International Environmental Politics*. Basingstoke: Palgrave Macmillan.

Finger, Matthias and James Kilcoyne. 1997. "Why Transnational Corporations are Organizing to 'Save the Global Environment'." *The Ecologist*, 27(4): 138–142.

Fuchs, Doris. 2005a. "Commanding Heights? The Strength and Fragility of Business Power in Global Politics." *Millennium: Journal of International Studies*, 33(3): 771–801. doi:10.1177/03058298050330030501.

Fuchs, Doris. 2005b. *Understanding Business Power in Global Governance*. Baden-Baden: Nomos.

Fuchs, Doris, Agni Kalfagianni, and Maarten Arentsen. 2009. "Retail Power, Private Standards, and Sustainability in the Global Food System." In *Corporate Power in Global Agrifood Governance*, ed. Jennifer Clapp and Doris Fuchs, 29–59. Cambridge, MA: MIT Press.

Gill, Stephen R. and David Law. 1989. "Global Hegemony and the Structural Power of Capital." *International Studies Quarterly*, 33(4): 475–499.

Haas, Peter M. 1992. "Banning Chlorofluorocarbons: Epistemic Community Efforts to Protect Stratospheric Ozone." *International Organization*, 46(1): 187–224. doi:10.1017/S002081830000148X.

Haufler, Virginia. 2001. *A Public Role for the Private Sector: Industry Self-Regulation in a Global Economy*. Washington, DC: Carnegie Endowment for International Peace.

Hoffman, Andrew J. 2001. *From Heresy to Dogma: An Institutional History of Corporate Environmentalism*. Stanford, CA: Stanford University Press.

Keck, Margaret E. and Kathryn Sikkink. 1998. *Activists beyond Borders: Advocacy Networks in International Politics*. Ithaca, NY: Cornell University Press.

Keohane, Nathaniel O., Richard L. Revesz, and Robert N. Stavins. 1998. "The Choice of Regulatory Instruments in Environmental Policy." *Harvard Environmental Law Review*, 22(2): 313–368. doi:10.1016/S0921-8009(97)00147-X.

Khagram, Sanjeev, James V. Riker, and Kathryn Sikkink, eds. 2001. *Restructuring World Politics: Transnational Social Movements, Networks, and Norms*. Minneapolis, MN: University of Minnesota Press.

Kolk, Ans, David Levy, and Jonatan Pinkse. 2008. "Corporate Responses in an Emerging Climate Regime: The Institutionalization and Commensuration of Carbon Disclosure." *European Accounting Review*, 17(4): 719–745. doi:10.1080/09638180802489121.

Krut, Riva and Harris Gleckman. 1998. *ISO 14001: A Missed Opportunity for Global Sustainable Industrial Development*. London: Earthscan.

Leighton, Michelle, Naomi Rhot-Arriaza, and Lyuba Zarsky. 2002. *Beyond Good Deeds: Case Studies and a New Policy Agenda for Corporate Accountability*. Berkeley, CA: Nautilus Institute for Security and Sustainable Development.

Levy, David L. and Daniel Egan. 1998. "Capital Contests. National and Transnational Channels of Corporate Influence on the Climate Change Negotiations." *Politics & Society*, 26(3): 337–361. doi:10.1177/0032329298026003003.

Levy, David L. and Daniel Egan. 2003. "A Neo-Gramscian Approach to Corporate Political Strategy: Conflict and Accommodation in the Climate Change Negotiations." *Journal of Management Studies*, 40(4): 803–829. doi:10.1111/1467-6486.00361.

Levy, David L. and Peter J. Newell. 2000. "Oceans Apart? Business Responses to Global Environmental Issues in Europe and the United States." *Environment*, 42(9):8–20. doi:10.1080/00139150009605761.

Levy, David L. and Peter J. Newell, eds. 2005. *The Business of Global Environmental Governance*. Cambridge, MA: MIT Press.

Levy, David L. and Maureen Scully. 2007. "The Institutional Entrepreneur as Modern Prince: The Strategic Face of Power in Contested Fields." *Organization Studies*, 28(7): 1–21. doi:10.1177/0170840607078109.

Litfin, Karen T. 1995. "Framing Science: Precautionary Discourse and the Ozone Treaties." *Millennium: Journal of International Studies*, 24(2): 251–277. doi:10.1177/0305829895 0240020501.

Mahoney, Christine. 2007. "Networking vs. Allying: The Decision of Interest Groups to Join Coalitions in the US and the EU." *Journal of European Public Policy*, 14(3): 366–383. doi:10.1080/13501760701243764.

Mattli, Walter and Ngaire Woods. 2009. *The Politics of Global Regulation*. Princeton, NJ: Princeton University Press.

McCormick, John. 2001. *Environmental Policy in the European Union*. Basingstoke: Palgrave Macmillan.

Meckling, Jonas. 2008. "Corporate Policy Preferences in the EU and the US: Emissions Trading as the Climate Compromise?" *Carbon and Climate Law Review*, 2(2): 171–180.

Meckling, Jonas. 2011a. "The Globalization of Carbon Trading: Transnational Business Coalitions in Climate Politics." *Global Environmental Politics*, 11(2): 26–50. doi:10.1162/ GLEP_a_00052.

Meckling, Jonas. 2011b. *Carbon Coalitions: Business, Climate Politics, and the Rise of Emissions Trading*. Cambridge, MA: MIT Press.

Morgera, Elisa. 2004. "From Stockholm to Johannesburg: From Corporate Responsibility to Corporate Accountability for the Global Protection of the Environment?" *Review of European Community & International Environmental Law*, 13(2): 214–222. doi:10.1111/j.1467-9388.2004.00398.x.

Mytelka, Lynn K. 2000. "Knowledge and Structural Power in the International Political Economy." In *Strange Power. Shaping the Parameters of International Relations and International Political Economy*, ed. Thomas C. Lawton, James N. Rosenau, and Amy C. Verdun, 49–56. Aldershot: Ashgate.

Neumayer, E. 2001. *Greening Trade and Investment. Environmental Protection without Protectionism*. London: Earthscan.

Newell, Peter. 2003. "Globalization and the Governance of Biotechnology." *Global Environmental Politics*, 3(2): 56–71. doi:10.1162/152638003322068218.

Newell, Peter. 2008. "The Political Economy of Global Environmental Governance." *Review of International Studies*, 34(3): 507–529. doi:10.1017/S0260210508008140.

Newell, Peter J. and David L. Levy. 2006. "The Political Economy of the Firm in Global Environmental Governance." In *Global Corporate Power*, ed. Christopher May, 157–181. Boulder, CO: Lynne Rienner.

Pattberg, Philipp. 2005. "The Institutionalization of Private Governance: How Business and Nonprofit Organizations Agree on Transnational Rules." *Governance: An International Journal of Policy, Administration and Institutions*, 18(4): 589–610. doi:10.1111/j.1468-0491.2005.00293.x.

Prakash, Aseem. 2000. "Responsible Care: An Assessment." *Business & Society*, 39(2): 183–209. doi:10.1177/000765030003900204.

Pulver, Simone. 2005. "Organising Business: Industry NGOs in the Climate Debates." In *The Business of Climate Change: Corporate Responses to Kyoto*, ed. Kathryn Begg, Frans Van der Woerd, and David Levy, 47–60. Sheffield: Greenleaf Publishing.

Raynolds, L. 2000. "Re-embedding Global Agriculture: The International Organic and Fair Trade Movements." *Agriculture and Human Values*, 17(1): 297–309. doi:10.1023/A:1007608805843.

Reifschneider, Laura M. 2002. "Global Industry Coalition." In *The Cartagena Protocol on Biosafety: Reconciling Trade in Biotechnology with Environment and Development?*, ed. Christoph Bail, Robert Falkner, and Helen Marquard, 273–281. London: RIIA/Earthscan.

Ronit, Karsten. 2007. *Global Public Policy. Business and the Countervailing Powers of Civil Society*. Abingdon: Routledge.

Schurman, Rachel. 2004. "Fighting 'Frankenfoods': Industry Opportunity Structures and the Efficacy of the Anti-Biotech Movement in Western Europe." *Social Problems*, 51(2): 243–268. doi:10.1525/sp.2004.51.2.243.

Sell, Susan K. and Aseem Prakash. 2004. "Using Ideas Strategically: The Contest between Business and NGO Networks in Intellectual Property Rights." *International Studies Quarterly*, 48(1): 143–175. doi:10.1111/j.0020-8833.2004.00295.x.

Skidmore-Hess, Daniel. 1996. "Business Conflict and Theories of the State." In *Business and the State in International Relations*, ed. Ronald W. Cox, 199–216. Boulder, CO: Westview Press.

Skjærseth, Jon Birger and Tora Skodvin. 2001. "Climate Change and the Oil Industry: Common Problems, Different Strategies." *Global Environmental Politics*, 1(4): 43–64. doi:10.1162/152638001317146363.

Strange, Susan. 1988. *States and Markets*. Oxford: Blackwell.

Tjernshaugen, Andreas. 2012. "Technological Power as a Strategic Dilemma: CO_2 Capture and Storage in the International Oil and Gas Industry." *Global Environmental Politics*, 12(1): 8–29.

UNCTAD (United Nations Conference on Trade and Development). 2006. *World Investment Report 2006: FDI from Developing and Transition Economies: Implications for Development*. Geneva: United Nations.

UNCTAD (United Nations Conference on Trade and Development). 2008. *World Investment Report 2008: Transnational Corporations and the Infrastructure Challenge*. Geneva: United Nations, http://unctad.org/en/docs/wir2008_en.pdf (accessed October 20, 2012).

UNCTAD (United Nations Conference on Trade and Development). 2011. *World Investment Report 2011: Non-equity Modes of International Production and Development*. Geneva: United Nations, http://www.unctad-docs.org/files/UNCTAD-WIR2011-Full-en.pdf (accessed October 20, 2012).

Utting, Peter and Jennifer Clapp, eds. 2008. *Corporate Accountability and Sustainable Development*. Delhi: Oxford University Press.

Vogel, David. 1986. *National Styles of Regulation: Environmental Policy in Great Britain and the United States*. Ithaca, NY: Cornell University Press.

Wapner, Paul. 1995. "Politics beyond the State: Environmental Activism and World Civic Politics." *World Politics*, 47(3): 311–341. doi:10.1017/S0022381600053962.

Williams, Marc. 2009. "Feeding the World? Transnational Corporations and the Promotion of Genetically Modified Food." In *Corporate Power in Global Agrifood Governance*, ed. Jennifer Clapp and Doris Fuchs, 155–185. Cambridge, MA: MIT Press.

Yandle, Bruce. 1983. "Baptists and Bootleggers: The Education of a Regulatory Economist." *Regulation*, May/June: 12–16.

Young, Stephen C. 2000. "Introduction: The Origins and Evolving Nature of Ecological Modernisation." In *The Emergence of Ecological Modernisation. Integrating the Environment and the Economy?*, ed. Stephen C. Young, 1–41. London: Routledge.

Chapter 18

International Regime Effectiveness

Steinar Andresen

Introduction

Unless you feel sure that international environmental regimes will have some effect, it does not make much sense to establish them. That realization motivated students of international regimes to embark on this strand of research some two decades ago. Hundreds of multilateral environmental agreements (MEAs) have been signed: indeed, in the 1990s some 20 to 30 multilateral and bilateral agreements were signed per year (Mitchell 2003: 438–439). When do these agreements solve the problems they were set out to deal with or at least contribute to a positive development – and when and why do they fail? Various large-scale projects were started in the 1990s to grapple with these questions. Through the many books and articles subsequently published we know much more about these questions today than we did some 20 years ago. Most of the research deals with "problem-solving effectiveness," less so with questions of distribution and fairness (Young 2003). This chapter therefore focuses on problem-solving effectiveness. The main message is positive, as research confirms that "regimes do matter." However, they do not make enough of a difference, as clearly shown in UNEP's recent *Global Environmental Outlook* (UNEP 2012).

The bulk of the empirical research on effectiveness was conducted in the 1990s. Several major works published in the last decade, such as Miles *et al.* (2002); Breitmeier *et al.* (2006); and Young *et al.* (2008), are also based mostly on data collected in the 1990s. To my knowledge no large-scale international research project focusing specifically on effectiveness has been launched since the turn of the millennium. This is not because this field of research is exhausted. It may be a coincidence, or it may be because attention has been diverted elsewhere, to studies on interplay between regimes and studies of partnerships. It may also reflect the relative stagnation in multilateral environmental diplomacy. There is a close connection between

The Handbook of Global Climate and Environment Policy, First Edition. Edited by Robert Falkner.
© 2013 John Wiley & Sons, Ltd. Published 2013 by John Wiley & Sons, Ltd.

the development on the environmental political arena and the research that gets conducted. Until the turn of the millennium there was rapid development and growth in the international environmental arena. This is no longer the case, as seen not least in the negotiations on the climate regime, which is where by far the most attention and resources have been invested (Victor 2011). I begin by briefly outlining the development of international environmental politics, to provide a platform for describing the subsequent development of effectiveness research. Then the focus shifts to the concept of effectiveness, what some regimes have achieved, and how this can be explained. Before concluding, there is a brief analysis of the climate regime, to shed light both on the relative standstill in multilateral environmental diplomacy and illustrate the methodological challenges.

The Development of International Environmental Politics

Although concern about nature and the natural environment arose in the nineteenth century and the first international instruments for nature protection were created at the turn of that century, it was the establishment of the United Nations that accelerated a more systematic push towards international environmental cooperation. Initially, such cooperation was quite narrow and technical, exemplified by the establishment of the UN Food and Agriculture Organization and the UN International Maritime Organization. The World Wildlife Fund, established in 1961, was the first non-governmental organization with an overtly international ambit. It marked a milestone, by introducing lobbying, campaigns, and active relations with the media. The 1960s also heralded a wider approach to environmental issues, as effects of the post-war rush to regenerate industry and manufacturing were beginning to take their toll in the form of polluted air and oceans, fanning public concern. This was, however, a typical expression of post-materialism, and was limited to the more affluent West (Andresen *et al.* 2011).

The 1972 Stockholm Conference on the Human Environment is generally seen as the watershed event that sparked a truly international approach. As a result of the Conference, the United Nations Environment Programme (UNEP) was created, important principles were adopted, and various MEAs emerged in the 1970s. Thus the Stockholm Conference scores high in terms of agenda-setting as well as institution-building (Andresen 2007a). The agreements created in the 1970s were typically *first-generation agreements*, framework conventions that essentially acknowledged the existence of a problem, without any demanding or specific obligations on the parties. The environment was fairly high on the agenda in the Western world in the 1970s, but this changed during the 1980s: the 1982 follow-up to the Stockholm Conference went virtually unnoticed by the public. And then two dramatic events changed this: the 1986 Chernobyl nuclear reactor accident, and the discovery of the hole in the ozone layer over Antarctica, attracting renewed public attention towards the environment. Both events occurred just as the UN was in the process of publishing *Our Common Future* in 1987, which introduced the concept of sustainable development. During the late 1980s, several new and more ambitious *second-generation agreements* came into being. They were more enterprising in setting numerical targets and deadlines for emission cuts. Targets were often quite random and not always well founded, but nevertheless represented a significant advance over the first generation, because now progress or lack of such could be measured (Andresen *et al.* 2011).

With the concept of *sustainable development,* a message was sent to developing countries that the developed world was ready to accommodate calls to approach the stewardship of the environment and development as two sides of the same coin. This contributed to the inclusion of the South in the process, and sustainable development was the key focus at the 1992 Rio Summit, which became the birthplace of the Climate Convention as well the Biodiversity Convention, Agenda 21, and the Commission for Sustainable Development. As the largest international conference ever launched with very high-level political representation, the Rio Summit also marked the breakthrough for green NGO participation. Its apparent success was due not least to the widespread optimism of the early 1990s: the East–West conflict was over, economic prospects looked good, and public pressure for ambitious environmental policy was strong (Andresen 2007a).

In many ways, the 1992 Summit marked the high point of international environmental enthusiasm. Later, the media spotlight shifted focus, as did the political will and capacity to follow up the pledges made in Rio. Although the "mega-conference approach" seemed to lose some of its relevance as the follow-up was so weak, efforts to make more sophisticated MEAs continued throughout the 1990s, notably for agreements regulating air pollution, paving the way for *third-generation agreements.* It was realized that the "one-size-fits-all" approach with equal targets and timetables for all parties was not always meaningful. Cost-effectiveness and fairness were the main motivations for this more nuanced approach. Cost-effectiveness implied that cuts should be made where it would cost least, and differentiated targets thereby emerged in some regimes. The principle of fair treatment also encouraged differentiation: it was recognized as fair to give developing countries more time to reach their targets than the developed world. There was also greater focus on the need for financing and assistance to enable developing countries to meet their obligations, through such mechanisms as the Global Environment Facility.

The process of arranging mega-conferences continued with the 2002 Johannesburg Summit. This event was heavily affected by the overall international political agenda, but now in a negative way. In the aftermath of the terrorist attacks on the USA in September 2001, the times were characterized by preoccupations with war and terrorism – not the environment and development. The agenda and approach continued to broaden as poverty was now a key issue, the idea of partnerships loomed large, and the business community set their mark on the event. Despite the original intention of focusing on implementation, the summit ended up recirculating some of the ambitious aims of the UN Millennium Declaration and adding some others. The Johannesburg Summit may have had some effects in terms of discourse, but on the whole it is hard to see any substantial influence on the subsequent development of international environmental politics (Andresen 2012).

In terms of MEAs, a chemicals cluster emerged, with the Basel Convention on Hazardous Waste Management (1989) and the Rotterdam Convention on the Trade in Hazardous Chemicals (1999) supplemented by the Stockholm Convention on POPS control in 2001 (Selin 2010), and negotiations on a mercury convention are currently (May 2012) underway. However, the last decade has been characterized more by concern over too many conventions than efforts to create new ones. The system has become incredibly complex, with demanding and often overlapping reporting requirements, particularly difficult for developing countries. There has also been concern over the conflicting commitments involved in the various regimes (Oberthür and Gehring 2006). In the run-up to the Johannesburg Summit, some policy-makers and academics called for a simplification of the system through a considerable strengthening of UNEP or even a World Environment Organization (WEO) (Biermann and Bauer 2005). The idea failed to gain momentum, and the discussion never surfaced at the Johannesburg Summit. On the other hand, this institutional architecture debate was resumed – or rather recirculated – as one of the main topics to be negotiated at the Rio+20 Conference in 2012.

In my view, the Rio+20 process as well as the global warming gridlock are further demonstrations of the recent stagnation in UN-based multilateral diplomacy. Some observers have also started to question the ability of UN multilateral diplomacy to deal effectively with the serious challenges to the environment and development, and have called for a more exclusive "club approach" (Victor 2011). Others have argued that there is no need for another global mega-conference, and the Rio+20 Conference is essentially a conference in search of a purpose (Andresen 2012).

Although the most recent period has been characterized by a rather stagnant global environmental diplomacy, tremendous progress has been achieved over a relatively short time in terms of *process and approach*: learning, knowledge, institutional design, scope of participation, as well as political effort. On this basis, the question is twofold: how and to what extent has this been translated into actual progress on the ground; and how do we measure progress or lack of such?

Regime Effectiveness: Conceptual Challenges and Solutions

The Link between the Policy Development and the Research Agendas

The focus of the International Relations (IR) community closely mirrors the political development outlined above. In the 1970s and 1980s, the establishment of international regimes was central. Why did some regimes emerge while others failed to materialize? In explaining the emergence of regimes some scholars followed the traditional IR schools of realism, liberalism, and social constructivism (Hasenclever *et al.* 1997). Others concentrated more on various types of leadership and their significance for regime creation (Young 1991). This line of research is still relevant as regimes continue to be established, although at a slower pace.

Since the early 1990s there has been a shift to studies of domestic implementation of international commitments and regime effectiveness. This reflects the fact that a significant number of MEAs have now reached maturity, having existed long enough to warrant investigation as to whether they have made a difference. Greatest attention has been paid to environmental regimes, probably because this is where the growth of regimes has been most pronounced. Some of the most important large-scale international study projects here are Brown Weiss and Jacobson (1998); Young (1999); Miles *et al.* (2002); and Breitmeier *et al.* (2006). Other significant contributions include Young (2001, 2003); Hovi *et al.* (2003a); Young *et al.* (2008); and Stokke (2012). In terms of analysis and methodology, these latter publications are very advanced, but overall they present no new empirical data on effectiveness.

In my opinion this reduces their value somewhat, as interaction between theoretical and empirical research is needed to move the field forward. In this regard, Stokke (2012) represents an interesting exception in his analysis of the effectiveness of the Barents Sea fisheries regime, combining a rigorous methodological approach with a comprehensive in-depth empirical investigation of the regime.

Most effectiveness studies have analyzed MEAs, but "soft-law" institutions have also been studied (Andresen 2007a, 2007b). As the approach has matured and developed, analysts have split regimes into different phases or particular units, in order to study changes in effectiveness over time or whether specific components were more important than others. This has enabled intra-regime comparison; comparisons between regimes and regime attributes have also been conducted (Miles *et al.* 2002). Breitmeier *et al.* (2006) have built up an International Regimes Database (IRD), in order to compare and test records on specific aspects of international regimes in relation to a range of aspects, including regime creation as well as effectiveness. Comparative quantitative analysis of these two major projects has also been conducted (Breitmeier *et al.* 2011). Another strand of effectiveness research has involved in-depth studies of domestic implementation of international regimes (Victor *et al.* 1998; Skjærseth 2000; Underdal and Hanf 2000).

Reflecting the new realities of "treaty congestion," analysis has also been conducted on regime interplay (Oberthür and Gehring 2006; Oberthür and Stokke 2011). The initial concern was that such "congestion" might reduce the effectiveness of international environmental governance. Research has revealed a more nuanced picture of conflicts as well as synergies created through regime interplay (Skjærseth *et al.* 2006). Oberthür and Stokke (2011) have concluded that interplay management serves to enhance the effectiveness of global environmental governance.

Scholars have also dealt with partnerships between various types of actors – partnerships often established because of failure to reach effective agreements between the states concerned (Pattberg 2007; Gulbrandsen 2010; see also Chapter 23 in this volume). Within this approach, the issue of certification of forests and fisheries has loomed large. On balance, certification seems to change some management practices and create better outcomes in some cases, but it has not been regarded an effective institution for addressing some of the most serious environmental challenges. Important from our perspective is Gulbrandsen's remark: "we still know too little about the environmental impact and efficacy of certification as a problem-solving instrument" (Gulbrandsen 2010: 180). As certification is a rather novel approach in politics and even more so in terms of academic studies, we know less about the effectiveness of this instrument compared to the more established MEAs.

Some Key Conceptual Challenges

In 1982 Oran Young wrote:

> [T]here are severe limitations to what we can expect from efforts to evaluate regimes... this suggests the importance of giving some consideration to non-consequentialist approaches to the evaluation of regimes (Young 1982: 138).

As noted, this cautionary note did not prevent Young and several others from embarking on this challenge a decade later. In view of the many uncertainties and shortcomings still existing, his warning may have been timely, but our insight, under-standing, and cumulative knowledge have also grown considerably over this rela-tively short period.

Initially the goal of the regime was used as a measuring rod to establish the effectiveness of the regime: the higher the level of goal achievement, the higher the effectiveness of the regime in question (Andresen and Wettestad 1995). Although some analysts still apply this measure, it is problematic to use, as goals are often vague and/or overly ambitious. Consider for example the official goal of the World Health Organization, "Health for all": not a very useful standard against which to measure progress. Moreover, goals may differ greatly in terms of ambitiousness as well as specificity, making comparison between regimes exceedingly difficult.

For quite some time there has been consensus within the "effectiveness com-munity" on three criteria for measuring the dependent variable: *output, outcome*, and *impact*.

Output: Output deals with the rules and regulations reflected in the relevant regime, so this indicator is often used by legal scholars. This has been labeled level 1 implementation; level 2 output is when *formal* domestic measures are taken to comply with the international commitments (Underdal 2002a: 7). As a point of departure, the more stringent and demanding the rules regulating the behavior of the parties, the higher the effectiveness of the regime is likely to be. For example we would expect second-generation agreements to have a higher potential for being effective than first-generation agreements. Similarly, the existence of an effective compliance mechanism would seem to indicate higher effectiveness than the absence of such rules. Still, it is an open empirical question whether this potential is realized. For example, from reporting from members, it seemed that the whaling nations were all in compliance with the quotas set in the 1950s by the International Whaling Com-mission (IWC). Later documentation has revealed massive cheating on the part of the Soviet Union in this period. Moreover, research conducted by the IWC Scientific Committee in the 1960s showed that the quota set at the time was five times higher than it should have been (Andresen 2000). Thus, merely being in compliance can be a far cry from being effective. The output indicator deals essentially with *potential* effectiveness. Also, given the severe methodological challenges in the causal substan-tiation of institutional effects as to interplay, most of the evidence concerns effects at the output level, and not the outcome or impact levels (Oberthür and Stokke 2011: 318). This indicator is therefore quite weak in terms of validity; on the other hand, it is easy to measure.

Outcome: The outcome indicator seeks to measure actual progress as shown by behavioral change among key target groups in the "right" direction (regime implementation) caused by the regime. This indicator scores high in terms of validity, but involves severe obstacles in methodology. The key challenge is to establish a causal link between the regime and behavior. Consider, for instance, the massive reductions in CO_2 emissions in states with economies in transition in the 1990s. This was not a result of the effectiveness or the "bite" of the climate regime, but was due to economic recession.

Impact: The methodological challenges are even more severe when it comes to the impact indicator, the link between the regime and the effect on the problem at hand. This is the ultimate question we want to answer – the extent to which the regime has been able to solve the problem it was set up to deal with. Unfortunately the influence of other drivers is so strong and difficult to measure that applying this indicator is usually exceedingly difficult. Consider, for example, all the many factors, apart from the regime, that are of significance for the status of given fish stock, or the level of air or ocean pollution. Consequently, this measuring rod must be used with great caution.

In short, studying outputs is usually a necessary starting point, but this needs to be supplemented with studies of outcomes to enable a better grasp on what is happening in practice. Impact indicators are so demanding in terms of methodology that they are difficult to apply in empirical studies.

We then turn to the question, and controversy, of measuring the dependent variable effectiveness. In the Miles *et al.* project, Arild Underdal (2002a) (of the "Oslo School") used three elements to construct two different measures of effectiveness. One compared the actual performance obtained under the regime as against the no-regime counterfactual. The other compared the actual state of affairs to the best solution that could be accomplished, "the collective optimum." The first question was used as a tool to investigate whether and to what extent regimes do matter – the relative improvement brought about by the regime. The second was designed to show whether and to what extent a particular problem is in fact solved by the regime (Hovi *et al.* 2003a). Helm and Sprinz (2000) (of the "Potsdam School") combine these components into one comprehensive measure, through a stepwise procedure arriving at "a simple coefficient of regime effectiveness that falls into a standardized interval (0, 1)" (Hovi *et al.* 2003a: 76). It gives a very "scientific" impression to have one firm figure as a precise indicator of the effectiveness of a given regime. However important this exercise may be, the point is not mathematics or quantification but the strategies identified for determining the no-regime counterfactual and the collective optimum. In their empirical research both Miles and colleagues as well as Helm and Sprinz use a structured expert-based scoring mechanism to arrive at estimates of the non-regime counterfactual. Various measures have been used to determine the collective optimum, using game theory and other approaches. In the Miles *et al.* project, key components were application of the best external sources available and ending up with the best judgment of the case-study specialist. There are other differences and similarities between the original Oslo and Potsdam Schools, but we shall not go into these here. More basic are the key challenge and dispute: how to measure the non-regime counterfactual and the collective optimum?

This discussion was taken up by Oran Young:

> How can we separate the signal of regime effects from the noise arising from the impacts of a wide range of other sources that operate simultaneously in our efforts to understand regime effectiveness? (Young 2001: 100)

This "noise" was identified as a number of driving forces including various demographic, economic, political, and technological forces that interact with each other in complex ways, producing far-reaching impacts on the problem quite apart from the dynamics of the regime. And so, "how can we determine the *proportion* of observed change in the target variables?" (Young 2001: 100; emphasis added). How precisely can causality be established? There is agreement that the specification of the no-regime counterfactual requires a causal judgment or a method of addressing what would have happened in the absence of a particular regime. Another hotly debated topic related to arriving at counterfactuals as well as the collective optimum is the use of "expert review teams," applied in the "Oslo School," the "Potsdam School," and even more extensively within the International Regimes Database (IRD) (Breitmeier *et al.* 2006).

Young (2001, 2003) raises serious doubts as to the suitability of this approach, pointing out that the different disciplines look at effectiveness through various lenses, and may thus reach divergent conclusions as to the locus of the no-regime counter-factual and thereby regime effectiveness in specific cases. Within the Miles *et al.* project the experts were essentially the case-study workers, but all authors were involved in coding all cases as a means of reducing the element of subjectivity. The process was also made easier within this project (as well as others) as the coders were mainly International Relations scholars. In the IRD, two coders were used on each case to reduce the problem of subjectivity. Still, Young's point – that other disciplines, NGOs, or policy-makers might arrive at very different conclusions – is certainly a valid one. The use of expert coders as a main benchmark for establishing effectiveness may in fact not be very scientific, as there are too many potential flaws. More generally:

> [F]rom the perspective of those having to decide whether the score on some complex dimension should be "two" or "three" on a five-point schedule, this can be a very painful process (Andresen and Wettestad 2004: 64).

Based on my experience (and that of others) from the Miles *et al.* project, scoring in terms of determining relative improvement was difficult, but still more straight-forward than when it came to determining the collective optimum, a rather elusive term in relation to our cases. Hovi *et al.* (2003b) note that although there may be weaknesses in the Oslo–Potsdam solution as well as more fundamentally in effective-ness studies, that should not lead us to abandon this strand of studies. The challenge to the effectiveness community now is to test out this approach as well as other potential candidates, through new and comprehensive empirical research projects.

Most debate within the research community has concerned these issues of mea-surement, with less discussion on the equally important question of how to explain effectiveness. Does this mean that there is agreement on how to approach this issue? No, here we find greater disparity within the research community. According to Mitchell (2010: 172) "there is, arguably, an 'embarrassment of riches' of variables, each with compelling logic and empirical support but which, collectively, lack logic." Brown Weiss and Jacobson (1998) cast the net wide, relating effectiveness charac-teristics to the participating countries, characteristics of the international environ-ment, as well as characteristics of the institution and the activity involved. Victor and colleagues focused particularly on systems of implementation review (Victor *et al.* 1998), while Young and colleagues highlighted various causal pathways by which institutions influence behavior (Young 1999). Breitmeier and colleagues con-cluded that behavior is not only shaped by consequences and appropriateness but also by discourse, legitimacy, and habit (Breitmeier *et al.* 2006: 234–235). Despite these differences, there is wide agreement that effectiveness is influenced by both institutional and non-institutional factors. Miles and colleagues (2002) used two main explanatory perspectives: the nature of the problem and the problem-solving capacity of the relevant institutions. The more politically and intellectually "malign" a problem, the less could be expected in terms of effectiveness. Problem-solving capacity was conceived of as a function of three variables: power, leadership, and institutional set-up.

In the following I rely heavily on the perspective of Miles and colleagues. Controlling for problem structure is important in order to be able to compare regimes. Thus it may be more of an accomplishment to achieve progress with an extremely malign problem, compared to solving an exceedingly benign one. The strength of the construction of the problem-solving capacity variable is that it includes all the major schools of IR thought. As we shall see, key elements from the realist school of thought need to be included, although this is often neglected by regime analysts.

Are Regimes Effective: When and Why?

Relatively few regimes have been analyzed, compared to the vast numbers that exist. Much is known about some favored regimes, but little or nothing about many others. Moreover, in many large-scale projects, pragmatism has often been a more important criterion for case selection than the objective principles for case selection laid down in textbooks on social science methodology. That is, the expertise and interest of those participating in the project has often been decisive for case selection. This further reduces possibilities for making general claims applicable to the wider universe of international environmental regimes (Andresen and Wettestad 2004).

In the following I present some general findings based primarily on Miles and colleagues, supplemented by other findings. Based on the scoring of the case-study workers, Underdal did the math and came up with some hard figures on the effectiveness of the 15 regimes (37 units of analysis) analyzed: "The short answer is that most of the regimes included in the sample make a positive difference but fall short of providing functionally optimal solutions" (Underdal 2002b: 435). This conclusion is in line with the findings of all the other main research projects that have been conducted. That is, there is support for the counterfactual argument that, if these regimes had not existed, things would have been worse. In more than half the cases the improvement brought about by the regime is shown to be significant. But there is also a flip side: in almost half the cases, these regimes did *not* make much of a difference, which illustrates the great variation in effectiveness.

More good news. According to Miles and colleagues, in about two out of three cases, regimes served as arenas for facilitating transnational learning and contributed to strengthening the knowledge base for policy-decisions. Moreover, we need not consider the distribution of power or the structure of the problem as given. The distribution of power may be changed through multiple maneuvers

including low-cost invasion of an existing organization (IWC), entry of new actors into the activity system itself (South Pacific Tuna), shift to another arena (radioactive waste) as well as a shift to a different regulatory approach (ship-generated oil-pollution) (Underdal 2002b: 457).

There were also instances where malign problems were dealt with rather effectively, although these were the exception. Most regimes had modest beginnings – even those that eventually achieved significant results, like the Oslo Commission on dumping in the North Sea and the ozone regime. This illustrates the positive practical effects of the more sophisticated approaches for some regimes over time. Particularly attractive are the fast-track options that have been applied in both the North Sea regime and the 1979 Convention on Long-Range Transboundary Air Pollution (LRTAP). What we find is essentially a combination of "soft law" and "hard law," where the most enthusiastic parties went ahead with ambitious soft-law commitments, which later trickled down to the more reluctant parties (Skjærseth *et al.* 2006). Creating forums for high-level leaders to meet may also have a positive effect when the problems are not too malign. However, when problems are truly malign, as with climate change, we have seen that this approach has its limitations.

Here the flip side to these positive messages should also be noted. Even though most regimes do make a positive difference, none of those studied, not even the ozone regime, have been able to solve the problems they were set up to deal with, and they have often fallen far short of the mark. Moreover, some of the overall improvements observed have been due to "fortunate circumstances" for which the regime itself can claim little or no credit. The most fundamental intervening factor was the general growth in public demand for and governmental supply of policies for environmental protection. This was clearly the case with overall developments from the early 1970s to the turn of the millennium. It may not be equally true for the more recent period as regards the main motor in this development, the OECD region. This makes the note of caution sounded by Underdal (2002b: 457) highly relevant: "the rate of success that we have observed for environmental regimes may not be easily replicated in a stagnant or declining policy field." The relative success achieved earlier may be hard to sustain in the current period of declining public concern over environmental issues and global economic recession.

Although malign problems can be dealt with quite effectively, this does not usually happen. The combination of high malignancy and high uncertainty is particularly lethal for problem-solving, as will be illustrated regarding the climate regime below. From these findings, power emerges as the most critical factor in dealing with strongly malign problems. Although there is little evidence that the hegemonic power approach has much significance for regime creation and effectiveness (Falkner 2005), that does not mean that power as such has no explanatory power. The shifting roles played by the USA in the ozone regime and the climate regime are illustrative. As to the negative impact of scientific uncertainty, this plagued the development of the work of the Paris Commission as well as LRTAP in the initial phases. When uncertainty was reduced, this was *one* important reason for the greater effectiveness that was achieved.

Also, although some key factors *can* be manipulated, the political engineering of institutional design of regimes tends to be a difficult exercise. For example, solid institutional capacity with an active secretariat is no guarantee of success: in most organizations, the secretariats depend more on powerful actors than vice versa. Good institutional capacity is likely to be most important when it interacts with other elements working in the same (positive) direction, like leadership and consensual science. Recent research offers new insights regarding the role of secretariats.

After studying the influence of 10 international environmental bureaucracies, mostly secretariats, Biermann and Siebenhuner (2009: 345) conclude that problem structure and what they label "people and procedures" are most important in explaining their influence, whereas institutional design makes less of a difference. Expectations as to the significance of institutional design should therefore be modest, but it is certainly not irrelevant.

The significance of various decision-making procedures is a potential candidate in this regard. The main premise in the Miles *et al.* project was the traditional one, that governance systems relying on consensus will tend to produce decisions in line with the preferences of the least ambitious member of the group. Interestingly, the Breitmeier *et al.* project reached a different conclusion: "But the evidence from the IRD does not support this conclusion. Problem improvement occurs in half the cases where regimes rely on consensus rules" (2006: 231). A further point can be added to this discussion, based on my research on the IWC (Andresen 2001). The IWC is one of the rare international organizations to practice majority voting, usually seen as a means to increase the potential effectiveness of the regime. With the IWC, however, the effect has been the opposite, as it has contributed to polarizing the already heated atmosphere in the Commission.

Finally, an observation on *exactly how much* of the change in behavior can be attributed to the regime compared to other drivers. In the studies that have been carried out, several factors are typically mentioned, of which the regime is one, but there is rarely any specification as to the proportion of change caused by the regime. Usually, the combination of these factors is said to explain the overall results achieved. For instance, Wettestad's recent study of the acid rain regime (Wettestad 2011) finds that achievements may have been the result of better scientific understanding, an advanced regime design, changed attitudes of key players, Germany's crucial policy turn-about, and the fall of communism, as well as the increased role of the EU. On this basis it is concluded: "It is hence clearly challenging to measure the exact impact of the multilateral mechanism on these achievements" (Wettestad 2011: 35). The more general fall-back is easier: "we are reasonably sure that emissions in Europe would have been considerably higher in such a counterfactual, no-regime situation" (Wettestad 2011: 35). This point will be elaborated further in the next section on the effectiveness of the climate regime.

The Lack of Effectiveness of the Climate Regime

Since the turn of the millennium, the environmental regime that has almost monopolized attention among policy-makers, the media, and the public has been the climate regime. By "climate regime" is here understood the political process under the UN umbrella. The main outputs have been the 1992 UN Framework Convention on Climate Change, the 1997 Kyoto Protocol, and the recent decisions taken by the Conference of the Parties (Cancun Agreement, 2010, and Durban Platform, 2011) on a future climate regime. The Framework Convention is a typical first-generation agreement with few specific or demanding commitments. Consequently it has been embraced by practically all nations of the world. Considering the short time needed to negotiate it, this convention represents a necessary first step, not least in terms of building up knowledge through the detailed reporting requirements. With the benefit of hindsight, perhaps its most important feature is the rather clear-cut – and static – division between developed and developing countries in terms of commitments (Bodansky 1993). More recently this has hampered progress during the climate negotiations.

In terms of treaty sophistication there was a significant development from the Framework Convention to the 1997 Kyoto Protocol, which was a typical third-generation agreement with both hard targets and differentiation. Moreover, three innovative market-based mechanisms were added to facilitate implementation. Thus from negotiations started in 1991 to the adoption of the Protocol the process was quite dynamic, at least compared to the lack of achievements over the last 15 years. It took four years to reach agreement on a much-diluted Marrakesh Accord (Hovi *et al.* 2003c). Another four years went by before the Protocol entered into force (2005), and in the meantime there was not much substance to negotiate over. Since the 2007 Bali COP, negotiations were intensified in order to reach agreement on a post-Kyoto regime, but progress has been modest. Postponing key decisions has characterized the process, as shown by the 2011 Durban Platform, where yet another "roadmap" has been set up in order to have a new agreement by 2015, to be in force by 2020. The Kyoto Protocol has been prolonged, but key parties have decided to leave it. Much time has also been spent on negotiating specific mechanisms like the Green Fund, technology transfer, and REDD+, but we do not yet know what significance these will have in terms of actual emissions reductions.

What about the effect of the specific measures agreed in the Protocol? Joint implementation has not been used much, but the spread of the Clean Development Mechanism (CDM) has accelerated significantly in recent years. From a methodology perspective, this mechanism has the advantage of flowing *directly* from the regime, which removes the problem of establishing causality. However, the good news stops there, as there is uncertainty and dispute over how much emissions reductions have been caused by this mechanism, in view of such factors as carbon leakage, additionality, and transaction costs.

The challenges in terms of determining causality are more difficult as regards emissions trading, as there is no trading based directly on the climate regime. Still, various national programs have obviously been *triggered* by the Protocol, of which by far the most important is the EU ETS (on emissions trading, see also Chapter 27 in this volume). With UN negotiations deadlocked, the stricter ETS over time seems likely to be more affected by internal EU dynamics than the climate regime will be. Rather, the lack of UN progress probably contributes more to provide ammunition to those forces within the EU that oppose a strong EU climate policy. There is also considerable disagreement on the extent to which the EU ETS has contributed to emission reductions that would otherwise not have occurred. Some reductions have probably been achieved, but this is linked only partially to the climate regime.

Turning to potential behavioral effects more broadly, in terms of emission reductions compared to a business-as-usual scenario, we find that the picture is even more uncertain. On the one hand global emissions are continuing to rise sharply, so the likelihood of meeting the goal of not more than a 2 °C rise in global temperature is almost non-existent. On the other hand, the increase in emissions would probably have been even higher without the regime. The best example is provided by the many measures taken in the EU and its member-states (Skjærseth and Wettestad 2008).

The climate regime has probably also had some effect for non-Annex I countries. For example, growing concerns in China with the climate issue as well as the intensity target adopted were no doubt spurred by the climate regime. However, given the recent stalemate internationally, it seems doubtful that the recent elevation of the issue on the political agenda can be causally linked to the climate regime (Stensdal 2012). In the USA, a country not noted for federal enthusiasm for climate change policies, it has been documented that there is considerable "bottom-up" climate action among various actors (Selin and VanDeveer 2009). These processes are probably driven largely by forces other than UN negotiations. In general, from emission figures and economic growth statistics, it seems that the development of the world economy has been a more important regulator than the climate regime. Thus, effectiveness is low – but equally important is the difficulty in deciding the precise effect of the climate regime.

In terms of explaining effectiveness, the extremely malign nature of the problem goes a long way towards providing an explanation. However, the problem-solving perspective also sheds light on the poor performance of the climate regime. During this process, the most powerful actor has been the USA – most of the time as a "laggard" and a "veto-power," blocking progress. The rising influence of China has not made progress easier. The EU has had leadership ambitions, but has been too weak to generate a sufficient number of followers to move the process forward (Andresen and Boasson 2011). In terms of institutional set-up, we do know that the massive global effort has yielded few results. The clearest example of failed multilateral UN diplomacy is probably the process leading up to COP 15 in Copenhagen 2009. Diplomats met during six negotiation sessions prior to COP 15 – but when the high-level segment arrived in Copenhagen there was no text on the table. In short, inefficiency and high transaction costs characterize the approach, not effective problem-solving.

Conclusion

The effectiveness community agrees that "regimes matter." We also know that there is significant variation as to how effective regimes are, and that few if any of them have managed to solve the problems that led to their being established in the first place. Although many explanatory perspectives have been used, there is consensus that institutional as well as non-institutional factors make a difference for regime performance. There is less agreement on how precisely regime effectiveness can be measured, and efforts to arrive at exact figures as an indicator of regime effectiveness remain mired in methodological difficulties. This I have sought to illustrate empirically through a brief analysis of the climate regime. Admittedly, however, my own shortcomings in game theory and quantitative methods may represent the real bottleneck in this regard.

Empirical studies of international environmental regime effectiveness peaked towards the end of the 1990s. Important analytical advances have been made since then, but as these have not been combined with new empirical research, their significance is reduced. Some may argue that effectiveness studies are no longer timely, as the research focus has shifted to other issue areas like regime interplay, influence, the role of partnerships, or regime resilience. However relevant and intriguing these approaches may be, they are no substitute for effectiveness studies.

It is now time to launch a new wave of empirical research on the effectiveness of regimes. Is it so, as I have assumed, that there has been a general lessening in effectiveness recently – or is the picture more nuanced? In any case, new research should not be restricted to replicas and updates of previous studies. Attention should also focus on non-environmental regimes, as we know that environmental regimes may not be the only, or even the most important, regimes that actually affect the environment. The scope should also be broader than problem-solving, and the significance of regimes in terms of legitimacy as well as fairness should be discussed. So far this research has been conducted almost exclusively by Western scholars, thereby reflecting Western ways of thinking. Not least because of new geopolitical realities, scholars from the emerging economies as well as more generally from the South should be included in future research. It is important also to bring in more extensive discussions on the significance of the new geopolitical realities represented by the new emerging economies for regime effectiveness.

References

Andresen, Steinar. 2000. "The Whaling Regime." In *Science and Politics in International Environmental Regimes: Between Integrity and Involvement*, ed. Steinar Andresen, Tora Skodvin, Arild Underdal, and Jørgen Wettestad, 35–69. Manchester: Manchester University Press.

Andresen, Steinar. 2001. "The International Whaling Regime: 'Good' Institutions but 'Bad' Politics?" In *Toward a Sustainable Whaling Regime*, ed. Robert Friedheim, 235–269. Seattle, WA: University of Washington Press.

Andresen, Steinar. 2007a. "The Effectiveness of UN Environmental Institutions." *International Environmental Agreements: Politics, Law and Economics*, 7(4): 317–336.

Andresen, Steinar. 2007b. "Key Actors in UN Environmental Governance: Influence, Reform and Leadership." *International Environmental Agreements: Politics, Law and Economics*, 7(4):457–468.

Andresen, Steinar. 2012. "Do We Need More Sustainable Development Conferences?" In *Handbook of Global Environmental Politics*, 2nd edn, ed. Peter Dauverge, 87–97. Cheltenham: Edward Elgar.

Andresen, Steinar and Elin Lerum Boasson. 2011. "International Climate Cooperation: Clear Recommendations, Weak Commitments." In *International Environmental Agreements: An Introduction*, ed. Steinar Andresen, Elin Lerum Boasson, and Geir Hønneland, 49–67. Abingdon: Routledge.

Andresen, Steinar, Elin Lerum Boasson, and Geir Hønneland, eds. 2011. *International Environmental Agreements: An Introduction*. Abingdon: Routledge.

Andresen, Steinar and Jørgen Wettestad. 1995. "International Problem-Solving Effectiveness: The Oslo Project Story So Far." *International Environmental Affairs*, 7(2): 127–149.

Andresen, Steinar and Jørgen Wettestad. 2004. "Case Studies of the Effectiveness of International Environmental Regimes." In *Regime Consequences: Methodological Challenges and Research Strategies*, ed. Arild Underdal and Oran Young, 49–70. Dordrecht: Kluwer Academic.

Biermann, Frank and Steffen Bauer, eds. 2005. *A World Environmental Organization: Solution or Threat for International Environmental Governance*. Aldershot: Ashgate.

Biermann, Frank and Berndt Siebenhuner, eds. 2009. *Managers of Global Change: The Influence of International Environmental Bureaucracies*. Cambridge, MA: MIT Press.

Bodansky, Daniel. 1993. "The United Nations Framework Convention on Climate Change: A Commentary." *Yale Journal of International Law*, 18: 451–558.

Breitmeier, Helmut, Arild Underdal, and Oran R. Young. 2011. "The Effectiveness of International Environmental Regimes: Comparing and Contrasting Findings from Quantitative Research." *International Studies Review*, 13(4): 579–605. doi:10.1111/j.1468-2486.2011.01045.x.

Breitmeier, Helmut, Oran R. Young, and Michael Zürn. 2006. *Analyzing International Environmental Regimes: From Case Study to Database*. Cambridge, MA: MIT Press.

Brown Weiss, Edith and Harold Jacobson, eds. 1998. *Engaging Countries: Strengthening Compliance with International Environmental Accords*. Cambridge, MA: MIT Press.

Falkner, Robert. 2005. "American Hegemony and the Global Environment." *International Studies Review*, 7(4): 585–598. doi:10.1111/j.1468-2486.2005.00534.x.

Gulbrandsen, Lars. 2010. *Transnational Environmental Governance: The Emergence and Effects of the Certification of Forests and Fisheries*. Cheltenham: Edward Edgar.

Hasenclever, Andreas, Peter Mayer, and Volker Rittberger. 1997. *Theories of International Regimes*. Cambridge: Cambridge University Press.

Helm, Carsten and Detlef Sprinz. 2000. "Measuring the Effectiveness of International Environmental Regimes." *Journal of Conflict Resolution*, 44(5): 630–652. doi:10.1177/0022002700044005004.

Hovi, Jon, Detlef Sprinz, and Arild Underdal. 2003a. "The Oslo–Potsdam Solution to Measuring Regime Effectiveness: Critique, Response, and the Road Ahead." *Global Environmental Politics*, 3(3):74–96. doi:10.1162/152638003322469286.

Hovi, Jon, Detlef Sprinz, and Arild Underdal. 2003b. "Regime Effectiveness and the Oslo–Potsdam Solution: A Rejoinder to Oran Young." *Global Environmental Politics*, 3(3): 105–107. doi:10.1162/152638003322469303.

Hovi, Jon, Tora Skodvin, and Steinar Andresen. 2003c. "The Persistence of the Kyoto Protocol: Why Other Annex I Countries Move On without the United States." *Global Environmental Politics*, 3(4): 1–23. doi:10.1162/152638003322757907.

Miles, Edward L., Arild Underdal, Steinar Andresen *et al.* 2002. *Environmental Regime Effectiveness: Confronting Theory with Evidence*. Cambridge, MA: MIT Press.

Mitchell, Ronald B. 2003. "International Environmental Agreements: A Survey of Their Features, Formation and Effects." *Annual Review of Environment and Resources*, 28, November: 429–461.

Mitchell, Ronald B. 2010. *International Politics and the Environment*. Thousand Oaks, CA: Sage.

Oberthür, Sebastian and Thomas Gehring, eds. 2006. *Institutional Interaction in Global Environmental Governance: Synergy and Conflict among International and EU Policies*. Cambridge, MA: MIT Press.

Oberthür, Sebastian and Olav Schram Stokke, eds. 2011. *Managing Institutional Complexity: Regime Interplay and Global Environmental Change*. Cambridge, MA: MIT Press.

Pattberg, Phillip. 2007. *Private Institutions and Global Governance: The New Politics of Environmental Sustainability*. Cheltenham: Edward Elgar.

Selin, Henrik. 2010. *Global Governance of Hazardous Chemicals: Challenges of Multilevel Management*. Cambridge, MA: MIT Press.

Selin, Henrik and Stacy D. VanDeveer. 2009. *Changing Climates in North American Politics: Institutions, Policymaking, and Multilevel Governance*. Cambridge, MA: MIT Press.

Skjærseth, Jon Birger. 2000. *North Sea Cooperation: Linking International and Domestic Pollution Control*. Manchester: Manchester University Press.

Skjærseth, Jon Birger, Olav Schram Stokke, and Jørgen Wettestad. 2006. "Soft Law, Hard Law, and Effective Implementation of International Environmental Norms." *Global Environmental Politics*, 6(3): 104–120. doi:10.1162/glep.2006.6.3.104.

Skjærseth, Jon Birger and Jørgen Wettestad. 2008. *EU Emission Trading: Initiation, Decision-Making and Implementation*. Aldershot: Ashgate.

Stensdal, Iselin. 2012. *China's Climate Policy 1998–2011: From Zero to Hero?* FNI Report 9/2012. Lysaker, Norway: Fridtjof Nansen Institute.

Stokke, Olav Schram. 2012. *Disaggregating International Regimes. A New Approach to Evaluation and Comparison.* Cambridge, MA: MIT Press.

Underdal, Arild. 2002a. "One Question Two Answers." In *Environmental Regime Effectiveness: Confronting Theory with Evidence*, ed. Edward L. Miles, Arild Underdal, Steinar Andresen *et al.*, 3–46. Cambridge, MA: MIT Press.

Underdal, Arild. 2002b. "Conclusions: Patterns of Regime Effectiveness." In *Environmental Regime Effectiveness: Confronting Theory with Evidence*, ed. Edward L. Miles, Arild Underdal, Steinar Andresen *et al.*, 433–466. Cambridge, MA: MIT Press.

Underdal, Arild and Kenneth Hanf, eds. 2000. *International Environmental Agreements and Domestic Politics: The Case of Acid Rain.* Aldershot: Ashgate.

UNEP. 2012. *GEO 5: Global Environment Outlook.* Nairobi: United Nations Environment Programme.

Victor, David G. 2011. *Global Warming Gridlock: Creating More Effective Strategies for Protecting the Planet.* Cambridge: Cambridge University Press.

Victor, David G., Karl Raustiala, and Eugene Skolnikoff, eds. 1998. *The Implementation and Effectiveness of International Environmental Commitments: Theory and Practice.* Cambridge, MA: MIT Press.

Wettestad, Jørgen. 2011. "Reducing Long-Range Air-Pollutants in Europe." In *International Environmental Agreements: An Introduction*, ed. Steinar Andresen, Elin Lerum Boasson, and Geir Hønneland, 23–37. Abingdon: Routledge.

Young, Oran R. 1982. *Resource Regimes: Natural Resources and Social Institutions.* Berkeley, CA: University of California Press.

Young, Oran R. 1991. "Political Leadership and Regime Formation: On the Development of Institutions in International Society." *International Organizations*, 45(3): 291–308.

Young, Oran R., ed. 1999. *The Effectiveness of International Environmental Regimes: Causal Connections and Behavioral Mechanisms.* Cambridge, MA: MIT Press.

Young, Oran R. 2001. "Inferences and Indices: Evaluating the Effectiveness of International Environmental Regimes." *Global Environmental Politics*, 1(1): 99–121. doi:10.1162/152638001570651.

Young, Oran R. 2003. "Determining Regime Effectiveness: A Commentary on the Oslo–Potsdam Solution." *Global Environmental Politics*, 3(3): 97–104. doi:10.1162/152638003322469295.

Young, Oran, Leslie King, and Heike Schroder, eds. 2008. *Institutions and Environmental Change: Principal Findings, Application, and Research Frontiers.* Cambridge, MA: MIT Press.

Chapter 19

Strengthening the United Nations

Steffen Bauer

Introduction

The environment arrived late on the United Nations' policy agenda, occupies limited institutional space within the convoluted United Nations system, and has tradition-ally been considered an issue of low politics in intergovernmental relations. It is thus unsurprising that the calls to strengthen the environmental mandate and, indeed, the corresponding capacities of the United Nations (hereafter UN) are as old as the UN's engagement in environmental policy itself.

The UN's involvement in environmental issues can be traced back to the emerging concern of the United Nations Educational, Scientific and Cultural Organization (UNESCO) for the "biosphere" in the 1960s. More pronounced engagement in the environmental realm is generally attributed to the United Nations Conference on the Human Environment (UNCHE). This was convened in Stockholm in June 1972 and brought about the United Nations Environment Programme (UNEP), which ultimately proved to be the conference's most visible and lasting achievement.

UNEP was formally established by Resolution 2997 (XXVII) of the UN General Assembly later in 1972 and took up its work in 1973, almost three decades after the establishment of the UN's principal organs and major specialized agencies (UNGA 1972). Deliberately designed as a subordinate program of the UN Economic and Social Council (ECOSOC) and with a "small secretariat" that is located in Kenya's capital Nairobi, that is, remote from UN hubs in New York and Geneva, UNEP epitomizes the late, limited, and low standing of environmental matters in the UN system. At the same time, it represents "the closest thing there is to an overarching global institution for the environment" (DeSombre 2006: 9).

There is a considerable gap between the high expectations that the UN's envi-ronmental institutions find themselves confronted with and their limited capability

The Handbook of Global Climate and Environment Policy, First Edition. Edited by Robert Falkner.
© 2013 John Wiley & Sons, Ltd. Published 2013 by John Wiley & Sons, Ltd.

to deliver. Still, the UN has evolved into the foremost arena of global environmental governance. At its center, UNEP has been reaffirmed time and again to be the international community's

> leading global environmental authority that sets the global environmental agenda, that promotes the coherent implementation of the environmental dimension of sustainable development within the UN system and that serves as an authoritative advocate for the global environment (Nairobi Declaration; see UNEP 1997: paragraph 2).[1]

Indeed, the dynamic growth in multilateral environmental institutions has largely occurred under the auspices of the UN, with many of the corresponding treaties being administered by UNEP. These include, *inter alia*, the 1992 Convention on Biological Diversity (CBD), the 1985 Vienna Convention for the Protection of the Ozone Layer and its 1987 Montreal Protocol on Substances that Deplete the Ozone Layer, the 1973 Convention on Trade in Endangered Species, and a host of chemicals- and waste-related conventions, as well as treaties pertaining to specific biodiversity-related issues and regional seas (Bauer 2009b).[2]

Moreover, UNEP has played a key role in raising governmental awareness around desertification and climate change. As a result, international efforts to tackle these issues are now both addressed by genuine UN conventions, that is, the United Nations Framework Convention on Climate Change (UNFCCC) and the United Nations Convention to Combat Desertification (UNCCD) respectively (Bauer 2009b). These environmental issues are thus provided with an elevated status within the UN system and a direct link to the UN Secretary General and General Assembly (Bauer *et al.* 2009). Many of these multilateral environmental agreements enjoy close to universal membership, which lends them global legitimacy and political clout, at least in theory.

In addition, several conventions relating specifically to the marine environment are governed through the International Maritime Organization (IMO), that is, one of the UN's oldest specialized agencies (Campe 2009). It is thus fair to conclude that the UN really is *the* arena for global environmental governance and international climate policy, yet still inadequate for solving the multiple environmental crises that humankind is facing.

A virtually "permanent state of reform" (Elliott 2005: 37) notwithstanding, the UN's expectations–capabilities gap has not been fundamentally addressed since the UNEP first took up its work. In fact, the apparent inefficacy of its environmental institutions has "in some respects served to erode the legitimacy of the UN and dilute its brand value because of the persisting gaps between rhetoric and commitments and implementation" (Weiss and Thakur 2010: 221). Meanwhile an increasing awareness of the ecological limits of growth-driven human development and a tendency to "securitize" key environmental problems such as water scarcity or climate change have created a sense of urgency, but have yet to prompt commensurate action. Indeed, it took 40 years for the 2012 United Nations Conference on Sustainable Development to seriously attempt to redress the UN's overall institutional framework at the nexus of environment and development (UN 2012a: section IV). To what end the ensuing reform decisions will eventually be implemented remains to be seen.

Any institutional reform will inevitably be appraised against the abundance of calls for a UN with stronger environmental capabilities that have amassed in the past decade alone (see *inter alia* Andresen 2001; WBGU 2001, 2011; Brack and Hyvarinen 2002; Tarasofsky 2002; Haas 2004; Biermann and Bauer 2005; Chambers and Green 2005; Najam *et al.* 2006; Swart and Perry 2007; Biermann 2012). From within the UN system itself the Secretary-General's High-Level Panel on System-Wide Coherence in the Areas of Development, Humanitarian Assistance, and the Environment acknowledges that:

> We possess fairly comprehensive knowledge and understanding of what we individually and collectively need to do to reverse these trends [of environmental degradation] – all spelled out in reports, declarations, treaties and summits since the early 1970s. While we have made significant advances within the UN framework, what is needed now is a substantially strengthened and streamlined international environmental governance structure, to support the incentives for change required at all levels (UN 2006: paragraph 31).

This chapter traces the evolution of institutions and concepts pertaining to global environmental governance by and through the UN, next discusses the structural obstacles that so far hinder a fundamental strengthening of environmental policy within the UN system, and concludes by sketching the prospects for environmental multilateralism in the wider context of shifting geopolitics and global aspirations for sustainable development.

Environmental Policy in the UN System: Key Concepts and Institutions

Societal concern for the environment naturally originated at local and national levels. It manifested itself in notions of nature conservation and ecological stewardship that are much older than the first inklings of international environmentalism. Yet they provided the basis on which political concern for the "human environment" was eventually elevated to the intergovernmental realm in the 1970s (see McCormick 1989; Caldwell 1996). As such, however, desires to conserve nature and to address transboundary pollution of air, land and water sources were perceived as a rather luxurious concern in large parts of the world that had only recently gained independence from colonial oppression and that were eager to advance on socio-economic scales by any means.

This sentiment was famously coined into a phrase by the then Indian prime minister Indira Gandhi (1972) when she rhetorically asked the United Nations Conference on the Human Environment "Are not poverty and need the greatest polluters?" It persevered as a quintessential point of reference around which the structural dichotomy between "the North," that is, rich industrialized countries, and "the South," that is, poor developing countries, revolved until the apparent contradiction between environment and development was formally overcome by aligning conflicting objectives under the conceptual formula of sustainable development.

Defined as "development that meets the needs of the present without compromising the ability of future generations to meet their own needs," the concept of sustainable development was first introduced by the World Commission on Environment and Development (WCED) in its 1987 report *Our Common Future* (also known as the Brundtland Report, after the commission's chairwoman, Gro Harlem

Brundtland; WCED 1987). To reflect both the socio-economic and the ecological context of development, the commission further emphasized

> the essential needs of the world's poor, to which overriding priority should be given; and the idea of limitations imposed by the state of technology and social organization on the environment's ability to meet present and future needs (WCED 1987).

The commission's definition of sustainable development was subsequently adopted in the Rio Declaration of the 1992 United Nations Conference on Environment and Development (UNCED). It has since proved paradigmatic for virtually all development activities of the UN, framing policy agendas in areas as diverse as multilateral aid, health care, food security, education, corporate social responsibility. and, not least, environmental regulation. The language of the UNFCCC and the CBD that were both adopted at the UNCED testify to the instant paradigmatic quality of the sustainable development concept. This is even more obvious in the case of the UNCCD, which was initiated at the Rio summit and eventually adopted in 1994 (see Bruyninckx 2005: 287–290; Johnson *et al.* 2006). Accordingly, any discussion of the UN's role in international environmental governance and, indeed, a strengthening of its corresponding institutions will need to be firmly placed in a context of sustainable development.

This is not to say that the principal North–South divide has been overcome. The inherent tension "between the sovereign prerogative of states to exploit, utilize, and develop resources within their jurisdiction … and the global impact of deforestation, stock depletions, desertification and atmospheric pollution" (Weiss and Thakur 2010: 208–209) prevails. Simply put, aspirations for inclusive growth tend to conflict with imperatives of sustainable resource use. Different interpretations of sustainable development thus continue to reflect the different priorities and interests of developed and developing countries and their ongoing struggle to balance the concept's three pillars of economic development, social development, and environmental protection (see also Chasek and Wagner 2012). Controversies about environmental governance within the UN thus continue to be characterized by mutual suspicion and a strong emphasis on the alleged trade-offs between developmental and environmental objectives.

If further proof was needed regarding the trade-offs that have dominated debates over sustainable development, the negotiations preceding the Rio+20 United Nations Conference on Sustainable Development in 2012 were marred by a stark North–South fault line. This prevailed throughout the summit, even as the South's traditional negotiating block of the "G77 and China" appeared more heterogeneous than ever. Its ubiquity and pervasiveness notwithstanding, sustainable development ultimately remains a contested concept. While virtually everybody can subscribe to sustainable development as a slogan and to the lofty aspirations attributed to it, it hardly lends itself to on-the-ground policy-making (see also Bruyninckx 2005; Weiss and Thakur 2010). Hence, whenever push comes to shove, negotiators utilize the conceptual vagueness of sustainable development as they see fit.

Strategic use of sustainable development's vagueness was particularly obvious at the World Summit on Sustainable Development (WSSD) that convened in Johannesburg in 2002. Rather than reaching substantive decisions to remedy the capabilities–expectations gap, governments evoked broad voluntary commitments that hurt no one, enthusiastically embraced public–private "Type II" partnerships as a solution to the problem of sustainable development, downplayed the relevance of environmental protection, and even watered down pre-existing multilateral outcomes in pertinent sectors such as water, health, and agriculture (see Andonova and Levy 2003; Pallemaerts 2003; Weiss and Thakur 2010).

As the UN now seeks to pick up the pieces after the Rio+20 United Nations Conference on Sustainable Development, it is helpful to recall that the WSSD was widely acknowledged to have failed to deliver, especially from an environmentalist viewpoint (e.g. Pallemaerts 2003; Speth 2003; Wapner 2003). To grapple with this realization the Rio+20 summit turned to notions of a "green economy" and their potential to facilitate sustainable development and poverty eradication (see UNEP 2011). In the event, however, the conference remained lukewarm about this latest addition to the global environmental governance lexicon and especially reflected developing countries' reluctance to subscribe to it (UN 2012a: section III).

In conjunction with the green economy debate, states were expected to finally reform the institutional framework for sustainable development to actually strengthen the UN at the nexus of environment and development and to complement the existing Millennium Development Goals with a new set of Sustainable Development Goals in the wake of the Rio+20 conference. On both counts some progress has been achieved even as its tangibility will be subject to intergovernmental interpretation and brinkmanship. However the implementation of corresponding decisions turns out, it will be vital to understand the key institutions that constitute the UN's environmental architecture so far and that have subsequently been the focus of reform debates. Of these institutions, the most notable are UNEP and the Commission on Sustainable Development.

United Nations Environment Programme

UNEP was conceived to provide the international community with leadership and guidance on global and regional environmental matters by assessing and monitoring the state of the environment and by serving as a norm-building catalyst for international environmental policy and law. Moreover, it was formally mandated to coordinate all of the UN's environmental activities (UNGA 1972; see also Ivanova 2007, 2010; Bauer 2009b). Its organizational set-up and governance structure, however, foresaw no autonomous role vis-à-vis other agencies and institutions within the UN system, many of which are superior to UNEP either hierarchically or politically. UNEP is thereby effectively prevented from living up to its assigned coordinating function.

UNEP's original mandate deliberately precluded it from having operative capacity beyond the promotion of environmental law. This has always been reflected by a relatively small budget and professional staff. The Environment Fund that technically determines UNEP's core budget relies on governments' voluntary pledges for replenishment and has over the years typically fluctuated somewhere between US\$30 million and 60 million per annum. This makes for a meek resource base, even as UNEP's overall budget has grown to some US\$200 million per annum (UNEP 2012). This is largely due to earmarked contributions from individual member-states, designated trust funds that are administered through UNEP, and increased access to operative funds, notably by virtue of UNEP being an implementing agency of the Global Environment Facility (GEF), and a partial expansion of its mandate through the Bali Strategic Plan on Technology Support and Capacity Building (UNEP 2004; see also Bauer 2009b and Ivanova 2012 for further details).

It is all the more remarkable against this background that UNEP, by and large, has been rather successful in acting as the UN's environmental consciousness and, indeed, as a catalyst for environmental action. As such it has played a proactive role in the facilitation of several pivotal multilateral environmental agreements, in the promotion of environmental law at international, regional, and even national levels, and in raising general awareness for the environmental challenges facing the international community. Ironically, UNEP's success in facilitating issue-specific multilateral environmental institutions is somewhat clouded by the concomitant proliferation of separate decision-making bodies, notably conferences of the parties (COPs) to distinct environmental treaties such as the CBD or the Montreal Protocol (see Andresen and Rosendal 2009; Bauer 2009a). Ultimately, the decision-making processes of these institutions are out of UNEP's reach, which further exacerbates the difficulties UNEP faces regarding the coordination of international environmental governance.

The dynamic proliferation of multilateral environmental agreements has intensified the pace, density, and complexity of international environmental governance (see also Depledge and Chasek 2012). Transaction costs are thereby multiplied not only for the UN, but also for member-states who are burdened with ever more meetings and concurrent reporting requirements. The meetings of the UNFCCC, CBD, and UNCCD and their respective subsidiary bodies alone consume up to 230 meeting days per year, triggering "treaty fatigue," and not only in developing countries whose capacities are easily stretched thin (Müller 2010: 164; see also Muñoz et al. 2009).

United Nations Commission on Sustainable Development

In 1992 the institutional landscape of the UN environmental architecture was expanded with the establishment of the Commission on Sustainable Development (CSD) and a corresponding Division for Sustainable Development at UN headquarters to serve as the commission's secretariat following the UNCED (UNGA 1992). Like UNEP, the CSD reports to the General Assembly via ECOSOC. It comprises 53 states that are elected by the General Assembly for three years and in accordance with a quota that warrants regional representation according to the UN's five regional groups. Moreover, the CSD stands out for "pioneering innovative arrangements for civil society participation" (Mingst and Karns 2007: 223) that facilitate input from nine Major Groups as called for in Agenda 21, the action plan for achieving sustainable development as established at the first Rio Conference in 1992 (UNCED 1992: chapter 23).

The CSD does not take legally binding decisions, but was tasked with monitoring and reporting on the implementation of the Rio decisions as spelled out in Agenda 21. Governments thus effectively mandated the CSD to engage with the environmental policy domain. While this reflected the paradigm shift towards sustainable development institutionally, it further blurred the delineation of competences between agencies dealing with environmental and development affairs without actually mainstreaming environmental and development objectives (Bauer 2009b: 184; see also Imber 1993; Elliott 2005). Rather than improving interagency coordination, the CSD evolved into a cumbersome and often politicized platform for fundamental North–South debates in which socio-economic concerns tend to take precedence over the interpretation of sustainable development, even as government delegates typically represent environmental ministries rather than ministries of finance, economy, or trade.

The CSD's status was formally reaffirmed when it was further tasked with monitoring the implementation of the Johannesburg Plan of Implementation after the 2002 World Summit on Sustainable Development. Its meeting format was concomitantly modified to address specific thematic complexes in biennial work cycles, such as water, sanitation, and human settlements in 2004/2005 and agriculture, rural development, land, drought, and desertification in 2008/2009 (see DeSombre 2006: 33 for an overview). The first year of the two-year cycles have been dedicated to evaluating the progress made regarding the respective policy themes. The ensuing year is then intended to galvanize implementation in the corresponding policy fields. While this has helped to streamline the CSD's agenda, it did not solve the controversies underlying much of its work, for instance, regarding reporting requirements of developed and developing countries.

On balance it seems fair to say that the CSD's major achievement as a legacy of the UNCED has been to increase multi-stakeholder dialogue through institutionalized participation of the Major Groups, even as their subsequent influence has remained ambiguous. Overall, however, the genesis and record of the CSD and its evident incapability to strengthen the UN's performance in environmental governance suggest "that it was created as a way to avoid, rather than institutionalize, action" (DeSombre 2006: 35).

International Environmental Governance at Country Level

With neither UNEP nor the CSD commanding significant operative capacities, an appraisal of the UN's role in environmental policy would be incomplete without considering the entities that do. Indeed, UNEP has shown a particular propensity to engage in interagency cooperation simply for the sake of increased involvement and visibility at country level, thereby circumventing to some extent the formal restrictions of its non-operative mandate (Bauer 2009b: 178).

Environment-related policies of specialized agencies and programs like the World Health Organization, the United Nations Industrial Development Organization (UNIDO), or UN-HABITAT notwithstanding, the United Nations Development Programme (UNDP) stands out as the UN's foremost operative actor at country level. As such, UNDP enjoys access to a wide range of multilateral environmental funds, including the Montreal Protocol's Multilateral Fund, the Global Environment Facility, and Capacity 21 funds that were specifically designated to build developing country capacities pertaining to the objectives of Agenda 21 (Murphy 2006: 270–271). Moreover, many issues in the UNDP's traditional portfolio, such as water management or energy provision, inherently relate to environmental policy.

The joint Poverty–Environment Initiative of UNDP and UNEP is arguably the most tangible undertaking so far regarding an operative UN program to support "country-led efforts to mainstream poverty–environment linkages into national development planning" (UNDP and UNEP 2011). Formally initiated in 2005, it was expanded after a five-year pilot phase (2004–2008) and is now up and running in 17 developing countries, providing "financial and technical assistance to government partners to set up institutional and capacity strengthening programmes" pertaining to each country's particular poverty–environment context (UNDP and UNEP 2011). The joint initiative mobilized an average of roughly US$4 million per annum in the pilot phase, and has since more than doubled its annual expenditure to US$8 million in 2009 and US$10 million in 2010 (UNDP and UNEP 2011).

Such efforts notwithstanding, the relationship between the UNDP and the UNEP – formally on equal footing in the UN hierarchy – has remained one of unequal siblings. In spite of numerous efforts to enhance constructive cooperation both in the field and at program level, joint initiatives are often marred by conflicting institutional interests and ensuing turf battles over competences, resources, and, not least, the attention of state principals (Biermann and Bauer 2004; see also Mee 2005). Any meaningful reform of the UN's environmental performance will also have to address the counterproductive side-effects of this uneasy competition and the general mode of cooperation between constituents of the United Nations Development Group, chaired by UNDP, and the Environmental Management Group, chaired by UNEP. With a reform of the UN development architecture representing an uphill struggle in its own right, this is of course easier said than done (see Weinlich 2011). It thus bodes poorly for the envisaged revamping of the Institutional Framework for Sustainable Development that the outcome of the United Nations Conference on Sustainable Development hardly considers these institutional realities and, in particular, the pivotal role of UNDP.

Another complex task facing, *inter alia*, the UN is the task of downscaling the global objectives negotiated under multilateral environmental agreements to meet the domestic needs of individual countries. Such a global–national-scale shift is intended to take place, for example, through the National Action Programmes of the UNCCD, or the National Adaptation Programmes of Action under the UNFCCC.[3] These instruments are critical to realizing effective multilevel governance and typically rely on the Global Environmental Facility and other multilateral funds that are often jointly administered by UNDP, UNEP, and other implementing agencies including the World Bank (e.g. Biermann 1997; Andler 2009; Horstmann and Chandani Abeysinghe 2011). Although considerable efforts are made to promote coherence and synergies regarding the implementation of multilateral environmental agreements, for instance in a Joint Liaison Group of the three "Rio conventions" – the CBD, UNCCD, and UNFCCC – the political reality seems more adequately characterized by institutional fragmentation and political competition.

Institutional fragmentation is not a direct outgrowth of UN institutions as such, but a result of member-states' readiness to create ever-new institutions rather than to redress or dissolve existing ones and their tendency to retain principal control through earmarked contributions. While institutionalist research suggests that inter-institutional disruption and conflict typically emerge as unintended side-effects rather than deliberate strategizing, they nonetheless inhibit efficient implementation (see also Gehring and Oberthür 2008).

Global Policy Dimensions: Ambitions and Obstacles Regarding a Substantive Strengthening of the UN's Environmental Governance Architecture

At the heart of the obstacles that have so far prevented a substantive strengthening of the UN's environmental architecture lie the very same fundamental issues that explain its current weakness: the North–South fault line that pervades global policy-making, the environment as a latecomer in international politics, and a limited institutional space within the UN that reflects governments' perception of a low, albeit growing, salience of global environmental problems.

Indeed, the UN Charter does not address the management of natural resources, even though the UN has always had a profound impact on how natural resources were perceived and addressed internationally (Schrijver 2007). Much less did the UN's founders conceive of concepts such as environment, ecology, or sustainability. In the absence of a "charter moment" that would correct this and other anachronisms of the UN system at the root, academic and public policy debate on the need for a specialized agency on the environment, indeed a world environment organization, has been thriving for some time.

In the mid-1990s this occurred at least partially in response to a management crisis at the helm of the UNEP secretariat, the emergence of the CSD in the wake of the UNCED, and an increasing awareness of the effects of economic globalization and the World Trade Organization's potential impact on environmental policy (Bauer and Biermann 2005). It has gained further momentum after the Global Ministerial Environment Forum's 2002 "Cartagena Package" decision on international environmental governance and the unsatisfactory outcomes of the WSSD. Together these factors have gradually galvanized intergovernmental consensus on incremental reform measures during the past decade. In 2005, the government of France even constituted a "Group of Friends" to push for a United Nations Environment Organization as a full-fledged specialized agency to replace UNEP. While this initiative was in itself unsuccessful, it did keep a potential "upgrade" of UNEP on the international agenda and provided a basis for a formalized intergovernmental consultation process in the context of Kofi Annan's broader reform agenda on system-wide coherence and the Rio+20 summit of 2012.

Named after the locations of key meetings, this consultative process is now referred to as the Belgrade process and the Nairobi–Helsinki process respectively (see Simon 2011, for further details). Compared to previous reform debates its outcomes are remarkable for their expedience and broad consensus on key areas of reform, which were also reflected in the original negotiating text that preceded the eventual outcome document of the 2012 United Nations Conference on Sustainable Development.[4] With an overall sense to make form follow function, the intergovernmental consultative process explicitly focused on pragmatic consensus-building regarding five functional objectives (see CGIEG 2010 and Simon 2011: 23 for further background):

- creating a strong, credible, and accessible science base and policy interface;
- developing a global authoritative and responsive voice for environmental sustainability;

- achieving effectiveness, efficiency, and coherence within the UN system;
- securing sufficient, predictable, and coherent funding;
- ensuring a responsive and cohesive approach to meeting country needs.

The politically delicate questions of whether and how the UN's environmental institutions should be strengthened in these ways were thus effectively circumvented. However, they are certain to resurface once tangible decisions on institutional competences, mandates, and resources pertaining to either of the desired functions are to be formally adopted.

At the Rio+20 summit, hopeful aspirations regarding a "double upgrade" that would transform UNEP into a UNEO specialized agency on the one hand and the CSD into an elevated Sustainable Development Council on the other hand were halfway met at best. The summit's general decisions to strengthen UNEP with universal membership and "secure, stable, adequate and increased financial resources from the [UN] regular budget" (UN 2012a: paragraph 88) and to replace the CSD with a "universal intergovernmental high-level political forum" (UN 2012a: paragraph 84) are yet to be implemented. Whether they actually strengthen the UN's "environmental pillar" ultimately depends on how this will be done. As always, the proof of the pudding will be in the eating.

Leaving the significance of pending institutional reforms aside, the predominance of socio-economic policy within the UN system and the concurrent North–South divide remain substantive obstacles to a major breakthrough on environmental institutions. This is true even as developing countries' interests appear far more heterogeneous than on previous occasions. Besides, political and economic powerhouses on both sides of the North–South divide, notably including the United States, China, India, and Russia, have always been skeptical regarding a substantive institutional strengthening of UNEP in particular. Thus far, it is hard to see what might prompt them to endorse more ambitious reform options or, for that matter, what price reform proponents like the European Union are actually willing to pay to get skeptics to align. Against this background, even adamant supporters of further-reaching institutional reform would seem well advised not to consider the formal status of a specialized agency as an end in itself, but to focus on strengthening the functional capacities required for effective international environmental governance (see also Najam 2005; Ivanova 2012).

Meanwhile, as any major reform of the UN system ultimately depends on the political will of its member-states, the UN can, to a certain extent, succeed in strengthening itself. As principals are generally hesitant to strengthen their agents, international organizations have often sought creative ways to improve their lot (Hawkins *et al.* 2006). Indeed, the role of international bureaucracies as drivers of incremental yet significant changes to the UN's performance in environmental governance increasingly warrants scholarly attention (see Bauer 2006; Biermann and Siebenhüner 2009).

The very decision to strengthen UNEP by granting its Governing Council universal membership (as opposed to the exclusive status quo with 58 members) provides a case study in endogenous institutional change (Bauer 2009b: 176–177). While states sought to avoid turning UNEP into a specialized agency, the UNEP secretariat managed to muster sufficient support to establish a Global Ministerial

Environment Forum (GMEF), thereby achieving *de facto* universal membership through the backdoor long before governments finally consented, at the Rio+20 summit, to accordingly expand the Governing Council's membership (see UN 2012a: paragraph 88(a)).

Since the GMEF was first invited to Malmö in 2000, it has become established practice to convene it as well as recurrent "special sessions" of the UNEP Governing Council in the intervals between the latter's biennial regular sessions. Moreover, the GMEF has since been routinely called to convene back to back with regular Council sessions. Though the Governing Council, with its restricted membership, continued to be *de jure* the decision-making body of UNEP, it could hardly ignore any substantive output from the GMEF. In the absence of a formal strengthening this was not insignificant as a measure of the UNEP's political clout, even though the GMEF cannot make formal decisions. The GMEF thus proves both the UN's internal potential to generate authority and the structural limits it is facing in redressing the fundamental parameters that ultimately determine its weakness.

The proliferation of ever more institutions and forums (like the GMEF) and the global interaction of ever more actors (notably non-state actors such as international bureaucracies, transnational civil society organizations, and multinational private businesses) demonstrate the growing fragmentation of international policy-making. While this is hardly exclusive to environmental governance, institutional fragmentation has proved particularly dynamic in international environment and climate policy (Biermann *et al.* 2009). Again, while synergetic and cooperative fragmentation are possible, empirical analysis suggests that fragmentation in global environmental policy, notably regarding climate change, is often conflictive and thereby undermines prospective organizational advantages of functional differentiation and redundancy (see also Keohane and Victor 2010; Zelli 2011). This is especially the case, since the complexity of institutional fragmentation and issue linkages in environmental governance is additionally enhanced by a recent phenomenon that has aptly been described as "climate change bandwagoning" (Jinnah and Conliffe 2012).

The dynamic development of environmental and climate institutions under the auspices of the UN alone, to say nothing of their inherent linkages with development policy, thus raises broader questions on the prospects for international cooperation and, indeed, organization. The 2012 United Nations Conference on Sustainable Development did not answer them, even as it extensively considered the institutional framework for sustainable development and corresponding means of implementation (UN 2012a). While it has reached some noteworthy decisions on long-pending issues of institutional reform – including the fate of the Commission on Sustainable Development, which is to be superseded, and the status of UNEP, which is to be strengthened, however half-heartedly – ensuing changes will yet again prove incremental rather than radical.

The prevailing sentiment in the wake of another lackluster UN summit thus is one of advanced skepticism. Even the most ardent advocates of the UN have to grapple with the notion that the current state of multilateralism seems to defy far-reaching institutional reform: "It is possible to add new organizations, forums or processes to the existing maelstrom, but it is impossible to shift what is already there in any fundamental way" (Halle 2012: 2).

Conclusions and Global Policy Implications: The Prospects for a Strengthened UN

The UN is often referred to as a proxy for "the international community" and, as such, it has taken to referring to "Our Common Future" (WCED 1987). If Earth Sciences are right, however, the international community's prospects in the nascent "anthropocene" are hardly commensurate with "The Future We Want" (UN 2012a; see also Biermann *et al.* 2012; Brito and Stafford Smith 2012). Indeed, the substance of the Rio+20 summit's outcome document and the tenacious stalemate of international climate negotiations reflect rather poorly on the ability of global institutions to tackle environmental challenges. Even as there have been considerable achievements since the environment first appeared on the UN's agenda, the overarching sense of the international community's response to global environmental change and the risks it entails for an already unstable world remains one of ineptitude, failure, and frustration (Falkner and Lee 2012).

As the glaring gap between the international community's expectations and the capabilities of global environmental institutions prevails, "the United Nations" is easily scapegoated. Nonetheless, a convincing alternative forum for addressing the dangerous trends of global environmental change in a manner that can claim universal legitimacy has yet to be found. In the absence of an effective and democratic world government, the UN with its universal membership remains unmatched for the provision of global vision and leadership, international legitimacy, and an indispensable convening power that is instrumental for mobilizing multilateral action. No issue could highlight these comparative advantages of the UN better than the quest for sustainable global development (see also Weiss and Thakur 2010).

If the international community is stuck with the UN to organize multilateral responses to universal challenges, how might it better incorporate the notion of "planetary boundaries" within the UN's historical domains of security, development, and human rights? One way forward in this respect would be to adjust the "uneven institutionalization" of environmental policy within the UN by standing it on four legs rather than its "two rear legs" (Conca 2010). That is to say that the normative and operative integration of environmental concerns must no longer be confined to the realm of sustainable development and international law, but also permeate the UN's activities regarding peace and human rights.

Indeed, the securitization of climate change and specific environmental issues, such as water scarcity or land degradation, and a discernible trend among non-governmental organizations and advocacy groups to pursue environmental agendas with human-rights-based approaches may point in that direction. It is no coincidence, for instance, that a proven and tested primer on the UN addresses environmental governance in a chapter on "Human Security" rather than in its chapter on "Economic Development and Sustainability" (Mingst and Karns 2007). Whether such approaches may help to further the environmental capacity of the UN is an open question, as both security policy and human rights law are highly politicized. Securitizing the environment, in particular, is a double-edged sword that may sideline precautionary approaches to environmental degradation and divert resources as much as it raises political awareness for socio-ecological interdependencies and the tangible problems caused by environmental degradation (see, for instance, Brock 1997; Detraz and Betsill 2009).

Any meaningful strengthening of the UN will have to respond to broader geopolitical trends. While the inner dynamics of the UN as we know it continue to be driven by a North–South antagonism largely fueled by global inequality, the terms of debate may be changing. The fragmentation of actor constellations in the environmental and climate policy arena, where the so-called BASIC countries and other emerging economies are ostensibly drifting away from their "traditional" G77 base, is suggestive of burgeoning tectonic shifts in world politics. As much as it has by now become a "commonplace" (Falkner and Lee 2012) to observe these shifts, it is not trivial to anticipate their implications for the UN. Whether and how states deal with the attendant material changes at the international level as well as at their respective domestic levels will determine the future role of the UN in world politics and, indeed, global environmental governance.

On the one hand, a realist reading of changes in North–South relations would suggest a renaissance of power politics with more conflictual fragmentation in which multilateral "forum shopping" further undermines the grasp of the UN and "reinforces the power of the strong" (Hurrell and Sengupta 2012). Indeed, the emergence of, and public interest in, the G20 may be seen as a case in point. Its global relevance was particularly evident in the handling of the financial crisis of 2008/2009 and, even in general terms, it may be expected to play a significant role in a globalized world. Still, it is unlikely the G20 can sideline the UN on matters of global environmental change and the development of effective and legitimate policy. For all of club governance's undeniable benefits, it also comprises considerable political disparities between its heavy-weight members, including China and the USA. While these same differences have so far also prevented progress on major global issues such as climate policy or, for that matter, UN reform, they also diminish the challenge that the G20's motley crew poses to multilateral efforts at tackling global environmental issues.

Rather than a return to power politics, a liberal reading of the global interdependencies that are as evident as they are complex suggests rather the possibility of synergetic polycentrism and enhanced multilateral cooperation. This could occur in a suitably strengthened UN with commensurate regulatory and coordinative competences as well as enhanced participatory mechanisms. This would arguably require a *de facto* qualification of sorts to the key principle of international law, that is, sovereignty, even as the latter has long been identified as a construct of "organized hypocrisy" (Krasner 1999). A deliberate qualification of sovereignty will be particularly hard to come by both for established powers that seek to preserve their status in world politics and for emerging powers that have long aspired to exploit the full potential of their sovereignty. Yet, the delegation of power and authority has been the essence of international organization ever since and was quintessential in the "constitutional moment" that brought about the UN after the Second World War. Ultimately, adjusting the UN to the geopolitical and socio-ecological realities of the anthropocene will require another constitutional moment.

Amending the UN Charter must not be off limits in the pursuit of a strengthened UN, although it appears politically prohibitive to negotiate any such amendment in the short run. A case in point, the Rio+20 summit failed to even commit to an explicit normative vision along these lines. Still, it can be argued that its "The

Future We Want" outcome document was consensually adopted by the world's governments at the summit and thereby provides at least a tangible vantage point and implicit legitimization for future intergovernmental consultations on that matter. A fundamentally restructured UN would have to acknowledge the planetary boundaries of human development as a guiding principle for all UN activities to finally place environmental sustainability on a par with the pursuit of peace, security, human rights, and welfare. It may then eventually overcome the expectations–capability gap that has thus far marred four decades of international environmental governance.

To sum up, as far as a genuine strengthening of the UN's environmental capabilities is concerned, the situation today is not entirely different from the assessment made more than a decade ago by the former administrator of UNDP and chair of the United Nations Development Group, James Gustave Speth, as the international community braced itself for the World Summit on Sustainable Development: "There is no great mystery about *what* must be done. What does remain a great mystery is *how* we get on that path" (Speth 2002: 26; original emphasis). That mystery in turn will hardly be solved unless "member states will recognize that they gain more than they lose by empowering the United Nations to carry out tasks that individually they have no prospect of fulfilling" (Kennedy 2006: 284–285).

Notes

1 For similar reaffirmations see *inter alia* the 2000 Malmö Declaration and 2010 Nusa Dua Declaration of the Global Ministerial Environment Forum (GMEF 2000; UNEP 2010) or the report of the Secretary-General's High-Level Panel on System-Wide Coherence in the Areas of Development, Humanitarian Assistance, and the Environment (UN 2006), and, most recently, the Rio+20 summit's outcome document (UN 2012: paragraph 88).

2 For detailed accounts of these and other multilateral environmental agreements and their relation to the UN see, for instance, Tolba and Rummel-Bulska (1998) or Chasek *et al.* (2010). For an overview of global multilateral environmental agreements, including their corresponding host institutions, number of member-states, and year of adoption, see Müller (2010: 165–166).

3 For a conceptual overview on scale and scaling in international environmental institutions see Gupta (2008); for illustrative case studies on the UNCCD's National Action Programmes see Bruyninckx (2004) and Pearce (2006).

4 Refer to the Co-chairs' "Zero Draft" Outcome Document of January 10, 2012 for the United Nations Conference on Sustainable Development, New York (UN 2012b).

References

Andler, Lydia. 2009. "The Secretariat of the Global Environment Facility: From Network to Bureaucracy." In *Managers of Global Change. The Influence of International Environmental Bureaucracies*, ed. F. Biermann and B. Siebenhüner, 203–224. Cambridge, MA: MIT Press.

Andonova, Liliana B. and Mark A. Levy. 2003. "Franchising Global Governance: Making Sense of the Johannesburg Type II Partnerships." In *Yearbook of International Co-operation on Environment and Development 2003/2004*, ed. O. Schram Stokke and O.B. Thommessen, 19–31. London: Earthscan.

Andresen, Steinar. 2001. "Global Environmental Governance: UN Fragmentation and Co-ordination." In *Yearbook of International Co-operation on Environment and Development 2001/2002*, ed. O.B. Thommessen and O. Schram Stokke, 19–26. London: Earthscan.

Andresen, Steinar and Kristin Rosendal. 2009. "The Role of the United Nations Environment Programme in the Coordination of Multilateral Environmenal Agreements." In *International Organizations in Global Environmental Governance*, ed. F. Biermann, B. Siebenhüner, and A. Schreyögg, 133–150. Abingdon: Routledge.

Bauer, Steffen. 2006. "Does Bureaucracy Really Matter? The Authority of Intergovernmental Treaty Secretariats in Global Environmental Politics." *Global Environmental Politics*, 6(1): 23–49.

Bauer, Steffen. 2009a. "The Ozone Secretariat: The Good Shepherd of Ozone Politics." In *Managers of Global Change. The Influence of International Environmental Bureaucracies*, ed. F. Biermann and B. Siebenhüner, 225–244. Cambridge, MA: MIT Press.

Bauer, Steffen. 2009b. "The Secretariat of the United Nations Environment Programme: Tangled Up in Blue." In *Managers of Global Change. The Influence of International Environmental Bureaucracies*, ed. F. Biermann and B. Siebenhüner, 169–202. Cambridge, MA: MIT Press.

Bauer, Steffen and Frank Biermann. 2005. "The Debate on a World Environment Organization: An Introduction." In *A World Environment Organization: Solution or Threat for Effective International Environmental Governance?*, ed. F. Biermann and S. Bauer, 1–23. Aldershot: Ashgate.

Bauer, Steffen, Per-Olof Busch, and Bernd Siebenhüner. 2009. "Treaty Secretariats in Global Environmental Governance." In *International Organizations in Global Environmental Governance*, ed. F. Biermann, B. Siebenhüner, and A.P. Schreyögg, 174–191. Abingdon: Routledge.

Biermann, Frank. 1997. "Financing Environmental Policies in the South: Experiences from the Multilateral Ozone Fund." *International Environmental Affairs*, 9(3): 179–219.

Biermann, Frank. 2012. "Greening the United Nations Charter: World Politics in the Anthropocene." *Environment*, 54(3): 6–17.

Biermann, Frank, Kenneth Abbott, Steinar Andresen *et al.* 2012. "Navigating the Anthropocene: Improving Earth System Governance." *Science*, 335, March 16: 1306–1307.

Biermann, Frank and Steffen Bauer. 2004. *United Nations Development Programme (UNDP) and United Nations Environment Programme (UNEP), Externe Expertise für das WBGU-Hauptgutachten "Welt im Wandel: Armutsbekämpfung durch Umweltpolitik."* Berlin: WBGU.

Biermann, Frank and Steffen Bauer, eds. 2005. *A World Environment Organisation. Solution or Threat to Effective International Environmental Governance?* Aldershot: Ashgate.

Biermann, Frank, Philipp Pattberg, Harro van Asselt, and Fariborz Zelli. 2009. "The Fragmentation of Global Governance Architectures. A Framework for Analysis." *Global Environmental Politics*, 9(4): 14–40.

Biermann, Frank and Bernd Siebenhüner. 2009. "The Influence of International Bureaucracies in World Politics: Findings from the MANUS Research Program." In *Managers of Global Change. The Influence of International Environmental Bureaucracies*, ed. F. Biermann and B. Siebenhüner, 319–350. Cambridge, MA: MIT Press.

Brack, D. and J. Hyvarinen, eds. 2002. *Global Environmental Institutions. Perspectives on Reform*. London: Royal Institute of International Affairs.

Brito, Lidia and Mark Stafford Smith. 2012. "State of the Planet Declaration. Planet under Pressure: New Knowledge towards Solutions." March 29, 2012. London: IGBP (International Geosphere-Biosphere Programme), DIVERSITAS, IHDP (International Human Dimensions Programme on Global Environmental Change), WCRP (World Climate Research Programme), and ICSU (International Council for Science).

Brock, Lothar. 1997. "The Environment and Security: Conceptual and Theoretical Issues." In *Conflict and the Environment*, ed. N.P. Gleditsch. 17–34. Dordrecht, the Netherlands: Kluwer.

Bruyninckx, Hans. 2004. "The Convention to Combat Desertification and the Role of Innovative Policy-Making Discourses: The Case of Burkina Faso." *Global Environmental Politics*, 4(3): 107–127.

Bruyninckx, Hans. 2005. "Sustainable Development: The Institutionalization of a Contested Policy Concept." In *International Environmental Politics*, ed. M. M. Betsill, K. Hochstetler, and D. Stevis, 265–298. Basingstoke: Palgrave Macmillan.

Caldwell, Lynton K. 1996. *International Environmental Policy: From the Twentieth to the Twenty-First Century*, 3rd edn. Durham, NC: Duke University Press.

Campe, Sabine. 2009. "The Secretariat of the International Maritime Organization: A Tanker for Tankers." In *Managers of Global Change. The Influence of International Environmental Bureaucracies*, ed. F. Biermann and B. Siebenhüner, 143–168. Cambridge, MA: MIT Press.

CGIEG (Consultative Group of Ministers or High-Level Representatives on International Environmental Governance). 2010. "Elaboration of Ideas for Broader Reform of International Environmental Governance." Information note from the Co-chairs of the Consultative Group of October 27, 2010. Nairobi: UNEP.

Chambers, Bradnee W. and Jessica F. Green, eds. 2005. *Reforming International Environmental Governance*. Tokyo: United Nations University Press.

Chasek, Pamela S., David L. Downie, and Janet Welsh Brown. 2010. *Global Environmental Politics*, 5th edn. Boulder, CO: Westview Press.

Chasek, Pamela S. and Lynn M. Wagner. 2012. "An Insider's Guide to Multilateral Environmental Negotiations since the Earth Summit." In *The Roads from Rio: Lessons Learned from Twenty Years of Multilateral Environmental Negotiations*, ed. P.S. Chasek and L.M. Wagner, 1–15. Abingdon: Routledge.

Conca, Ken. 2010. "Standing on Its Two Rear Legs: The Uneven Institutionalization of Environmental Concerns within the UN System." Presented at 51st Annual Convention of the International Studies Association, February 17–20, New Orleans, LA.

Depledge, Joanna and Pamela S. Chasek. 2012. "Raising the Tempo. The Escalating Pace and Tempo of Environmental Negotiations." In *The Roads from Rio: Lessons Learned from Twenty Years of Mulitlateral Environmental Negotiations*, ed. P.S. Chasek and L.M. Wagner, 19–38. Abingdon: Routledge.

DeSombre, Elizabeth R. 2006. *Global Environmental Institutions*. Abingdon: Routledge.

Detraz, Nicole and Michele M. Betsill. 2009. "Climate Change and Environmental Security: For Whom the Discourse Shifts." *International Studies Perspectives*, 10(3): 303–320.

Elliott, Lorraine. 2005. "The United Nations' Record on Environmental Governance: An Assessment." In *A World Environment Organization: Solution or Threat for Effective International Environmental Governance?*, ed. F. Biermann and S. Bauer, 27–56. Aldershot: Ashgate.

Falkner, Robert and Bernice Lee. 2012. "Introduction. Special Issue on 'Rio+20 and the Global Environment: Reflections on Theory and Practice'." *International Affairs*, 88(3): 457–462.

Gandhi, Indira. 1972. "Life is One and the World is One. Prime Minister Indira Gandhi Speaks to Plenary." In *Only One Earth. United Nations Conference on the Human Environment, Stockholm 5–16 June*, ed. Center for Economic and Social Information at United Nations European Headquarters, 18. Geneva: United Nations.

Gehring, Thomas and Sebastian Oberthür. 2008. "Interplay: Exploring Institutional Interaction." In *Institutions and Environmental Change. Principal Findings, Applications, and Research Frontiers*, ed. O.R. Young, L.A. King, and H. Schroeder, 187–224. Cambridge, MA: MIT Press.

GMEF (Global Ministerial Environment Forum). 2000. "Malmö Ministerial Declaration. Declaration of the Global Ministerial Environment Forum and the United Nations Environment Programme Governing Council." Doc. UNEP/GCSS VI/L.3, May 31. Nairobi: UNEP, www.unep.org/malmo/malmo_ministerial.htm (accessed October 20, 2012).

Gupta, Joyeeta. 2008. "Global Change: Analyzing Scale and Scaling in Environmental Governance." In *Institutions and Environmental Change. Principal Findings, Applications, and*

Research Frontiers, ed. O.R. Young, L.A. King, and H. Schroeder, 225–257. Cambridge, MA: MIT Press.

Haas, Peter M. 2004. "Addressing the Global Governance Deficit." *Global Environmental Politics*, 4(4): 1–15.

Halle, Mark. 2012. *Life after Rio*. IISD Commentary, June 2012. Winnipeg, Manitoba: International Institute for Sustainable Development.

Hawkins, Darren G., David A. Lake, Daniel L. Nielson, and Michael J. Tierney. 2006. "Delegation under Anarchy: States, International Organizations and Principal-Agent Theory." In *Delegation and Agency in International Organizations*, ed. D.G. Hawkins, D.A. Lake, D.L. Nielson, and M.J. Tierney, 3–37. Cambridge: Cambridge University Press.

Horstmann, Britta and Achala Chandani Abeysinghe. 2011. "The Adaptation Fund of the Kyoto Protocol: A Model for Financing Adaptation to Climate Change?" *Climate Law*, 2(3): 415–437.

Hurrell, Andrew and Sandeep Sengupta. 2012. "Emerging Powers, North–South Relations and Global Climate Politics." *International Affairs*, 88(3): 463–484.

Imber, Mark F. 1993. "Too Many Cooks? UN Reform after the Rio United Nations Conference on Environment and Development." *International Affairs*, 69: 55–70.

Ivanova, Maria. 2007. "Designing the United Nations Environment Programme: A Story of Compromise and Confrontation." *International Environmental Agreements: Politics, Law and Econmics*, 7(3): 337–361.

Ivanova, Maria. 2010. "UNEP in Global Environmental Governance: Design, Leadership, Location." *Global Environmental Politics*, 10(1): 30–59.

Ivanova, Maria. 2012. "Institutional Design and UNEP Reform: Historical Insights on Form, Function And Financing." *International Affairs*, 88(3): 565–584.

Jinnah, Sikina and Alexandra Conliffe. 2012. "Climate Change Bandwagoning: Climate Change Impacts on Global Environmental Governance." In *The Roads from Rio. Lessons Learned from Twenty Years of Multilateral Environmental Negotiations*, ed. P.S. Chasek and L.M. Wagner, 199–221. Abingdon: Routledge.

Johnson, Pierre Marc, Karel Mayrand, and Marc Paquin. 2006. "The United Nations Convention to Combat Desertification in Global Sustainable Development Governance." In *Governing Global Desertification. Linking Environmental Degradation, Poverty and Participation*, ed. P.M. Johnson, K. Mayrand, and M. Paquin, 1–10. Aldershot: Ashgate.

Kennedy, Paul. 2006. *The Parliament of Man. The Past, Present, and Future of the United Nations*. London: Allen Lane.

Keohane, Robert O. and David G. Victor. 2010. *The Regime Complex for Climate Change*. Discussion Paper 10–33, The Harvard Project on International Climate Agreements. Cambridge, MA: Harvard Kennedy School.

Krasner, Stephen D. 1999. *Sovereignty: Organized Hypocrisy*. Princeton, NJ: Princeton University Press.

McCormick, John. 1989. *The Global Environmental Movement. Reclaiming Paradise*. Bloomington, IN: Indiana University Press.

Mee, Laurence D. 2005. "The Role of UNEP and UNDP in Multilateral Environmental Agreements." *International Environmental Agreements: Politics, Law and Economics*, 5: 227–263.

Mingst, Karen A. and Margaret P. Karns. 2007. *The United Nations in the 21st Century*, 3rd edn. Boulder, CO: Westview Press.

Müller, Joachim, ed. 2010. *Reforming the United Nations. The Challenge of Working Together*. Leiden: Martinus Nijhoff.

Muñoz, Miquel, Rachel Thrasher, and Adil Najam. 2009. "Measuring the Negotiation Burden of Multilateral Environmental Agreements." *Global Environmental Politics*, 9(4): 1–13.

Murphy, Craig N. 2006. *The United Nations Development Programme. A Better Way?* Cambridge: Cambridge University Press.

Najam, Adil. 2005. "Neither Necessary, nor Sufficient: Why Organizational Tinkering Will Not Improve Environmental Governance." In *A World Environment Organization: Solution or Threat for Effective International Environmental Governance?*, ed. F. Biermann and S. Bauer, 235–256. Aldershot: Ashgate.

Najam, Adil, Mihaela Papa, and Naada Taiyab. 2006. *Global Environmental Governance. A Reform Agenda*. Winnipeg, Manitoba: International Institute for Sustainable Development.

Pallemaerts, Marc. 2003. "International Law and Sustainable Development: Any Progress in Johannesburg?" *Review of European Community & International Environmental Law*, 12(1): 1–11.

Pearce, Richard. 2006. "Decentralisation and Sustainable Resources Management in West Africa: A Line of Action for Revising National Action Programmes." In *Governing Global Desertification. Linking Environmental Degradation, Poverty, and Participation*, ed. P.M. Johnson, K. Mayrand, and M. Paquin, 147–162. Aldershot: Ashgate.

Schrijver, Nico. 2007. "Natural Resource Management and Sustainable Development." In *The Oxford Handbook on the United Nations*, ed. T.G. Weiss and S. Daws, 592–610. Oxford: Oxford University Press.

Simon, Nils. 2011. *International Environmental Governance for the 21st Century. Challenges, Reform Processes and Options for Action on the Way to Rio 2012*. Berlin: Stiftung Wissenschaft und Politik/German Institute for International and Security Affairs.

Speth, James Gustave. 2002. "The Global Environmental Agenda: Origins and Prospects." In *Global Environmental Governance. Options and Opportunities*, ed. D.C. Esty and M.H. Ivanova, 11–30. New Haven, CT: Yale School of Forestry and Environmental Studies.

Speth, James Gustave. 2003. "Perspectives on the Johannesburg Summit." *Environment*, 45(1): 24–29.

Swart, Lydia and Estelle Perry, eds. 2007. *Global Environmental Governance. Perspectives on the Current Debate*. New York: Center for UN Reform Education.

Tarasofsky, Richard G. 2002. *International Environmental Governance: Strengthening UNEP*. UNU Working Paper. Tokyo: United Nations University.

Tolba, Mostafa K. and Iwona Rummel-Bulska. 1998. *Global Environmental Diplomacy. Negotiating Environmental Agreements for the World, 1973–1992*. Cambridge, MA: MIT Press.

UN (United Nations). 2006. "Delivering as One. Report of the Secretary-General's High-Level Panel on System-Wide Coherence in the Areas of Development, Humanitarian Assistance, and the Environment." New York: United Nations.

UN (United Nations). 2012a. "The Future We Want. Outcome of the United Nations Conference on Sustainable Development, Rio de Janeiro, Brazil, 20–22 June 2012." UN Doc. A/CONF.216/L.1* of June 22, 2012. New York: United Nations.

UN (United Nations). 2012b. "The Future We Want." Co-chairs' Zero Draft Outcome Document of January 10, 2012 for the United Nations Conference on Sustainable Development. New York: United Nations.

UNCED (United Nations Conference on Environment and Development). 1992. "Agenda 21. Report of the United Nations Conference on Environment and Development, Annex II, 12 August." Doc. UN A/CONF.151/26 (vol. I–III). New York: United Nations.

UNDP (United Nations Development Programme) and UNEP (United Nations Environment Programme). 2011. "PEI Annual Progress Report 2010." Nairobi: United Nations.

UNEP (United Nations Environment Programme). 1997. Governing Council. "Nairobi Declaration of the Heads of Delegation." Nairobi: UNEP.

UNEP (United Nations Environment Programme). 2004. "Bali Strategic Plan for Technology Support and Capacity-Building." UN Doc. EP/GC.23/6/Add. 1 of December 23, 2004. Nairobi: United Nations.

UNEP (United Nations Environment Programme). 2010. "Proceedings of the Governing Council/Global Ministerial Environment Forum at Its Eleventh Special Session, 24–26 February 2010." UN Doc. EP/GCSS.XI/11 of March 3. Nairobi: UNEP.

UNEP (United Nations Environment Programme). 2011. *Towards a Green Economy. Pathways to Sustainable Development and Poverty Eradication*. Paris: UNEP.

UNEP (United Nations Environment Programme). 2012. *Annual Report 2011*. Nairobi: UNEP.

UNGA (United Nations General Assembly). 1972. "Resolution 2997 (XXVII). Institutional and Financial Arrangements for international Environmental Cooperation." UN Doc. A/RES/27/2997 of December 15. New York: United Nations.

UNGA (United Nations General Assembly). 1992. "Resolution 47/191. Institutional Arrangements to Follow Up the United Nations Conference on Environment and Development." UN Doc. A/RES/47/191 of December 22. New York: United Nations.

Wapner, Paul. 2003. "World Summit on Sustainable Development: Toward a Post-Jo'burg Environmentalism." *Global Environmental Politics*, 3(1): 1–10.

WBGU (German Advisory Council on Global Change). 2001. *New Structures for Global Environmental Policy*. World in Transition Series. London: Earthscan.

WBGU (German Advisory Council on Global Change). 2011. *A Social Contract for Sustainability*. World in Transition Series. Berlin: WBGU.

WCED (World Commission on Environment and Development). 1987. *Our Common Future*. Oxford: Oxford University Press.

Weinlich, Silke. 2011. *Reforming Development Cooperation at the United Nations: An Analysis of Policy Positions and Actions of Key States on Reform Options*. Bonn: Deutsches Institut für Entwicklungspolitik/German Development Institute.

Weiss, Thomas G. and Ramesh Thakur. 2010. *Global Governance and the UN. An Unfinished Journey*. Bloomington, IN: Indiana University Press.

Zelli, Fariborz. 2011. "The Fragmentation of the Global Climate Governance Architecture." *Wiley Interdisciplinary Reviews: Climate Change*, 2(2): 255–270.

Chapter 20

International Negotiations

Radoslav S. Dimitrov

Introduction

The proliferation of global environmental institutions is a distinct development in modern international relations. In recent decades, states have negotiated over 700 multilateral policy agreements and over 1000 bilateral agreements on ecological issues (Mitchell 2003). International policy-making is accelerating as governments negotiate new agreements and renegotiate existing ones. Climate change alone was the subject of 20 rounds of formal negotiations between 2007 and the end of 2011. At any given time, a multilateral environmental meeting of government representatives is taking place somewhere in the world, with Geneva, New York, Bonn, Bangkok, and New Delhi being among the most common venues for diplomacy. Between 1992 and 2007, major conferences related to only 10 of the existing multilateral environmental agreements filled 115 days per year (Muñoz *et al.* 2009). When we add other environmental issues as well as the plethora of technical workshops and pre-negotiations, we observe an international community of states in perpetual negotiation over environmental policy.

Multilateral negotiations have been described as "a process of mutual persuasion and adjustment of interests and policies which aims at combining non-identical actor preferences into a single joint decision" (Rittberger 1998: 17). In a more recent definition, negotiation is

> purposeful communication consisting of strategies developed and implemented by two or more actors to pursue or defend their interests. The entire pattern of interaction constitutes a *process* played against a *structure* of background factors that change slowly over the long term (Avenhaus and Zartman 2007: 5).

The Handbook of Global Climate and Environment Policy, First Edition. Edited by Robert Falkner.
© 2013 John Wiley & Sons, Ltd. Published 2013 by John Wiley & Sons, Ltd.

the process through which human communities make collective decisions in the governance of public affairs is shaped by many factors that have preoccupied the academic literature.

The negotiation process unfolds in analytically distinct stages. I. William Zartman (1994) described three phases: problem diagnosis, invention of the bridging formula, and negotiation on the details. Oran Young (1994) more usefully distinguished between pre-negotiation, negotiating, and implementation of international regimes and showed that each stage is affected by different political factors. Pamela Chasek (2001) borrows this insight and provides perhaps the most elaborate discussion of stages in various environmental negotiations. Pre-negotiation, for instance, involves agenda-setting where countries choose the negotiating forum, decision-making procedures, relevant actors, and which policy issues to include and exclude from discussions. Negotiations typically consist of years of formal and informal discussions on the rules of a treaty, including targets, timetables, policy implementation options, and compliance procedures. The third stage, implementation, consists of domestic treaty ratification and policy development. Chasek charts the unfolding of negotiations with an elaborate "phased process model" that includes precipitants, issue definition, statements of initial positions, bargaining, and turning points.

Analytical Perspectives on Negotiations

Negotiations are the principal means of constructing international environmental institutions (Haas *et al.* 1993; Levy *et al.* 1995; Young 1998; Goldstein *et al.* 2000). Logically, the intellectual roots of the negotiations literature are in neoliberal institutionalism, a school of thought that focuses on the role of institutions in world politics and posits that institutions affect state behavior by creating incentives for cooperation and reducing transaction costs. Scholarship on environmental regime formation, in particular, is prolific and has strengthened neoliberal institutionalism in IR theory (see for instance Hasenclever *et al.* 1997 and the work of Oran Young, Scott Barrett, and Peter Haas, to name a few). Today the academic literature on environmental negotiations can be divided in three primary realms: rationalist, constructivist, and descriptive work by insiders, the latter driven mostly by interest in policy-making.

Rationalism

Early research sought to explain why some negotiations succeed while others fail to produce policy agreements. To take one instance, a project funded by the Ford Foundation attempted to identify the "determinants of success" through a comparison of five empirical cases of successful regime formation (Young and Osherenko 1993). The authors concluded that none of the independent variables under consideration could explain the outcomes. Subsequent scholarship has scaled down its ambition and desisted from broad theoretical explanations of negotiation outcomes. In another classic example of rationalism, Detlef Sprinz and Tapani Vaahtoranta (1994) stress cost–benefit analysis and explain country positions in negotiations with their expected policy costs and vulnerability to ecological problems.

In other empiricist-rationalist scholarship, explorations of multilateral negotiations examine configurations of interests and changes in national positions (Andresen
and Agrawala 2002; Vogler and Bretherton 2006), coalition-building (Hampson
1995; Dupont 1996), the role of leadership (Young 1991; Underdal 1994), and the
role of issue linkage (Zartman 1994; Hopmann 1996; Jinnah 2011). The intellectual
roots of this work can be found in game theory.

Game Theory   Game theory focuses on formal modeling of negotiations and uti-
lizes formal logic to derive probable outcomes from fixed actor preferences. Models
such as the prisoner's dilemma, chicken, or stag hunt usually portray the situation as
a matrix indicating the choices facing negotiators and consequences for each strategy,
without describing the negotiation process (Avenhaus and Zartman 2007). Howard
Raiffa and others have utilized elaborate models and decision analysis to calcu-
late optimal solution outcomes given a particular configuration of state preferences
(Raiffa 1982, 2002).

Scott Barrett (1998, 2003) has built a body of work that consistently uses game
theory to clarify the obstacles to global environmental cooperation. Another pioneer
in this realm is Hugh Ward, who used the game of chicken to illuminate climate
negotiations (1993) and showed that iterative prisoner's dilemma games can yield
cooperation if states do not discount the future too heavily (Ward 1996). Related
work developed a model of climate negotiations incorporating divergent national
positions of dragger and pusher countries (Ward *et al.* 2001). Rational choice models
have been used to explain both the domestic sources of national policy positions
and the dynamics of international negotiations, and generate recommendations for
promising political strategies (Grundig 2009).

Kaitala and Pohjola (1995) developed a dynamic model of global climate change
negotiations that differentiates between countries depending on their vulnerability to
climate impacts. In their model, countries negotiate international transfer payments
to address the asymmetric effects of global warming but concrete negotiations are
not described. Akira Okada (1999) applied a cooperative market model to illuminate
international trading of carbon emission permits. This study is not empirical either; it
evaluates hypothetical allocation rules for the United States, Russia, and Japan. Bruce
Bueno de Mesquita (2009) declared with great confidence that predicting the future
is possible and used a computer to state that the 2009 Copenhagen conference would
fail and that global climate policy would gain momentum over several decades, then
steadily decline between 2050 and 2100.

Three observations are in order. First, game theory is the most elegant, parsimo-
nious approach to the study of negotiations. It brings major insights into bargaining
and is indispensable in clarifying strategic choices that political actors face, identifying zones of agreement, and explaining failure to reach agreement. Second, formal
models of bargaining[1] have rarely been applied to actual cases of environmental
negotiations (Avenhaus and Zartman 2007). A collection of essays, for instance, used
extended game theoretic methods to speculate on potential agreements on the reduction of greenhouse gases (Carraro 1997). Heterogeneity of state actors was theorized
to benefit the prospects for burden-sharing arrangements and coalition-building,
while issue linkage is believed to improve the chances of agreements. Whether this

actually occurs in negotiations is unknown since existing studies do not compare formal models with actual negotiations.

Third, because of their focus on a priori preferences and outcomes, game theorists skip the entire *process* of negotiations. This undermines their position, particularly given recent findings that the process of communicating policy preferences has a pronounced impact on the prospects for agreement – independent of distributional issues and concerns about cheating (Earnest 2008). Finally, assumptions used in modeling are rarely applicable in the real world of environmental negotiations: the number of actors is rarely only two (there are more than 190 in climate change negotiations); actors are rarely unified; information about the positions and preferences of other countries is far from perfect; and preferences of a country change, sometimes dramatically, as in the case of Australia's turnabout in ratifying the Kyoto Protocol in 2007.

Oran Young sought to correct these well-known shortcomings with his seminal model of integrative bargaining (1994). He noted that power theorists overemphasize the role of hegemons, rational-choice theorists use models of bargaining that are simplistic, and cognitivists have not modeled the process through which social learning leads to convergence of policy preferences. Young calls for a model of institutional bargaining that captures the role of multiple actors, consensus rules of decision-making, the veil of uncertainty about future costs and benefits, and evolving configurations of interests, among other factors. His model is commonly recognized as influential in the discipline but, curiously, has not been applied in empirical studies of negotiations.

Leadership Hegemonic power appears rarely to determine outcomes in environmental negotiations, because military or even economic power is not fungible, and the academic study of environmental diplomacy features few studies in the realist tradition. Scholars of global environmental politics agree that structural power matters little in environmental diplomacy (Young 1991; Underdal 1994; Falkner 2005). In a thorough treatment of the topic, Robert Falkner (2005) reminds us of the role of American hegemony but shows that hegemony provides an incomplete perspective that can explain neither the direction of US policy nor international outcomes such as regime formation. Furthermore, even small countries can exercise strong influence in negotiations. The Netherlands has used initiative and shrewd diplomacy to influence both European and global negotiations (Kanie 2003). The Alliance of Small Island States has been an active participant in the climate change negotiations and has influenced the process and outcomes considerably. Politically weak countries such as Tuvalu, Micronesia, Barbados, and the Maldives have shaped climate negotiations by "borrowing external power" (Betzold 2010).

The weak relevance of power hierarchy has led to a vibrant body of research on leadership. There are three principal types of leadership: structural, directional, and instrumental (Gupta and Grubb 2000).[2] Structural leadership derives from material resources a state possesses that give it power in the structure of the game, including a share of polluting emissions. Deborah Davenport (2005), for instance, argues that US policy preferences are the principal explanatory factor behind the failure of negotiations to produce a global forest convention. Directional leaders such as the European Union in climate change or the United States in the ozone negotiations lead by example through unilateral domestic policies that demonstrate feasible solutions to other countries (Underdal 1994). Instrumental leadership is a function of political initiative, skill, and creativity in the process of negotiations, including submission of policy proposals and persuasive arguments.

Instrumental leadership can be subdivided into two types: entrepreneurial and intellectual (Young 1991; Kanie 2003). One entrepreneurial leader is the small island nation of Tuvalu, whose delegation has been remarkably influential in climate discussions by providing concrete proposals, including a full-fledged, elaborate treaty text tabled in 2009 that they insisted be the basis of negotiations in Copenhagen. Intellectual leaders introduce innovative policy solutions to the ecological problem at hand. The United States played an intellectual leadership role in the 1990s by introducing the idea of emission trading into the Kyoto Protocol negotiations.

Some of the most sophisticated scholarship explores the causal mechanisms through which leaders emerge. Norichika Kanie (2003), for instance, provides us with a rich empirical study of the Netherlands' leadership in climate talks. Replete with concrete facts from the negotiation of the Kyoto Protocol and an extensive account of domestic policy formation, his article shows that Dutch leadership was made possible by domestic political processes as well as intense cooperation between the government delegation and Dutch NGOs during the international game. The study of environmental diplomacy could greatly benefit from more such multilevel work that straddles both state–society interactions and the domestic–international interface.

Explaining European Leadership The European Union has provided strong leadership in environmental negotiations on various issues (Gupta and Grubb 2000; Vogler 2005; Harris 2007; Oberthür and Kelly 2008) and generates an academic debate on how to explain it. Some scholars argue that EU environmental leadership is a product of norms and identity of Europe as an ideational leader (Manners 2002; Krämer 2004). Others caution against idealism and argue that political economy and material considerations can explain EU positions (Falkner 2007). When the USA abandoned the Kyoto Protocol in 2001, some IR scholars predicted the end of the global climate regime. Theorists expected other states to abandon the regime out of concerns with relative gains: why stay to pay high policy costs and give competitive advantage to America? The facts interfered with that theory as well: Europe did precisely the opposite of what scholars and pundits expected. The EU not only stayed in Kyoto but adopted unilateral policies for steep emission reductions. They emerged as the international leader, whose followers included Canada, Japan, and Russia, who also ratified the treaty, and the Kyoto Protocol entered into force in February 2005.

Vogler (2005) considers carefully institutionalist hypotheses and finds evidence of "normative entrapment." European leadership is a product of a normative stance on climate change and remains part of an enduring self-image that continues to propel strong policies. Jon Hovi and his colleagues compare four alternative explanations and argue that the EU move is a product of the combined effects of domestic institutional inertia and a power-seeking desire for international leadership (Hovi *et al.* 2003). By pulling out of Kyoto, the USA offered the EU and other actors an opportunity to gain political power in one of the most important current negotiations.

Domestic–International Connections The interplay between domestic politics and international discussions is another fruitful area of study in the rationalist framework. Robert Putnam's seminal work established that each state actor in negotiations plays two "games" simultaneously with domestic constituents and foreign counterparts (Putnam 1988). His concept of the two-level game continues to inform scholars in understanding state behavior. In her award-winning work, Beth DeSombre (2000) shows the domestic sources of foreign environmental policy that can indirectly illuminate negotiations, too. Aslaug Asgeirsdottir (2008) examines bargaining between Iceland and Norway over fish stocks, and her findings confirm Putnam's view that powerful domestic interest groups actually strengthen the negotiating position of states vis-à-vis other countries. Iceland's strong fishing industry exerted pressures on the government that helped its delegation win concessions from Norway, whose weaker internal pressures left the delegation with more maneuvering space and therefore more openness for compromise. Other empirical studies cast doubt on the theory and suggest that state leaders may choose to ignore domestic constraints and may pursue international strategies without paying close attention to the "domestic game." In a study of the Kyoto Protocol, McLean and Stone argue that the European Union has made a principled commitment to climate cooperation and subordinates its domestic politics to the international level regardless of negotiation outcomes (2012).

Issue Linkage Negotiations on a specific environmental problem rarely develop in isolation from international discussions on other ecological problems. Tapping into the literature on institutional interplay (Young 2002), studies of issue linkage have enriched our understanding of its impacts on the construction of agreements (Jinnah 2011). State and non-state actors make deliberate decisions to affect policy outcomes by drawing linkages between climate change, forestry, desertification, ozone depletion, biodiversity, and other issues. These strategies have inundated UNFCCC conferences, making the climate problematique a central hub of global environmental politics at large:

> Indeed, with over 1,200 NGO and IGO observers now accredited to attend the UNFCCC negotiations, representing over 22 issue areas, and drawing over 20,000 observers, it seems that everyone from McDonald's to the Vatican is jumping on the proverbial climate change bandwagon (Jinnah 2011: 2).

The particular effects of such issue linkage are still open for debate. Linking environmental and trade issues made negotiations on ozone depletion easier and is credited with contributing to the success of the Montreal Protocol (Barrett 1997). Bandwagoning has the potential to facilitate more effective policy outcomes on climate change (Jinnah 2011). At the same time, linkages increase issue complexity that is already overwhelming in climate politics and presents an obstacle to productive negotiations (Victor 2011).

Constructivism

Norms and Trust Constructivist scholars argue that shared global norms affect international environmental policy. Ozone treaties resulted from social discourse tailored to favor the precautionary principle (Litfin 1994). A norm of environmental multilateralism explains the creation of the impotent UN Forum on Forests and global state participation in it (Dimitrov 2005). And outcomes of the 1991 Earth Summit reflect a broad normative paradigm of liberal environmentalism (Bernstein 2001).

In the same intellectual tradition, John Vogler (2010) offers constructivist advice on how to strengthen the global climate policy-making process. He calls for building trust between states as a key ingredient in the kitchen of environmental institution-building. Trust can be developed not only through strict compliance mechanisms or long-term institutional interactions but also through the development of shared understanding of the problem and domestic policy action that signals commitment. Ultimately, Vogler sees the development of trust as inextricable from the evolution of identity and perceptions of national interests. His inspiring work is future-oriented rather than empirical as it provides important recommendations for future political efforts. Systematic observations on the actual behavior of actors in building or undermining trust would be an important follow-up in this line of research.

Persuasion and Argumentation in Negotiations

Despite the widespread recognition that "in essence, international negotiation is communication" (Stein 1988: 222), communication is the *terra incognita* of negotiation studies. Sweeping literature reviews conclude that the exchange of arguments is the least-explored topic in this field of research (Jönsson 2002; Zartman 2002). There is an academic tendency to treat international politics as a series of strategic policy moves, hence our traditional focus on state "behavior" and action rather than words. Talk is cheap indeed (because the supply exceeds the demand, one might quip), yet international relations occur through speech acts as well as policy actions. Besides, we have empirical evidence that cheap-talk diplomacy can diffuse international crises and prevent war between countries in bargaining games with multiple equilibria (Ramsay 2011). Listening to intergovernmental conversations is important and also interesting.

What do delegations actually say to one another? "The back-and-forth communication . . . the dynamics of mutual persuasion attempts that we usually associate with negotiations are insufficiently caught" (Jönsson 2002: 224). Thomas Risse (2000) and Harald Müller (2004) cogently argued for the need to study communicative behavior, but the few scholars who tried to follow up admitted failure to produce conclusive results, partly due to a lack of verbatim records of negotiations (Deitelhoff and Müller 2005). Scholars rarely have access to international negotiations, particularly those held behind closed doors. Important books by Farhana Yamin and Joanna Depledge rectify the general neglect of process and provide detailed descriptions of the logistical and bureaucratic organization of climate negotiations but also leave out the discursive exchange among delegations (Yamin and Depledge 2004; Depledge 2005).

In a complex marriage between rationalism and constructivism, Christian Grobe advances a rationalist theory of argumentative persuasion. He claims that changes in bargaining positions are motivated by new causal knowledge about the problem at hand. After a thoughtful review of the relevant literature, Grobe makes a compelling case for the study of persuasion and sketches two empirical cases: negotiations on the International Convention for the Prevention of Pollution from Ships (MARPOL) and the Commission for the Conservation of Antarctic Marine Living Resources (CCAMLR). Paradoxically, his work dismisses the role of arguments made during negotiations in changing policy preferences:

> In the MARPOL negotiations, where the parties were perfectly informed about the situation at hand, argumentative talk was without effect on the outcome... On the other hand, states were highly receptive to arguments in the CCAMLR case. But these arguments did not lead to a reformulation of preferences (Grobe 2010: 22).

Notably, Grobe does not examine actual argumentation during the negotiations. The two brief case studies underlying his "functional persuasion theory" draw on secondary sources and include no information about the actual conversation between delegations during the negotiations.

Others argue, alternatively, that persuasion and discourse do alter policy preferences (Dimitrov 2012). European arguments during the climate negotiations induced fundamental change in many countries' views on the economic benefits of climate policy. International discussions during the 1990s were dominated by the premise that climate policy is expensive and countries must choose between economic and environmental interests. In the early 2000s, the European Union introduced the concept of "win-win solutions" to the climate discourse (Dimitrov 2012). Their new argument was contrary to conventional wisdom at the time: climate policy can bring economic *benefits* and there is no juxtaposition between economic and environmental interests. States can reduce emissions through energy savings and renewable energy. The benefits of such action are multiple: financial savings, increased economic competitiveness, improved energy security, increased political independence from unstable regions such as the Middle East, improved public health – as well as mitigating climate change and its devastating impacts.

The EU pounded this argument tirelessly over many years of discussions. They also backed their words with actions and unilaterally adopted the ambitious 2007 "Energy and Climate Package" that is binding on all 27 members (Morgera *et al.* 2010; Oberthür and Pallemaerts 2010). In March 2011, after extensive continent-wide public consultations, the European Commission publicized a "Roadmap for Moving to a Competitive Low-Carbon Economy in 2050." The roadmap envisions emission reductions up to 95% by 2050. The transition would cost €270 billion per year, or 1.5% of GDP, but would save up to €320 billion per year on fuel costs. The "win-win" rationale was embraced by other countries, including South Korea, who adopted "Green Growth" as the paradigm underlying their current economic development (see also Chapter 12 in this volume).

Insider Perspectives and Empirical Accounts

Empirical studies based on direct observation of negotiations are relatively few. Much published work offers recycled information that can be derived without negotiations actually having been observed. Typically studies of international regime formation produce a chronological list of conferences and their main outcomes (agreements whose text can be obtained online), and select dramatic moves by particular countries, such as Canada's withdrawal from the Kyoto Protocol (which one can learn by following the newspaper headlines). The dynamics around the negotiation table often remain hidden. What is the verbal exchange? What are the offers and responses made during informal consultations? Relevant literature tends to avoid these questions and gravitate toward related topics such as theorizing about the creation of institutions and their impact on state behavior (Young 1994; Barrett 2003) or future policy options (Victor 2011).

This tendency is understandable and perhaps unavoidable. Lack of direct access to negotiations is the likely main reason for leaving the process out. Few scholars attend UN environmental conferences or carry out extensive interviews with key actors. The very few who do are observers without access to what goes on behind closed doors. They attend as non-governmental participants (typically with accreditation through environmental groups) and are barred from sessions of the "working groups" and informal consultations where most of the strategic political exchange takes place.

One distinct body of literature comes from participants and rectifies the problem of data shortage. Detailed accounts of negotiation processes offer an insider view, based either on authors' direct involvement (Benedick 1998; Depledge 2005; Rajamani 2008, 2010; Kulovesi and Gutiérrez 2009; Smith 2009; Bodansky 2010; Dimitrov 2010) or interviews with key actors (Falkner 2000). A recent compendium offers intimate perspectives on various environmental negotiations from expert writers for the Earth Negotiations Bulletin with extensive exposure to actual negotiations (Chasek and Wagner 2012). These and other works offer a palpable taste of environmental diplomacy and an in-depth expertise that can inform both theory and practice.

Climate Change Negotiations

Global climate negotiations have attracted considerable academic attention. Matthew Paterson and Daniel Bodansky have documented the early efforts to formulate a global response to climate change in 1980s and the 1990s (Paterson 1996; Bodansky 2001). Participants in the UN political process have documented more recent negotiations on post-Kyoto policy (Fry 2008; Kulovesi and Gutiérrez 2009; Dimitrov 2010; Sterk et al. 2010; Oberthür 2011). These comprehensive guides to global climate change negotiations clarify the notoriously complex policy issues on the table and the positions of key countries and coalitions. Many studies analyze existing climate agreements and discuss future prospects for cooperation (Paterson 1996; Ott 2001; Victor 2001; Betsill 2004; Yamin and Depledge 2004; Depledge 2006; Clémençon 2008; Ott et al. 2008; Watanabe et al. 2008). Others focus on national policies and negotiation positions of particular actors (Hovi et al. 2003; Kanie 2003; Najam 2005; Oberthür and Kelly 2008; Betzold 2010) and study domestic policy discourse (Pettenger 2007; McCright and Dunlap 2008; Harrison and Sundstrom 2010). Finally, another important body of literature debates future policy options, offers policy recommendations, and discusses issues of justice and equity (Agrawala and Andresen 2001; Aldy et al. 2003; Najam et al. 2003; Bodansky 2004; Victor 2004; Adger et al. 2006; Roberts and Parks 2007; Hare et al. 2010; Müller 2011).

Current international negotiations on climate change are an example of *post-agreement negotiation* defined by Bertram Spector as "dynamic and cooperative processes, systems, procedures and structures that are institutionalized to sustain dialogue on issues that cannot, by their nature, be resolved by a single agreement" (Spector 2003: 55). Countries disagree on a splendid variety of contentious issues (Dimitrov 2010). One disagreement pertains to "the legal architecture" of the future climate policy regime: whether to extend the Kyoto Protocol that places the onus on industrialized countries, or create a new global agreement with obligations for all major emitters – or both. In addition, the method of determining national targets for emission reductions is disputed. The European Union and the Alliance of Small Island States advocated a classic "top-down approach" of determining global targets based on science-based goals (e.g. 25–40% global emission cuts needed to keep temperature rise to below the critical threshold of 2 °C). Others such as Australia, the USA, and China fought for a "bottom-up" approach allowing every country to determine its national goals regardless of global environmental results. Other key debates pertain to obligations for developing countries; level and mechanisms of international funding for climate policy in poor countries; the role of agriculture and forestry in calculating emission levels (LULUCF, or land-use and land-use change and forestry); the transfer of environmentally friendly technologies; and the creation of an Adaptation Framework.

Twenty rounds of formal negotiations occurred in the four years between Bali and Durban (December 2011). In a historic breakthrough, the Cancun Agreements of 2010 established for the first time an official global goal of limiting temperature rise to below 2 °C, and stipulated that developing countries "will" take nationally appropriate mitigation actions. The deal also included a principled agreement to establish a Global Adaptation Framework; an international registry for developing country policies; and a Green Climate Fund to provide up to US$100 billion per year for climate policy by 2020.

The negotiations suffered a major blow at Durban 2011. After two weeks of discussions, including three days of intense high-level talks between environment ministers, states decided to postpone a globally binding climate treaty for at least nine years. Only three countries openly supported this outcome (Australia, Canada, and the United States), while others accepted it in exchange for a continuation of the Kyoto Protocol. The EU privately considered boycotting the conference and island nations described the outcome as a form of hara-kiri that "places entire nations on death row." The collective decision is to continue negotiations with a new deadline of 2015 for finalizing an agreement for *after* 2020. This constituted an open admission that the Bali mandate had failed, and turned the famed "post-2012 policy" into a post-2020 possibility. A second major decision was to extend the Kyoto Protocol, with a second commitment period. Two stipulations weaken Kyoto 2: first, the duration of the new commitment period will be decided at a later, unspecified date (five or eight years, until 2017 or 2020). Second, Kyoto 2 relies on voluntary national commitments to be determined by countries domestically. The text merely "invites countries" to report internationally their policy goals. Thus, the original Kyoto Protocol with its binding absolute emissions reductions was replaced by a bottom-up approach and voluntary goals, without even obliging countries to communicate those goals internationally.

Today the global negotiations have been placed on hold, and prospects for change over the next several years are bleak. The climate case outcomes confirm the pessimistic views of game theorists who argue that policy agreements tend to work only in situations of simple coordination and are likely to fail in real collaboration, where countries face strong incentives to defect from a collective agreement. The case also appears to obey the "law of the least ambitious program" formulated by Arild Underdal (1980) that remains foundational in mainstream scholarship on environmental diplomacy. Underdal observes that negotiations involving multiple actors tend to produce outcomes that reflect the lowest common denominator. Indeed, the large number of actors (194 to be more precise) makes effective climate agreements difficult. The requirement of global political consensus as a basis for decision-making at the UNFCCC creates major obstacles to effective multilateralism on climate change. Consensus weakens prospective international agreements by giving every actor veto power. If every single government must endorse a collective policy, the emerging agreement is likely to reflect the preferences of the most obstructionist player.

Future Prospects and Policy Recommendations

There is striking convergence of academic views on the poor prospects for climate talks. David Victor (2006) and Bruce Bueno de Mesquita (2009) state with certainty that failure of the current global approach to climate change is guaranteed, given the enormous issue complexity of climate policy combined with highly diverse national interests and conflicting country assessments of the climate danger. In a particularly thoughtful and extensively researched piece, Røgeberg et al. (2010) bring charts and numbers to prove that the international community of states cannot solve the climate problem. A veteran diplomat, Richard Smith (2009), considers the climate negotiations process as a manual for how not to negotiate agreements. He worries about the absence of domestic support and national policies in key countries as an important precondition for productive international negotiations.

Academic observers share skepticism on the prospects but disagree on how to improve them. The subject of a cottage industry of academics and think tanks, the proposals for international climate policy are numerous and diverse (Aldy et al. 2003; Bodansky 2004; Aldy and Stavins 2010; Vogler 2010; Victor 2011). Falkner and his colleagues stress the need for redesigning the current international approach to tackling the problem. They caution against the dangers of a decentralized bottom-up approach and advocate a "building-blocks" strategy of negotiating a broad global legal framework with firm binding commitments, in an incremental fashion (Falkner et al. 2010).

David Victor (2011) recommends the exact opposite: negotiating a narrow non-binding agreement on key issues among a few key players and a bottom-up approach to country commitments. He advocates abandoning the UNFCCC approach that relies on political consensus and seeks legally binding treaties (Victor 2011). The alternative approach he proposes involves: negotiating a non-binding climate agreement among a small group of major emitters, who retain freedom to determine their national policies (a bottom-up approach), and reciprocal country commitments that create the incentives for participation. Essentially, he advocates replacing the binding model of international law and creating a global oligarchy of powerful countries to provide global climate governance reflecting their national interests and abilities.

Promising Research Directions

Rethinking the Link between Institutions, Negotiations, and Governance

The failure of the UN talks to produce a climate treaty is clear, but scholars draw different conclusions and disagree on the implications of this outcome for governance theories. Some dismiss the intergovernmental realm as unimportant and focus on non-state initiatives (Hoffman 2011). Others draw a causal connection between inter-state negotiations and multilevel climate governance by both state and non-state actors (Dimitrov 2010). Still others expand academic definitions of regimes and argue that the climate regime encompasses multiple institutions and non-governmental initiatives (Keohane and Victor 2011).

The disappointment that climate negotiations have failed to produce a treaty is understandable but it need not create skepticism about the importance of international discussions. In my view, UN negotiations have affected state behavior and fostered the development of domestic policies in the absence of a formal treaty (Dimitrov 2010). The conversations have helped state and corporate actors alike to recalculate their interests in green policies. The last four years of formal negotiations have seen major policy shifts in China, India, Australia, Japan, Korea, and many other countries. These policy shifts converge in one direction: a low-carbon economy based on alternative energy and energy efficiency. The discourse has therefore changed the perception of national interests and today governments behave differently.

Negotiations scholars need to reconsider the meaning of "outcome" and recognize the diverse impacts of negotiations on state behavior apart from treaty-making. In a rich empirical study, for instance, Antto Vihma argues that India's domestic climate discourse as well as decision-making processes have changed as a result of the country's engagement in UN talks (2010). Peter Haas has argued that the most important effect of United Nations environmental conferences is the growth of global environmental norms (2002). Depledge and Yamin would agree: "The negotiating environment of a regime enmeshes delegations in a dense web of meetings, practices, processes, and rules, generating an inherent motivation among negotiators to advance the issue" (2009: 439; cited in Falkner *et al.* 2010: 255). Indeed, "the Kyoto Protocol" is now a household phrase in communities around the world and raises awareness of climate change.

Many diplomats describe China's new five-year plan (2011–2015) as the most progressive legislature toward a low-carbon economy in history. Influenced by European arguments about the economic benefits of green action, in 2008 South Korea officially embraced a "Green Growth" paradigm of economic development, committed to 30% cuts by 2020 below business-as-usual, and established a Global Green Growth Institute to systematize the green growth theory and spread it to developing countries. Countries are establishing new branches of government dedicated to climate policy such as Australia's Department of Climate Change and Energy. Norway plans to slash its emissions by 40% by 2020 and be carbon neutral by 2030. Japan's decision to cut its emission by 25% by 2020 is also remarkable.

Hence, international discussions have helped change the world despite their failure to produce a treaty. Global climate governance is dramatically different today compared to the 1990s, and is now a remarkably vibrant realm of policy development and implementation. Aggregate climate governance comprising regional, national, sub-national, and local policies as well as non-state initiatives worldwide is thriving (Schreurs 2008; Selin and VanDeveer 2009; Hoffman 2011).

Changes in Policy Preferences

The academic discipline would benefit from research on the evolution of policy preferences. We know that governments change their mind in the course of negotiations. Germany dropped its opposition to international regulations in the acid rain case, a crucial breakthrough that turned the tide in constructing one of the most effective environmental regimes. The United States took a U-turn in forestry negotiations and became an active opponent of a forest treaty. In climate change negotiations of the 1990s, developing countries were adamant in refusing to take any action on emissions reductions; such a notion was taboo until the Bali Conference in 2007, when India led the G77 coalition to endorse the concept of "nationally appropriate mitigation actions" (NAMAs) in the South. China and others then refused to subject their NAMAs to international monitoring, reporting, and verification but later accepted this at Cancun 2010. Finally, a major historic milestone was marked in Durban in 2011 when China signaled a willingness to accept binding commitments under a future global treaty.

These changes in policy preferences are milestones in every story of negotiations and should constitute a key research topic. How do countries come to embrace policy options they previously opposed? When and why do changes in national positions occur over time? Domestic politics is an obvious influence that can explain policy changes. Elections sometimes lead to new country positions, as in Australia, when ratification of the Kyoto Protocol was the first act in office of Kevin Rudd, the new prime minister. Alternatively, constructivism is particularly well positioned to pursue this research, by virtue of its interest in the evolution of ideas and interest formation. Vogler suggests briefly that the British government made efforts to change other countries' perceptions of the climate problem as well as their national economic interests in mitigating it:

> [Emission reductions] are now claimed to constitute an economic benefit and a necessary investment, rather than a burden to be borne. From a constructivist perspective, this is an audacious move to subvert accepted meanings and constructions of self-interest (Vogler 2010: 2685–2686).

Argumentation and Persuasion in Negotiations

There is now evidence that specific arguments made during negotiations help persuade countries and change their calculations of self-interest. The European argument regarding economic benefits of climate policy has persuaded countries to change domestic policies (Dimitrov 2012). Today 90 states have considerable domestic plans for clean energy and emission reductions. While establishing a strict causal connection between particular arguments and state behavior would be premature at this early stage, the extensive global discussions on climate change over the last 10 years correlate with a global pattern of national policy developments.

Further research on argumentation can make valuable contributions to theory and practice. First, it would enable conclusions on the effectiveness of negotiating strategies that can be useful to practitioners and policy-makers. Building an inventory of argumentative approaches can uncover the foundations of discursive strategies and allows us to compare the effectiveness of different approaches to persuasion. In the long run, such research can generate recommendations to policy-makers on designing effective negotiating strategies. Second, research on policy change and persuasion can facilitate the development of a future theory of interest formation. Political scientists of all theoretical stripes agree that social actors pursue their perceived interests. Yet, we know little about how interests and policy preferences emerge and change (Moravcsik 1997; Finnemore and Sikkink 1998). Argumentation studies can illuminate the role of dialogue in the evolution of policy preferences and help clarify sociological processes of interest creation, reconstitution, and change.

Acknowledgments

I am grateful to Robert Falkner for his excellent suggestions, which greatly improved this chapter, and to Carla Lee, for her dedicated research assistance.

Notes

1   Game theorists use the term "bargaining" while political scientists prefer "negotiation"; the two terms are often used synonymously.
2   Young (1991) offers an alternative typology and lists three leadership types: structural, entrepreneurial, and intellectual.

References

Adger, W. Neil, Jouni Paavola, Saleemul Huq, and M.J. Mace, eds. 2006. *Fairness in Adaptation to Climate Change*. Cambridge, MA: MIT Press.

Agrawala, Shardul and Steinar Andresen. 2001. "Two-Level Games and the Future of the Climate Regime." *Energy and Environment*, 12(2–3): 5–11.

Aldy, Joseph E., John Ashton, Richard Baron *et al.* 2003. *Beyond Kyoto: Advancing the International Effort against Climate Change*. Arlington, VA: Pew Center on Global Climate Change.

Aldy, Joseph E. and R.N. Stavins, eds. 2010. *Post-Kyoto Climate International Climate Policy: Implementing Architectures for Agreements*. Cambridge: Cambridge University Press.

Andresen, Steinar and Shardul Agrawala. 2002. "Leaders, Pushers and Laggards in the Making of the Climate Regime." *Global Climate Change*, 12: 41–51.

Asgeirsdottir, Aslaug. 2008. *Who Gets What: Domestic Influences on International Negotiations Allocating Shared Resources*. Albany, NY: State University of New York Press.

Avenhaus, Rudolf and I. William Zartman. 2007. "Formal Models of, in and for International Negotiations." In *Diplomacy Games: Formal Models and International Negotiations*, ed. Rudolf Avenhaus and I. William Zartman, 1–22. Berlin: Springer.

Barrett, Scott. 1997. "The Strategy of Trade Sanctions in International Environmental Agreements." *Resource and Energy Economics*, 19(4): 345–361.

Barrett, Scott. 1998. "On the Theory and Diplomacy of Environmental Treaty-Making." *Environmental and Resource Economics*, 11(3–4): 317–333.

Barrett, Scott. 2003. *Environment and Statecraft: The Strategy of Environmental Treaty-Making*. Oxford: Oxford University Press.

Benedick, Richard Elliot. 1998. *Ozone Diplomacy: New Directions in Safeguarding the Planet*. Cambridge, MA: Harvard University Press.

Bernstein, Steven. 2001. *The Compromise of Liberal Environmentalism*. New York: Columbia University Press.

Betsill, Michele. 2004. "Global Climate Change Policy: Making Progress or Spinning Wheels?" In *The Global Environment: Institutions, Law and Policy*, ed. Regina S. Axelrod, David L. Downie, and Norman J. Vig, 103–124. Washington, DC: CQ Press.

Betzold, Carola. 2010. "Borrowing Power to Influence International Negotiations: AOSIS in the Climate Change Regime, 1990–1997." *Politics*, 30(3): 131–148.

Bodansky, Daniel. 2001. "The History of the Global Climate Change Regime." In *International Relations and Global Climate Change*, ed. Urs Luterbacher and Detlef F. Sprinz, 23–40. Cambridge, MA: MIT Press.

Bodansky, Daniel. 2004. *International Efforts on Climate Change beyond 2012: A Survey of Approaches*. Arlington, VA: Pew Center on Global Climate Change.

Bodansky, Daniel. 2010. "The Copenhagen Climate Change Conference: A Postmortem." *American Journal of International Law*, 104(2): 230–240.

Carraro, Carlo. 1997. *International Environmental Negotiations: Strategic Policy Issues*. Cheltenham: Edward Elgar.

Chasek, Pamela S. 2001. *Earth Negotiations: Analyzing Thirty Years of Environmental Diplomacy*. Tokyo: United Nations University Press.

Chasek, Pamela S. and Lynn M. Wagner, eds. 2012. *The Roads from Rio: Lessons Learned from Twenty Years of Multilateral Environmental Negotiations*. New York: Routledge.

Clémençon, Raymond. 2008. "The Bali Road Map: A First Step on the Difficult Journey to a Post-Kyoto Protocol Agreement." *Journal of Environment & Development*, 17(1): 70–94.

Davenport, Deborah. 2005. "An Alternative Explanation of the Failure of the UNCED Forestry Negotiations." *Global Environmental Politics*, 5(1): 105–130.

Deitelhoff, Nicole and Harald Müller. 2005. "Theoretical Paradise – Empirically Lost? Arguing with Habermas." *Review of International Studies*, 31: 167–179.

de Mesquita, Bruce Bueno. 2009. "Recipe for Failure." *Foreign Policy*, 175, November/December: 76–88.

Depledge, Joanna. 2005. *The Organization of Global Negotiations: Constructing the Climate Change Regime*. London: Earthscan.

Depledge, Joanna. 2006. "The Opposite of Learning: Ossification in the Climate Change Regime." *Global Environmental Politics*, 6(1): 1–22.

DeSombre, Elizabeth R. 2000. *Domestic Sources of International Environmental Policy: Industry, Environmentalists, and U.S. Power*. Cambridge, MA: MIT Press.

Dimitrov, Radoslav S. 2005. "Hostage to Norms: States, Institutions and Global Forest Politics." *Global Environmental Politics*, 5(4): 1–24.

Dimitrov, Radoslav S. 2010. "Inside UN Climate Change Negotiations." *Review of Policy Research*, 27(6): 795–821.

Dimitrov, Radoslav S. 2012. "Persuasion in World Politics: The UN Climate Change Negotiations." In *Handbook of Global Environmental Politics*, 2nd edn, ed. Peter Dauvergne, 72–86. Cheltenham: Edward Elgar.

Dupont, Christophe. 1996. "Negotiation as Coalition Building." *International Negotiation*, 1(1): 47–64.

Earnest, David. 2008. "Coordination in Large Numbers: An Agent-Based Model of International Negotiations." *International Studies Quarterly*, 52: 363–382.

Falkner, Robert. 2000. "Regulating Biotech Trade: The Cartagena Protocol on Biosafety." *International Affairs*, 76(2): 299–313.

Falkner, Robert. 2005. "American Hegemony and the Global Environment." *International Studies Review*, 7(4): 585–599.

Falkner, Robert. 2007. "The Political Economy of 'Normative Power' Europe: EU Environmental Leadership in International Biotechnology Regulation." *Journal of European Public Policy*, 14(4): 507–526.

Falkner, Robert, Hannes Stephan, and John Vogler. 2010. "International Climate Policy after Copenhagen: Towards a 'Building Blocks' Strategy." *Global Policy*, 1(3): 252–262.

Finnemore, Martha and Kathryn Sikkink. 1998. "International Norms Dynamics and Political Change." *International Organization*, 52(4): 887–917.

Fry, Ian. 2008. "Reducing Emissions from Deforestation and Forest Degradation: Opportunities and Pitfalls in Developing a New Legal Regime." *Review of European Community & International Environmental Law*, 17(2): 166–182.

Goldstein, Judith, Miles Kahler, Robert O. Keohane, and Ann-Marie Slaughter. 2000. "Legalization and World Politics." *International Organization*, 54(3): 385–399

Grobe, Christian. 2010. "The Power of Words: Argumentative Persuasion in International Negotiations." *European Journal of International Relations*, 16(1): 5–29.

Grundig, Frank. 2009. "Political Strategy and Climate Policy: A Rational Choice Perspective." *Environmental Politics*, 18(5): 747–764.

Gupta, Joyeeta and Michael Grubb. 2000. *Climate Change and European Leadership: A Sustainable Role for Europe?* Berlin: Springer.

Haas, Peter M. 2002. "UN Conferences and Constructivist Governance of the Environment." *Global Governance*, 8(1): 73–91.

Haas, Peter M., Robert O. Keohane, and Marc Levy, eds. 1993. *Institutions for the Earth: Sources of Effective International Environmental Protection*. Cambridge, MA: MIT Press.

Hampson, Fen Osler. 1995. *Multilateral Negotiations: Lessons from Arms Control, Trade, and the Environment*. Baltimore, MD: Johns Hopkins University Press.

Hare, William, Claire Stockwell, Christian Flachsland, and Sebastian Oberthür. 2010. "The Architecture of the Global Climate Regime: A Top-Down Perspective." *Climate Policy*, 10: 600–614.

Harris, Paul G., ed. 2007. *Europe and Global Climate Change: Politics, Foreign Policy and Regional Cooperation*. Cheltenham: Edward Edgar.

Harrison, Kathryn and Lisa M. Sundstrom, eds. 2010. *Global Commons, Domestic Decisions: The Comparative Politics of Climate Change*. Cambridge, MA: MIT Press.

Hasenclever, Andreas, Peter Mayer, and Volker Rittberger. 1997. *Theories of International Regimes*. Cambridge: Cambridge University Press.

Hoffman, Matthew J. 2011. *Climate Governance at the Crossroads: Experimenting with a Global Response after Kyoto*. New York: Oxford University Press.

Hopmann, P. Terrence. 1996. *The Negotiation Process and the Resolution of International Conflict*. Columbia, SC: University of South Carolina Press.

Hovi, Jon, Tora Skodvin, and Steinar Andresen. 2003. "The Persistence of the Kyoto Protocol: Why Annex I Countries Move On without the United States." *Global Environmental Politics*, 3(4): 1–23.

Jinnah, Sikkina. 2011. "Climate Change Bandwaggoning: The Impacts of Strategic Linkages on Regime Design, Maintenance, and Death." *Global Environmental Politics* (special issue), 11(3): 1–9.

Jönsson, Christer. 2002. "Diplomacy, Bargaining and Negotiation." In *Handbook of International Relations*, ed. Walter Carlsnaes, Thomas Risse, and Beth A. Simmons, 212–234. London: Sage.

Kaitala, V. and M. Pohjola. 1995. "Sustainable International Agreements on Greenhouse Warming: A Game Theoretic Study." In *Control and Game Theoretic Models of the Environment*, ed. Carlo Carraro and J.A. Filar, 67–87. Basel: Birkhäuser.

Kanie, Norichika. 2003. "Leadership in Multilateral Negotiation and Domestic Policy: The Netherlands at the Kyoto Protocol." *International Negotiation*, 8(2): 339–365.

Keohane, Robert O. and David G. Victor. 2011. "The Regime Complex for Climate Change." *Perspectives on Politics*, 9(1): 7–23.

Krämer, L. 2004. "The Roots of Divergence: A European Perspective." In *Green Giants? Environmental Policies of the United States and the European Union*, ed. Norman J. Vig and Michael G. Faure, 53–72. Cambridge, MA: MIT Press.

Kulovesi, Kati and Maria Gutiérrez. 2009. "Climate Change Negotiations Update: Process and Prospects for a Copenhagen Agreed Outcome." *Review of European Community & International Environmental Law*, 18(3): 229–243.

Levy, Marc A., Oran R. Young, and Michael Zürn. 1995. "The Study of International Regimes." *European Journal of International Relations*, 1: 267–330.

Litfin, Karen T. 1994. *Ozone Discourses: Science and Politics in Global Environmental Cooperation*. New York: Columbia University Press.

Manners, I. 2002. "Normative Power Europe: A Contradiction in Terms?" *Journal of Common Market Studies*, 40(2): 235–258.

McCright, Aaron M. and Riley E. Dunlap. 2008. "Defeating Kyoto: The Conservative Movement's Impact on U.S. Climate Change Policy." *Social Problems*, 50(3): 348–373.

McLean, Elena V. and Randall W. Stone. 2012. "The Kyoto Protocol: Two-Level Bargaining and European Integration." *International Studies Quarterly*, 56: 1–15.

Mitchell, Ronald. 2003. "International Environmental Agreements: A Survey of Their Features, Formation and Effects." *Annual Review of Environment and Resources*, 28: 429–461.

Moravcsik, Andrew. 1997. "Taking Preferences Seriously: A Liberal Theory of International Politics." *International Organization*, 51(4): 513–553.

Morgera, Elisa, Kati Kulovesi, and Miquel Muños. 2010. *The EU's Climate and Energy Package: Environmental Integration and International Dimensions*. Edinburgh Europa Paper Series 2010/07; University of Edinburgh School of Law Working Paper No. 2010/38, http://ssrn.com/abstract=1711395 (accessed April 15, 2011).

Müller, Benito. 2011. *UNFCCC: The Future of the Process*. Climate Strategies Paper. Cambridge: Climate Strategies, http://www.oxfordclimatepolicy.org/publications/documents/UNFCCC-TheFutureoftheProcess.pdf (accessed April 15, 2011).

Müller, Harald. 2004. "Arguing, Bargaining and All That: Communicative Action, Rationalist Theory and the Logic of Appropriateness in International Relations." *European Journal of International Relations*, 10(3): 395–435.

Muñoz, Miquel, Rachel Thrasher, and Adil Najam. 2009. "Measuring the Negotiation Burden of Multilateral Environmental Agreements." *Global Environmental Politics*, 9(4): 1–13.

Najam, Adil. 2005. "Developing Countries and Global Environmental Governance: From Contestation to Participation to Engagement." *International Environmental Agreements: Politics, Law and Economics*, 5: 303–321.

Najam, Adil, Saleemul Huq, and Youba Sokona. 2003. "Climate Negotiations beyond Kyoto: Developing Countries Concerns and Interests." *Climate Policy*, 3: 221–231.

Oberthür, Sebastian. 2011. "Global Climate Governance after Cancun: Options for EU Leadership." *International Spectator*, 46(1): 5–13.

Oberthür, Sebastian and Claire Roche Kelly. 2008. "EU Leadership in International Climate Policy: Achievements and Challenges." *International Spectator*, 43(3): 35–50.

Oberthür, Sebastian and Marc Pallemaerts. 2010. *The New Climate Policies of the European Union: Internal Legislation and Climate Diplomacy*. Brussels: Brussels University Press.

Okada, A. 1999. *A Cooperative Game Analysis of CO_2 Emission Permits Trading: Evaluating Initial Allocation Rules*. Discussion Paper No. 495, Institute of Economic Research, Kyoto University, Japan.

Ott, Herman E. 2001. "Climate Change: An Important Foreign Policy Issue." *International Affairs*, 7(2): 277–296.

Ott, Herman E., Wolfgang Sterk, and Rie Watanabe. 2008. "The Bali Roadmap: New Horizons for Global Climate Policy." *Climate Policy*, 8: 91–95.

Paterson, Matthew. 1996. *Global Warming and Global Politics*. New York: Routledge.

Pettenger, Mary E., ed. 2007. *The Social Construction of Climate Change*. Toronto: Ashgate.

Putnam, Robert D. 1988. "Diplomacy and Domestic Politics: The Logic of Two-Level Games."
    International Organization, 42(3): 427–460.
Raiffa, Howard. 1982. *The Art and Science of Negotiation.* Cambridge, MA: Harvard University Press.
    versity Press.
Raiffa, Howard. 2002. *Negotiation Analysis: The Art and Science of Collaborative Decision
    Making.* Cambridge, MA: Harvard University Press.
Rajamani, Lavanya. 2008. "From Berlin to Bali: Killing Kyoto Softly." *International and
    Comparative Law Quarterly,* 57(4): 909–939.
Rajamani, Lavanya. 2010. "The Making and Unmaking of the Copenhagen Accord." *International and Comparative Law Quarterly,* 59(3): 824–843.
    national and Comparative Law Quarterly,* 59(3): 824–843.
Ramsay, Kristopher W. 2011. "Cheap Talk Diplomacy, Voluntary Negotiations, and Variable
    Bargaining Power." *International Studies Quarterly,* 55: 1003–1023.
Risse, Thomas. 2000. "Let's Argue! Communicative Action in World Politics." *International
    Organization,* 54(1): 1–39.
Rittberger, V. 1998. "International Conference Diplomacy: A Conspectus." In *Multilateral
    Diplomacy: The United Nations System at Geneva,* ed. M.A. Boisard and E.M. Chossudovsky, 15–28. The Hague: Kluwer.
    dovsky, 15–28. The Hague: Kluwer.
Roberts, J. Timmons and Bradley C. Parks. 2007. *A Climate of Injustice: Global Inequality,
    North–South Politics, and Climate Policy.* Cambridge, MA: MIT Press.
Røgeberg, Ole, Steinar Andresen, and Bjart Holtsmark. 2010. "International Climate Treaties:
    The Case for Pessimism." *Climate Law,* 1(1): 177–197.
Schreurs, Miranda A. 2008. "From the Bottom Up: Local and Subnational Climate Change
    Politics." *Journal of Environment & Development,* 17(4): 343–355.
Selin, Henrik and Stacy D. VanDeveer. 2009. *Changing Climates in North American Politics:
    Institutions, Policymaking, and Multilevel Governance.* Cambridge, MA: MIT Press.
Smith, Richard J. 2009. *Negotiating Environment and Science: An Insider's View of International Agreements, from Driftnets to the Space Station.* Washington, DC: Resources for the
    tional Agreements, from Driftnets to the Space Station.* Washington, DC: Resources for the
    Future.
Spector, Bertram I. 2003. "Deconstructing the Negotiations of Regime Dynamics." In *Getting
    it Done: Post-agreement Negotiation and International Regimes,* ed. B.I. Spector and I.
    William Zartman, 51–88. Washington, DC: United States Institute of Peace Press.
Sprinz, Detlef, and Tapani Vaahtoranta. 1994. "The Interest-Based Explanation of International Environmental Policy." *International Organization,* 48(1): 77–105.
    tional Environmental Policy." *International Organization,* 48(1): 77–105.
Stein, Janice Gross. 1988. "International Negotiation: A Multidisciplinary Perspective." *Negotiation Journal,* 4(3): 221–231.
    tiation Journal,* 4(3): 221–231.
Sterk, Wolfgang, Christof Arens, Sylvia Borbonus *et al.* 2010. *Something was Rotten in the
    State of Denmark: Cop-out in Copenhagen.* Munich: Wuppertal Institute for Climate,
    Environment and Energy.
Underdal, Arild. 1980. *The Politics of International Fisheries Management: The Case of the
    Northeast Atlantic.* New York: Columbia University Press.
Underdal, Arild. 1994. "Leadership Theory: Rediscovering the Art of Management." In *International Multilateral Negotiations: Approaches to the Management of Complexity,* ed. I.
    national Multilateral Negotiations: Approaches to the Management of Complexity,* ed. I.
    William Zartman, 178–200. San Francisco: Jossey-Bass.
Victor, David G. 2001. *The Collapse of the Kyoto Protocol and the Struggle to Slow Global
    Warming.* Princeton, NJ: Princeton University Press.
Victor, David G. 2004. *Climate Change: Debating America's Policy Options.* New York:
    Council on Foreign Relations.
Victor, David. 2006. "Toward Effective International Cooperation on Climate Change: Numbers, Interests and Institutions." *Global Environmental Politics,* 6(3): 90–103.
    bers, Interests and Institutions." *Global Environmental Politics,* 6(3): 90–103.
Victor, David. 2011. *Global Warming Gridlock: Creating More Effective Strategies for Protecting the Planet.* Cambridge: Cambridge University Press.
    tecting the Planet.* Cambridge: Cambridge University Press.
Vihma, Antto. 2010. *Elephant in the Room: The New G77 and China Dynamics in Climate Talks.* Finnish Institute of International Affairs Briefing Paper 62. Helsinki: Finnish
    mate Talks.* Finnish Institute of International Affairs Briefing Paper 62. Helsinki: Finnish

Institute of International Affairs, www.upi-fiia.fi/en/publication/118/elephant_in_the_room (accessed April 15, 2011).

Vogler, John. 2005. "The European Contribution to Global Environmental Governance." *International Affairs*, 81: 835–850.

Vogler, John. 2010. "The Institutionalisation of Trust in the International Climate Regime." *Energy Policy*, 38: 2681–2687.

Vogler, John and Charlotte Bretherton. 2006. "The European Union as a Protagonist to the United States on Climate Change." *International Studies Perspectives*, 7(1): 1–22.

Ward, Hugh. 1993. "Game Theory and the Politics of the Global Commons." *Journal of Conflict Resolution*, 37(2): 203–235.

Ward, Hugh. 1996. "Game Theory and the Politics of Global Warming: The State of Play and Beyond." *Political Studies*, 44(5): 850–871.

Ward, Hugh, Frank Grundig, and Ethan R. Zorick. 2001. "Marching at the Pace of the Slowest: A Model of International Climate-Change Negotiations." *Political Studies*, 49(3): 438–461.

Watanabe, Rie, Christof Arens, Florian Mersmann *et al.* 2008. "The Bali Roadmap for Global Climate Policy: New Horizons and Old Pitfalls." *Journal of European Environmental and Planning Law*, 5(2): 139–158.

Yamin, Farhana and Joanna Depledge. 2004. *The International Climate Change Regime: A Guide to Rules, Institutions and Procedures*. Cambridge: Cambridge University Press.

Young, Oran R. 1991. "Political Leadership and Regime Formation: On the Development of Institutions in International Society." *International Organization*, 45(3): 281–308.

Young, Oran R. 1994. *International Governance. Protecting the Environment in a Stateless Society*. Ithaca, NY: Cornell University Press.

Young, Oran R. 1998. *Creating Regimes: Arctic Accords and International Governance*. Ithaca, NY: Cornell University Press.

Young, Oran R. 2002. *The Institutional Dimensions of Environmental Change: Fit, Interplay, and Scale*. Cambridge, MA: MIT Press.

Young, Oran R. and Gail Osherenko, eds. 1993. *Polar Politics: Creating International Environmental Regimes*. Ithaca, NY: Cornell University Press.

Zartman, I. William, ed. 1994. *International Multilateral Negotiations: Approaches to the Management of Complexity*. San Francisco: Jossey-Bass.

Zartman, I. William. 2002. "What I Want to Know about Negotiations." *International Negotiation*, 7(1): 5–15.

Chapter 21

Regionalism and Environmental Governance

Miranda Schreurs

Introduction

The European Union is widely seen as a global leader in environmental protection. In areas ranging from the control of chemicals, the reduction of packing waste, the promotion of e-waste recycling, and the development of renewable energy to the reduction of greenhouse gas emissions, the EU sets examples that are looked to by many other countries in the world.

There are many examples of areas where EU standards are among the leading ones globally. The EU's policy regulating chemicals and their safe use (Registration, Evaluation, Authorisation, and Restriction of Chemical Substances (REACH)) requires industry to register chemicals they use, provide information about potential hazards, and reduce the use of the most hazardous chemicals. REACH regulations are setting new global standards, and other states are choosing to adopt similar national chemical control approaches. EU recycling requirements not only for glass, paper, packaging, and metals, but also batteries and electronic components, are among the most demanding in the world. Here too the EU is setting high internal standards that are having a global reach (Selin and VanDeveer 2006; Schreurs *et al.* 2009). The EU is also very active in relation to the promotion of renewable energy and the setting of greenhouse gas emission reduction targets. Under the Kyoto Protocol, the EU took on a target to reduce its greenhouse gas emissions by 8% of 1990 levels by 2012 (Harris 2007). This target has been met, with EU emissions for 2010 estimated to be 10.6 % below their 1990 level (European Environment Agency 2011). The EU has set goals to obtain 20% of its primary energy from renewable sources by 2020, albeit with different targets for individual member-states, and introduced an international carbon emissions trading system (Skjærseth and Wettestad 2008). The EU has established a target to reduce its greenhouse gas emissions by 20% of 1990 levels

The Handbook of Global Climate and Environment Policy, First Edition. Edited by Robert Falkner.
© 2013 John Wiley & Sons, Ltd. Published 2013 by John Wiley & Sons, Ltd.

by 2020 and pushed for the establishment of a global climate treaty with binding targets to follow the first phase of the Kyoto Protocol (Schreurs and Tiberghien 2007, 2010; Jordan *et al.* 2012). A 2050 roadmap sets a target for reducing greenhouse gas emissions by 80–95% relative to a 1990 base year.

How is it that a supranational entity composed of 27 (and soon to be more) states with diverse economic, cultural, and geographic conditions has managed to achieve this? What factors have driven the EU to become an environmental leader? In addition to understanding what factors have driven EU efforts to promote high environmental standards across the Union, it is important to consider what shortcomings there are to the EU's approach. It is also important to consider to what extent other regions are trying to emulate the EU's approach to dealing with regional and global environmental problems.

Institutionalizing Environmental Protection

Environmental protection is one of the most advanced areas of cooperation in Europe. In comparison, taxation, military security, and energy are areas where national sovereignty remains strong and European regionalism is less well developed.

One reason environmental protection is relatively advanced in Europe is because it has been incorporated into the EU's governing treaties and institutionalized in its governing structures. No other regional governance structure has institutionalized environmental protection as deeply as has the EU.

Environmental protection has become an increasingly important area of community activity. The EU is based on a series of treaties. The founding treaties of the Union made almost no mention of the environment, but today environmental protection is considered a key aspect of the Union's activities.

Initially, European integration was about promoting peace across the continent through economic integration. In its first decades, the community focused strongly on promoting trade and creating a single market. Slowly, beginning in the 1970s, greater attention began to be paid to environmental protection matters. Harmonization of environmental standards was considered important in order to eliminate barriers to trade. With major differences in the environmental standards of different member-states, there was concern that uneven environmental rules in different states could result in competitive disadvantages for industry. Growing international attention to environmental problems and the first United Nations Conference on the Human Environment in Stockholm in 1972 contributed to new understandings of the importance of pollution control and nature conservation that went beyond mere trade coordination. Over time the importance of environmental protection in its own right came to be recognized as critical for the protection of human health and ecosystems and for the quality of life in Europe.

The European Community launched its first five-year environmental action plan in 1972 and established a Directorate-General for the Environment (better known as DG Environment) the following year. In 1986, the Single European Act, the first major modification of the 1957 Treaty of Rome establishing the European Coal and Steel Community (the predecessor to the European Union), elevated environmental protection to a Community responsibility. The Treaty of Maastricht, establishing the

European Union, made environmental protection a central element of Community policy (European Communities 1992). Article 130(r) of the treaty stated that Community policy on the environment shall contribute to "preserving, protecting, and improving the quality of the environment" and promote steps at the "international level to deal with regional or worldwide environmental problems." It moreover states that environmental protection will be

> based on the precautionary principle and on the principles that preventive action should be taken, that environmental damage should as a priority be rectified at source and that the polluter should pay.

Importantly, it also requires that environmental protection "be integrated into the definition and implementation of other Community policies." The Treaty of Lisbon that came into effect in 2009, abolishing the Community and replacing it with the Union, maintained these basic principles and added new provisions related to climate change and renewable energy (European Communities 2007). The precautionary principle, climate change, and sustainable development are central elements of EU environmental policy (Vogel 2012).

Environmental Leadership through Multi-level Reinforcement

Beyond the treaties, the three main EU institutions – the European Council, the European Commission, and the European Parliament – all play important roles in promoting environmental protection.

The European Council and the Council of Ministers bring together heads of government and ministers of the member-states. They provide an avenue for states to push environmental issues onto the European agenda. On different environmental issues, various states tend to be pioneers or trend setters in terms of establishing national environmental regulations. On issues with a European dimension, these states often try to put these issues on the agenda of the European Council (Liefferink and Andersen 1998; Schreurs and Tiberghien 2007). Examples of this include Germany's push in the mid-1980s to have similar controls introduced on emissions from large combustion plants at the EU level as it had introduced domestically (Ramus 1991) or Denmark's efforts to promote sustainable development policy with the EU (OECD 2007: 123). Often efforts to promote new policies are made by the country holding the rotating presidency of the Council. Thus, during its presidency in the first half of 2012, Denmark prioritized establishment of an Energy Efficiency Directive, which is to increase energy efficiency by more than 17% by 2020 (Danish Presidency of the Council of the European Union 2012: 19). As a world leader in deployment of wind energy, Denmark also has substantially influenced EU renewable energy goals.

In the European Commission in Brussels, the Directorate General for the Environment is charged with formulating environmental regulations, enforcing member-state compliance with environmental regulations, representing member-states in some international negotiations, and overseeing programs to promote environmental protection within the Union. The Directorate General Climate Action established in 2010 has taken over responsibility for promoting climate change action, negotiating on climate, reaching the EU's 2020 goals, and implementing the carbon emissions trading system. As the European Commission is responsible for drafting and implementing laws, it has considerable ability to influence both environmental agenda-setting and policy implementation. This can be seen, for example, in the Commission's active role in promoting a carbon emissions trading system and greenhouse gas emission targets (Skjærseth and Wettestad 2008).

Decision-making rules pertaining to environmental protection have also been altered so that most environmental regulations no longer need unanimous support but can be passed by qualified majorities (a voting procedure that takes the population of a member-state into account when assigning a weight to its vote). Efforts to make decision-making procedures more democratic have resulted in a re-evaluation of the roles of the main EU institutions. Whereas in the past, the European Parliament had limited ability to influence the shape of European regulations, it now shares co-decision authority with the European Council, meaning that the Parliament can now require substantial modifications to regulations. The Parliament, for example, has issued numerous resolutions demanding tighter regulations on oil and gas drilling, energy efficiency, pesticides, and nanomaterials, as well as higher targets for greenhouse gas emission reductions.

The multiple decision points in the relatively loose structure of the European Union provide many avenues for influencing the EU agenda (Schreurs and Tiberghien 2007, 2010). While certainly there are many cases of veto players attempting to block EU environmental policy formation – such as Poland's veto of efforts to raise the EU's greenhouse gas emission reduction target from 20% to 25% at the spring 2012 summit of the EU Council or the German auto industry's efforts in 2007 to block tighter emission standards for automobiles – there are many avenues by which European environmental policy can be moved forward. The drivers may be progressive member-states within the European Council, technocrats working in the European Commission, members of the European Parliament, or the holder of the EU presidency. The entrepreneurial push for policy change tends to pass between and among these actors depending on the issue and circumstances. The relatively loose structure of the EU has allowed for a kind of multilevel reinforcement of European environmental leadership. When a state or actor that led in the past can no longer lead, they may pass the baton on to another state or actor to take over. Alternatively, other actors may themselves seize the baton and run with it when they feel no other actor is playing this role.

The Normative Dimension to Environmental Protection

Calls for stricter environmental regulations in Europe are also tied to relatively high levels of environmental awareness throughout Europe (albeit with differences among member-states). A June 2011 Eurobarometer survey found that respondents considered climate change to be the second most serious problem facing the world today (poverty, hunger, and lack of drinking water being the first). There were some differences among member-states, with countries hardest hit by the economic recession putting economic concerns higher on the list. Sixty-eight percent of respondents ranked climate change as a very serious problem (with an average score of 7.4 out of a possible most serious score of 10) (Directorate General Climate 2011).

The reality of living in a densely populated region where little truly natural environment remains certainly has heightened Europeans' appreciation of protecting what nature does still exist. Sensitized by the wide-scale biodiversity loss that has already occurred as a result of many centuries of human settlement and development, Europe has in recent years tried to expand its protected areas and connect them through the Natura 2000 initiative. Natura 2000 is the centerpiece initiative of the EU in its efforts to protect the survival of Europe's most important species and habitats.[1] The EU also has proposed measures for halting the loss of biodiversity in Europe in connection with the biodiversity strategy that came out of the Nagoya Conference of the Parties to the Biodiversity Convention in 2010. Europe's many environmental non-governmental organizations have demanded stronger protections for nature at the European level.

Numerous environmental crises – the *Torrey Canyon* oil spill off the coast of England in 1967, the accident at the Seveso chemical plant in Italy that resulted in a toxic vapor cloud contaminating the region around the plant in 1976, the Chernobyl nuclear accident in 1986, the Baia Mare cyanide spill in Romania that polluted the Danube in 2000, food safety problems like mad cow disease, among countless others, have played their part in sensitizing the European population to the importance of preventive action and precaution. The severe pollution of the former Eastern bloc states, which first became fully apparent after the fall of the Iron Curtain, was also of major concern to the original members of the EU given that the air and water pollution affected wide regions. A central activity of the EU has been to strengthen the environmental standards throughout Europe in order to reduce the likelihood of such future environmental catastrophes.

Efforts to deepen environmental awareness in Europe have been made by the environmental community, environmental-leaning political parties, the more environmentally progressive member-states, the Commission, and the Parliament. Various European instruments have been established to support environmental projects. The LIFE+ Programme, for example, supports best practice and demonstration projects that contribute to the implementation of the Natura 2000 network, the Birds and Habitats Directives, and biodiversity preservation goals. It also promotes awareness-raising tied to nature protection and biodiversity.[2] Pre-accession funding is made available to EU candidate and potential candidate countries to help bring their national laws, including those tied to the environment, into compliance with EU laws. The 12 states that joined the EU between 2004 and 2007 were beneficiaries of such funding, with the funding beginning prior to actual accession (Andonova 2004).

Yet, as hinted above, it would be wrong to suggest that there is not also considerable conflict regarding environmental standards and policy direction. One way the EU deals with the different environmental and financial capacities of member-states as well as their different levels of environmental concern is to establish burden-sharing arrangements that assign different targets to member-states but preserve a common EU target. This has been used in relation to greenhouse gas emission reduction targets, renewable energy targets, and emissions trading. In the case of the 2020 renewable energy target, for example, the EU common target is 20% but with different targets taken on by each member-state. The highest target is held by Sweden with

49% and the lowest by Malta with 10%. Similarly, although there is a 20% CO_2 emission reduction target for Europe by 2020, there are different targets for member-states. The highest reduction targets of 20% are shared by Denmark, Ireland, and Luxembourg, and the lowest is a 20% growth in emissions target set by Bulgaria.[3]

This flexibility in approach may have stymied opposition to policies that would otherwise have been blocked by various member-states (Jordan et al. 2012).

Framing Environmental Leadership as an Opportunity

David Vogel et al. (2010: 36) argue that one reason EU institutions have been eager to harmonize environmental standards across the member-states is in order to support the still relatively young single market. The single market functions better when environmental standards are harmonized.

The push for environmental leadership, however, goes beyond simple harmonization of standards. In relation to global environmental issues, the European Union has pursued a prominent role. This is seen as important in terms of promoting long-term planetary sustainability, addressing growing resource scarcities, exporting European environmental norms and standards abroad, creating greater international avenues of cooperation, and even enhancing potential export markets for Europe's green technologies.

EU environmental leadership is increasingly portrayed as critical to the EU's future and as a key means of assuring Europe's long-term economic competitiveness. The EU is highly dependent on imported fuel and mineral resources. To the extent that the EU can become highly energy and resource efficient, production costs can decline substantially. Expanding the use of renewable energy will reduce the need to import fossil fuels and have other positive environmental consequences (such as reducing greenhouse gas emissions and pollutants from the burning and extraction of fossil fuels). The EU argues that this can be an important way of keeping money within Europe rather than sending money to regimes with political systems that do not share many EU values and approaches. Beyond the environmental benefits to be derived from efficiency improvements, environmental leadership is seen as a way of stimulating new jobs and potential new export industries. Individual European states are already leaders in many environmental technologies and processes. In 2011, the EU had over 1 million jobs in the renewable energy sector and sales worth €127 billion (Observ'ER 2011).

The concept of sustainable development has become increasingly prominent in European policy documents. Much of the EU's effort at the United Nations Conference on Sustainable Development (Rio+20) focused on winning greater international support for the concept of green growth – the idea that there are many win-wins that can be achieved for environmental protection and economic stability through more efficient use of resources and more environmentally sensitive forms of production (Clémençon 2012; Schreurs 2012).

The EU and Environmental Norm Diffusion

EU environmental leadership is also understood as an opportunity for the EU to influence environmental developments in would-be accession states, neighboring countries, other regions, and at the global level. Within European decision-making circles, reference is often made to the idea of Europe serving as a model that other countries or regions can follow. There is evidence that many European environmental norms are diffusing regionally and globally (Busch and Jörgens 2012).

Certainly the strongest influence the EU has is on its own member-states' environmental policies. Member-states are required to transpose EU regulations into national law and can be punished for non-compliance.

For Europe, promotion of environmental protection in neighboring states has been a means of supporting cooperation and diffusing European norms and values. States that accede to the EU are required to transpose the Acquis Communautaire, the complete body of EU law. They are aided in preparing this transition in the years prior to accession and are usually accorded additional years to come into compliance with EU regulations and directives after accession. This has been one of the most powerful and rapid ways the EU has influenced policy change (Carmin and VanDeveer 2005).

The EU also uses it neighborhood policy to try to promote environmental policies, programs, and norms in closer and more distant neighbors, and integrates environmental protection into its overseas activities. Finally, the EU has tried – with different levels of success and also considerable failure – to upload its environmental standards, norms, and approaches to the international level and in this way to influence environmental negotiations.

Monitoring, Enforcement, and Compliance

One of the differences between the European Union and many other regional groupings is that the EU has the authority to enforce compliance with EU regulations.

Member-states do not always manage or choose to comply with EU regulations in a timely fashion. DG Environment is responsible for ensuring compliance with EU laws. Citizens and non-governmental organizations (NGOs) can lodge complaints about member-state non-compliance with environmental regulations with DG Environment. DG Environment is then expected to evaluate the situation, warn member-states that are not in compliance, and initiate infringement procedures against member-states that remain in non-compliance. In May 2012, for example, the Commission initiated infringement procedures against Romania and Slovakia, urging them to bring their national laws on end-of-life vehicles banning hazardous metals in materials and components of vehicles into line with EU legislation. Hungary and Romania were warned to do more to protect natural habitats and ensure that environmental impact assessments are conducted as required by habitats protection legislation. Italy was urged to ensure adequate pre-treatment of waste that is landfilled as stipulated in EU landfill legislation. Greece has been told it needs to improve its treatment of waste water. In each of these cases, the member-states were given two months to come into compliance with the EU regulations. Member-states can be brought before the European Court of Justice and penalized financially for failure to comply with EU laws. Thus in May 2012, Germany was referred to the EU Court of Justice for allegedly not fully applying the principle of cost recovery for water services in order to promote efficiency as stipulated in the Water Framework Directive. In April 2012 the European Commission referred Bulgaria, Hungary,

Poland, and Slovakia to the EU Court of Justice for not meeting the December 2010 deadline for establishing national laws bringing the countries in line with the EU's Waste Framework Directive and requested the court to impose penalty payments.[4] Research suggests, however, that member-states tend to delay in complying with court orders and that the Commission is under-resourced, hindering its ability to fully carry out its mandate to ensure compliance (Jack 2011).

Stimulating Sub-national Civil Society Cooperation at the Regional Level

Regional cooperation can occur at multiple levels of government. Stimulated by the success of the US Conference of Mayors Climate Protection Agreement that has led over a thousand US cities to agree to take action to address climate change and share best-practice information, the European Commission launched the Covenant of Mayors in 2009. The Covenant of Mayors now has over 4000 cities as members; they have committed to take action on energy efficiency and the promotion of renewable energies in order for the EU to meet and exceed its goal to reduce greenhouse gas emissions by 20% of 1990 levels by 2020.[5]

Regionalism and Environmental Cooperation in the International Context

The EU has certainly attracted much interest in other regions of the world that are themselves experimenting with greater regional cooperation. The EU has also actively sought to export its own model of cooperation to other regions of the world. Efforts to expand regional environmental cooperation are becoming more numerous.

In other regions, too, states have entered into multilateral arrangements that have as one of their goals the promotion of environmental protection. These include broader associations, such as the Association of Southeast Asian Nations (ASEAN), the North American Free Trade Agreement (NAFTA), the South Asian Association for Regional Cooperation (SAARC), the Union of South American Nations (UNASUR), and the African Union (AU), as well as more focused forums, such as the Northeast Asian Sub-Regional Programme on the Environment or the Mekong River Commission. Regional approaches to environmental governance are also forming at the sub-national level, such as the European Covenant of Mayors mentioned previously, as well as the Regional Greenhouse Gas Initiative and the Western Climate Initiative, greenhouse gas emission trading systems that have formed between US states and Canadian provinces. What factors are driving the formation of these arrangements and how effective are they?

Bilateral Environmental Agreements

The cross-border nature of many environmental matters has led to the birth of many bilateral environmental agreements. In the North American context, there is a long history of bilateral environmental agreements. In 1909 the Boundary Waters Treaty was established. Advanced for its time, it not only called for free navigation between Canada and the United States along its rivers, tributaries, bays, and lakes, it stipulated that any activity that would change the water levels of a boundary water resulting in adverse impacts in the other country would require the approval of an International Joint Commission established between the two countries. It also stated that the boundary waters "shall not be polluted on either side to the injury of health or property on the other" (International Joint Commission 1909). The Great Lakes Water Quality Agreement of 1972 (amended in 1978 and with a protocol established in 1987) was formed to address the increasingly severe pollution of the Great Lakes Basin ecosystem. It aims to restore and maintain the chemical and biological health of the Great Lakes through the control of the release of toxic substances, the abatement, control, and prevention of pollution from municipal and industrial sources and shipping, the promotion of waste water treatment, and the control of agricultural runoff (International Joint Commission 1989).

In terms of air pollution and acid rain, Canada and the United States were pushed into greater cooperation due to early transboundary air pollution disputes. In 1941, the two countries' governments settled a transboundary water and air pollution dispute caused by a mine and smelting company in Trail, British Columbia that was impacting agriculture and forestry in Washington. Eventually, the arbitration tribunal found the Canadian side responsible and compensation was paid by the Canadian government to the US government (which then distributed the funds to the landowners). The case was important in terms of establishing the polluter pays principle in transboundary contexts and in promoting the development of international environmental law (Wirth 1996). Decades later new disputes arose between the two countries related to acid rain. Canada's lakes and forests were being adversely impacted by the burning of coal for the production of electricity and other industrial purposes in the US Midwest. Prevailing winds carried the acidic compounds produced by coal-burning from the Midwest northward toward New England and Canada. After almost a decade of cross-national tensions tied to the issue, the 1991 Agreement between Canada and the United States on Air Quality was formed. The agreement promoted joint scientific research and domestic measures to reduce sulfur dioxide and nitrogen oxides, the primary precursors to acid rain. Later, the agreement was extended to include measures to address transboundary air pollution leading to ground-level ozone (smog) problems (International Joint Commission 1991; Munton 2007).

North American Agreement on Environmental Cooperation

One of the striking features of the growing number of regional governance structures is that all have deemed it necessary to incorporate environmental protection concerns into their structures. When the North American Free Trade Agreement, which has as its main aim the removal of trade barriers for the promotion of cross-border trade, was being negotiated, there was considerable opposition from labor and environmental groups. Environmentalists argued strongly for the inclusion of environmental protection standards in order to prevent downward pressure on states to remove environmental legislation as a barrier to free trade and to prevent a flight of US industry to Mexico due to its weaker environmental standards. Their pressure succeeded in persuading the three governments to conclude an environmental side agreement in parallel to the free trade agreement, the North American Agreement on Environmental Cooperation (NAAEC).

The NAAEC seeks to ensure that domestic environmental standards are protected and improved, that environmental protection and trade are better integrated, and that environmental cooperation be strengthened among Canada, Mexico, and the United States (Commission for Environmental Cooperation n.d.). The three member-states are expected to submit reports on the state of their environment, make use of environmental impact assessments, and cooperate on regional environmental matters.

The NAAEC does not seek to harmonize environmental laws, as is done in the EU, nor does it create supranational institutions with authority to develop and implement environmental laws. It is an example of a more decentralized form of regional cooperation.

The NAFTA environmental side agreement led to the establishment of a Commission for Environmental Cooperation in 1994 that includes a Council of Ministers (the three environmental ministers), a Secretariat (based in Montreal), and a Joint Public Advisory Committee (JPAC) that provides for civil society input to the Council. The Council of Ministers is at cabinet level and meets at least once a year. It must approve the Commission's budget and the work of the Secretariat; assess the environmental effects of the NAFTA; and develop recommendations on public access to information, limits for pollutants, and transboundary environmental assessments. It has some agenda-setting capacity through the development of strategic plans. The 2010–2015 strategic plan is focused on climate change (improving comparability of emissions data, establishing an interactive online platform with information on climate change), greening the economy and healthy communities and the environment (e.g. green building, sound management of electronic wastes across the continent) (Podhora 2011).[6]

The Secretariat is charged with implementing the agreement and issuing reports. NGOs can submit complaints to the Secretariat regarding the failure of any of the three countries to enforce their existing national environmental regulations (Markell and Knox 2003; Hufbauer and Schott 2005), and if the Secretariat makes a recommendation that a factual report be produced and the Council approves the recommendation, the Secretariat is expected to issue a neutral report on the matter. The Secretariat is not, however, a court and can issue no opinions or decisions. Apparently, even this relatively weak oversight capacity can have its effects, pressuring the states in question to review their own situations. Many cases are submitted to the Secretariat. Examples include a complaint that the government of Quebec is failing to enforce its environmental law with regard to vapor emissions from service stations in suburban Montreal; another that alleges that tailing ponds tied to the extraction of bitumen from oil sands in northern Alberta are polluting groundwaters, soil, and surface water; a third, that the United States government is failing to enforce the federal Clean Water Act against coal-fired power plants for mercury emissions that are degrading water bodies; and a fourth that charges that the Mexican government is failing to enforce environmental legislation related to a hazardous waste landfill in Sonora, Mexico. The submissions tend to be made by environmental NGOs as well as individuals.[7]

In cases where there is a consistent pattern of non-enforcement of a domestic law, the Council may oversee an arbitration process that is performed by an expert arbitration board appointed from a list of candidates that is consensually established. The arbitration panel can recommend that the party complained against adopt and implement an action plan to remedy the situation. Panel decisions are not, however, binding. Countries are not mandated to implement the decisions of the panel. They can instead choose to forgo trade concessions established by the NAFTA (Abbott 1993).

The NAAEC is a very different model from the regional environmental cooperation that is found in Europe. It provides a model for greater regional environmental cooperation in cases where countries are not prepared to pool their sovereignty to the degree that is being done in the European Union. Still, it must be recognized that as environmental protection is in this case still primarily seen as a responsibility of national authorities and no efforts are made to harmonize environmental laws, there are substantial limitations to what can be achieved in comparison to the situation in Europe. Michelle Betsill (2009), for example, argues that there are few prospects that regional carbon emissions trading will be embedded into the NAFTA environmental side agreement, as is the case with carbon emissions trading in the EU.

Regional Approaches to Addressing Climate Change in North America

Considering the difficulty of achieving global agreements on climate change, and the reality that in 2001 the United States pulled out of the Kyoto Protocol and Canada followed suit in 2011, more regional approaches may at least in the shorter term play a critical role in addressing the continent's major contributions to global greenhouse gas emissions. Canada, Mexico, and the United States combined account for about 22% of global carbon dioxide emissions and are respectively the seventh, eleventh, and second largest emitters globally. Per capita emissions in Canada are particularly high (Marland *et al.* 2008).

One of the more interesting developments related to climate change in the North American context has been the development of regional carbon emissions trading regimes. The Regional Greenhouse Gas Initiative (RGGI) links nine eastern US states in a carbon emissions trading system that is to reduce emissions from power plants. Several Canadian provinces are observers to the RGGI. More ambitious is the Western Climate Initiative that links California and several Canadian provinces in a greenhouse gas emissions system covering seven greenhouse gases beginning in 2013. Partners in the Western Climate Initiative have agreed to reduce their greenhouse gas emissions by 15% of 2005 levels by 2020.[8] It is far more ambitious than the RGGI as it is to cover not just emissions from the power sector but also from industry, transportation, and residential and commercial fuel use (Western Climate Initiative; Selin and VanDeveer 2009).

The Association of Southeast Asian Nations

Other regions of the world are also developing new regional environmental structures. The Association of Southeast Asian Nations (ASEAN) includes 10 member-states: Brunei Darussalam, Cambodia, Indonesia, the People's Democratic Republic of Lao, Malaysia, Myanmar, the Philippines, Singapore, Thailand, and Vietnam. ASEAN has developed an ASEAN free trade area (AFTA) and set a goal to establish an ASEAN Community by 2015 (Schreurs 2010). The ASEAN Community, which is already beginning to take shape, is composed of three pillars: a Political Security Community, an Economic Community, and a Socio-Cultural Community. The Socio-Cultural Community addresses a wide variety of issues including education, health, labor, rural development, women, and environment. ASEAN is developing an approach to environmental protection that has taken many lessons from the EU model while still retaining greater degrees of national sovereignty in decision-making. Cooperation within ASEAN is guided by the ASEAN Way, the principle of non-interference in internal affairs of other states but the idea that cooperation can be achieved through common norm development. Thus, unlike in the EU, where qualified majorities can pass environmental legislation, in ASEAN consensus is required.

Still, as ASEAN becomes increasingly institutionalized greater degrees of sovereignty may be pooled in ASEAN institutions. Along with the main goals of creating a liberal trade arrangement and enhancing regional security, environmental protection and nature conservation are important aspects of the region's growing cooperative governance structures (Elliott 2003). The impacts of the Indian Ocean tsunami in December 2004 and regional haze problems have highlighted to the region the importance of deepened regional cooperation and harmonization of standards and rules.

An ASEAN Ministerial Meeting on the Environment was formed in 1981 and brings together the environmental ministers of the region on a periodic basis to establish joint goals and policy directions. Ten priority areas guide the community's work: global environmental issues, managing transboundary environmental pollution (and especially transboundary haze and the movement of hazardous waste), sustainable development through public participation, environmentally sound technology, quality living standards in urban areas, harmonizing environmental policies and databases, sustainable use of coastal and marine environment, sustainable management of natural resources and biodiversity, sustainability of freshwater resources, and climate change.[9]

ASEAN has issued numerous joint declarations on key environmental and sustainable development matters. One example includes the Singapore Declaration on Climate Change, Energy and the Environment and the ASEAN Declaration on Environment Sustainability. The Singapore Declaration, which was issued by the ASEAN member-states plus Australia, China, India, Japan, South Korea, and New Zealand, called for the long-term stabilization of greenhouse gas emissions at a level that will not cause dangerous anthropogenic interference with the climate system while also stressing that developed countries need to play a leading role in greenhouse gas mitigation and adaptation and that action should follow the principle of common but differentiated responsibilities.

Working groups are expected to carry out the priorities spelled out in declarations. ASEAN working groups have been formed to address nature conservation and biodiversity, the coastal and marine environment, and environmentally sustainable cities, among other issues. In January 2011, representatives of ASEAN cities and national governments plus the ASEAN Secretariat met in Jakarta to exchange information on best practices and lessons learned (ASEAN 2011).[10]

A handful of international environmental agreements have been formed as well. After years of tensions due to the transboundary haze that is caused by the burning of forests, ASEAN member-states established the ASEAN Agreement on

Transboundary Haze Pollution in 2002. The goal of the agreement is to address the transboundary haze through national efforts and regional and international cooperation. The agreement has entered into force and is the first legally binding international agreement among the ASEAN member-states. Indonesia, the source of most of the haze and the last hold-out on ratification of the agreement, has indicated that it will work to ratify the agreement soon (Maruli 2011).

In the area of natural-disaster management, painful lessons tied to the loss of human life that could conceivably have been reduced with better regional cooperation have resulted in a reassessment of regional cooperation. After the Indian Ocean tsunami, the ASEAN Agreement on Disaster Management and Emergency Response was formed to enhance coordination on disaster response and prevention.

Within ASEAN, environmental governance is less well institutionalized and environmental protection a less well developed area than is the case in the European Union. Yet, signs are pointing towards growing concern about population and development pressures that are threatening the region's highly biologically diverse and rich ecosystems and the impacts of severe pollution on human health. Both as a result of bottom-up initiatives and assistance provided from outside, the region is strengthening national environmental laws and slowly embracing more regional approaches to environmental governance.

Conclusion

Several factors are pushing states towards greater use of regional governance strategies for dealing with environmental matters. One obvious reason is that many environmental problems have transboundary and international impacts. Addressing transboundary environmental problems from a regional perspective can reduce conflicts among neighbors while improving regional environmental quality.

A second reason appears to be that harmonization of environmental standards is considered useful for improving trade relations and limiting industrial flight to regions of lower environmental standards, although this may be more true for Europe, which has aggressively pursued harmonization of standards, than North America. It may also be the case that in North America, Canada and the USA have already converged in many of their environmental standards and the expectation is that with greater interaction and the development of the Mexican economy, it too will raise its environmental standards. ASEAN appears to be considering the benefits that could derive from greater harmonization of standards.

A third factor may be related to norm diffusion and a growing recognition of the importance of environmental protection. In regions where there is close interaction due to economic and cultural exchange, environmental norms may diffuse from environmentally more advanced countries or regions to other states.

In some instances it may also be the case that regional environmental governance is being pursued in reaction to the lack of national leadership on environmental matters. Pursuit of environmental protection at the regional level is not just occurring at the state level, but also among sub-national actors. This is the case with the regional greenhouse gas emissions trading initiatives in the United States.

Other regions of the world not discussed in this chapter are also taking steps towards greater environmental cooperation. It may well be that as global environmental agreements become increasingly cumbersome and hard to advance attention will turn to the possibilities for promoting sustainable development and environmental conservation at the regional level.

Notes

1   Natura 2000, http://ec.europa.eu/environment/nature/natura2000/index_en.htm (accessed July 19, 2012).
2   Life+ Programme, http://ec.europa.eu/environment/life/about/index.htm (accessed July 15, 2012).
3   Europe 2020 Targets, http://ec.europa.eu/europe2020/pdf/targets_en.pdf. (accessed July 20, 2012).
4   European Commission, Environment Infringement Cases, http://ec.europa.eu/environment/legal/law/press_en.htm (accessed July 14, 2012).
5   Covenant of Mayors, http://www.eumayors.eu/index_en.html (accessed July 22, 2012).
6   Commission for Environmental Cooperation, http://www.cec.org/Page.asp?PageID=751&SiteNodeID=1008&BL_ExpandID=155 (accessed July 22, 2012).
7   Commission for Environmental Cooperation, http://www.cec.org/Page.asp?PageID=751&SiteNodeID=1008&BL_ExpandID=155 (accessed July 22, 2012).
8   Western Climate Initiative, http://www.westernclimateinitiative.org (accessed July 19, 2012).
9   ASEAN, http://environment.asean.org/index.php?page=overview (accessed July 22, 2012).
10 ASEAN, http://environment.asean.org/index.php?page=overview (accessed July 22, 2012).

References

Abbott, Frederick M. 1993. "The NAFTA Environmental Dispute Settlement System as a Prototype for Regional Integration Arrangements." *Yearbook of International Environmental Law*, 3(4): 3–29.

Andonova, Liliana. 2004. *Transnational Politics of the Environment: The European Union and Environmental Policy in Central and Eastern Europe*. Cambridge, MA: MIT Press.

ASEAN. 2011. "ASEAN Cities Explore Methods and Tools to Address Climate Change," http://environment.asean.org/index.php?page=media:jps:aseancities (accessed July 22, 2012).

Betsill, Michelle M. 2009. "NAFTA as a Forum for CO_2 Permit Trading?" In *Changing Climates in North American Politics: Institutions, Policymaking, and Multilevel Governance*, ed. Henrik Selin and Stacy D. VanDeveer, 161–180. Cambridge, MA: MIT Press.

Busch, Per Oluf and Helge Jörgens. 2012. "Europeanization through Diffusion? Renewable Energy Policies and Alternative Sources for European Convergence." In *European Energy Policy: An Environmental Approach*, ed. Francesc Morata and Slorio Sandoval, 66–82. Cheltenham: Edward Elgar.

Carmin, Joann and Stacy D. VanDeveer. 2005. *EU Enlargement and the Environment: Institutional Change and Environmental Policy in Central and Eastern Europe*. New York: Routledge.

Clémençon, Raymond. 2012. "From Rio 1992 to Rio 2012 and Beyond: Revisiting the Role of Trade Rules and Financial Transfers for Sustainable Development." *Journal of Environment & Development*, 21: 5–14.

Commission for Environmental Cooperation. 2012. "North American Agreement on Environmental Protection, 1993," http://www.cec.org/Page.asp?PageID=1226&SiteNodeID=567 (accessed October 20, 2012).

Commission for Environmental Cooperation. n.d. "Registry of Submissions," http://www.cec.org/Page.asp?PageID=751&ContentID=&SiteNodeID=250&BL_ExpandID=156 (accessed July 22, 2012).

Danish Presidency of the Council of the European Union. 2012. *Europe at Work: The Results of the Danish Presidency of the Council of the European Union in the First Half of 2012.* Copenhagen: Ministry of Foreign Affairs, http://eu2012.dk/en/NewsList/Juni/Uge-26/~/media/702749D703AB4F1790A110951A22E8FE.pdf (accessed October 20, 2012).

Directorate General Climate. 2011. *Climate Change.* Special Eurobarometer 372, Wave EB754, October. Brussels: European Commission, http://ec.europa.eu/public_opinion/archives/ebs/ebs_372_en.pdf (accessed October 20, 2012).

Elliott, Lorraine. 2003. "ASEAN and Environmental Cooperation." *Pacific Review*, 16(1): 29–52. doi:10.1080/0951274032000043235.

European Communities. 1992. "Treaty of Maastricht: Provisions Amending the Treaty Establishing the European Economic Community with a View to Establishing the European Community," February 7, 1992, http://www.eurotreaties.com/maastrichtec.pdf (accessed October 20, 2012).

European Communities. 2007. "Treaty of Lisbon amending the Treaty on European Union and the Treaty establishing the European Community," December 13, 2007. *Official Journal of the European Union*, C 306 (50), http://eur-lex.europa.eu/JOHtml.do?uri=OJ:C:2007:306:SOM:EN:HTML (accessed October 20, 2012).

European Environment Agency. 2011. *Greenhouse Gas Emission Trends and Projections in Europe 2011: Tracking Progress towards Kyoto and 2020.* EEA Report No. 4. Copenhagen: European Environment Agency, http://www.eea.europa.eu/publications/ghg-trends-and-projections-2011 (accessed October 20, 2012).

Harris, Paul G. 2007. *Europe and Global Climate Change: Politics, Foreign Policy and Regional Cooperation.* Cheltenham: Edward Elgar.

Hufbauer, Gary Clyde and Jeffrey J. Schott. 2005. *NAFTA Revisited: Achievements and Challenges.* Washington, DC: Institute for International Economics.

International Joint Commission. 1909. "Treaty between the United States and Great Britain relating to Boundary Waters, and Questions Arising between the United States and Canada," http://bwt.ijc.org/index.php?page=Treaty-Text&hl=eng (accessed October 20, 2012).

International Joint Commission. 1989. "Great Lakes Water Quality Agreement of 1978 as amended by Protocol," http://www.ijc.org/rel/agree/quality.html (accessed October 20, 2012).

International Joint Commission. 1991. "Air Quality Agreement," http://www.ijc.org/rel/agree/air.html (accessed October 20, 2012).

Jack, Brian. 2011. "Enforcing Member State Compliance with EU Environmental Law: A Critical Evaluation of the Use of Financial Penalties." *Journal of Environmental Law*, 23(1): 73–95.

Jordan, Andrew, Dave Huitema, Harro van Asselt, and Tim Rayner, eds. 2010. *Climate Change Policy in the European Union: Confronting the Dilemmas of Mitigation and Adaptation?* Cambridge: Cambridge University Press.

Jordan, Andrew, Harro van Asselt, Frans Berkhout *et al.* 2012. "Understanding the Paradoxes of Multi-level Governing: Climate Change Policy in the European Union." *Global Environmental Politics*, 12(2): 43–66.

Liefferink, Duncan and Michael Skou Andersen. 1998. "Strategies of the 'Green' Member State in EU Environmental Policy-Making." *Journal of European Public Policy*, 5(2): 254–270.

Markell, David L. and John H. Knox, eds. 2003. *Greening NAFTA: The North American Commission for Environmental Cooperation.* Stanford, CA: Stanford University Press.

Marland, G., T.A. Boden, and R.J. Andres. 2008. "Global, Regional, and National Fossil Fuel CO_2 Emissions." In *Trends: A Compendium of Data on Global Change.* Oak Ridge, TN: US Department of Energy, Carbon Dioxide Information Analysis Center, Oak Ridge National Laboratory, http://cdiac.ornl.gov/trends/emis/tre_tp20.html (accessed October 20, 2012).

Maruli, Aditia. 2011. "Indonesia to Ratify ASEAN Agreement on Trans-boundary Haze Pollution," Antaranews.com, March 7, http://www.antaranews.com/en/news/68888/indonesia-to-ratify-asean-agreement-on-trans-boundary-haze-pollution (accessed July 23, 2012).

Munton, Donald. 2007. "Acid Rain Politics in North America: Conflict to Cooperation to Collusion." In *Acid in the Environment: Lessons Learned and Future Prospects*, ed. Gerald R. Visgillio and Diana M. Whitelaw, 175–202. New York: Springer.

Observ'ER. 2011. *The State of Renewable Energies in Europe*. EurObserv'ER Report 11. Paris: Observ'ER, http://www.eurobserv-er.org/pdf/barobilan11.pdf (accessed October 20, 2012).

OECD (Organisation for Economic Co-operation and Development). 2007. *Environmental Performance Reviews: Denmark*. Paris: OECD.

Podhora, Aranka. 2011. "Environmental Assessment and the Greening of NAFTA: Das Instrument der Umweltprüfung in der Ökologisierung des NAFA." PhD dissertation, Technishe Universität Berlin, http://opus.kobv.de/tuberlin/volltexte/2011/3013/ (accessed October 20, 2012).

Ramus, Catherine. 1991. *The Large Combustion Plant Directive: An Analysis of European Environmental Policy*. Oxford Institute for Energy Studies EV7. Oxford: Oxford Institute for Energy Studies, http://www.oxfordenergy.org/wpcms/wp-content/uploads/2011/03/EV7-TheLargeCombustionPlantDirectiveAnAnalysisofEuropeanEnvironmentalPolicy-CA Rasmus-19911.pdf (accessed October 20, 2012).

Schreurs, Miranda A. 2010. "Multi-level Governance the ASEAN Way." In *Handbook on Multi-level Governance*, ed. Michael Zürn, Sonja Wälti, and Henrik Everlein, 308–322. Cheltenham: Edward Elgar.

Schreurs, Miranda A. 2012. "Rio+20: Assessing Progress to Date and Future Challenges." *Journal of Environment & Development*, 21: 19–23.

Schreurs, Miranda A., Henrik Selin, and Stacy D. VanDeveer, eds. 2009. *Transatlantic Environment and Energy Politics: Comparative and International Perspectives*. Farnham: Ashgate.

Schreurs, Miranda A. and Yves Tiberghien. 2007. "Multi-level Reinforcement: Explaining European Union Leadership in Climate Change Mitigation." *Global Environmental Politics*, 7(4): 19–46.

Schreurs, Miranda A. and Yves Tiberghien. 2010. "European Union Leadership in Climate Change: Mitigation through Multi-level Reinforcement." In *European Union Leadership in Climate Change: Mitigation through Multilevel Reinforcement*, ed. Kathryn Harrison and Lisa McIntosh Sundstrom, 23–66. Cambridge, MA: MIT Press.

Selin, Henrik and Stacy D. VanDeveer. 2006. "Raising Global Standards: Hazardous Substances and E-waste Management in the European Union." *Environment*, 48(10): 6–17.

Selin, Henrik and Stacy D. VanDeveer, eds. 2009. *Changing Climates in North American Politics: Institutions, Policymaking, and Multilevel Governance*. Cambridge, MA: MIT Press.

Skjærseth, Jon Birger and Jørgen Wettestad. 2008. *EU Emissions Trading: Initiation, Decision-making, and Implementation*. Aldershot: Ashgate.

Vogel, David. 2012. *The Politics of Precaution: Regulating Health, Safety, and Environmental Risks in Europe and the United States*. Princeton, NJ: Princeton University Press.

Vogel, David, Michael W. Toffel, Diahanna Post, and Nazli Uludere. 2010. *Environmental Federalism in the European Union and the United States*. Harvard Business School Working Paper 10-085, March 17, http://www.hbs.edu/research/pdf/10-085.pdf (accessed October 20, 2012).

Wirth, John D. 1996. "The Trail Smelter Dispute: Canadians and Americans Confront Trans-boundary Pollution, 1927–41." *Environmental History*, 1(2): 34–51.

Part IV Global Economy and Policy

Chapter 22

Globalization

Peter Newell

Introduction

Global environmental change is as much a product and manifestation of globalization as trade, production, and finance. Indeed, what gets financed, produced, and traded in the global economy is, in many cases, goods and products made up of, or embodying, natural resources. Even the production of synthetic and artificial components or the operation of service industries, while seemingly not so dependent on direct extraction, still consume vast amounts of water and energy. Hence while those of us who live in more affluent parts of the world may be less attentive to the value of environmental resources, compared to low-income countries for whom close to one third of their wealth comes from their "natural capital" that includes forests, protected areas, agricultural lands, energy, and minerals (World Bank 2011), the circuits of capital that underpin globalization impact hugely, though highly unevenly, on the environment that we all share. Yet discussion of either the environmental dimensions of globalization or the environment as a form of globalization itself is often subsumed by other "high political" concerns, even if "the environment" sustains our collective ability to produce, trade, and consume.

Globalization can of course mean many things, and I have reproduced some definitions below (see Box 22.1). The use of the term often combines an attempt to describe shifts in the scale, speed, or intensity of flows (of capital, information, technology, people, and pollution) with changes in political power and authority (supranationalization, decentralization, private governance, civil regulation, among others), while not losing sight of the social and cultural aspects of globalization such as identity, politics, nationalism, and cultures of knowledge and consumption. Within this broad canvas, I will focus on those aspects of the slippery phenomena we call globalization that interact most directly and clearly with the world of (global)

The Handbook of Global Climate and Environment Policy, First Edition. Edited by Robert Falkner.
© 2013 John Wiley & Sons, Ltd. Published 2013 by John Wiley & Sons, Ltd.

environmental politics. In this regard what is perhaps most significant ecologically is the rescaling of capitalist relations and strategies of accumulation to secure access to new resources and markets made possible through technological advance, internationalized production strategies, liberalized trade, and the globalization of finance that underpins this. It is this that has *intensified* and *globalized* environmental harm in a way that has brought about the multiple crises we now currently face around the availability of food, water, and energy. It is also what has enabled capital to overcome the limits imposed by national regulation, including pressures from environmental groups and trade unions for stronger forms of environmental protection on the one hand, and more stringent forms of legislation regarding occupational health and safety legislation on the other. For globalizing capital, accessing new markets and consumers has also been imperative to addressing domestic crises of overaccumulation and the lack of viable domestic investment opportunities (Harvey 2003).

Box 22.1   Definitions of Globalization

[T]he growing interconnectedness and interrelatedness of all aspects of society (Jones 2006: 2).

[G]lobalization refers to the widening and deepening of the international flows of trade, capital, technology and information within a single integrated global market (Petras and Veltmeyer 2001: 11).

[G]lobalization is a transformation of social geography marked by the growth of supraterritorial spaces (Scholte 2000: 8).

[G]lobalization is what we in the Third World have for several centuries called colonization (Khor 1995).

[A] process (or set of processes) which embodies a transformation in the spatial organization of social relations and transactions – assessed in terms of their extensity, intensity, velocity and impact – generating transcontinental or interregional flows and networks of activity, interaction and the exercise of power (Held *et al.* 1999: 16).

Globalization, from this more historical and critical perspective, is better understood as a deepening and globalizing of earlier patterns of capitalist development, rather than a decisive and tangible break with previous economic relations. The inequalities, disruptions, and patterns of environmental injustice that are apparent throughout the history of the world economy take on increasingly transnational dimensions in this latest era of capitalist development. The exploitation of resources through uneven development can be traced back to colonialism, where "accumulation through dispossession" was more obvious and the use of force more prominent, but continues to manifest itself in uneven terms of trade, the debt crisis, and structural adjustment programs that prise economies open to foreign investors as restless capital seeks new outlets for investment and return to avert crises of overaccumulation (Harvey 2003). The process of moving crises around geographically, spatially, or temporally, rather than resolving them, which David Harvey has high-lighted so eloquently in his work, explains, in part, phenomena such as the export of resource-intensive forms of production overseas while rich countries continue to capture the benefits through consumption. It accounts for the export of toxic and hazardous wastes to poor countries so that richer countries are not faced with the consequences of their consumption (Clapp 2001), and the double standards that many TNCs employ when they operate overseas, allowing them to produce more cheaply but still sell their wares to richer consumers (Madeley 1999). The inter-national division of labor upon which this model is premised, which has enabled wealthier countries in the core of the global economy to enjoy a comparative advan-tage in service and high-tech sectors while benefiting from industrial production reliant on cheap labor in the global South explains, for example, why emissions of greenhouse gases have risen so sharply in "rising powers" such as China, India, and Brazil that are now home to more energy- and pollution-intensive stages of global production chains.

Such strategies both feed upon and reproduce global inequalities, even if they clearly bring tangible benefits to some social groups within host countries. As Roberts and Parks show, these forms of ecologically uneven exchange also mean that the responsibility for pollution, as well as the pollution itself, is redistributed globally such that: "Emissions are increasing sharply in developing countries as wealthy countries 'offshore' the energy and resource intensive stages of production" (2008: 169). While moving things around makes sense for richer countries or social groups able to do so (out of sight, out of mind), it serves to disperse rather than resolve environmental problems.

The chapter is structured as follows. First, I review the evidence of the rela-tionship between globalization and the environment, looking at debates about the impact of different globalizing trends, largely economic ones, on the environment. The way in which this occurs reflects and is mediated by social relations of class, race, and gender that help us to determine who wins and who loses from the way resources are exploited and distributed in today's global economy: the glob-ally uneven distribution of burden and benefit (Newell 2005). Second, I look at attempts to date to govern and mediate the relationship between globalization and the environment. Numerous institutions, public and private, have been created seek-ing to contain the worst ecological and social effects of globalization and build upon the benefits it delivers. This is true of global and regional trade agreements and institutions, as well as of private governance by and for private actors that seek to reduce their pollution voluntarily in preference to state-based regulation. Third, the limits of these forms of governance and the institutions that oversee them have prompted widespread mobilizations and contestation from a range of social movements and civil society actors that have questioned the orientation and per-formance of these institutions: the model of growth they pursue and their limited effectiveness in squaring this with rising levels of environmental damage. This section looks at the strategies adopted by these groups to try and re-embed globalization in an altogether different set of values and institutions guided as much by con-cerns with sustainability and social justice as the pursuit of profit. The fourth and final section offers some conclusions and global policy implications based on the previous discussion.

Globalization and the Environment: Exploring the Connections

This section looks at evidence of the "nature" of the relationship between what has come to be called "globalization," notably trade, production, and finance and different environmental domains. It documents key trends and controversies regarding evidence of the impact of globalization upon, and its relationship to, different socio-ecological systems.

Trade

There is a vast literature on trade liberalization and the environment that arrives at an array of competing conclusions about whether, when, how, and why trade liberalization can be compatible with the goal of environmental protection (for an overview, see Chapter 24 in this volume). Opinion is divided, for example, over whether the lowering of trade barriers enables a "trading up" of environmental standards as companies and countries seek to export their products to the richest regions in the global economy (Europe, North America, and East Asia), where environmental standards also tend to be highest – creating a positive incentive for upgrading (Vogel 1997). In its World Development Report for 1992, the World Bank claimed, more broadly, that: "Liberalized trade fosters greater efficiency and higher productivity and may actually reduce pollution by encouraging the growth of less polluting industries and the adoption and diffusion of cleaner technologies" (World Bank 1992: 67).

Critics claim, however, that more open markets allow investors to play countries off against one another in the pursuit of "pollution havens": zones where environmental regulation is lower, ignored, or unenforced. For years activists have berated the mining industry for its poor track record on environmental pollution, human rights violations, and displacement of indigenous peoples in its overseas operations (Evans *et al.* 2002). The oil industry too has been accused of double standards when it operates in developing countries (Okanta and Douglas 2001). The activities of firms such as Shell in the Niger Delta, Nigeria and Texaco in Ecuador have attracted global attention as a result of activist exposure and high-profile legal actions against those companies (Newell 2001; Garvey and Newell 2005; Frynas 2009). Copeland (2008: 68) claims, nevertheless:

> Fears that trade liberalization will cause an exodus of polluting industry to poorer countries with weak environmental policy appear to be unfounded. Although there is evidence that stringent environmental policy does reduce competitiveness in industries intensive in production-generated pollution, there is no evidence that it is the most important factor affecting trade and investment flows.

Rather than an active downgrading of regulations to attract mobile capital, many have observed a "chilling" or even a "deepfreeze" effect on countries' environmental regulations whereby reforms are not undertaken, or new policies either not introduced or not implemented for fear of deterring investors. Zarsky (2006: 395) finds, moreover, that:

While there is little evidence that MNCs select investment sites on the basis of lower environmental standards, it seems safe to conclude that many perform below standards of global best practice once they get there. They do not, in other words, actively seek out a "pollution haven" but, if the local environmental regulation is weak, create one through their operations.

The answer to whether trade liberalization is compatible with environmental protection in many ways seems to be "it depends." It depends on the country and region in question (how much power they have to negotiate terms with investors); the power of the corporation (how much the country needs their investment and what other rival locations are really viable); and the sector (how resource-intensive it is and what the global distribution of those resources is – how concentrated they are in particular jurisdictions).

What this narrow discussion on whether and under what conditions formally ascribed environmental standards are revised, lowered, or ignored altogether often overlooks, however, is the bigger and more fundamental question of the sustainability of the current organization of the world trading system, where principles of comparative advantage are held sacrosanct and the desirability of export-led growth strategies, which require concentrated production, often leading to monocultures and the subsequent loss of biodiversity and the intensive use of chemicals in agriculture to boost production, are left unquestioned. A broader developmental critique is the idea that export-led growth is often at the expense of meeting basic needs. For example, cash crops are grown in countries where people are starving because they generate more revenue, when that land could be better used to grow subsistence crops from which the poor are more likely to benefit directly. Icke (1990: 63–64) argues:

> [T]he poorest countries in the world grow cash crops on land that could be growing food for their own people. That's why Ethiopia was still exporting food at the height of the famine... in Ghana half their farming land is not growing food for the malnourished, but cocoa for western chocolate bars... 40 per cent of the food-growing land in Senegal is growing peanuts for western margarine... during the great drought in the Sahel the production of peanuts for export increased there while tens of thousands starved... in Colombia where malnutrition is common, fertile land is used to grow cut flowers for the rich in the west.

Such calls underpin social movement calls for "food sovereignty," for example (Borras *et al.* 2008), whereby greater efforts are made by producers to regain control over what they produce and on what terms. Increasingly powerful incentives exist for countries to be more self-sufficient in food, water, and energy given the uncertainties associated with securing these resources globally, where high dependency on oil has led powerful countries to war, and concerns about food and water provision have led to land grabs in parts of the developing world (Borras *et al.* 2011). Despite this, there is little near-term prospect that the global economy will be reoriented around shorter circuits of production and consumption, rather than driven by the economic rather than ecological logic of where things can be most cheaply produced.

Production

The debate about the impact of production very much mirrors that about trade and the environment, since agreements to liberalize trade enable transnational corporations to enter new markets. Those who take the view that new entrants bring cleaner technologies, employment opportunities, and revenue for governments and communities see this is a good thing, while critics suggest that trade agreements between unequal partners tilt the gains towards richer countries while opening up poorer ones to exploitation. As with the trade debate, however, the answer to the question of whether business can be a force for greening seems to be: "it depends." Patterns of "greening" appear to reflect the size of companies: their environmental footprint, their ability to demand change from their suppliers, and their exposure to pressure from consumers and shareholders. This means most focus remains on the activities of transnational corporations rather than small and medium-sized enterprises. There are also important regional and national differences, however, which reflect different regulatory cultures (Levy and Newell 2000; Utting 2002) and levels of integration within global markets (see the "trading up" argument above). There are also important sectoral differences, and while resource-intensive sectors such as oil and mining attract most attention, assumptions that "lighter" industries tend to pollute less need to be subject to critical scrutiny, as examples of toxic contamination from the computer and electronics industry make clear (Pellow and Park 2002).

It is easy enough to locate examples of corporate irresponsibility and environmental negligence (Karliner 1997; Madeley 1999; Okanta and Douglas 2001), just as organizations such as the Business Council for Sustainable Development and the World Bank can identify competing examples of business leadership on environmental issues based on the "business case" for sustainable development (Schmidheiny 1992; Holliday *et al.* 2002). The more interesting debate in a way is to identify the conditions in which it is possible to harness the power of businesses to improve their environmental performance as well as the measures necessary to deter and penalize corporate irresponsibility. The debate then moves on to which policy tools are most effective, efficient, and equitable: regulation or voluntary responses, partnerships or litigation. Unsurprisingly the evidence is mixed, but combinations of tools, tailored to particular national needs and sectoral circumstances, often end up being advocated (World Bank 2000; Newell 2001).

Again, what this framing of the relationship between production and the environment as being about the greening of existing businesses serves to obscure is a bigger debate about the viable and legitimate ways to generate wealth in a resource-constrained world. In other words, are the goals of producing more and more goods year on year, and seeking to create consumer demand for ever more products, compatible with deeper notions of sustainability? The debate comes back to basic notions such as what we mean by growth, progress, and wealth (see also Chapter 12 in this volume). Corporations are given charters and a license to operate based on an assumption that they serve a legitimate public need: they produce things we need, employ people, and pay taxes. But what if they fail to serve that need? Should governments use their powers more forcefully to revoke the charters and licenses of corporations that are found guilty of repeated social and environmental misconduct, as some people claim (Korten 1995)? Can we imagine an economy in which rather than just producing differently (in a more sustainable manner) businesses actually produce less, but where production is oriented towards meeting the basic needs that remain unmet for the majority of the world's people rather than fueling the overconsumption of 1% of the world's population? For obvious reasons, given the interests at stake, these broader issues and concerns struggle to get a hearing in the debate about the greening of business in a context of globalization.

Finance

There are numerous ways in which public and private finance interacts with the environment. Aid, debt, and private finance are among the vehicles that both fund environmental degradation through support to large-scale infrastructural projects, for example, and are also expected to pay for environmental measures (see also Chapters 25 and 28 in this volume).

Debt, for example, has been seen both as a driver of environmental degradation (George 1992) and as a potential opportunity to connect debt relief with conservation measures through debt-for-nature swaps (Jakobeit 1996). These schemes initially involved NGOs, such as leading conservation NGOs Conservation International, Nature Conservancy, and WWF, and then latterly governments. They involved soliciting donations for the purchase of a foreign debt title of a developing country at a discount on the face value from a commercial bank. The debt title is then converted into domestic currency, reducing the foreign debt, and freeing up an agreed fraction of the debt title to be used to finance a conservation project. US$128 million was raised for environmental projects, while developing countries reduced their stock of foreign debt by US$177 million (Jakobeit 1996: 134). The logic of the schemes continues today in relation to approaches for the payment for ecosystem services such as REDD (Reduced Emissions from Deforestation and Forest Degradation).

The aid lending of bilateral and multilateral institutions, meanwhile, has come under fire for its failure to take into account environmental impacts (Young 2002; Goldman 2005), even where key global actors such as the World Bank present themselves as leading players in financing climate mitigation efforts. For example a report by the Washington-based group the World Resources Institute (WRI) found that between 2005 and 2008 less than 30% of the World Bank's lending to the energy sector integrated climate considerations into project decision-making. As late as 2007, more than 50% of the Bank's US$1.8 billion energy-sector portfolio did not include climate-change considerations at all (WRI 2008).

Given the growing importance of private finance relative to public money, there has been an understandable shift in emphasis towards the ways in which private capital can be levered for environmental goals. The context in which this is most apparent is climate change, where it is already very clear that large amounts of private money will need to be raised if governments are to get even close to meeting the obligations they agreed to in Copenhagen and Cancun regarding the delivery of up to US$100 billion a year by 2020 through the Green Climate Fund. As well as being an important source of money for environmental goods, however, the more difficult and important issue perhaps is trying to green existing flows of private finance, which, in many cases, underpin environmentally damaging investments. The US$1 trillion that changes hands every day in private financial markets, mainly through currency speculation and investments in stocks and bonds, has an obvious, if disputed and difficult to quantify, effect on the global environment and patterns of resource use. As Helleiner (2011: 51) suggests: "If the global economy is to be made more environmentally sustainable, this powerful 'electronic herd' of global money will need to be steered in greener directions."

Globalization and the Governance of the Environment

The interface between trade, production, finance, and the environment described in the section above does not occur in a vacuum. The nature of the relationships described is mediated by institutions, power, and social relations. These then are critical to our understanding of the governance of the environment in a context of globalization: which interventions aimed at safeguarding the environment are likely and possible; who wins, who loses, how and why from the prevailing global distribution of benefits and burdens from existing (natural) resource allocations; and what spaces and opportunities might exist to contest these, an issue I address in the final section of the chapter.

This section looks not only then at the extent to which and the ways in which globalization is subject to new modes of environmental governance in the arena of trade, production, and finance, but also explores the way in which structures of global environmental governance reflect, embody, and are themselves part of globalization. This trend is traced, amongst other things, through the growth of private governance and regulation, the turn towards market-based solutions, and the growth of markets in environmental services for water, carbon, and forests.

First, trade. Here much of the literature has focused on actual and potential instances of conflicts between environmental regulations and trade rules (Vogel 1997; Lieberman and Gray 2008), as well as the broader governance arrangements in place within global and regional trade institutions such as the WTO and NAFTA, which enable or inhibit the adoption of environmental protection measures (Conca 2000; Audley *et al.* 2003; Newell 2007). Despite moves towards the acceptance of trade restrictive measures for environmental ends, where norms exist internationally for protection measures (Barkin 2008), environmentalists remain concerned about the ongoing resistance to acceptance of process-based environmental measures that would allow countries to discriminate between products on the basis of the extent to which goods have been produced in sustainable ways. As LeQuesne (1996: 81) notes:

> [F]rom an environmental point of view, there is no meaningful distinction to be drawn between environmental harm which is generated by a product, or the harm generated by its process and production methods.

Second, regarding production, the issue is less the power of existing global institutions for regulation, but rather the near-total absence of them. An international code of conduct to regulate the activities of TNCs has been on the international agenda since the 1970s. The UN Centre for TNCs (UNCTC) was set up in 1973, largely at the request of developing country governments, amid concern about the power of TNCs, but was unable to conclude negotiations on a code of conduct. This failure was explained by conflicts of interest between developed and developing countries and the opposition of the United States, in particular, and in 1993 the CTC was restructured to become the Commission on International Investment and Transnational Corporations, housed within the United Nations Conference on Trade and Development.

Guidelines and standards promoted by bodies such as the International Labour Organization (ILO) (Tripartite Declaration of Principles Concerning Multinational Enterprises (MNEs) and Social Policy) and the OECD (such as the OECD Guidelines on MNEs) are not widely known and rarely used, are entirely voluntary and without sanction, and are outdated, compared even with companies' own codes of conduct (MacLaren 2000). In the environmental domain the issue of TNC regulation was dropped from the UNCED agenda amid sustained efforts on the part of organizations such as the Business Council for Sustainable Development to present themselves as the solution to environmental problems (Schmidheiny 1992), and while Agenda 21 includes recommendations that affect TNCs, it does not take the form of a code of conduct. Instead of business regulation, the overriding preference has been to view business as a partner in promoting sustainability, as demonstrated with the type 2 partnerships agreed at the World Summit on Sustainable Development in 2002 (Bäckstrand 2008; Pattberg et al. 2012).

Concern remains, however, about the perceived imbalance between the rights and responsibilities of TNCs. The history of business regulation reveals an imbalance between the promotion and protection of investor rights over investor responsibilities (Muchlinski 1999): regulation for business rather than regulation of business (Newell 2001). Protection of investor rights can include provisions such as those contained in the NAFTA agreement, which permit companies to challenge governments and local authorities about restrictions on their activities and set a precedent for later Free Trade Agreements (FTAs), such as the Central America–Dominican Republic Free Trade Agreement (CAFTA-DR). It also includes the creation of bodies to address investor disputes such as the International Centre for Settlement of Investment Disputes (ICSID) of the World Bank, as well as generic investment treaties, of which by 2010 there were 5900 (UNCTAD 2010). Attempts not only to protect the exit and entry options of TNCs, but also confer upon them rights to challenge and reverse the public policies of sovereign governments in the ways noted above, have provoked particular ire. These trends provide evidence of what Gill calls the "new constitutionalism," which refers to efforts

> to develop a politico-legal framework for the reconstitution of capital on a world scale and thus the intensification of market forms of discipline . . . The new constitutionalism seeks to reinforce a process whereby government policies are increasingly accountable to (international) capital and thus to market forces (1995: 78–79).

Third, regarding finance, even more so than the domain of production, the issue is one of un-governance and active neglect. Despite repeated calls, not least in the wake of the latest financial crisis to infect the world from 2008, to reregulate aspects of finance capital or impose taxes on short-term and volatile financial transactions, and recognition even within the neoliberal heartlands of the World Bank and IMF that the use of capital controls may, on occasion, be appropriate, global finance remains the least regulated pillar of the global economy. The extent to which this is so depends on which aspect of finance we are talking about. There have been growing pressures on bilateral agencies and multilateral development banks to screen their lending for potential environmental impacts, often coming from Washington-based NGOs as well as social movements in the global South (Fox and Brown 1998; Goldman 2005), which have produced an array of reforms, though not ones their critics would consider adequate. There is also increasing focus on export credit agencies that provide public money to private firms to encourage investment in overseas infrastructural projects. This credit is provided in the form of government-backed loans, investment guarantees, and risk insurance. Official Export Credit and Investment Insurance Agencies (ECAs) have become the largest source of public international finance, accounting for 24% of all developing country debt and 56% of the debt owed to official governmental agencies (ECA Watch 2011). What is most significant perhaps from an environmental point of view is the support that ECAs provide to high-risk ventures often associated with resource extraction in environmentally sensitive and socially vulnerable areas of the world. Most ECAs are not subject to social and environmental standards or assessment procedures, and operate in a highly secretive manner, rarely disclosing information about the projects they finance or evaluating the impacts of such projects.

In spite of (or perhaps because of) the weak regulation of most aspects of finance, there has been a series of initiatives to govern finance, led either by public international institutions such as the United Nations Environment Programme (UNEP) or the International Finance Corporation (IFC) of the World Bank or by businesses themselves, such as the Carbon and Water Disclosure Projects.

Contesting Globalization

This section shows how some of the ecological impacts associated with the different dimensions of globalization discussed in the first section, as well as the structures of governance and un-governance of globalization explored in the second, are being contested by a range of environmental NGOs and social movements concerned about the environmental and social impacts of globalization and engaged in campaigns for institutional reform or resistance to attempts to "privatize" and commodify natural resources. It provides examples of prominent campaigns to green trade, production, and finance. These help to show that globalization is not a uniform or linear process, nor is it apolitical in its outcomes or neutral in terms of the interests it serves.

Despite general consensus among many environmental activists that the current organization of the global economy is unsustainable, views differ about why this is and, therefore, which strategies are most appropriate to bring about change. There has been a series of campaigns targeted at trade institutions globally and regionally, aimed either at securing environmental side agreements to trade treaties, as in the case of the North American Free Trade Area, highlighting the environmental consequences of attempts to liberalize key sectors such as energy and agriculture, as with the Free Trade Area of the Americas agreement currently under negotiation, or challenging the privatization of resources such as water through agreements such as the World Trade Organization (WTO) General Agreement on Trade in Services (Newell 2007; Icaza et al. 2010). There have also been debates about whether environmental standards should be incorporated into trade agreements to establish basic floors above which investors should operate, or whether the more critical battle is to keep the WTO out of environmental policy such that its rules do not trump those of other multilateral environmental agreements (MEAs) or countries' national regulations that appear to impede trade. As noted above, however, more radical groups also question the sustainability of the model of trade liberalization being promoted in terms, for example, of the intensification of resource use that is often required, or the impacts associated with transporting more goods over longer distances around the world (NEF 2003; Acción Ecológica 2004).

Around production there has been a rising tide of what is sometimes referred to as "civil regulation": civil-society-based "regulation" of the corporate sector (Bendell 2000; Newell 2000; Zadek 2001). It takes as its point of departure and justification for action the lack of effective regulation by states unwilling or unable (or both) to address corporate irresponsibility, or by corporations themselves, who may promote acts of corporate responsibility but have few incentives or collective means to confront corporate wrong-doing. These strategies respond to a perceived "governance deficit," in that the global power of TNCs is not adequately matched by existing regulatory instruments. The term incorporates a range of "liberal" strategies of engagement with business that seek to work with and through the market to achieve reform, examples being the negotiation of codes of conduct, shareholder activism, and the creation of standards of certification. Examples of liberal strategies include the creation of certification schemes in the forestry and marine sectors or project-specific collaborations between companies such as McDonald's and the environmental NGO Environmental Defense (Murphy and Bendell 1997). More "critical" modes of engagement, meanwhile, aim to contest and restrict corporate power. These include the organization of consumer boycotts (such as those organized against Exxon or Shell), the creation of "watchdog" groups that monitor the activities of TNCs (such as Corporate Watch or Oilwatch), as well as traditional protest strategies of naming and shaming and resistance (Newell 2001).

Finally, activism around finance has taken a number of forms, from issue-specific protests regarding particular projects sponsored by the World Bank, for example, through to general campaigns for the reform of international financial institutions (Fox and Brown 1998; Edwards and Gaventa 2001; Scholte and Schnabel 2002) or ECAs (ECA Watch 2011). With respect to private finance, we see among environmental groups a divide between those that are interested in locating and activating levers that exist within the current financial system that can be used to engineer positive change, and those that engage in a more full-frontal attack on the basic principles and means by which the financial system operates. This strategic difference separates groups such as BankTrack (2011), engaged in monitoring and exposing acts of environmental negligence enabled by the support of banks (such as the involvement of commercial banks sponsoring damaging environmental investments in projects like the tar-oil sands in Canada), from coalitions of activist investors aiming to work with financial investors to sensitize them to the importance of environmental risks to their investments, to disclose their investments, and to use their power to disinvest from polluting activities and invest in sustainable projects and sectors of the economy. An example of the latter would be the Interfaith Centre for Corporate Responsibility, a coalition of 275 faith-based institutional investors that use their financial muscle to hold firms to account for their performance on climate change (Newell 2008).

Conclusions

I have argued in this chapter that while contemporary globalization has many historical precedents and essentially derives from established patterns and tendencies
of (uneven) capitalist development, the way in which specific policy tools such as
trade agreements and financial deregulation and corporate strategies such as globalized forms of production through networks and vertical integration have removed
barriers to accumulation has generated a specific set of environmental challenges.

Some of these relate to the spatial and temporal fixes employed to simultaneously
shift responsibility and benefit from ecologically uneven exchange. We observe this
in relation to the offshoring or outsourcing of the most resource-intensive parts of the
production process to the developing world (where labor costs are lower and where
in the case of climate change emissions reductions obligations do not yet apply) and
in the export of hazardous materials or the use of lower environmental and worker
health standards in poorer countries. We also see it in innovations such as carbon
offsets that pay poorer countries to reduce emissions on the part of richer countries:
sold as a win-win situation that generates capital for poorer countries while relieving
pressure on capitalists in the global North to reduce their own emissions.

Others relate to the political challenges of holding powerful corporate actors to
account for their social and environmental performance when the distance between
sites of production and consumption is so large, or when their power outstrips
their responsibilities. A mixture of private certification schemes as well as watchdog
activism from groups such as Corporate Watch has sought to address this potential
governance gap and to expose the use of double standards, but can clearly go only
so far. It raises the issue of whether, and if so how, universal, even if very minimal,
standards of environmental conduct can be applied to corporations wherever in the
world they operate. This is certainly the ambition behind repeated calls from many
civil society organizations for a new UN legally binding corporate accountability
convention to provide clearer and more enforceable forms of protection for workers
and their environment than are currently afforded by the existing patchwork of
voluntary agreements, self-regulation, and weak international law.

The extent and nature of these challenges reflect different understandings of where
the problems lie, and lead to different ideas about what the solutions should be. For
market liberals, those that take a favorable view of the ability of markets to deliver
positive environmental outcomes, the issue is pricing: internalizing the externalities
of environmental pollution that producers are currently able to pass on to society
(Clapp and Dauvergne 2011). Nicholas Stern's (2006) claim that climate change
represents the world's greatest market failure then becomes a call to introduce more
wide-ranging carbon taxes or to strengthen carbon markets that put a price on
carbon and incentivize its reduction. For those who place more faith in institutions,
such measures need to be complemented by strong institutions that establish clear
rules, coordinate cooperative outcomes among states, and produce international law
that imposes obligations on states to address environmental problems. In this view,
if more effective treaties could be negotiated for forests, climate change, and water,
as they have been for ozone depletion and to a lesser extent the trade in endangered
species, this would go a long way to setting in place responsible collective stewardship
in an era of globalization (Young 1998). For others adopting a more critical view,

neither market reforms nor institutional innovation alone will go far enough in confronting the basic reality that a global economy organized around notions of endless year-on-year increases in growth and increases in the throughput of natural resources is fundamentally incompatible with any serious notion of sustainability. In other words, it cannot sustain such patterns of production and consumption indefinitely in a world where there are limits to growth (Meadows *et al.* 1972). Where this leads in terms of solutions is deglobalization: a conscious attempt to de-link economies through a greater emphasis on self-sufficiency and the prioritization of meeting basic needs rather than creating and then serving manufactured "wants" and "desires" (Trainer 1996).

It should be obvious which reading of globalization and its relationship to the environment currently prevails, despite the efforts described in this chapter's section on "Contesting Globalization" by a growing array of activists to question the orientation and organization of the current global economy. A combination of material, institutional, and discursive power coheres around the idea that environmental problems occur because there is a lack of something: growth, technology, cleaner production, or capital. The solution then becomes an opportunity to accumulate capital by providing these things, generating demand for more goods and services and creating entrepreneurs for environmental services. The implication of those very things in causing problems in the first place is then airbrushed from the picture.

It follows then that the implications for policy of the body of research summarized here depend on which research is considered to be credible, reliable, and applicable to the policy needs of a diversity of policy-makers who themselves do not agree in many instances on which aspects of globalization require reform, or the extent to which abundant evidence of worsening environmental conditions relates to the current organization of the global economy. There may nevertheless be some compatibilities between these approaches, or at least in the idea that no one set of strategies is likely to deliver the scale or speed of change required. Most evidence points to the need for market-based mechanisms to be embedded within strong institutions and rule-based frameworks if they are to deliver effective and equitable outcomes. Likewise, for them to gain traction with investors, prices need to be high and scarcity is a precondition for that. That takes us back to clear targets that drive interest in reducing pollution in the first place (Newell 2012). At the same time, resistance to market mechanisms and the commodification of everything often has the effect of creating problems and legitimacy crises for market actors that they then have to address through improved standards and governance in order to maintain the credibility of the market as a whole. We see these dynamics clearly at work in carbon markets, where doubts about the "additionality" and authenticity of claimed emissions savings, as well as about the development benefits they claim to deliver, have given rise to a series of voluntary standards aiming to address these issues as well as a greater use of third-party verification and other tools (Paterson 2009; Newell and Paterson 2010).

That there can be mutually reinforcing dynamics between regulation, markets, and resistance is not the same as saying that these can be relied upon to adequately detect and address the range of environmental problems we face, which are clearly about more than inefficient markets and rogue traders within them. Since environmental problems derive from everyday practices of production and consumption in every domain of human life, some of which citizens have direct control over themselves, but many over which they do not, change is clearly required at all levels from personal behavior to structural change in the organization of the global economy. This is so even for those who firmly believe that capitalism can grow its way out of environmental crisis through innovation, finance, and technology, because ultimately every aspect of the global economy depends on a sustainable supply of resources to preserve itself. The consequences of global market forces being allowed to reign without serious social or ecological restraint were reflected upon with hindsight and foresight by Karl Polanyi almost 70 years ago:

> To allow the market mechanism to be sole director of the fate of human beings and their natural environment... would result in the demolition of society... Nature would be reduced to its elements, neighbourhoods and landscapes defiled, rivers polluted, military safety jeopardized, the power to produce food and raw materials destroyed... [T]he commodity fiction disregarded the fact that leaving the fate of soil and people to the market would be tantamount to annihilating them (1944: 73).

References

Acción Ecológica. 2004. "Acción Ecológica," http://www.accionecologica.org/alca.htm (accessed October 5, 2004).

Audley, John, Demetrios G. Papademetriou, Sandra Polaski, and Scott Vaughan. 2003. *NAFTA's Promise and Reality. Lessons from Mexico for the Hemisphere*. Washington, DC: Carnegie Endowment for International Peace.

Bäckstrand, Karin. 2008. "Accountability of Networked Climate Governance: The Rise of Transnational Climate Partnerships." *Global Environmental Politics*, 8(3): 74–102.

BankTrack. 2011. "About BankTrack," www.banktrack.org/show/pages/about_banktrack (accessed July 3, 2012).

Barkin, Samuel. 2008. "Trade and Environment Institutions." In *Handbook on Trade and the Environment*, ed. Kevin Gallagher, 318–326. Cheltenham: Edward Elgar.

Bendell, Jem, ed. 2000. *Terms for Endearment: Business, NGOs and Sustainable Development*. Sheffield: Greenleaf Publishing.

Borras, Jun, Marc Edelman, and Cristóbal Kay, eds. 2008. *Transnational Agrarian Movements Confronting Globalization*. Oxford: Wiley-Blackwell.

Borras, Jun, Ruth Hall, Ian Scoones *et al.* 2011. "Towards a Better Understanding of Land Grabbing." *Journal of Peasant Studies*, 38(2): 209–217.

Clapp, Jennifer. 2001. *Toxic Exports: The Transfer of Hazardous Wastes from Rich to Poor Countries*. Ithaca, NY: Cornell University Press.

Clapp, Jennifer and Peter Dauvergne. 2011. *Paths to a Green World: The Political Economy of the Global Environment*, 2nd edn. Cambridge, MA: MIT Press.

Conca, Ken. 2000. "The WTO and the Undermining of Global Environmental Governance." *Review of International Political Economy*, 7(3): 484–494.

Copeland, Brian. 2008. "The Pollution Haven Hypothesis." In *Handbook on Trade and the Environment*, ed. Kevin Gallagher, 60–71. Cheltenham: Edward Elgar.

ECA Watch. 2011. "Jakarta Declaration for Reform of Official Export Credit and Investment Insurance Agencies," www.eca-watch.org/goals/jakartadec.html (accessed January 27, 2012).

Edwards, Michael and John Gaventa, eds. 2001. *Global Citizen Action*. Boulder, CO: Lynne Rienner.

Evans, Geoff, James Goodman, and Nina Lansbury. 2002. *Moving Mountains: Communities Confront Mining and Globalization*. London: Zed Books.

Fox, Jonathan and David Brown, eds. 1998. *The Struggle for Accountability: The World Bank, NGOs, and Grassroots Movements*. Cambridge MA: MIT Press.

Frynas, George. 2009. *Beyond Corporate Social Responsibility: Oil Multinationals and Social Challenges*. Cambridge: Cambridge University Press.

Garvey, Niamh and Peter Newell. 2005. "Corporate Accountability to the Poor? Assessing the Effectiveness of Community-Based Strategies." *Development in Practice*, 15(3–4): 389–405.

George, Susan. 1992. *The Debt Boomerang*. London: Pluto Press.

Gill, Stephen. 1995. "Globalization, Market Civilisation and Disciplinary Neo-liberalism." *Millennium: Journal of International Studies*, 24(3): 399–423.

Goldman, Michael. 2005. *Imperial Nature: The World Bank and Struggles for Social Justice in an Age of Globalization*. New Haven, CT: Yale University Press.

Harvey, David. 2003. *The New Imperialism*. Oxford: Oxford University Press.

Held, David, Anthony McGrew, David Goldblatt, and Jonathan Perraton. 1999. *Global Transformations: Politics, Economics, and Culture*. Stanford, CA: Stanford University Press.

Helleiner, Eric. 2011. "The Greening of Global Financial Markets?" *Global Environmental Politics*, 11(2): 51–53.

Holliday, Charles, Stephen Schmidheiny, and Phillip Watts. 2002. *Walking the Talk: The Business Case for Sustainable Development*. Sheffield: Greenleaf Publishing.

Icaza, Rosalba, Peter Newell, and Marcelo Saguier. 2010. "Citizenship and Trade Governance in the Americas." In *Globalizing Citizens: New Dynamics of Inclusion and Exclusion*, ed. John Gaventa and Rajesh Tandon, 163–185. London: Zed Books.

Icke, David. 1990. *It Doesn't Have to Be like This: Green Politics Explained*. London: Green Print.

Jakobeit, Carl. 1996. "Non-state Actors Leading the Way: Debt-for-Nature Swaps." In *Institutions for Environmental Aid*, ed. Robert Keohane and Marc Levy, 127–167. Cambridge, MA: MIT Press.

Jones, Andrew. 2006. *Dictionary of Globalization*. Cambridge: Polity.

Karliner, Joshua. 1997. *The Corporate Planet: Ecology and Politics in the Age of Globalization*. San Francisco: Sierra Club Books.

Khor, Martin. 1995. "Address to the International Forum on Globalization." New York City, November.

Korten, David. 1995. *When Corporations Rule the World*. West Hartford, CT: Kumarian Press.

LeQuesne, Caroline. 1996. *Reforming World Trade: The Social and Environmental Priorities*. Oxford: Oxfam Publishing.

Levy, David and Peter Newell. 2000. "Oceans Apart? Comparing Business Responses to the Environment in Europe and North America." *Environment*, 42(9): 8–20.

Lieberman, Sarah and Tim Gray. 2008. "The WTO's Report on the EU's Moratorium on Biotech Products: The Wisdom of the US Challenge to the EU in the WTO." *Global Environmental Politics*, 8(1): 33–52.

MacLaren, Duncan. 2000. "The OECD's Revised 'Guidelines for Multinational Enterprises': A Step towards Corporate Accountability?" February. London: Friends of the Earth.

Madeley, John. 1999. *Big Business, Poor Peoples*. London: Zed Books.

Meadows, Donella, Dennis Meadows, Jorgen Randers, and William Behrens. 1972. *The Limits to Growth*. London: Pan Books.

Muchlinksi, Peter. 1999. "A Brief History of Business Regulation." In *Regulating International Business: Beyond Liberalization*, ed. Sol Picciotto and Ruth Mayne, 47–60. Basingstoke: Macmillan, in association with Oxfam.

Murphy, David and Jem Bendell. 1997. *In the Company of Partners*. Bristol: Policy Press.

NEF (New Economics Foundation). 2003. *Collision Course: Free Trade's Free Ride on the Global Climate*. London: New Economics Foundation.

Newell, Peter. 2000. "Environmental NGOs and Globalization: The Governance of TNCs." In *Global Social Movements*, ed. Robin Cohen and Shirin Rai, 117–134. London: Athlone Press.

Newell, Peter. 2001. "Managing Multinationals: The Governance of Investment for the Environment." *Journal of International Development*, 13(7): 907–919.

Newell, Peter. 2005. "Race, Class and the Global Politics of Environmental Inequality." *Global Environmental Politics*, 5(3): 70–94.

Newell, Peter. 2007. "Trade and Environmental Justice in Latin America." *New Political Economy*, 12(2): 237–259.

Newell, Peter. 2008. "Civil Society, Corporate Accountability and the Politics of Climate Change." *Global Environmental Politics*, 8(3): 124–155.

Newell, Peter. 2012. "Of Markets and Madness: Whose Clean Development Will Prevail at Rio+20?" *Journal of Environment & Development*, 21: 40–43.

Newell, Peter and Matthew Paterson. 2010. *Climate Capitalism: Global Warming and the Transformation of the Global Economy*. Cambridge: Cambridge University Press.

Okanta, Ike and Oronto Douglas. 2001. *Where Vultures Feast: Shell, Human Rights, and Oil in the Niger Delta*. New York: Sierra Club Books.

Paterson, Matthew. 2009. "Resistance Makes Carbon Markets." In *Upsetting the Offset: The Political Economy of Carbon Markets*, ed. Steffen Böhm and Siddhartha Dabhi, 244–255. London: May Fly Books.

Pattberg, Philipp, Frank Biermann, Ayşem Mert, and Sander Chan, eds. 2012. *Public–Private Partnerships for Sustainable Development: Emergence, Influence, and Legitimacy*. Cheltenham: Edward Elgar.

Pellow, David and Lisa Sun-Hee Park. 2002. *The Silicon Valley of Dreams: Environmental Injustice, Immigrant Workers, and the High-Tech Global Economy*. New York: New York University Press.

Petras, James and Henry Veltmeyer. 2001. *Globalization Unmasked: Imperialism in the 21st Century*. Delhi: Madhyam Books.

Polanyi, Karl. 1944. *The Great Transformation*. Boston, MA: Beacon Press [reissued 1980].

Roberts, J. Timmons and Bradley Parks. 2008. "Fuelling Injustice: Globalization, Ecologically Unequal Exchange and Climate Change." In *The Globalization of Environmental Crises*, ed. Jan Ooshthoek and Barry Gills, 169–187. Abingdon: Routledge.

Schmidheiny, Stephen. 1992. *Changing Course*. Cambridge, MA: MIT Press.

Scholte, Jan Aart. 2000. *Globalization: A Critical Introduction*. Basingstoke: Palgrave Macmillan.

Scholte, Jan Aart and Annabel Schnabel, eds. 2002. *Civil Society and Global Finance*. London: Routledge.

Stern, Nicholas. 2006. *The Stern Review on the Economics of Climate Change*. London: HM Treasury.

Trainer, Ted. 1996. *Towards a Sustainable Economy: The Need for Fundamental Change*. Oxford: Jon Carpenter.

UNCTAD (United Nations Conference on Trade and Development). 2010. *World Investment Report: Investing in a Low Carbon Economy*. Geneva: UNCTAD.

Utting, Peter. 2002. *The Greening of Business in Developing Countries*. London: Zed Books.

Vogel, David. 1997. *Trading Up: Consumer and Environmental Regulation in the Global Economy*, 2nd edn. Cambridge, MA: Harvard University Press.

World Bank. 1992. *World Development Report 1992: Development and the Environment*. New York: Oxford University Press.

World Bank. 2000. *Greening Industry: New Roles for Communities, Markets and Governments*. New York: Oxford University Press.

World Bank. 2011. *The Changing Wealth of Nations: Measuring Sustainable Development in the New Millennium*. Washington, DC: World Bank

WRI (World Resources Institute). 2008. *Correcting the World's Greatest Market Failure: Climate Change and Multilateral Development Banks.* Washington, DC: WRI, www.wri.org/publication/correcting-the-worlds-greatest-market-failure (accessed October 20, 2012).

Young, Oran. 1998. *Global Governance: Learning Lessons from the Environmental Experience.* Cambridge, MA: MIT Press.

Young, Zoe. 2002. *A New Green Order? The World Bank and the Politics of the Global Environment Facility.* London: Pluto Press.

Zadek, Simon. 2001. *The Civil Corporation: The New Economy of Corporate Citizenship.* London: Earthscan.

Zarsky, Lyuba. 2006. "From Regulatory Chill to Deepfreeze?" *International Environmental Agreements: Politics, Law and Economics,* 6(4): 395–399.

Chapter 23

Private Regulation in Global Environmental Governance

Graeme Auld and Lars H. Gulbrandsen

Introduction

Over the past decades, private regulations – instances where non-state actors set rules to govern their behavior and/or the behavior of others – have emerged as a vibrant source of global environmental governance. They are diverse in form. From the individual actions of companies to enforce social and environmental performance requirements within their supply chains to industry-wide codes of conduct or multi-stakeholder bodies setting environmental and social standards with third-party compliance audits, these private regulatory efforts are governing the practices of global production, distribution, and consumption.

While reviewing the broad landscape of private initiatives, this chapter focuses on social and environmental certification initiatives. This form of private regulation exists in many sectors, but has particularly deep roots in natural resource management and agriculture. Hence, we use the forest, fishery, and agricultural sectors as focal points to draw attention to broader trends. The chapter discusses demand and supply factors which contributed to the emergence of these initiatives. It then turns attention to the consequences of these processes for both the institutional evolution of private regulators and the effects these initiatives have for problem amelioration. Though a rich body of research attends to the emergence and, to a lesser degree, the evolutionary questions, it is the on-the-ground impacts that current work is increasingly assessing. Taking the insights from these lead sectors, we review the private regulatory activities on climate change, drawing parallels and noting differences that emerge.

We proceed in four parts. First, the chapter details three overlapping analytical perspectives on private regulation. Second, it presents factors associated with the emergence of private regulation in the focal sectors, and it reviews the evolving

The Handbook of Global Climate and Environment Policy, First Edition. Edited by Robert Falkner.
© 2013 John Wiley & Sons, Ltd. Published 2013 by John Wiley & Sons, Ltd.

institutional and problem-oriented effects of these initiatives. Next it turns to climate change, where we present a preliminary review of the emergence and effects of a wide array of private regulatory efforts. The final section builds from the review to discuss options for global public policy.

Private Regulation

While International Relations (IR) theories for many years focused on states and tended to dismiss corporations, NGOs, and civil society networks as insignificant actors in world politics, this has notably changed. Cutler *et al.*'s (1999) edited volume on private authority in fields from technical standards to credit rating provided an important foundation for a now-sizable research agenda built on the premise that IR underestimated the role of private actors. These actors, Cutler and colleagues maintained, are "increasingly engaged in authoritative decision-making that was previously the prerogative of sovereign states" (1999: 16).

Particularly relevant to this chapter, several studies have investigated the emergence and proliferation of voluntary codes of conduct, standards, and governance programs, conceptualized as "private authority" (Cutler *et al.* 1999; Hall and Biersteker 2002), "civil regulation" (Bendell 2000; Zadek 2001), and "regulatory standard-setting schemes" (Abbott and Snidal 2009). These studies, following Rosenau and Czempiel (1992), generally agree that we may talk of a shift from government to *governance* in global environmental politics. Across issue areas, we now observe a diversity of collaborative partnerships between states and non-state actors, shared rule-making authority, and private authority supplementing and sometimes supplanting traditional multilateral treaty-making.

The widespread emergence of these initiatives has captured the attention of scholars from various traditions. We outline three overlapping analytic perspectives that capture important threads of existing work. From here, we proceed with an institutional perspective, seeking to outline common features of extant initiatives.

Three Analytic Perspectives

The first analytic perspective – a governance or institutional perspective – comprises work focused on the rules, procedures, and bureaucracies created by actors other than states to regulate, steer, or nudge activities in particular directions and away from others (Meidinger 2006; Abbott and Snidal 2009). As Dingwerth and Pattberg (2006) point out, this strand has an explanatory and prescriptive agenda. Networked governance, greater civil society activism, or business self-regulation are seen as solutions to the complex social and environmental challenges posed by our globalized world (Conroy 2006), or private regulation may be transformative if it develops along particular paths (Bernstein and Cashore 2007; Auld *et al.* 2009; Auld *et al.* 2010).

A second perspective takes businesses – and sometimes business associations – as the unit of analysis. One vein, grounded in neo-institutionalism and business strategy, examines how firms act individually and collectively to regulate markets without direct state involvement (King and Lenox 2000). Businesses adopt codes of conduct and voluntary standards to protect their reputations, provide credible information to consumers, and gain competitive advantages (Reinhardt 2000; Prakash and Potoski

2006). Peer pressure from within the industry may facilitate adoption of standards because environmental and social reputations often reflect on the entire industry, not just individual companies (Gunningham and Rees 1997; Gulbrandsen 2006). The collective action problem is, in many cases, lessened because companies join associations and are able to monitor one another's behavior. An industry response of this kind occurred when the US chemical industry developed the Responsible Care code of conduct following the 1984 Bhopal disaster in India (Prakash 2000). In other cases, individual companies or smaller groups of companies seek to protect their reputation or reap market benefits by making claims about social and environmental responsibility (Auld *et al.* 2008a). Certification can credibly verify such claims, separating responsible companies from free riders. According to a club theory approach, certification systems provide excludable reputation benefits which are non-rival among participating companies (Prakash and Potoski 2006). Companies may also adopt voluntary standards to prevent enactment of more demanding regulations, hoping that adherence to voluntary standards will preempt or soften present and future public regulations (Segerson and Miceli 1998; Vogel 2005). But adoption of standards is just as often caused by pressure from activists and advocacy coalitions that target companies through coordinated campaigns (Cashore *et al.* 2004; Gulbrandsen 2006; Bartley 2007b).

Another vein of this research adopts a critical perspective to understand how corporations exert political power in international environmental politics, both inside intergovernmental negotiations and outside the traditional arena for multilateral treaty negotiations (Levy and Newell 2005; Fuchs 2007; Falkner 2008). Scholarship from this perspective also seeks to subvert and counter the claims of certain scholars in the governance and institutional strand. Rather than noting the transformative potential of private regulation in the social and environmental field, these new initiatives are often, on balance, seen to *reinforce* neoliberal globalization. For instance, Guthman (2007) examines agro-food labels to assess how likely they represent the vanguard of a Polanyian countermovement to the negative effects of neoliberal globalization. She finds that the project of creating a label is in fact consistent with many facets of the neoliberal project. Klooster (2010) investigates similar concerns with forest-sector private regulation, while other studies note instances where private regulations reinforce, rather than work against, distributional inequity (Mutersbaugh 2005; Taylor 2005) and power imbalances among domestic interests (Ponte 2008) and undermine the enforcement of an existing or an incipient regulatory regime (Besky 2011).

A third perspective, related to the critical business scholarship, comprises work that applies various sociological lenses to understand the processes by which private authority emerges and spreads. Bartley (2003), for instance, shows how neoliberal ideas and institutions served as preconditions for the emergence of certification systems in the forest and labor sectors. Bartley and Smith (2007) and Dingwerth and Pattberg (2009) have used the concept of an organizational field to understand the emerging constellations of transnational private regulators. Isomorphic pressures have also been used to explain the commonalities and differences in the programmatic form private regulation takes in different settings (Gulbrandsen 2008). Overdevest's (2005, 2010) work in the forest sector examined the convergence of standards between programs as a product of competitive benchmarking. This analysis has informed subsequent work on the emergence of experimental governance whereby various interventions are treated as experiments to feed into reasoned public debate with the aim of collectively learning something from each (Overdevest *et al.* 2010). The application of these ideas is beginning to take root in several areas where private regulation is burgeoning (for an application to climate governance, see Hoffmann 2011).

Characterizing Private Regulation

What, then, are the key characteristics of private regulation in global environmental governance? Numerous classification systems exist, each with different theoretical assumptions about important design features. The inclusiveness of private regulatory initiatives, for example, has received considerable attention, with scholars noting different degrees of multi-stakeholderism across programs that have implications for the legitimacy, accountability, and stringency of these initiatives (Gulbrandsen 2004, 2010; Fransen and Kolk 2007; Raynolds *et al.* 2007; Tollefson *et al.* 2008). Others home in on the systems of monitoring, particularly whether an initiative involves independent verification in the form of third-party audits and sanctions for those failing to meet program standards (King and Lenox 2000; Prakash and Potoski 2006). IR and policy scholars have, in particular, given notable attention to the relationship between private regulators and the state (Cashore 2002; Falkner 2003; Börzel and Risse 2005). According to Cutler and colleagues (1999), three features of "private authority" render their rule-making authority distinct. First, those subject to the private rules must accept them as legitimate. Second, there must be a high degree of compliance with rules and decisions being made by private actors. Third, non-state actors "must be empowered either explicitly or *implicitly* by governments and international organizations" granting them the authority to make decisions for others (Cutler *et al.* 1999: 19; emphasis in original). This latter assumption has been questioned by Cashore (2002), who argues that it is precisely the lack of government delegation of rule-making authority which is one of the defining features of market-based certification programs, termed "non-state market-driven" governance. States may influence non-state governance systems in several ways, but they do not use their sovereign authority to require compliance with rules. Indeed, the main claim of much of the literature on private authority is not that states do not contribute to the governance processes, but that private regulatory programs do not derive rule-making authority from governments (Cashore 2002; Bartley 2007b).

Following the governance and institutional perspective, private regulation can be defined as voluntary standards, rules, and practices that are created by non-state actors and govern the behavior of participants in an issue area. There are two aspects of this definition that should be noted. First, since private regulatory programs are created by non-state actors there is no use of legal coercion to force companies to adopt the standards. As participation is voluntary, operators have to be convinced that the benefits of standard adoption will outweigh the costs, or that standard adoption is appropriate and justified in terms of their commitment to corporate social responsibility (Cashore *et al.* 2007; Gulbrandsen 2010). Second, the fact that private regulatory programs govern the behavior of participants means they require companies to undertake behavioral changes they would otherwise not be required to implement (Cashore *et al.* 2004).

Beyond this minimum definition, there is significant variation in the type, design, and requirements of private regulatory programs. Requirements vary from disclosure rules to mandatory prohibitions on certain activities (e.g. no use of genetically modified organisms). When disclosure is required, the accuracy or value of the information may not be verified, nor is it necessarily tied to required behavioral changes. Examples include information disclosure initiatives such as the Global Reporting Initiative, the Carbon Disclosure Project, and the Forest Footprint Disclosure Project. In these cases, the hope is that information disclosure will enable stakeholders to demand certain performance levels, compare performance across companies, and exert pressure on non-disclosing companies and poor performers (Kolk *et al.* 2008).But without verification or required behavioral benchmarks, they represent a "soft" mode of governance, leaving it up to civil society stakeholders to demand better environmental performance levels based on the disclosed information.

Other programs, such as the Forest Stewardship Council (FSC), the Marine Stewardship Council (MSC), and Fairtrade Labelling Organizations (FLO) International, have created environmental and/or social performance standards and mechanisms to verify compliance with the standards. Compliance verification is usually done by independent auditors. Operators that pass the inspection audit are awarded a certificate attesting to compliance. Although most environmental certification programs involve on-the-ground inspections, the number and types of issues addressed by auditors vary. Depending on the seriousness of a compliance failure, but also on the program's rules, penalties range from minor to major requests to correct practices to revocation of the certificate.

Though varied, most private regulatory programs have membership rules and/or stakeholder bodies. Membership rules sometimes favor industry and business interests; in other cases the rules balance decision-making powers across a broader array of stakeholders. The powers given to stakeholder bodies also vary. Some programs grant ultimate decision-making authority to their membership as a whole. Other programs have granted this authority a board of directors while giving stakeholders an advisory role.

Private Regulation in Practice

Drawing on the analytic perspectives noted above, this section reviews what we know about the emergence of private regulation, the evolution of various initiatives, and their current and future impacts on environmental and social problems. It focuses on prominent forest, fisheries, and agriculture programs, which operate in sectors where the development of private regulation has been vibrant and extensive.

Emergence of Private Regulation

Demand- and supply-side factors underlie the emergence of numerous private regulatory programs in the focal sectors. Two demand-side factors stand out: public policy failures and balancing consistency against demands for choice. Each serves as motive for the creation of private regulation. On the supply side, institutional entrepreneurs are critical.

Demand for Private Regulation   Rather than operating alone and in isolation from governmental processes, private regulatory programs often emerge in response to governance failures or inadequate public regulations. With forest certification, for instance, a series of shortcomings with intergovernmental processes served as an impetus for NGOs and businesses to form private regulatory alternatives. The limited effects of the International Tropical Timber Organization – created in 1986 to implement the first International Tropical Timber Agreement – on tropical defor-estation and the failed attempts to produce a binding forest convention at the 1992 UN Conference on Environment and Development nurtured demand for alternative solutions (Humphreys 1996). One outcome was the 1993 launch of the FSC, a part-nership between environmental and social NGOs, retailers, manufacturers, forest companies, and professional certification bodies. By circumventing intergovernmen-tal forest policy negotiations, the hope was that forest certification would offer an alternative, fast-track route to improved global forest practices (Elliott 2000; see also Chapter 5 in this volume).

Challenges with ocean governance also served as a motivator for the creation of private fisheries regulation (Gulbrandsen 2005). Multilateral fisheries agreements are more extensive and have greater teeth than those in the forest sector. Still, in the early 1990s, the dramatic collapse of the cod fishery off the east coast of Canada and similar concerns with stocks in the North Sea helped motivate the creation of a private certification system, the MSC (Auld 2009). In 1996, WWF teamed up with Unilever – at the time one of the world's largest seafood buyers – to establish the MSC as a seafood certification scheme. The MSC was formally established in 1997 as a non-profit organization (Fowler and Heap 2000).

Demand for choice and harmonization, too, have underpinned the formation and evolution of several private regulators. The FSC was not the first organization formed to certify responsible forest practices or practices in other sectors. Before it launched, and critical to its formation, a number of private certification organizations and newly founded non-profits had begun building a certification industry. The Rainforest Alliance, founded in 1987 to advance rainforest protection by means other than boycotts, was a key player in this process. It created the SmartWood program in 1989, which certified its first forest operation in Indonesia in 1990. SmartWood and another early certifier – Scientific Certification Services (SCS) – were important contributors to the formation of the FSC. Their prior operations were a key reason the FSC focused on standard-setting and accreditation as opposed to offering certification services itself. The FSC's role was to bring better consistency and harmony to emerging certifiers. SmartWood and SCS were among the first to be accredited as third-party certifiers by the FSC (Auld 2009).

The motivation for harmonization is clearer still in organic agriculture. The Inter-national Federation of Organic Agriculture Movements (IFOAM) was founded in 1972 to build and share knowledge about the practices of organic farming. By the 1980s, however, many private certifiers had formed and were loosely attached to IFOAM, raising concerns about consistency. That inspection practices varied across different organic certifiers helped motivate IFOAM to create an accreditation unit, which eventually became an independent organization, Organic Accreditation Services International. Even before this, IFOAM had been providing international guidance on the basic standards for organic practices in different sectors, taking on the role of promoting consistency within the sector (Auld 2009).

Counter to the drive for consistency and harmonization, demand for private regulatory choice has been an equally important factor. Sometimes the demand has been for alternatives which offer a different approach to accomplishing the same goal. Many of the forest certification programs now housed under the umbrella scheme, the Program for the Endorsement of Forest Certification (PEFC), fit this category. Proponents of national schemes endorsed by the PEFC call their standards equivalent to those of the FSC for what ends they accomplish, even if the means are different. Another example is the apparel industry labor rights scheme – Worldwide Responsible Accredited Production (WRAP) – that emerged in response to the NGO-sponsored Fair Labor Association (Abbott and Snidal 2009: 76). Demand in other cases has been for more complementary alternatives, which tackle different, often emerging problems in a given sector. The Bird Friendly program created by the Smithsonian Migratory Bird Center, for instance, requires that operators which meet its shade-coffee standards must also be certified organic. Only those operators with both certifications can use the program's "Bird Friendly" label (Auld 2009; Auld et al. 2009).

Supply of Private Regulation The demand for alternative governance approaches only partly explains the surge in private regulatory programs. Institutional entrepreneurs have played their own critical role in supplying private regulation to an increasing number of issues, including: sustainable tourism, the aquarium trade, palm oil production, soy production, and parks management (Honey 2002; Conroy 2006; Auld et al. 2007). Some certification initiatives have largely independent roots; labor standards and forestry standards emerged roughly at the same time, for example, but those working on the respective schemes had little knowledge of what was happening in the other sector (Bartley 2003). In other cases, entrepreneurs have worked to spread the certification idea across sectors and industries. As explained by Auld et al. (2007), three entrepreneurial groups are particularly important: environmental NGOs, professional certification bodies, and philanthropic foundations.

First, environmental NGOs have created or supported a range of certification initiatives. The WWF, for instance, was central in launching the FSC, which it copied in modified form when establishing the MSC. The WWF has since helped to form certification schemes for the marine aquarium trade, sustainable palm oil, and sustainable soy oil (Auld et al. 2007). The Aquaculture Stewardship Council, a certification program to promote sustainable fish farming, is one the WWF's most recent projects.

Second, certifiers have been key entrepreneurs for the certification idea. Some certifiers were operating well before the advent of social and environmental certification programs, but the growth of sustainability certification initiatives has presented a new business opportunity. Certification bodies like SGS have a long history of auditing technical standards. Established in 1878 to offer agricultural inspection services to European grain traders, SGS was among the first certifiers to be FSC accredited. In

1997, by helping form the labor standards program, Social Accountability International (SAI), SGS facilitated the spread of certification to the apparel industry (Auld *et al.* 2007). SGS, SCS, and a few other professional certifiers have become accredited to certify operations for numerous certification schemes.

Another example is the Rainforest Alliance. It has applied its SmartWood model to the production of various commodities affecting the integrity of tropical forests (Taylor and Scharlin 2004; Auld 2009). In 1994, the first two Chiquita-owned banana farms in Costa Rica were certified, followed the next year by the first coffee farms to be certified in Guatemala.[1] According to the Rainforest Alliance, more than 15% of the bananas in international trade currently come from farms it has certified.[2] The program now certifies a range of tropical commodities, including cocoa, tea, citrus, and cut flowers (Auld 2009).

Finally, philanthropic foundations have provided financial support to certification schemes. Bartley (2007a) details the role of US foundations in the formation of forest certification. The FSC was significant, he explains, because it provided foundations with a project they could jointly support and demonstrated that certification was a potential solution for several environmental and social problems. Some foundations that supported the FSC then supported the MSC; other foundations observed the success of forest certification and decided to support the nascent fisheries certification program. The Packard Foundation was vital in supporting MSC's transformation from a WWF–Unilever partnership to a fully independent, multi-stakeholder certification program (Gulbrandsen 2010). As with FSC, foundation grants remain MSC's most important source of income. Foundations have also supported a range of other social and environmental certification initiatives.

Evolving Effects of Private Regulation

The demand and supply of private regulation within the focal sectors are closely entwined and continue to evolve as the number and diversity of programs expands. Two questions arise about this growing field. A first concerns how the interaction of programs shapes their own institutional development. A second turns attention to the effects of any and all private regulatory programs for problem amelioration. That is, are they addressing the environmental and social problems facing different economic sectors?

Institutional Development   The role of entrepreneurs in spreading certification to new problem areas is one part of a larger set of interactions among existing and still-to-develop programs. Copying has been widespread. The FSC modeled its chamber system after those of IFOAM and IUCN (Elliott 2000). The FSC, in turn, has become an organizational model for other certification programs, including the MSC. And recently, the International Social and Environmental Accreditation and Labelling (ISEAL) Alliance has begun actively promoting greater consistency among a growing group of certification systems. Studying a broader set of cases, Dingwerth and Pattberg (2009) argue that the logic of organizational fields and mimetic processes helps to explain convergence to certain design principles for transnational rule-making. However, a closer look at private regulatory programs reveals *persistent variation* in design and organizational characteristics.

The spread of certification from forestry to fisheries illustrates how strategic design choices can result in different organizational features across programs (Auld 2009; Gulbrandsen 2009; Auld and Gulbrandsen 2010). Although the founders of MSC mimicked some of FSC's features, they avoided other features. First, while FSC gives its members ultimate authority, the MSC granted ultimate decision-making authority to a Board of Trustees, which is self-recruiting and functions much like a corporate board of directors. Second, MSC's founders chose not to give national affiliates a role in developing locally appropriate standards, as had the FSC. The localization of standards was instead controlled by certification bodies that were to assess individual applicant fisheries (Auld and Gulbrandsen 2010). Third, the founders of MSC decided not to address social issues in the program's principles and criteria. Several commentators argued in favor of standards covering both environmental and social issues, but MSC decided to keep them narrower, focusing primarily on environmental issues. While MSC's assessment methodology and procedures have been modified over the course of its development, the principles and criteria have remained the same (Auld and Gulbrandsen 2010).

Hence, while mimetic processes have resulted in convergence among some programs, specialization and selective mimicry allow diversity to persist. In this respect, we see that imitation of a specific governance model is likely to be mixed with innovation as a result of strategic design choices, adaptation to a different context, and power struggles over whose interests the model is to serve.

Problem Amelioration   The effect of private regulation for problem amelioration is a growing issue of interest to practitioners and academics. In the early stages of many certification programs, proxy measures such as the number of companies participating, areas certified, or values of certified products traded or consumed were cited to capture effects. Attention has since evolved to focus on the behavioral changes operators have undertaken as a consequence of participating in certification programs. As noted above, the voluntary nature of certification programs has led most analysts to identify factors which allow some operators to more easily achieve certification than others. First, evidence suggests that companies facing relatively low costs of standards adoption tend to participate more frequently in schemes with stringent standards than do companies that face high adoption costs (Gulbrandsen 2004, 2010; Cashore et al. 2007; Auld et al. 2008b). In other words, it may be less costly for companies in countries with relatively stringent environmental regulations to join voluntary programs than it is for companies in countries with lax regulations. Second, large companies in developed countries may find it easier to certify operations than do small operations in developing countries, owing to the benefits of economies of scale (Cashore et al. 2004; Cashore et al. 2007). Third, several practical barriers impede adoption in developing countries, including lack of technical information, shortcomings of scientific data, or inadequate legal and administrative systems (Gulbrandsen 2004, 2010; Pattberg 2005, 2006; Ponte 2008; Ward and Phillips 2008). Although several certification programs have introduced specialized arrangements to reduce entry barriers for small producers from developing countries, patterns of adoption continue to raise questions about the global effectiveness of private regulatory programs (Auld 2010, Gulbrandsen 2010). Having a patchwork of support for a certification program, particularly ones attempting to promote the ecological integrity of forests or marine ecosystems, raises questions about landscape-level or ecosystem impacts.

The character of the rules also affects what the programs mean for on-the-ground performance. According to the Rainforest Alliance, its farm certification program creates social and environmental benefits, including decreased water pollution and soil erosion, reduction of pesticides, protection of wildlife, and improved conditions for farm workers.[3] Unlike Fairtrade certification, however, it does not guarantee producers a minimum price, nor does it seem to have a strong impact on working conditions and wages (Daviron and Ponte 2005; Conroy 2006: 251). Because retailers pay less than the Fairtrade price for certified commodities, Rainforest Alliance certification has been tremendously successful in increasing market adoption, thereby allowing multinational corporations like Chiquita, Unilever, and Kraft Foods to capture a large share of the ethical consumer market (Conroy 2006: 251).

The demand-side factor leading to the creation of "choice" in the private regulatory market has also led to competition between programs. In some cases, this is seen as a force for downward pressure on standards. Such a charge has been leveled at the Rainforest Alliance's coffee and banana programs, for instance (Bacon *et al.* 2008). In other cases, competition has resulted in some convergence and upward change of standards, increasing the average stringency of certification systems. In forestry, for example, criteria-by-criteria comparisons of FSC and industry-backed schemes have found substantial differences in environmental ambitiousness, demonstrating that the latter schemes were the least stringent (Overdevest 2005). Such comparisons have placed upward pressure on industry-backed programs, narrowing the gap between their approach and that of FSC, although differences do remain (McDermott *et al.* 2008).

Even this focus on standards, however, leaves some questions unanswered. Particularly, we do not know if compliance alone, even if it is widespread, will mean the environmental or social conditions actually improve. This assumes certification programs have the right standards in place. Examining ultimate performance is beyond the scope of this chapter, but it is one which researchers are attending to with greater energy.

Climate-Related Private Regulation

Against the backdrop of the expanding array of private regulatory programs, it is no surprise that climate change has attracted its share of attention. Each of the three factors we described as affecting the emergence of private regulation is clearly in play.

First, on the demand side, the absence of effective government action has been critical. Following the US withdrawal from the Kyoto Protocol in 2001, and lack of substantial progress in negotiating a post-2012 multilateral climate agreement, myriad climate governance experiments have been initiated by corporations, civil society actors, sub-national governments, cities, and municipalities (Bulkeley and Betsill 2003; Newell 2008; Hoffmann 2011; Meckling 2011). Hoffmann (2011), for example, explains that urgency about climate change and frustration with the lack of progress by intergovernmental processes motivated the development of climate governance experiments involving a wide range of sub-national actors. This observation is equally important for understanding the emergence of private regulation relevant to climate change.

Additional to this governance failure, the possibility that a new regulatory regime would form further enticed private action. Certain private regulations have emerged to supplement the "flexible mechanisms" established by the Kyoto Protocol. The clean development mechanism (CDM) and joint implementation allow countries with emissions reduction commitments (Annex I countries) to meet their commitments by purchasing credits, through approved emission reduction projects in other countries (non-Annex I countries). Their establishment has provided opportunities for carbon offset standard-setters, verifiers, and traders that could make a profit from regulatory carbon markets (Newell and Paterson 2010).

The regulatory market is the tip of the iceberg, however. On the supply side, institutional entrepreneurs have developed a sizable voluntary, over-the-counter carbon offset market, established to take advantage of the expansion of the climate change policies beyond the initial phase of the Kyoto Protocol. These entrepreneurs have created a variety of offset projects, including wind power, renewable biomass, agricultural methane, landfill gas, small- and large-scale hydro, energy efficiency, avoided deforestation, industrial gas destruction, biofuels, biogas, and fuel switching.[4] The links to forest and agricultural certification programs, and the role of entrepreneurs in spreading certification, is also clearly illustrated in the climate case. The FSC has been working on forest carbon accounting since 2009. The Rainforest Alliance is also in the business of carbon-offset verification and validation. It has been accredited by the American National Standards Institute to the ISO standard (ISO 14065) for greenhouse gas validation and verification bodies,[5] allowing it to provide auditing services to forest managers and landowners.

While the number of offset providers, validators, and verifiers has exploded since the early 2000s (Hoffmann 2011: 132), the standard-setters that develop or approve project methodologies are fewer but serve important regulatory functions. By establishing protocols and methods for measuring, verifying, and recording greenhouse gas reductions, these actors are seen to provide the foundational infrastructure for voluntary carbon markets (Hoffmann 2011). One example is the WWF's Gold Standard, which "essentially applies an extra set of screens to CDM or voluntary projects" (Newell and Paterson 2010: 119), requiring that projects: employ renewable energy or energy efficiency technologies, adhere to the strictest standards on additionality, and create positive effects for local communities (Auld et al. 2009).[6] Other well-known standard-setters include Voluntary Carbon Standard, Carbon Fix, Climate Action Reserve, the Climate, Community and Biodiversity standards, Plan Vivio, Voluntary Offset Standard, and Social Carbon. There are also several standards for voluntary emissions reporting, including the Greenhouse Gas Protocol and the Carbon Trust's Carbon Footprint Measurement Methodology.

The Kyoto Protocol also helped foster the global diffusion of marked-based instruments, particularly private experiments with emission trading (Meckling 2011). The first private emission trading systems were innovative but short-lived internal corporate systems implemented by the European oil majors BP and Shell to gain practical experience in anticipation of an international trading system under the Kyoto Protocol. Launched in 2000, BP's internal emission trading system ceased to exist by the end of 2001, when the company had achieved a 10% reduction in greenhouse gas emissions (Victor and House 2006). Internal trading quickly became dated as national schemes (Denmark and the UK) and regional schemes (the EU Emissions Trading System) were developing in Europe. Meanwhile, in the USA, a private cap-and-trade system got under way in 2001, drawing on the experiences of Shell and BP and emissions trading in Denmark and the UK. This led to the 2003 establishment of the Chicago Climate Exchange (CCX) – the world's first private cap-and-trade system.

Another private regulatory program in the field of climate governance is the Carbon Disclosure Project (CDP). Founded in 2000, the CDP is a London-based, independent non-profit foundation representing a consortium of investors concerned about climate change – and their investments. It is backed by several major blue-chip investors, including HSBC, JPMorgan Chase, Bank of America, Merrill Lynch, and Goldman Sachs. Guided by the aphorism "what gets measured can be managed,"[7] the CDP works with some of the world's biggest corporations, including Walmart, Tesco, Procter and Gamble, Dell, and PepsiCo. It annually surveys these companies on a wide range of climate-related activities and provides this information to investors seeking to reduce exposure to climate-related risks. By publicly disclosing individual corporate responses and summary analyses, the CDP provides valuable information on what companies are and are not doing to address climate change in their own operations and their supply chains. However, unlike certification programs, the CDP stops short of setting standards that require companies to improve performance.

The initiatives reviewed above illustrate how climate-related private regulation has compensated for weak or lacking government regulation, supplemented government regulation, and, sometimes, facilitated the development of government regulation. Some regulations that now are mandatory, for instance, began as voluntary initiatives. As Hoffmann (2011: 52) observes, voluntary programs are often "stepping stones toward mandatory climate policy – voluntary reporting gives way to voluntary action which leads to mandatory action with low targets and finally mandatory action with high targets." The voluntary California Climate Action Registry, for instance, gave way to California's landmark climate legislation signed in 2006, and re-emerged as the Climate Registry. This registry has become the major climate inventory initiative in North America; voluntary reporting to the Climate Registry has, in turn, developed into mandatory reporting for large emitters (Hoffmann 2011: 88–89). Private regulations should thus not only be evaluated on the basis of their direct effects but also on the basis of their broader consequences, including demonstration effects, spillover effects, and educational effects (Auld et al. 2008b). How and whether private regulation will act synergistically with government rules is a key question for considering the evolutionary potential of these initiatives (Auld et al. 2009; Gulbrandsen 2010).

Options for Global Policy

Just as private regulation has evolved in the past several decades, the attention of research and practitioners has too. Increasingly, scholars are asking how private regulation interacts with public policy domestically and internationally to create hybrid, synergistic, or other relationships which do or do not help address the ultimate global problems of concern (Schneiberg and Bartley 2008; Auld *et al.* 2009). Based on our review, what conclusions can we initially draw about these interactions and their implications for global public policy?

First, in the domestic context, states can influence private regulatory programs at all stages of the regulatory process from agenda-setting to negotiation of standards and on to implementation, monitoring, and enforcement (Abbott and Snidal 2009). The most obvious example is that existing rules and norms provide the framework for private regulatory schemes (Cashore *et al.* 2004: 20). Other examples are government-controlled or -owned operations acting as clients for certification, governments covering auditing costs for clients, and public procurement policies stipulating the purchase of certified products (Cashore *et al.* 2004: 20–22; Klooster 2006).

Second, in the transnational context, states have fewer possibilities to stimulate, strengthen, or regulate private programs, but they can influence such programs through intergovernmental organizations (IGOs) such as UNEP or international regimes such as the trade regime (Abbott and Snidal 2009: 67). According to Abbott (2012), two types of engagement with private regulatory programs are especially promising for IGOs. The first is what he calls "regulatory cooperation," in which IGOs engage directly with private regulatory programs and the targets of regulation to influence their behavior. Through regulatory cooperation, IGOs can encourage firms to adopt standards, "stimulate and focus public demand," "reduce fragmentation by promoting industry-wide standards," "encourage business schemes to become more participatory and deliberative," and "facilitate learning across firms and industries" (Abbott 2012).

The second mode of engagement is "orchestration." This is where IGOs bring private regulatory programs into the governance arrangement to act as intermediaries between the IGOs and the targets of regulation, such as UNEP's engagement with the Global Reporting Initiative, where UNEP chaired its planning committee, endorsed its sustainability reporting guidelines, and supported it financially (Abbott 2012). The World Bank's alliance with the WWF to promote forest protection and certification is another example. Launched in 1998 and renewed in 2005, the alliance seeks to increase the area of protected and certified forests, particularly in developing countries. The World Bank's commitment to forest certification demanded that it take a clear position on acceptable standards. Although the Bank has not formally endorsed a specific program, the requirements in its operational policies on forests are remarkably similar to the FSC principles (Humphreys 2006: 173–174). The operational policies are officially an internal reference guide for World Bank managers, but the Bank can transmit its policy to countries to which it lends, thereby promoting the FSC.

Third, specific to climate change, scholars have called for "leadership in an experimental world" (Hoffmann 2011: 158); embedding distinct "institutional building blocks" in an international political framework (Falkner *et al.* 2010); and "a light coordination mechanism" for a highly decentralized system (Pattberg 2010: 285). Yet, their rather general prescriptions for global policies leave a lot to be answered. Indeed, government engagement with private regulatory schemes is not always desirable or productive from an environmental or social point of view. Several private schemes, such as the FSC, emerged precisely because of stalemate in intergovernmental negotiations and because IGOs, such as the International Tropical Timber

Organization, were seen as dominated by states that promoted industry interests at the expense of environmental interests (Gulbrandsen 2004; Humphreys 2006). Likewise, states seeking to protect their fishing industries responded to the MSC by urging the UN Food and Agriculture Organization to regulate fisheries certification (Gulbrandsen 2009). Public engagement with private regulatory schemes – and especially those backed by environmental NGOs – thus runs the risk of regulatory capture by the industries those schemes seek to regulate.

Fourth, removing macro-institutional constrains for transnational environmental regulation, both private and public, may do more to facilitate transnational governance by private authorities than would be possible via IGO backing. A much-discussed obstacle to trade-related eco-labeling requirements is the international trade regime. Hence, modifying multilateral trade rules in ways that facilitate rather than hinder eco-certification and eco-labeling could encourage wider adoption of private regulatory programs. In the long term, building an environmentally friendly trade regime could be an effective way of stimulating private regulation in global environmental governance.

Our review has shown that private regulatory programs have become vibrant and dynamic institutions for environmental governance across sectors and countries. These programs represent a remarkable policy innovation by non-state actors, but we have seen that their evolutionary potential depends critically on synergies with government regulations. Future research should examine how the dynamic interactions between private regulatory programs and public policies influence the ultimate performance of these programs in ameliorating pressing environmental and social problems.

Notes

1 www.rainforest-alliance.org/about/documents/ra_timeline.pdf (accessed October 20, 2012).
2 www.rainforest-alliance.org/agriculture.cfm?id=fruits, August 6, 2009 (accessed October 20, 2012).
3 www.rainforest-alliance.org/agriculture.cfm?id=main, August 12, 2009 (accessed October 20, 2012).
4 For an overview of credit and project types, see http://www.endscarbonoffsets.com/ (accessed October 20, 2012).
5 www.rainforest-alliance.org/climate.cfm?id=international_standards, August 6, 2009 (accessed October 20, 2012).
6 http://www.cdmgoldstandard.org/about-us/who-we-are (accessed November 16, 2011).
7 Quotation from Lord (Adair) Turner, Chairman, UK Financial Services Authority at the CDP web site: https://www.cdproject.net/en-US/WhatWeDo/Pages/overview.aspx (accessed November 11, 2011).

References

Abbott, K.W. 2012. "Engaging the Public and Private in Global Sustainability Governance." *International Affairs*, 88(3): 543–565.

Abbott, K.W. and D. Snidal. 2009. "The Governance Triangle: Regulatory Standards Institutions and the Shadow of the State." In *The Politics of Global Regulation*, ed. W. Mattli and N. Wood, 44–88. Princeton, NJ: Princeton University Press.

Auld, G. 2009. "Reversal of Fortune: How Early Choices Can Alter the Logic of Market-Based Authority." PhD dissertation, Yale University.

Auld, G. 2010. "Assessing Certification as Governance: Effects and Broader Consequences for Coffee." *Journal of Environment & Development*, 19(2): 215–241.

Auld, G., C. Balboa, T. Bartley *et al.* 2007. "The Spread of the Certification Model: Understanding the Evolution of Non-state Market-Driven Governance." Presented at the 48th Convention of the International Studies Association, February 27–March 3, Chicago, IL.

Auld, G., C. Balboa, S. Bernstein, and B. Cashore. 2009. "The Emergence of Non-state Market-Driven (NSMD) Global Environmental Governance: A Cross-Sectoral Assessment." In *Governance for the Environment: New Perspectives*, ed. M.A. Delmas and O.R. Young, 183–218. Cambridge: Cambridge University Press.

Auld, G., S. Bernstein, and B. Cashore. 2008a. "The New Corporate Social Responsibility." *Annual Review of Environment and Resources*, 33(1): 413–435.

Auld, G., B. Cashore, C. Balboa *et al.* 2010. "Can Technological Innovations Improve Private Regulations in the Global Economy?" *Business and Politics*, 12(3): Article 9. doi:10.2202/1469-3569.1323.

Auld, G. and L.H. Gulbrandsen. 2010. "Transparency in Nonstate Certification: Consequences for Accountability and Legitimacy." *Global Environmental Politics*, 10(3): 97–119.

Auld, G., L.H. Gulbrandsen, and C.L. McDermott. 2008b. "Certification Schemes and the Impact on Forests and Forestry." *Annual Review of Environment and Resources*, 33(1): 187–211.

Bacon, C.M., V.E. Méndez, S.R. Gliessman et al., eds. 2008. *Confronting the Coffee Crisis: Fair Trade, Sustainable Livelihoods and Ecosystems in Mexico and Central America*. Cambridge, MA: MIT Press.

Bartley, T. 2003. "Certifying Forests and Factories: States, Social Movements, and the Rise of Private Regulation in the Apparel and Forest Products Fields." *Politics & Society*, 31(3): 433–464.

Bartley, T. 2007a. "How Foundations Shape Social Movements: The Construction of an Organizational Field and the Rise of Forest Certification." *Social Problems*, 54(3): 229–255.

Bartley, T. 2007b. "Institutional Emergence in an Era of Globalization: The Rise of Transnational Private Regulation of Labor and Environmental Conditions." *American Journal of Sociology*, 113(2): 297–351.

Bartley, T. and S. Smith. 2007. *"The Evolution of Transnational Fields of Governance: A Network Analytic Approach."* Bloomington, IN: Department of Sociology, Indiana University.

Bendell, J. 2000. "Civil Regulation: A New Form of Democratic Governance for the Global Economy?" In *Terms for Endearment: Business, NGOs and Sustainable Development*, ed. J. Bendell, 239–254. Sheffield: Greenleaf Publishing.

Bernstein, S. and B. Cashore. 2007. "Can Non-state Global Governance Be Legitimate? An Analytical Framework." *Regulation & Governance*, 1(4): 347–371.

Besky, S. 2011. "Colonial Pasts and Fair Trade Futures: Changing Modes of Production and Regulation on Darjeeling Tea Plantations." In *Fair Trade and Social Justice: Global Ethnographies*, ed. S. Lyon and M. Moberg, 97–122. New York: New York University Press.

Börzel, T.A. and T. Risse. 2005. "Public–Private Partnerships: Effective and Legitimate Tools of Transnational Governance." In *Complex Sovereignty: Reconstituting Political Authority in the Twenty-First Century*, ed. E. Grande and L.W. Pauly, 195–216. Toronto: University of Toronto Press.

Bulkeley, H. and M.M. Betsill. 2003. *Cities and Climate Change: Urban Sustainability and Global Environmental Governance*. London and New York: Routledge.

Cashore, B. 2002. "Legitimacy and the Privatization of Environmental Governance: How Non-state Market-Driven (NSMD) Governance Systems Gain Rule-Making Authority." *Governance*, 15(4): 503–529.

Cashore, B., G. Auld, S. Bernstein, and C.L. McDermott. 2007. "Can Non-state Governance 'Ratchet Up' Global Environmental Standards? Lessons from the Forest Sector." *Review of European Community & International Environmental Law*, 16(2): 158–172.

Cashore, B., G. Auld, and D. Newsom. 2004. *Governing through Markets: Forest Certification and the Emergence of Non-state Authority*. New Haven, CT: Yale University Press.

Conroy, M.E. 2006. *Branded: How the "Certification Revolution" is Transforming Global Corporations*. Gabriola Island, BC: New Society Publishers.

Cutler, C., V. Haufler, and T. Porter, eds. 1999. *Private Authority in International Politics*. Albany NY: State University of New York Press.

Daviron, B. and S. Ponte. 2005. *The Coffee Paradox: Global Markets, Commodity Trade and the Elusive Promise of Development*. London and New York: Zed Books, in association with CTA Wageningen.

Dingwerth, K. and P. Pattberg. 2006. "Global Governance as a Perspective on World Politics." *Global Governance*, 12(2): 185–203.

Dingwerth, K. and P. Pattberg. 2009. "World Politics and Organizational Fields: The Case of Transnational Sustainability Governance." *European Journal of International Relations*, 15(4): 707–743.

Elliott, C. 2000. *Forest Certification: A Policy Network Perspective*. Bogor, Indonesia: Center for International Forestry Research (CIFOR).

Falkner, R. 2003. "Private Environmental Governance and International Relations: Exploring the Links." *Global Environmental Politics*, 3(2): 72–87.

Falkner, R. 2008. *Business Power and Conflict in International Environmental Politics*. Basingstoke: Palgrave Macmillan.

Falkner, R., H. Stephan, and J. Vogler. 2010. "International Climate Policy after Copenhagen: Towards a 'Building Blocks' Approach." *Global Policy*, 1(3): 252–262.

Fowler, P. and S. Heap. 2000. "Bridging Troubled Waters: The Marine Stewardship Council." In *Terms for Endearment: Business, NGOs and Sustainable Development*, ed. J. Bendell, 135–148. Sheffield: Greenleaf Publishing.

Fransen, L.W. and A. Kolk. 2007. "Global Rule-Setting for Business: A Critical Analysis of Multi-stakeholder Standards." *Organization*, 14(5): 667–684.

Fuchs, D.A. 2007. *Business Power in Global Governance*. Boulder, CO: Lynne Rienner.

Gulbrandsen, L.H. 2004. "Overlapping Public and Private Governance: Can Forest Certification Fill the Gaps in the Global Forest Regime?" *Global Environmental Politics*, 4(2): 75–99.

Gulbrandsen, L.H. 2005. "Mark of Sustainability? Challenges for Fishery and Forestry Eco-Labeling." *Environment*, 47(5): 8–23.

Gulbrandsen, L.H. 2006. "Creating Markets for Eco-Labelling: Are Consumers Insignificant?" *International Journal of Consumer Studies*, 30(5): 477–489.

Gulbrandsen, L.H. 2008. "Accountability Arrangements in Non-state Standards Organizations: Instrumental Design and Imitation." *Organization*, 15(4): 563–583.

Gulbrandsen, L.H. 2009. "The Emergence and Effectiveness of the Marine Stewardship Council." *Marine Policy*, 33(4): 654–660.

Gulbrandsen, L.H. 2010. *Transnational Environmental Governance: The Emergence and Effects of the Certification of Forests and Fisheries*. Cheltenham: Edward Elgar.

Gunningham, N. and J. Rees. 1997. "Industry Self-Regulation: An Institutional Perspective." *Law & Policy*, 19(4): 363–414.

Guthman, J. 2007. *The Polanyian Way? Voluntary Food Labels as Neoliberal Governance*. Oxford: Wiley-Blackwell.

Hall, R.B. and T.J. Biersteker. 2002. *The Emergence of Private Authority in Global Governance*. Cambridge Studies in International Relations 85. Cambridge: Cambridge University Press.

Hoffmann, M.J. 2011. *Climate Governance at the Crossroads: Experimenting with a Global Response after Kyoto*. New York: Oxford University Press.

Honey, M., ed. 2002. *Ecotourism and Certification: Setting Standards in Practice*. Washington, DC: Island Press.

Humphreys, D. 1996. *Forest Politics: The Evolution of International Cooperation*. London: Earthscan.

Humphreys, D. 2006. *Logjam: Deforestation and the Crisis of Global Governance*. London: Earthscan.

King, A.A. and M.J. Lenox. 2000. "Industry Self-Regulation without Sanctions: The Chemical Industry's Responsible Care Program." *Academy of Management Journal*, 43(4): 698–716.

Klooster, D. 2006. "Environmental Certification of Forests in Mexico." *Annals of the Association of American Geographers*, 96(3): 541–565.

Klooster, D. 2010. "Standardizing Sustainable Development? The Forest Stewardship Council's Plantation Policy Review Process as Neoliberal Environmental Governance." *Geoforum*, 41(1): 117–129.

Kolk, A., D. Levy, and J. Pinkse. 2008. "Corporate Responses in an Emerging Climate Regime: The Institutionalization and Commensuration of Carbon Disclosure." *European Accounting Review*, 17(4): 719–745.

Levy, D.L. and P.J. Newell. 2005. *The Business of Global Environmental Governance*. Cambridge, MA: MIT Press.

McDermott, C.L., E. Noah, and B. Cashore. 2008. "Differences that 'Matter'? A Framework for Comparing Environmental Certification Standards and Government Policies." *Journal of Environmental Policy and Planning*, 10(1): 47–70.

Meckling, J. 2011. *Carbon Coalitions: Business, Climate Politics, and the Rise of Emissions Trading*. Cambridge, MA: MIT Press.

Meidinger, E. 2006. "The Administrative Law of Global Private–Public Regulation: The Case of Forestry." *European Journal of International Law*, 17(1): 47–87.

Mutersbaugh, T. 2005. "Fighting Standards with Standards: Harmonization, Rents, and Social Accountability in Certified Agrofood Networks." *Environment and Planning A*, 37(11): 2033–2051.

Newell, P. 2008. "Civil Society, Corporate Accountability and the Politics of Climate Change." *Global Environmental Politics*, 8(3): 122–153.

Newell, P., and M. Paterson. 2010. *Climate Capitalism: Global Warming and the Transformation of the Global Economy*. Cambridge: Cambridge University Press.

Overdevest, C. 2005. "Treadmill Politics, Information Politics and Public Policy toward a Political Economy of Information." *Organization & Environment*, 18(1): 72–90.

Overdevest, C. 2010. "Comparing Forest Certification Schemes: The Case of Ratcheting Standards in the Forest Sector." *Socio-Economic Review*, 8(1): 47–76.

Overdevest, C., A. Bleicher, and M. Gross. 2010. "The Experimental Turn in Environmental Sociology: Pragmatism and New Forms of Governance." In *Environmental Sociology: European Perspectives and Interdisciplinary Challenges*, ed. M. Gross and H. Heinrichs, 279–294. New York: Springer.

Pattberg, P.H. 2005. "The Forest Stewardship Council: Risk and Potential of Private Forest Governance." *Journal of Environment & Development*, 14(3): 356–374.

Pattberg, P.H. 2006. "Private Governance and the South: Lessons from Global Forest Politics." *Third World Quarterly*, 27(4): 579–593.

Pattberg, P.H. 2010. "Public–Private Partnerships in Global Climate Governance." *Wiley Interdisciplinary Reviews: Climate Change*, 1(2): 279–287.

Ponte, S. 2008. "Greener than Thou: The Political Economy of Fish Ecolabeling and Its Local Manifestations in South Africa." *World Development*, 36(1): 159–175.

Prakash, A. 2000. "Responsible Care: An Assessment." *Business & Society*, 39(2): 183–209.

Prakash, A. and M. Potoski. 2006. *The Voluntary Environmentalists: Green Clubs, ISO 14001, and Voluntary Regulations*. Cambridge: Cambridge University Press.

Raynolds, L.T., D. Murray, and A. Heller. 2007. "Regulating Sustainability in the Coffee Sector: A Comparative Analysis of Third-Party Environmental and Social Certification Initiatives." *Agriculture and Human Values*, 24(2): 147–163.

Reinhardt, F.L. 2000. *Down to Earth: Applying Business Principles to Environmental Management*. Boston, MA: Harvard Business School Press.

Rosenau, J.N. and E.O. Czempiel, eds. 1992. *Governance without Government: Order and Change in World Politics*. Cambridge: Cambridge University Press.

Schneiberg, M. and T. Bartley. 2008. "Organizations, Regulations, and Economic Behavior: Regulatory Dynamics and Forms from the Nineteenth to Twenty-First Century." *Annual Review of Law and Social Science*, 4: 1–31.

Segerson, K. and T.J. Miceli. 1998. "Voluntary Environmental Agreements: Good or Bad News for Environmental Protection?" *Journal of Environmental Economics and Management*, 36(2): 109–130.

Taylor, J.G. and P.J. Scharlin. 2004. *Smart Alliance: How a Global Corporation and Environmental Activists Transformed a Tarnished Brand*. New Haven, CT: Yale University Press.

Taylor, P.L. 2005. "In the Market but Not of It: Fair Trade Coffee and Forest Stewardship Council Certification as Market-Based Social Change." *World Development*, 33(1): 129–147.

Tollefson, C., F.P. Gale, and D. Haley. 2008. *Setting the Standard: Certification, Governance, and the Forest Stewardship Council*. Vancouver: UBC Press.

Victor, D.G. and J.C. House. 2006. "BP's Emissions Trading System." *Energy Policy*, 34(15): 2100–2112.

Vogel, D. 2005. *The Market for Virtue: The Potential and Limits of Corporate Social Responsibility*. Washington, DC: Brookings Institution Press.

Ward, T.J. and B. Phillips. 2008. "Anecdotes and Lessons of a Decade." In *Seafood Labelling: Principles and Practice*, ed. T.J. Ward and B. Phillips, 416–436. Oxford: Wiley-Blackwell.

Zadek, S. 2001. *The Civil Corporation: The New Economy of Corporate Citizenship*. London: Earthscan.

Chapter 24

International Trade, the Environment, and Climate Change

Nico Jaspers and Robert Falkner

Trade liberalization has been a key driving force behind global economic growth since the Second World War. During this period, global environmental degradation reached new heights. Three sets of questions arise from this. First, is the liberalization of international trade responsible for the global ecological crisis, or do freer trade, increased global competition and greater wealth help to promote environmental protection and a more efficient use of scarce resources? Second, do the rules of the international trading system (mainly the World Trade Organization – WTO) help or hinder efforts to protect the environment, and are international environmental agreements consistent with the rules and obligations of the WTO order? Third, with regard to the threat of global warming, does free trade undermine the climate policies of more ambitious countries because of the threat of industrial relocation to laggard countries, and should trade measures be employed as a tool of international climate policy?

These and other questions about the trade–environment nexus have been intensely debated for some time and remain critical to the future of the trading system, particularly with regard to climate change (for a general overview of the debate, see Sampson 2005). This chapter reviews the trade–environment debate and recent scholarship. It opens with a brief discussion of the general relationship between trade, the environment, and climate change; then focuses on the institutional and jurisdictional context for trade and environmental policy-making; and concludes with an analysis of the trade implications of recent developments in climate change policy.

Links between Trade and the Environment

Are international trade and environmental protection compatible or in conflict? Two types of causal links between trade and environment can be identified: the

The Handbook of Global Climate and Environment Policy, First Edition. Edited by Robert Falkner.
© 2013 John Wiley & Sons, Ltd. Published 2013 by John Wiley & Sons, Ltd.

first concerns the effect that trade liberalization has on environmental quality in a given country or worldwide; the second reverses the perspective and addresses the impact that environmental protection policies have on international trade. In general, free-trade supporters argue that liberalizing trade has a positive effect on the environment while some environmental measures pose a protectionist threat (Bhagwati and Srinivasan 1996; Hettige *et al.* 1998; Bhagwati 2004). In contrast, environmentalists see free trade as one of the main causes of environmental pollution and advocate that environmental policy should limit free trade where it harms the environment (Daly 1993; Goldsmith and Mander 2001). This second argument has gained new prominence in the context of climate change, where it is sometimes argued that unilateral efforts to reduce carbon emissions might shift industrial activity from countries with strict regulation to those with laxer regulations. This so-called "carbon leakage" is widely regarded to undermine the effect of climate change mitigation policies (Frankel 2009; Gros *et al.* 2010).

Closer examination of the empirical evidence behind these claims reveals a more nuanced picture (Neumayer 2001). Free trade can lead to more polluting production and greater consumption of natural resources, as is the case in countries that specialize in the production of pollution-intensive goods in response to trade liberalization, such as China, which has seen a dramatic rise in air and water pollution caused by the expansion of export-oriented manufacturing (Economy 2004). Free trade can also promote greater efficiency in production and the diffusion of environmental technologies and standards worldwide, as can be seen in more globally oriented companies such as the chemical and steel industries (Reppelin-Hill 1999; Garcia-Johnson 2000). The empirical record is also mixed when it comes to the impact of environmental policies on trade. Environmental protection efforts can disrupt international trade and give rise to disguised protectionism, an accusation often leveled by developing countries at advanced economies (OECD 2005). Other measures, however, can be compatible with the international trading system. Abolishing subsidies for fossil-fuel use, for example, would not only help in the fight against global warming; it would also promote a level playing field in international energy markets (Anderson and McKibbin 2000). Overall, therefore, generalizations about the trade–environment nexus are problematic. Trade liberalization and environmental protection can, but need not, be in conflict, and much depends on specific circumstances and policies under consideration.

A more useful way to think about these connections is, therefore, to focus on particular mechanisms by which trade impacts on the environment. Grossman and Krueger (1993) propose three such mechanisms: The *scale* effect occurs when liberalized trade stimulates economic growth, which in turn increases pollution and resource consumption. The *composition* effect leads to greater specialization between countries and differential rates of environmental degradation, as countries with lower environmental standards will see an expansion of environmentally harmful activity in response to trade-induced specialization. The *technique* effect involves efficiency improvements in the technologies for production and resource extraction, which can raise the level of environmental protection worldwide. Environmentalists add two further mechanisms that tend to be neglected by economic models. The first is *cultural change* in society caused by trade liberalization, which creates shifts not only in production technologies but also in consumption patterns. The spread of consumerist values and greater availability of goods leads to rising consumption, which may outstrip any efficiency gains from freer trade (Princen *et al.* 2002). The second is the *distancing* effect that is the result of ever longer and more complex chains between geographically dispersed economic actors, from resource extraction and manufacturing to international trade and retailing. This weakens the ability of consumers to identify and accept responsibility for the consequences that their decisions have on the environment in ever more distant locations (Princen 1997).

While the debate over the right way to conceptualize the linkages between trade and the environment continues, international policy-makers are keen to stress the mutual supportiveness of trade and environmental policies. Whether trade and environmental policy-making support each other or clash depends to some extent on how existing international norms and rules are to be interpreted. We thus need to consider how the rules of the GATT/WTO trade system affect environmental policy, and vice versa. Other bilateral and regional trade agreements (e.g. NAFTA) also affect the trade–environment relationship (Gallagher 2004; Heydon and Woolcock 2009: 123–142), but the subsequent analysis focuses on the relationship between multilateral trade rules and environmental policies and regimes.

International Trade Rules and Environmental Protection

At the time of the creation of the GATT (General Agreement on Tariffs and Trade) in the late 1940s, there was no international environmental agenda to speak of. Understandably, therefore, the GATT did not include any special provisions on the relationship between trade and environmental policy. Still, it recognized that some trade restrictions might be needed in the interest of public health or nature conservation.[1]

The GATT's main objective is to reduce the overall level of tariffs and other trade barriers through a series of multilateral negotiations. Its legal structure is based on a number of fundamental norms, most importantly reciprocity and non-discrimination. Reciprocity in the GATT system is evident from the conduct of negotiations. Rather than lower trade barriers unilaterally, GATT members have only agreed to reduce their levels of protection in return for reciprocal concessions from others. Non-discrimination is expressed in two principles in the GATT agreement: the most-favored-nation (MFN) principle (Article I), which requires each GATT member to accord to all other members the same privileges it has granted to its "most-favored nation"; and the national treatment principle (Article III), which demands that GATT members treat "like products" imported from foreign producers in the same way as those of domestic producers. The concept of "like products" is an important and controversial one in the trade–environment context, even though no definite interpretation of it exists in GATT/WTO law and jurisdiction (Sampson 2005: 82). Internationally traded goods may reflect different designs or production techniques, but are to be considered as "like products" if they share important physical characteristics or are functionally equivalent (e.g. cars by different manufacturers).

Article XX, the only specific environmental provision in the GATT, sets out the conditions for restricting international trade in the interests of human, animal, or plant life or health (Art. XX(b)) and the conservation of natural resources (Art. XX(g)). Such measures are allowed if they do not arbitrarily and unjustifiably discriminate between countries with similar conditions or constitute a disguised protectionist measure; if (in the case of subclause (b)) they can be considered necessary, that is, no other, less trade-intrusive, measures are available; and if (in the case of subclause (g)) equivalent domestic restrictions are imposed as well. The GATT thus allows environmental exceptions from its trade disciplines but seeks to prevent "green" discrimination or protectionism (Neumayer 2001: 24–25).

As discussed below, the GATT contains provisions that are bound to come into conflict with environmental policies. This is most clearly the case with the non-discrimination rule for "like products," which prohibits member-states from restricting trade based on the way in which goods have been produced (process and production methods – PPMs). From an environmental perspective, it is often the production process that needs to be regulated (e.g. to reduce greenhouse gas emissions from manufacturing) and many international environmental agreements seek to restrict such environmentally damaging side-effects. Indeed, environmentalists have long complained about the GATT's "chilling" effect on taking out trade measures focused on polluting production methods (Eckersley 2004).

More recently, the creation of the WTO at the end of the Uruguay Round has signaled a greater willingness in the trading system to recognize the legitimacy of environmental policies (Charnovitz 2007). Thus, the preamble of the Marrakesh Agreement Establishing the WTO lists sustainable development and environmental protection as explicit objectives for the trading system. Although not legally binding, the preamble represents an important departure from the GATT's previous philosophy of a strict separation of trade and environmental policy. Furthermore, because the WTO also strengthened the GATT's dispute settlement mechanism and made its rulings legally binding, the evolving WTO jurisdiction on cases involving environmental trade measures has assumed greater importance in balancing the trade–environment relationship.

Other notable achievements of the Uruguay Round include the Agreements on Technical Barriers to Trade (TBT) and on the Application of Sanitary and Phytosanitary Measures (SPS). The TBT agreement sets rules for the use of technical regulations and standards with a view to minimizing their trade-distorting effect (Stein 2009). It recognizes the right of countries to impose such measures to protect human health and the environment, but stipulates that these should be not more trade-restrictive than necessary. For example, an environmental label that informs consumers about the potential health risks associated with a particular product could be considered acceptable under WTO rules, if applied in a non-discriminatory manner, but not if it aims solely at PPM characteristics of a product (e.g. carbon intensity of car manufacturing). However, voluntary measures such as eco-labels created by private actors do not fall under WTO jurisdiction.

The SPS agreement, which deals with measures to protect human, animal, or plant life or health, similarly allows states to take such measures where they do not lead to discrimination or disguised trade restrictions (Charnovitz 1999). The TBT and SPS Agreements both encourage the harmonization or creation of international standards. Article 2.2 of the Agreement further specifies that SPS measures are to be based on scientific principles of risk assessment and sufficient scientific evidence. This requirement can be temporarily suspended where "relevant scientific evidence is insufficient," but additional scientific information is to be obtained to carry out a full risk assessment "within a reasonable period of time" (Article 5.7). The SPS Agreement is the only trade agreement that formally recognizes precaution as a justification for taking trade measures where there is scientific uncertainty but some evidence of potential harm. The question that has repeatedly pitted the WTO against environmentalists is whether such uncertainty is only a temporary phenomenon or a more persistent problem that pervades many areas of environmental policy-making, such as food safety and genetically modified organisms (Post 2006; Isaac and Kerr 2007).

Multilateral Environmental Agreements, Trade Measures, and the WTO

Of the over 500 multilateral environmental agreements (MEAs) that have been created in the last four decades, a small but growing proportion includes trade measures among their regulatory instruments. As trade restrictions become more popular in global environmental policy-making, concern is rising that these measures will increasingly come into conflict with WTO rules. The definition of trade measures in MEAs is fairly wide and often imprecise. It most commonly refers to various forms of restrictions on trade for environmental purposes, such as bans on the trade of certain polluting substances or embargoes of specific countries that are in breach of environmental obligations. It may also include other measures that have an indirect trade impact, such as reporting requirements, labeling systems, prior consent requirements, or fiscal instruments (e.g. taxes, subsidies) (Brack and Gray 2003: 5–6). Some MEAs are designed to regulate trade, such as the Convention on Trade in Endangered Species (CITES), while others use trade restrictions as one of several instruments to support their main environmental goal (e.g. the Montreal Protocol on Substances that Deplete the Ozone Layer).

Trade measures have become popular instruments in MEAs for a number of reasons. They broadly serve three purposes (see Brack and Gray 2003: 13–15):

- *Target environmental harm.* Most trade measures in MEAs seek to tackle environmental problems by restricting the international movement of products or species that are potentially harmful or endangered.
- *Promote compliance and regime effectiveness.* Some MEAs use trade measures to ensure the effective operation of an environmental regime. For example, restrictions may be imposed to punish countries that do not fully comply with a regime's provisions, or to prevent industrial flight to non-parties, so-called "leakage."
- *Encourage participation in environmental regimes.* Trade restrictions are also seen as a form of pressure on countries that are reluctant to join an environmental regime. For example, the Montreal Protocol's prohibition of trade with non-parties encouraged some countries to join the agreement to prevent being excluded from the international trade in regulated substances and products containing them.

Trade experts have raised several concerns about trade measures in MEAs. WTO rules require environmental trade measures to be non-discriminatory, that is they should not discriminate between "like products" from different WTO members or between domestic and international production. Where environmental treaties target products because of the underlying process and production methods rather than the environmental quality of the product itself, any resulting trade interference could be seen to be in breach of WTO obligations. A further area where MEAs and the WTO rules could clash is where one party to a MEA uses trade sanctions against a non-party, but both parties are members of the WTO. In such cases, the party that suffers a trade sanction could take action under the WTO alleging breach of trade rules. As yet, no WTO member has challenged an MEA in the WTO's dispute-settlement mechanism, but with growing use of MEA-based trade measures a future conflict over their WTO compatibility cannot be ruled out. The next section considers what recent WTO dispute settlement cases tell us about the evolution of WTO jurisdiction on trade–environment conflicts.

Trends in WTO Jurisprudence

Only a very small fraction of the over 500 disputes that have been considered under the GATT/WTO dispute-settlement mechanism relate to environmental issues, even though environment-related trade disputes have attracted a great deal of public attention. A closer examination of the most important cases reveals important developments in trade jurisdiction.[2]

Tuna–Dolphin

One of the earliest and most controversial trade–environment disputes concerned a US ban on certain tuna imports as part of a wider effort to protect dolphins. The 1972 Marine Mammal Protection Act (MMPA) required US fishermen to use dolphin-safe fishing methods to prevent the unwanted trapping of dolphins in purse seine nets. In 1984, the US Congress allowed the US to impose import bans on tuna from countries that did not employ dolphin-safe fishing methods. This trade measure was designed to prevent foreign competition from circumventing the MMPA's provisions and gaining an unjustified competitive advantage over US fishermen. When the USA implemented an embargo on tuna imports from Mexico and a few other countries in 1990, Mexico filed a complaint with the GATT, arguing that the US ban was illegal as it was focused on process and production methods (the type of nets that trap dolphins), rather than the product itself (tuna). Mexico further argued that the USA was not allowed to use GATT Article XX to force other countries to abide by its domestic environmental laws (extraterritoriality). The GATT panel that heard the case decided in Mexico's favor in 1991, but the ruling never became legally binding. In light of the upcoming negotiations on the North American Free Trade Agreement (NAFTA), Mexico decided not to demand the formal adoption of the decision. In any case, the GATT rules gave any party, such as the USA, the right to veto a panel decision. The decision caused uproar among environmentalists and led to a protracted debate in the 1990s about whether the GATT was fundamentally hostile to environmental concerns (Esty 1994).

USA–Gasoline

In 1990, the USA amended the Clean Air Act (CAA) in an effort to improve air quality by reducing adverse emissions from gasoline use. The law mandated the sale of "reformulated" (i.e. cleaner) gasoline in heavily populated urban areas but permitted the continued sale of "conventional" gasoline in more rural areas. To prevent a shift in inexpensive but highly polluting gasoline ingredients from urban to rural areas, the law also stipulated that conventional gasoline must remain as clean as it was in 1990 (the "baseline"). By and large, domestic refiners were allowed to use individual baselines that were actually in use in 1990, while foreign producers had to follow an average baseline set by the Environmental Protection Agency (EPA). This, Venezuela and Brazil argued, was in conflict with Article III of the GATT as it discriminated against imported products. In 1996, the WTO Appellate Body decided that the baseline establishment methods were indeed inconsistent with Article III and could not be justified by Article XX, as the US had claimed. However, the Appellate Body found that the US measures were aimed at the conservation of natural resources, and that WTO members were free to set their own environmental objectives, provided they do so in conformity with WTO rules, in particular with regard to the treatment of domestic and foreign products. The dispute settlement body, now operating under the strengthened rules of the WTO agreement, thus took a broader view of the environmental purpose of the trade measure and did not focus solely on the discriminatory nature of the measure (Trebilcock and Howse 2005: 526–528).

Shrimp–Turtles

A similar case to the *Tuna–Dolphin* dispute emerged in 1997, when India, Malaysia, Pakistan, and Thailand filed complaints at the WTO against a US decision to force foreign shrimp trawlers to use so-called "turtle excluder devices" (TEDs) when fishing in areas where sea turtles are present. The plaintiffs argued that this measure, which was based on America's Endangered Species Act of 1973, was in breach of WTO rules as it threatened foreign producers with a trade ban if they did not comply with US environmental law. Again, the case was decided under the enhanced powers of the WTO agreement and in the context of the WTO's greater emphasis on balancing free trade with environmental sustainability. In 1998, the dispute settlement body ruled that the US import ban was generally a legitimate policy with regard to provisions under Article XX related to "exhaustible natural resources." However, it also found that the way the ban operated, and the fact that the USA had previously negotiated treaties on sea-turtle protection with some but not all affected countries, constituted "arbitrary and unjustifiable discrimination" between WTO members. The USA subsequently changed its rules so that they were targeted at individual shipments rather than at countries – a practice that the WTO decided was justified under Article XX. While the USA technically lost the initial case, the decision marked an important shift in WTO jurisdiction as it acknowledged that in certain circumstances, countries can use trade measures with the aim of protecting natural resources. The USA lost the case not because it aimed to protect the environment but because it had designed the measure in a discriminatory way – similar to the above gasoline case. Critically for the debate on whether the WTO and environmental policies are compatible, the ruling also pointed to the possibility that trade restrictions can be based on process and production methods in another country if these restrictions do not arbitrarily and unjustifiably discriminate between different countries (Howse 2002).

EU–Biotech

A series of food and feed safety scares in Europe in the late 1980s and in the 1990s created considerable public pressure for more stringent food-safety measures at the European level. In the second half of the 1990s, NGO campaigns and consumer hostility against genetically modified organisms (GMOs) led the EU to impose a *de facto* moratorium on GMO approvals and imports. Under pressure from their farming and biotechnology sectors, the USA, Canada, and Argentina in 2003 brought a WTO case against the EU's restrictions on the marketing of GMOs. At the heart of the dispute was the question of whether the EU was entitled to act in a precautionary manner even though a high degree of scientific uncertainty surrounded the GMO safety debate. The use of the WTO as a forum to settle a dispute over the appropriate use of precaution in environmental risk regulation proved controversial, not least since the Cartagena Protocol on Biosafety had been adopted in 2000 in the face of US resistance (Falkner 2007). In 2006, the WTO ruled against the EU on procedural grounds, finding that the *de facto* GMO moratorium was in violation of WTO law, but did not pass a substantive judgment on the WTO consistency of the EU's precautionary GMO legislation as such. By the time the ruling was announced, the EU had already revised its regulations on GMOs and lifted its moratorium at least partially, even though its GMO approval process remains complex and prone to substantial delays due to domestic resistance to agricultural biotechnology (Lieberman and Gray 2008).

Brazil–Retreaded Tires

In late 2004, Brazil decided to strengthen its import restrictions on retreaded tires (reconditioned old tires for further use) from non-Mercosur countries, arguing that the disposal of such tires creates environmental and human health problems. A year after Brazil imposed these restrictions, the EU asked for a WTO panel to consider whether they conformed to WTO rules. Brazil claimed that its import restrictions were justified under Article XX and that it was obliged to exclude Mercosur countries from the restrictions according to the rules of the customs union. The EU countered that the exemption of Mercosur countries from the import restriction constituted a breach of the WTO's non-discrimination rule, among others. Both the Panel and the Appellate Body ruled in 2007, albeit for different reasons, that Brazil's import restrictions were inconsistent with WTO rules and could not be justified by Article XX. As in earlier rulings such as *US–Gasoline* or *Shrimp–Turtles*, the Appellate Body argued that import bans can be justified on environmental grounds, but that the chapeau (introductory provisions) of Article XX stipulates that they must not lead to "arbitrary and unjustifiable discrimination between countries." Brazil complied with the DSB's request to revise its laws to make them conform to WTO rules.

Overall Trends in WTO Jurisdiction

Over the past two decades, GATT/WTO jurisdiction on environment-related trade measures has changed considerably. While earlier rulings (*Tuna–Dolphin* case) rejected trade restrictions aimed at process and production methods (PPMs) outside a country's own jurisdiction, the *US–Gasoline* case marked the cautious beginning of a less restrictive interpretation of environmental measures. In this case, the WTO panel stressed that trade measures must not discriminate among countries but acknowledged that they can be based on grounds of environmental protection. The *Shrimp–Turtle* case further strengthened this shift in the WTO's interpretation of environmental trade measures. The decision almost reversed the earlier *Tuna–Dolphin* decision by arguing that a trade measure based on PPMs *can* be directed at other countries under Article XX, and that animals can qualify as an "exhaustible natural resource" that may be protected through trade bans. In the *EC–Biotech* case, the WTO Panel reinforced the importance of non-discrimination and the proper application of regulatory procedures, but acknowledged the importance of scientific uncertainty in justifying trade restrictions, arguing that a moratorium amidst scientific uncertainty need not necessarily violate international trade law. Thus, WTO jurisdiction has gradually come to accept that trade-restricting measures under Article XX can be justified for environmental reasons, but continues to insist that they must not constitute an arbitrary and/or unjustifiable discrimination. Indeed, the primary reason why environmental measures in *Gasoline*, *Shrimp–Turtle*, and *Retreaded Tires* were found to be in breach of WTO rules was not the ultimate objective of these measures but the way in which they had been applied (DeSombre and Barkin 2002).

Climate Change and International Trade

Climate change has added a new and urgent dimension to the debate on trade and the environment: whereas previous trade–environment conflicts usually focused on only a limited number of industries or countries, global warming affects virtually every country and all aspects of economic life. Addressing climate change creates fundamental questions of current and inter-generational fairness, equity, and freedom, and involves global collective action combined with unprecedented degrees of market failure and scientific uncertainty (Stern 2007). Rich industrialized countries are largely responsible for causing global warming in the past, but the majority of future greenhouse gas emissions will come from rapidly industrializing emerging economies such as China and India. In 2007, China overtook the USA as the world's largest emitter, and in 2008, China and India together produced almost twice as much CO_2 as the 27 European Union countries combined.[3]

If some countries decide to reduce CO_2 emission without similar commitments from others, international trade allows industrial activity to simply shift from the former to the latter. This so-called "carbon leakage" can occur in three ways: energy-intensive industries physically relocate to countries with less stringent regulations; domestic producers lose market share to foreign competitors that increase production; or a lower demand for fossil fuels in high-regulation countries decreases the overall price for these fuels and thus leads to increased consumption in low-regulation countries (Frankel 2009; Weber and Peters 2009). Other factors such as transportation costs, local market conditions, and the cost of capital and labor are often equally or more important reasons behind industrial relocation (World Bank 2008; Weber and Peters 2009). However, where industries move to avoid carbon regulations, such "leakage" undermines the goal of reducing global emissions and discourages ambitious climate policies. Depending on the type of emission reduction scheme, leakage rates (i.e. the increase in emissions in low-regulation countries as a share of reductions in high-regulation countries) have been estimated at as high as one fourth for the iron and steel sector and up to one third for the cement industry (Reinaud 2008). When European countries ratified the Kyoto Protocol and introduced the EU Emissions Trading Scheme (ETS), for instance, there was great concern over whether the refusal of the USA to ratify the Kyoto Protocol would allow it to be a "free rider" on Europe's climate policy (Biermann and Brohm 2005), with some arguing that the US rejection of Kyoto can be interpreted as a hidden subsidy for its industry and may thus conflict with trading rules (Stiglitz 2007).[4]

Against this background, some have called for trade measures to be used to discourage carbon leakage and free-riding (Stern 2007; Stiglitz 2007). As Barrett (2010: 3) put it bluntly: "If trade measures can enforce trade agreements, why not use trade measures to enforce climate agreements?" A common line of argument is that since the WTO has the strongest compliance system of any international regime, it could strengthen international climate policies (Biermann and Brohm 2005; Frankel 2009). Two specific proposals for climate-related trade measures have been put forward. A first proposal involves taxing imports from countries that apply less stringent carbon emission limits. A so-called "border tax adjustment" (BTA) forces importers to pay a fee that reflects the costs of carbon emissions while exporters may obtain a tax credit to avoid double taxation (Frankel 2009; Kaufmann and Weber 2011). A second proposal envisions requirements to purchase emission permits in a cap-and-trade system so that foreign and domestic producers pay the same price for emitting a ton of CO_2. France, the USA, and the EU have already tabled proposals for how to incorporate trade measures into climate change legislation in the form of BTAs or the mandatory purchase of emission permits (Biermann and Brohm 2005; Cosbey 2008; Tarasofsky 2008).

Ideally, since all current economic activity in one way or another produces carbon emissions, trade measures ought to cover the largest possible range of products. One way to do this would be to consider the carbon footprint of each individual product, possibly with the help of a standard developed by the International Organization for Standardization (ISO 14067; see also Gros et al. 2010). While this would ensure that a large share of international trade is covered, developing a comprehensive method to quantify the carbon content of every traded product for tax purposes is difficult: national authorities may not have the capacity to collect data, producers have an incentive to underestimate carbon content, and complex international supply chains make tracing the carbon content of each individual component of a product cumbersome (Cosbey 2008). It would certainly be easier to target only a limited range of internationally traded energy-intensive materials such as aluminum, cement, steel, paper, glass, iron, and chemicals. This approach, however, may disproportionately harm manufacturers in technologically advanced countries who import these materials for further processing without significantly reducing emissions of energy-intensive manufacturing in heavily polluting countries (Cosbey 2008; Weber and Peters 2009). Moreover, the quantity of emissions associated with basic materials very much depends on the source of energy with which they were processed (e.g. fossil fuels versus hydropower) (Cosbey 2008).

Measuring carbon content poses further difficulties. A system that seeks to count the emissions of individual firms may be too complicated to work, given the complexity of international supply chains. Establishing carbon content on the basis of industry averages would be easier to achieve but raises the question of whether industrialized or emerging-economy standards are used as a point of reference or whether actual emissions count. Any aggregate national measure raises the question of how to quantify and compare different types of emission reduction policies. Should the carbon intensity per capita or per unit of economic output count? How will changes in emission policies be reflected in trade measures? How can a situation be avoided where trade is simply redirected via a third country with nominally stricter emission targets (Weber and Peters 2009)?

Finally, a critical question concerns the relationship between such measures and international trade law. On this, there seems to be agreement that many of these climate policies could, *in principle*, be seen to conform to WTO rules (Bhagwati and Mavroidis 2007; Frankel 2009). In a joint report, the WTO and UNEP (2009: xix) argue that international trade rules permit, "under certain conditions, the use of border tax adjustments on imported and exported products." The conditions, however, are crucial, as are the trade rules under which conformity is claimed. For instance, aiming trade measures at a country as such is likely to fall foul of the most-favored-nation (MFN) principle, and the WTO has in the past been skeptical of measures that seek to directly influence policies in another country (Tarasofsky 2008; Messerlin 2012). Equally controversial is the idea of trade measures aimed at process and production methods (PPMs). Although the WTO Appellate Body found in the asbestos case that "consumer preferences" are a valid consideration for distinguishing products that would normally be treated as "like" products (Kaufmann and Weber 2011), it is questionable whether products can be treated differently based on the energy and emissions profile of production processes which do not affect the final product as such (Weber and Peters 2009). Furthermore, growing reliance on national or regional emissions trading schemes has led to situations in which governments provide direct or indirect subsidies to domestic companies (e.g. by allocating low-cost permits or offering subsidies to compensate for the costs of permits), which may violate provisions of the WTO's Agreement on Subsidies and Countervailing Measures (Henschke 2012). Alternatively, the compulsory inclusion of foreign companies in such emissions trading schemes can be seen as a unilateral act that falls foul of the WTO's restrictions on the extraterritorial application of domestic environmental laws, as can be seen in the international spat over the EU's plan to include foreign airlines in its emissions trading scheme.

On the other hand, it has been argued that many climate measures would conform to Article XX of the GATT on the depletion of natural resources (Biermann and Brohm 2005; Kaufmann and Weber 2011). The outcome of the *Shrimp–Turtle* case can be interpreted to have "legalized" trade measures aimed at process and production methods, and the decision in the *US–Gasoline* case defined clean air as an exhaustible natural resource. Together, these two decisions may have paved the way for trade-related climate measures to be consistent with world trade law (Bhagwati and Mavroidis 2007). Indeed, since climate change also affects biodiversity, the "clean air" decision might not even be a necessary condition for a trade measure to conform to WTO statutes (Wiers 2008). Thus, while CO_2-related trade measures are technically difficult to implement,[5] they may not necessarily conflict with world trade law. Still, imposing climate-related trade restrictions would be politically controversial, which may explain why climate leaders such as the EU have been reluctant to introduce BTAs or the mandatory purchase of emission allowances for importers.

Since any long-term solution to the challenge of climate change requires close international cooperation, there is considerable concern that unilateral trade measures would undermine international political processes and could prove counterproductive (Cosbey 2008; Weber and Peters 2009; Barrett 2010). After all, the UNFCCC stipulates that emerging economies have "common *but differentiated* responsibilities," and in the run-up to the 2009 Copenhagen climate conference, a group of developing countries warned that:

> Parties shall not resort to any form of unilateral measures, including fiscal and non-fiscal border measures, against goods and services imported from other Parties, in particular from developing country Parties, on grounds of stabilization and mitigation of climate change (Ad Hoc Working Group 2009).

Indeed, Charnovitz (2003) argues that political concern over trade measures has already led to a "chilling effect" in environmental negotiations (see also Cosbey 2008: 6). Proposals for compensating developing countries by transferring the revenue from such trade measures (Biermann and Brohm 2005; Weber and Peters 2009) may go some way to assuage their concerns but may not solve the underlying problem of carbon leakage and shifts in competitiveness. Furthermore, trade experts warn that unilateral measures would lead to a "slippery slope" towards an abuse of climate change for protectionist purposes (Bhagwati and Mavroidis 2007; Frankel 2009). Still, climate leaders will be tempted to use trade measures as a "stick" in negotiations to put pressure on other countries to join a global agreement.

Conclusion: Global Policy Implications

The trade-environment nexus remains a controversial and challenging issue on the international trade agenda. Some progress has been made in identifying the circumstances in which international trade and environmental protection can be mutually compatible, but several areas of contention and conflict remain.

The first area relates to the WTO's general approach to environmental policy. Some observers call on the WTO to become more engaged with environmental issues, not least since the WTO already adjudicates cases that involve conflicts between environmental measures and international trade law. Given the WTO's *de facto* impact on global environmental policy, they argue that the WTO should take on more formal environmental responsibilities, even though details of such a closer engagement with the global environmental agenda remain sketchy. On the other hand, concerns have been raised that environmental protection might actually take a back seat on the international trade agenda due to an increasing use of bilateral agreements instead of multilateral ones and a generally low interest among some countries in issues related to environmental protection (Neumayer 2004). The WTO has so far trod a careful path through this debate, stating repeatedly that, while it aims to contribute to sustainable development, it does not consider itself as an environmental protection agency (WTO 2004).

The second area relates to the interpretation of existing legal provisions. Despite an evolving mandate and institutional framework, the WTO has had significant impact on certain environmental measures, as outlined above. Past decisions have clarified what a "necessary" environmental measure is; what is meant by "exhaustible natural resource"; whether measures can extend extraterritorially; and how "arbitrary" and "unjustifiable" should be interpreted under the chapeau of Article XX. Disagreement still exists, however, with regard to environmental measures aimed at PPMs, especially when they are "unincorporated," that is, when they cannot be detected in the final product. The definition and use of precaution remains equally contested, as has been illustrated by the *EC–Biotech* case and the question of "sound" science as a criterion for policy-making versus a broader interpretation of the evidence basis for risk assessment.

The third area relates to the question of inclusiveness and transparency of decision-making. While the CTE has been tasked with addressing the relationship between MEAs and the WTO, both in institutional and jurisdictional terms, there remains considerable debate on how to integrate the two, especially when the former continue to employ trade-restricting measures that remain vulnerable to challenges under WTO law (Eckersley 2004; Palmer and Tarasofsky 2007). Another contentious point is the access of external stakeholders, especially civil society and NGOs, to WTO decision-making processes. While the WTO has promoted dialogue with interested organizations, NGOs continue to raise concerns about the lack of transparency in the WTO's deliberations and negotiations, especially with regard to environmental issues.

The fourth and final area relates to the increasingly important impact of the climate-change debate on international trade. As states explore different options for reducing greenhouse gas emissions and global climate governance becomes increasingly fragmented (Falkner *et al.* 2010), it is becoming clear that trade measures will be part of the international effort to combat global warming. This could be in the form of border tax adjustment to address international competitiveness issues, preferential treatment of climate-friendly goods and services, renewable energy subsidies and product labels indicating carbon content, among others (Brewer 2010). Efforts to enforce international climate policy through trade measures may test the scope of Article XX (Frankel 2009), and a push to target carbon content in internationally traded goods may test the WTO's willingness to accept unilateral trade measures that are based on PPMs (Hufbauer and Kim 2009). The WTO itself recognizes its responsibility in the international community to address climate change as part of its sustainable development agenda, but sees its role primarily as an arbiter of conflicts. The challenge will be to avoid the trap of green protectionism, where general trade restrictions are used to seek compliance with quite distinct climate goals. Climate policy may yet prove to be the biggest challenge for the WTO's ability to manage the trade–environment relationship.

Acknowledgments

This chapter builds on but extends Falkner and Jaspers (2012). Research towards this chapter was supported by a grant from the Kolleg-Forschergruppe (KFG), "The Transformative Power of Europe" at Freie Universität Berlin. The KFG is funded by the German Research Foundation (DFG) and brings together research on the diffusion of ideas in the EU's internal and external relations. For further information please consult www.transformeurope.eu.

Notes

1   A comprehensive guide to WTO law and jurisdiction in relation to environmental matters can be found in Bernasconi-Osterwalder *et al.* (2006).
2   An overview of these and other environment-related cases, as well as panel and appellate body reports, can be found at: http://www.wto.org/english/tratop_e/dispu_e/dispu_status_e.htm (accessed October 20, 2012).
3   United Nations Statistics Division (UNSTATS), available at http://mdgs.un.org/unsd/mdg/SeriesDetail .aspx?srid=749 & crid= (accessed 20 October, 2012).
4   Bhagwati and Mavroidis (2007), however, disagree and argue that for a subsidy in the form of a tax rebate to conflict with world trade law, a country must first signal that it intends to impose a tax but then refrain from doing so. In the case of the Kyoto Protocol, this was not the case for the USA.
5   Gros *et al.* (2010) argue that there are no "insurmountable practical obstacles" to introducing a CO_2 border tax.

References

Ad Hoc Working Group on Long-Term Cooperative Action under the Convention. 2009. "Report of the Ad Hoc Working Group on Long-Term Cooperative Action under the Convention on its Seventh Session, Held in Bangkok from 28 September to 9 October 2009, and Barcelona from 2 to 6 November 2009," http://unfccc.int/resource/docs/2009 /awglca7/eng/14.pdf (accessed October 20, 2012).

Anderson, K. and W.J. McKibbin. 2000. "Reducing Coal Subsidies and Trade Barriers: Their Contribution to Greenhouse Gas Abatement." *Environment and Development Economics*, 5(4): 457–481.

Barrett, S. 2010. "Climate Change and International Trade: Lessons on their Linkage from International Environmental Agreements." Background paper for "Thinking Ahead on International Trade (TAIT)," 2nd Conference Climate Change, Trade and Competitiveness: Issues for the WTO, Geneva, June 16–18.

Bernasconi-Osterwalder, N., D. Magraw, M.J. Oliva *et al.* 2006. *Environment and Trade: A Guide to WTO Jurisprudence*. London: Earthscan.

Bhagwati, J. 2004. *In Defense of Globalization*. Oxford: Oxford University Press.

Bhagwati, J. and P.C. Mavroidis. 2007. "Is Action against US Exports for Failure to Sign Kyoto Protocol WTO-Legal?" *World Trade Review*, 6(2): 299–310.

Bhagwati, J. and T.N. Srinivasan. 1996. "Trade and Environment: Does Environmental Diversity Detract from the Case for Free Trade?" In *Fair Trade and Harmonization: Prerequisites for Free Trade?*, ed. by J. Bhagwati and R. Hudec, 159–223. Cambridge, MA: MIT Press.

Biermann, F. and R. Brohm. 2005. "Implementing the Kyoto Protocol without the USA: The Strategic Role of Energy Tax Adjustments at the Border." *Climate Policy*, 4: 289–302.

Brack, D. and K. Gray. 2003. *Multilateral Environmental Agreements and the WTO*. London: Royal Institute of International Affairs.

Brewer, T.L. 2010. "Trade Policies and Climate Change Policies: A Rapidly Expanding Joint Agenda." *World Economy*, 33(6): 799–809.

Charnovitz, S. 1999. "Improving the Agreement on Sanitary and Phytosanitary Standards." *Trade, Environment, and the Millennium*, ed. G.P. Sampson and W.B. Chambers, 171–194. Tokyo: United Nations University Press.

Charnovitz, S. 2003. *Beyond Kyoto: Advancing the International Effort against Climate Change*. Arlington, VA: Pew Center on Global Climate Change.

Charnovitz, S. 2007. "The WTO's Environmental Progress." *Journal of International Economic Law*, 10(3): 685–706.

Cosbey, A. 2008. *Border Carbon Adjustment*. Winnipeg, Manitoba: International Institute for Sustainable Development.

Daly, H.E. 1993. "The Perils of Free Trade." *Scientific American*, 269(5): 50–57.

DeSombre, R. and J.S. Barkin. 2002. "Turtles and Trade: The WTO's Acceptance of Environmental Trade Restrictions." *Global Environmental Politics*, 2(1): 12–18.

Eckersley, R. 2004. "The Big Chill: The WTO and Multilateral Environmental Agreements." *Global Environmental Politics*, 4(2): 24–50.

Economy, E.C. 2004. *The River Runs Black: The Environmental Challenge to China's Future*. Ithaca, NY: Cornell University Press.

Esty, D. 1994. *Greening the GATT*. Washington, DC: Institute for International Economics.

Falkner, R. 2007. "The Political Economy of 'Normative Power' Europe: EU Environmental Leadership in International Biotechnology Regulation." *Journal of European Public Policy*, 14(4): 507–526.

Falkner, R. and N. Jaspers. 2012. "Environmental Protection, International Trade and the WTO." In *The Ashgate Research Companion on International Trade*, ed. Kenneth Heydon and Stephen Woolcock, 245–260. Aldershot: Ashgate.

Falkner, R., H. Stephan, and J. Vogler. 2010. "International Climate Policy after Copenhagen: Towards a 'Building Blocks' Approach." *Global Policy*, 1(3): 252–262.

Frankel, J.A. 2009. "Addressing the Leakage/Competitiveness Issue in Climate Change Policy Proposals." In *Climate Change, Trade, and Competitiveness: Is a Collision Inevitable?*, ed. I. Sorkin and L. Brainard, 69–82. Washington, DC: Brookings Institution Press.

Gallagher, K.P. 2004. *Free Trade and the Environment: Mexico, NAFTA, and Beyond*. Palo Alto, CA: Stanford University Press.

Garcia-Johnson, R. 2000. *Exporting Environmentalism: U.S. Multinational Chemical Corporations in Brazil and Mexico*. Cambridge, MA: MIT Press.

Goldsmith, E. and J. Mander, eds. 2001. *The Case against the Global Economy: And for a Turn Towards Localization*. London: Earthscan.

Gros, D., C. Egenhofer, N. Fujiwara *et al.* 2010. *Climate Change and Trade*. Brussels: Centre for European Policy Studies.

Grossman, G.M. and A. Krueger. 1993. "Environmental Impacts of a North American Free Trade Agreement." *The US–Mexico Free Trade Agreement*, ed. P. Garber, 13–56. Cambridge, MA, MIT Press.

Henschke, L. 2012. "Going It Alone on Climate Change. A New Challenge to WTO Subsidies Disciplines: Are Subsidies in Support of Emissions Reductions Schemes Permissible under the WTO?" *World Trade Review*, 11(01): 27–52.

Hettige, H., M. Mani, and D. Wheeler. 1998. *Industrial Pollution in Economic Development: Kuznets Revisited*. World Bank Development Research Group Working Paper, No. 1876. Washington DC: World Bank.

Heydon, K. and S. Woolcock. 2009. *The Rise of Bilateralism: Comparing American, European and Asian Approaches to Preferential Trade Agreements*. Tokyo: United Nations University Press.

Howse, R. 2002. "The Appellate Body Rulings in the *Shrimp/Turtle* Case: A New Legal Baseline for the Trade and Environment Debate." *Columbia Journal of Environmental Law*, 27(2): 491–521.

Hufbauer, G.C. and J. Kim. 2009. *The World Trade Organization and Climate Change: Challenges and Options*. Washington, DC: Peterson Institute for International Economics.

Isaac, G.E. and W.A. Kerr. 2007. "The Biosafety Protocol and the WTO: Concert or Conflict?" In *The International Politics of Genetically Modified Food: Diplomacy, Trade and Law*, ed. R. Falkner, 195–212. Basingstoke: Palgrave Macmillan.

Kaufmann, C. and R.H. Weber. 2011. "Carbon-Related Border Tax Adjustment: Mitigating Climate Change or Restricting International Trade?" *World Trade Review*, 10(4): 497–525.

Lieberman, S. and T. Gray. 2008. "The World Trade Organization's Report on the EU's Moratorium on Biotech Products: The Wisdom of the US Challenge to the EU in the WTO." *Global Environmental Politics*, 8(1): 33–52.

Messerlin, P.A. 2012. "Climate and Trade Policies: From Mutual Destruction to Mutual Support." *World Trade Review*, 11(01): 53–80.

Neumayer, E. 2001. *Greening Trade and Investment: Environmental Protection without Protectionism*. London: Earthscan.

Neumayer, E. 2004. "The WTO and the Environment: Its Past Record is Better than Critics Believe, but the Future Outlook is Bleak." *Global Environmental Politics*, 4(3): 1–8.

OECD. 2005. *Environmental Requirements and Market Access*. OECD Trade Policy Studies. Paris: Organisation for Economic Co-operation and Development.

Palmer, A. and R. Tarasofsky. 2007. *The Doha Round and Beyond: Towards a Lasting Relationship between the WTO and the International Environmental Regime*. London: Chatham House.

Post, D.L. 2006. "The Precautionary Principle and Risk Assessment in International Food Safety: How the World Trade Organization Influences Standards." *Risk Analysis*, 26(5): 1259–1273.

Princen, T. 1997. "The Shading and Distancing of Commerce: When Internationalization is Not Enough." *Ecological Economics*, 20: 235–253.

Princen, T., M. Maniates, and K. Conca, eds. 2002. *Confronting Consumption*. Cambridge, MA, MIT Press.

Reinaud, J. 2008. *Issues behind Competitiveness and Carbon Leakage: Focus on Heavy Industry*. IEA Information Paper. Paris: International Energy Agency.

Reppelin-Hill, V. 1999. "Trade and Environment: An Empirical Analysis of the Technology Effect in the Steel Industry." *Journal of Environmental Economics and Management*, 38(3): 283–301.

Sampson, G.P. 2005. *The WTO and Sustainable Development*. Tokyo: United Nations University Press.

Stein, J. 2009. "The Legal Status of Eco-Labels and Product and Process Methods in the World Trade Organization." *American Journal of Economics and Business Administration*, 1(4): 285–295.

Stern, N. 2007. *The Economics of Climate Change: The Stern Review*. Cambridge: Cambridge University Press.

Stiglitz, J.E. 2007. *Making Globalization Work*. New York: W.W. Norton.

Tarasofsky, R.G. 2008. "Heating Up International Trade Law: Challenges and Opportunities Posed by Efforts to Combat Climate Change." *Carbon and Climate Law Review*, 1: 7–17.

Trebilcock, M.J. and R. Howse. 2005. *The Regulation of International Trade*, 3rd edn. New York: Routledge.

Weber, L.C. and G.L. Peters. 2009. "Climate Change Policy and International Trade: Policy Considerations in the US." *Energy Policy*, 37: 432–440.

Wiers, J. 2008. "French Ideas on Climate and Trade Policies." *Carbon Climate Law Review*, 1: 18–32.

World Bank. 2008. *International Trade and Climate Change: Economic, Legal, and Institutional Perspectives*. Washington, DC: World Bank.

WTO. 2004. *Trade and Environment at the WTO*. Geneva: World Trade Organization.

WTO and UNEP. 2009. *Trade and Climate Change. WTO–UNEP Report*. Geneva: World Trade Organization.

Chapter 25

Global Finance and the Environment

Christopher Wright

Introduction

This chapter considers the relationship between global finance and the environment. The health of the global environment is increasingly vulnerable. The principal drivers of environmental degradation have been the exponential increase in demand for natural resources – including fossil fuels, fresh water, basic minerals, and food – and the widespread pollution associated with their consumption, including greenhouse gas emissions. The scale of natural resource use has been greatly facilitated by the transformation of finance from being organized within a tapestry of interdependent national markets to being an increasingly integrated global system. Yet, research in international political economy has only recently begun to consider how the globalization of financial markets may impact the state of the environment and global environmental governance (Helleiner and Clapp 2012).

This chapter aims to contribute to this emerging area of research by providing an overview of the ways in which global finance impacts the environment, emphasizing a core set of institutional rules and financial actors. With reference to the latter, the analysis does not focus on government-owned financing mechanisms explicitly mandated to promote environmental sustainability (see Matz 2005; van Putten 2008; Park 2012). Instead, it considers the most important mainstream financial institutions with long-term investment mandates. The first section identifies three pathways of influence – systemic, institutional, and instrumental. The second section examines the current state of corporate environmental accounting, reporting, and stock-listing requirements. The third and fourth sections consider the extent to which a set of long-term investors – pension funds, sovereign wealth funds, project finance banks, and export credit agencies – address environmental issues in their investment practices. The concluding section summarizes the analysis and looks ahead.

The Handbook of Global Climate and Environment Policy, First Edition. Edited by Robert Falkner.
© 2013 John Wiley & Sons, Ltd. Published 2013 by John Wiley & Sons, Ltd.

Finance and the Environment: Three Pathways of Influence

The state of the environment today cannot be understood without reference to how the global economy is organized and the patterns of resource production and consumption it encourages (Richardson 2008; Newell and Paterson 2010). Finance plays a significant role in sustaining the global economy by mobilizing capital for energy, infrastructure, manufacturing, and technological innovation. The cost of capital is an important determinant for whether different economic activities are commercially viable and, by implication, whether governments and companies are able and willing to pursue them. Economic booms and busts in industries that have a positive or adverse impact on the environment – whether it is renewable energy generation or open-pit coal mining – are often preceded by shifts in investor sentiments that move capital in or out of particular companies and projects. Overall, patterns of capital allocation across time and space shape the choices governments, companies, and consumers make, and thereby greatly influence the environmental consequences of economic life.

Systemic Impacts

Three systemic features of the global financial system are particularly relevant to global environmental governance. First, financial markets in the current wave of globalization are even more integrated across geographic regions, financial products, and financial services than in previous periods of global economic convergence (Porter 2005; Cerny 2010). Global financial integration has facilitated greater international capital flows but also increased the exposure of governments and companies to financial volatility and shocks. The Latin American debt crisis in the 1980s put pressure on debtor countries to rapidly increase exports of natural resources to an unsustainable rate in order to quickly generate hard currency to service debt and make up for shortfalls in public finances (WCED 1987). The Asian financial crisis in 1997–1998, while reducing regional demand for natural resources, induced a rise in illegal fishing, logging, mining, waste disposal, and clearing of degraded forests for plantations, in part as a result of cuts to government budgets that weakened environmental law enforcement (Dauvergne 2005: 180). Since the global financial crisis in 2008–2009, a number of European governments have scaled back fiscal subsidies and research funding that supported the development and deployment of renewable energy (Bloomberg New Energy Finance 2011: 23).

Second, the structure of financial markets is increasingly shaped by financial innovation. As an example, the creation of global voluntary and regulated carbon markets has allowed companies in the clean technology and renewable energy sectors to obtain financing from investors in disparate places beyond what they otherwise would have access to (Lederer 2010). The emergence of an environmental bond market – underpinned by financial relationships between public and private financial institutions and project developers – provides an additional source of green financing (OECD 2011: 16). However, financial innovations in securitization, coupled with technological developments in investment execution, have also encouraged investors to adopt shorter investment time horizons when deciding on investment strategies and making investment decisions (MacKenzie 2011). Investors with short time horizons – such as hedge funds – and the use of short-term financial instruments – such as derivatives – have become much more widespread than previously. While adding liquidity to capital markets, short-term finance has been found to generate adverse impacts beyond the financial system itself. For example, derivatives markets in food commodities over the past decade have led to a "financialization" of agriculture and greater volatility in food prices, undermining sustainable land use and access to affordable food (Helleiner and Clapp 2012).

And third, structural changes in global stock markets have changed the relationship between companies and shareholders. Whereas individuals – advised by brokers – held most company stock five decades ago, institutional investors such as pension funds, mutual funds, and insurance companies dominate stock markets today. The globalization of national stock markets entails that a much smaller share of stock is now held by domestic investors. In their place, financial leverage in stock markets is concentrated among fewer financial institutions that invest across national markets. However, their power to change corporate behavior is mediated by increasingly lengthy investment chains with multiple intermediaries, whereby sources of capital rarely interact directly with their ultimate beneficiaries. In addition, large companies rely less on stock issuance for their financing needs than previously. While these developments entail that the relationship between companies and shareholders is becoming less consequential to the financing of corporate activities, the latter still retain leverage through the rights they have as owners.

Institutional Impacts

We can distinguish between three types of regulatory institutions that mediate the relationship between finance and the environment. First, there are financial regulations that have implications for how companies account for and report on their environmental impacts to shareholders and the investment community more broadly. Financial statements summarize the profits generated by a company within a given period by making a tally of assets, less liabilities. International financial accounting standards govern how companies prepare financial statements, including how company impacts on the environment are recorded (Thistlethwaite 2011). Corporate disclosure rules define the reporting obligations of companies, while corporate governance standards determine the responsibilities of company boards relative to shareholders. The extent to which such rules mandate companies to account for and report on their environmental impacts and future risks will not only influence corporate environmental management practices, but also whether and how investors account for environmental impacts when attributing value to a financial asset, such as a company stock. Stock exchanges also influence the corporate accounting, management, and reporting practices of listed companies through their indices, listing requirements, and monitoring activities (Siddy 2009; Morales and van Tichelen 2010).

Second, there are environmental regulations that are implemented through financial markets. Given the apparent failure of the Kyoto Protocol to achieve its environmental objectives, its most important legacy may be the emergence of regulated carbon markets as an institutional mechanism for mitigating global climate change.

The three flexible mechanisms under the Kyoto Protocol – emissions trading, joint implementation, and the clean development mechanism – each created new financial markets facilitating trades in carbon-related commodities and helping market participants manage transaction risks (Lederer 2010; Newell and Paterson 2010). These markets and the politics around them produced new spaces for financial actors to engage with environmental issues and shape regulatory responses to environmental problems. Looking ahead, financial transfers will likely remain a cornerstone of any grand bargain struck between states with the largest historical responsibilities for greenhouse gas concentrations in the atmosphere and states that are most rapidly increasing their emissions.

And finally, there are investor-led networks and associations that promote environmental objectives. In general terms, they fall into three categories. First, there are self-regulatory associations established by financial actors to develop and diffuse environmental commitments and standards in particular financial markets or among themselves and their peers (Pattberg 2007; Park 2012). Second, there are networks supported by investors that produce reporting standards and encourage companies to annually disclose how they manage environmental risks, including strategies, action plans, data on environmental impacts, and targets for reducing them (Kolk *et al.* 2008; Newell and Paterson 2010: 65–67). And third, there are investor-driven governance networks that are created to facilitate collective shareholder engagement and influence business-relevant environmental policy and regulation (MacLeod and Park 2011). They pursue their goals through shareholder activism, petitioning, open letters, and policy statements, as well as participating in public hearings and roundtables with regulatory agencies.

Instrumental Impacts

The evolution of the global financial system has coincided with a proliferation of financial institutions that differ in their ownership, mandates, activities, and size. Whereas it can be easy to identify companies that contribute to environmental harm – such as oil companies causing spills, power plants emitting air toxins, or timber companies clear-cutting virgin forest – it can be much more difficult to map the myriad of investors that provide them with capital and benefit from their profits (Richardson 2008: 3). While a comprehensive overview of this complex landscape is beyond the scope of this chapter, it is instructive to identify the main groups of financial actors that influence how financial markets interact with the environment. First, there are large funds that invest capital owned by others, mostly in stocks and bonds. This group includes pension funds, mutual funds, hedge funds, insurers, and sovereign wealth funds (SWFs). Second, there are banking institutions that issue loans and guarantees to companies, projects, and governments. These include commercial banks, government-owned banks, such as national infrastructure banks and multilateral development banks (MDBs), and export credit agencies (ECAs). Third, there are insurers and reinsurers that assist companies, financial institutions, and individuals to manage risks related to the physical environment (Paterson 2001). And finally, there are a large number of financial intermediaries that operate as brokers and advisors between those offering a financial asset – such as a stock or a bond – and those looking to invest in it.

Financial Regulatory Institutions and the Environment

During the last three decades, the deliberate facilitation among governments of the emergence of an integrated global financial system has made it more difficult for national financial regulatory agencies to provide effective regulation (Cerny 2010: 262). As there are no supranational institutions with legal authority akin to the World Trade Organization (WTO) to govern the global financial system, the enforcement of uniform national implementation of international standards and codes has been weak (Helleiner and Pagliari 2010: 3). The policy space left open by governments has been filled by numerous public and private networks and standard-setting organizations, many of which have overlapping mandates and jurisdictions (see also Chapter 23 in this volume). Compared to other policy fields, finance is notable for the significant influence exerted by businesses and industry associations over international regulatory structures central to the functioning of markets (Porter 2005). For example, the four largest accounting firms in the world hold powerful positions in the International Accounting Standards Board (IASB), the international body accepted by most governments as having the legitimate authority to set international financial accounting standards. In some cases, this balance between private and public interests has undermined rule-making that encourages financial institutions to make decisions in favor of the public good (Richardson 2008; Nölke 2010).

It is beyond the scope of this chapter to comprehensively review the challenges associated with regulating the systemic risks embedded in global finance. Instead, the analysis will focus on three sets of rules that are particularly relevant for understanding how finance impacts the environment: corporate environmental accounting, corporate environmental reporting, and stock-listing requirements.

Corporate Environmental Accounting

Financial accounting influences whether and how a company's impact on the environment is given a monetary value and reflected in corporate accounts and statements. International financial reporting standards (IFRS) – defined by the IASB – currently employ a high threshold for companies to record environmental liabilities and risks in their financial statements (Ascui and Lowell 2011; Thistlethwaite 2011). By implication, companies that have caused environmental harm are only in limited cases being expected to report these as either liabilities or risks. This means investors are not being encouraged by international financial accounting standards to divest from companies that cause environmental harm or invest more in those that avoid harm or generate environmental benefits.

There is limited institutional interaction between the policy community promoting corporate environmental accounting and reporting and the established accounting profession and its formal institutions. Private governance initiatives created to expand and standardize environmental accounting and disclosure – such as the Greenhouse Gas Protocol, the Global Reporting Initiative (GRI), and the Carbon Disclosure Project (CDP) – have emerged and evolved with only limited input from financial regulatory institutions. Furthermore, the growing field of carbon accounting – which can be defined as the measurement, collation, and communication of carbon emissions data (Bowen and Wittneben 2011) – has yet to significantly inform financial accounting rules. Full-cost accounting and life-cycle analysis, each of which would facilitate the greater inclusion of environmental externality costs into corporate planning, are not accepted by current international accounting standards as legitimate methods of accounting (Prakash 2000: 26). Since the accounting profession tends to use existing accounting entities, such as taxes, leases, subsidies, and commodities, as references when addressing environmental risks and liabilities, it finds it difficult to accommodate the unique complexity of environmental causes and consequences (Ascui and Lovell 2011).

Corporate Environmental Reporting

There has been a proliferation of corporate environmental reporting during the last decade. Today, virtually all large companies in Europe and North America, and increasingly in other regions as well, report on their environmental risks and activities in annual sustainability reports, sections of their corporate web site, and in submissions to voluntary reporting initiatives. This has been encouraged and facilitated by investor-led reporting organizations. Each year since 2003, the CDP has sent a questionnaire to hundreds of listed companies – on behalf of hundreds of investors – asking them to publicly report on their climate risk strategies, risk assessments, actions plans, greenhouse gas emissions, and reduction targets (Kolk *et al.* 2008; Harmes 2011). The CDP encourages companies to report emissions in accordance with the Greenhouse Gas Protocol developed jointly by the World Resources Institute (WRI) and the World Business Council on Sustainable Development (WBCSD). In 2010, CDP created a similar annual survey for promoting corporate reporting on water risk management. The Forest Footprint Disclosure project and the Global Real Estate Sustainability Benchmark are two other investor-led organizations that work towards improving corporate environmental reporting through annual surveys.

These institutional developments have increased the volume of corporate environmental reporting from individual companies, and across geographic markets. Among a sample of 458 companies from the FTSE All-Share index, 99% referred to environmental topics in their 2009–2010 annual reports, and 67% published quantitative metrics on their environmental impacts (Environment Agency 2010). The most significant improvements in reporting quality have been in the area of climate change risks and energy use. Among the 458 companies on the FTSE All-Share index, 62% produced quantitative information on greenhouse gas emissions and/or energy consumption in 2009–2010, a 112% increase compared to 2006 (Environment Agency 2010). In 2003, the CDP was backed by 35 investors, and 221 companies chose to complete the questionnaire. By 2011, support had grown to 551 investors, and more than 2 124 companies responded, including 81% of the Global 500, 68% of the Standard & Poors 500, and 83% of the South Africa 100 companies (Carbon Disclosure Project 2011).

The growth of voluntary reporting seems to have encouraged some governments to embed environmental disclosure requirements in mandatory disclosure rules. All companies with high emissions covered by the EU's emissions trading system have been required to report verified carbon emissions figures since 2005 (Ascui and Lovell 2011). EU member-states' implementation of the EU Accounts Modernisation Directive (AMD) and the EU Transparency Directive has further encouraged the strengthening of environmental reporting requirements at the national level. For example, the UK Companies Act of 2006 requires companies to include in their annual reports an assessment of environmental risks and uncertainties, corporate policies and their effectiveness, and key performance indicators (Environment Agency 2010). The Danish government introduced a mandatory requirement in 2009 that all large companies publicly report on their corporate social responsibility commitments and activities. In Sweden, all state-owned enterprises are required to report on their environmental and social performance according to GRI guidelines. In the USA, the Securities and Exchange Commission (SEC) has issued rules that require companies to disclose how environmental laws may influence their capital expenditures, earnings, and competitive position, and also to disclose information on ongoing legal proceedings involving environmental liabilities.

Notwithstanding these developments, corporate environmental reporting remains uneven, reflecting weak and poorly coordinated institutions. First, companies retain significant discretion in deciding what and how to report. Voluntary standards are not monitored and enforced. National regulations on environmental reporting are much less prescriptive than those addressing financial reporting, and enforcement is often weak. When information is not collected and reported according to standardized metrics, investors will find it cumbersome to determine whether the environmental performance of one company is better or worse than another (Kolk *et al.* 2008; Solomon *et al.* 2011). In aggregate, this results in information that is insufficiently comprehensive to understand and compare the environmental impacts of different companies, and how they may be impacted by environmental change. Second, corporate environmental reporting varies significantly by region, reflecting variations in commitments among companies to corporate social responsibility, and demand among investors and other stakeholders for such information. In general, European companies tend to be the most transparent, followed by North American companies, with those in Asia and Latin America the least transparent.

And third, much of the information on how companies impact the environment is either not integrated into traditional financial reporting, or done so in an inconsistent manner. For example, among companies covered by the EU ETS, some have chosen not to disclose their carbon liabilities on their balance sheets, whereas others are charging them to their income statements (Solomon *et al.* 2011). Within companies, the annual process of writing a report covering environmental and social issues – commonly known as a sustainability report – is often undertaken separately from the process of writing financial reports. As a result, companies often fail to communicate whether the analysis and results in one report may affect or be affected by the analysis and results in the other. Among the sample of 458 companies from the FTSE All-Share index surveyed for their environmental disclosures, only 36% included some environmental information in audited sections of their respective annual reports (Environment Agency 2010). This is a concern given that many companies may have strong incentives to withhold or manipulate information about environmental harms that may undermine their reputation (Vogel 2006).

Stock-listing Requirements

Stock exchanges act as gatekeepers for public equity markets by setting the terms and conditions for companies to raise capital from investors by issuing shares. Exchanges that are members of the World Federation of Exchanges (WFE) transacted over US$80 trillion in 2008, facilitating millions of transactions between buyers and sellers of corporate stocks (Morales and van Tichelen 2010). Each stock exchange has listing requirements that are meant to ensure that listed companies provide basic financial information about themselves to the investing public. They are designed to build confidence among investors of the quality of company management underpinning all stocks listed on the exchange. Institutional investors, such as pension funds, mutual funds, and sovereign wealth funds, account for a significant share of investment activity at the largest stock exchanges. Many of them systematically invest capital in all companies listed on the exchange according to a weighted index. The prevalence of index-based investing entails that any company added to a stock exchange or issuing new stock is guaranteed an immediate demand for its newly listed stocks.

In recent years, many stock exchanges have included references to environmental reporting in their listing requirements, commonly in the form of voluntary guidance. Perhaps the most far-reaching is the Shanghai Stock Exchange, which requires companies in highly polluting industries wishing to issue stocks to obtain a permit from the Chinese Ministry of Environmental Protection (Siddy 2009). The permitting process subjects the company to an environmental assessment, including a short public consultation process. According to the rules, all listed companies are required to disclose information relating to environmental protection, and breaches of compliance may be subject to investigations, fines, and public outing. In South Africa, the King Codes on Corporate Governance require companies listed on the Johannesburg Stock Exchange to integrate sustainability disclosures with annual reports (Morales and van Tichelen 2010). Other stock exchanges have issued guidance that encourages voluntary environmental reporting, with some including a "report, or explain" requirement. Furthermore, there has been a proliferation of sustainability indices that use environmental, social, and governance criteria to identify a subset of listed companies on an exchange. Some comprise only clean technology and renewable energy companies, whereas others track companies with superior environmental performance across industrial sectors. Between 1999 and 2009, the number of such indices grew from fewer than 5 to more than 50 (Siddy 2009).

Institutional Investment and the Environment

Institutional investment refers to capital allocation that is carried out by investment organizations on behalf of one or more investors. Among the most important institutional investors in global stock markets by size are pension funds investing capital on behalf of pension beneficiaries, and sovereign wealth funds mandated to invest in global capital markets in accordance with a variety of national policy objectives. In OECD countries alone, pension funds are estimated to hold US$28 trillion in financial assets, with annual in-flows from new contributions of approximately US$850 billion (OECD 2011: 10). As much of this capital is typically invested in stock markets, pension funds collectively own a large share of listed companies in the major economies, potentially giving them significant influence across industries (Woods 2009). Meanwhile, sovereign wealth funds are government-owned investment vehicles mainly created to safeguard and augment national wealth. In general terms, they either derive capital from the production and sale of a commodity, for instance oil, gas, or minerals (commodity funds), or from sovereign budget surpluses, trade surpluses, and/or central bank currency reserves. They are estimated to collectively hold at least US$3–4 trillion in capital, projected to rise as much as US$10 trillion by 2015 (Monk 2009), which is more than hedge funds, private equity funds, and all of official development assistance combined. Compared to pension funds, sovereign wealth funds are generally less transparent with their investment strategies and portfolios (Bahgat 2008).

Many institutional investors have publicly committed to address environmental issues in their investment practices, and some report how they have implemented this commitment. Among the 539 signatories to the UN Principles for Responsible Investment (UN PRI) that chose to fill out its annual survey in 2011, 94% of fund owners indicated that they have a responsible investment policy, 79% said they integrated ESG factors to "some" or "a large" extent into internally managed (active) investments in developed market listed equities, whereas 30% of investment managers had invested in clean technology funds (UN PRI 2011). An industry survey of 12 institutional investors, representing almost US$2 trillion in assets under management, found that half had undertaken or had plans to make changes to their actual asset allocations in order to respond to climate change, whereas more than half had or would increase their engagement on climate change with companies and policy-makers (Mercer 2012). Solomon et al. (2011) interviewed professionals in 20 leading investment institutions in the UK with responsible investment roles and found that many asked companies to disclose how they managed climate change risk in order to encourage changes in corporate behavior.

Institutional investors committed to socially responsible investment (SRI) have typically engaged in some or all of four types of investment practices (Vogel 2006: 60–65). First, they have divested from companies operating in certain sectors associated with unethical products or companies that have engaged in acts of corporate misconduct. The practice of negative screening dates back to at least the 1970s, when religious groups first asked fund managers to invest their capital in a way that conformed to their moral codes (Richardson 2008: 73–79). Today, the most common sector exclusions are tobacco, alcohol, pornography, and weaponry. Some investors also divest from companies in other industry sectors that are complicit in major environmental damages or human rights abuses. Second, investors have established special funds that invest only in companies operating in industrial sectors that generate public goods. Such positive screening has benefited companies in the clean technology and renewable energy sectors, as well as companies operating in poor countries or promoting social development. Third, investors have used environmental criteria alongside more conventional financial criteria when designing stock portfolios and weighted capital towards companies with superior environmental performance. For example, they may invest in companies in the Dow Jones Sustainability Index, which tracks the financial performance of multinational companies identified by Dow Jones as leaders in sustainability. The emergence of such indices has provided incentives for companies to improve their environmental performance and report positive results so as to raise capital from investors committed to overweight such companies in their portfolios. And fourth, investors have used their rights as shareholders in companies to engage with company boards on environmental issues and put forward and vote on shareholder resolutions at annual general meetings (Clark *et al.* 2008). In recent years, a number of institutional investors have collaborated to demand that shale gas companies listed in the USA would publicly report on the environmental risks associated with the process of hydraulic fracturing. In 2011, 315 investors joined engagement activities coordinated by the UN PRI through its web-based clearing-house, including a joint initiative by 33 investors to send letters to 92 companies in energy-intensive sectors asking them to disclose plans for reducing their greenhouse gas emissions (UN PRI 2011: 12).

Pension funds have also defined and advanced shared environmental objectives and public policy positions through collaboration (MacLeod and Park 2011). UN PRI has encouraged nearly 1000 asset owners, fund managers, and financial services firms to publicly endorse six principles of responsible investment (UN PRI 2011). CERES, the sustainable business coalition, has drafted a document that outlines specific actions that institutional investors expect companies to undertake in the area of corporate climate risk management (CERES 2012). Investor associations have also emerged as influential actors in public policy-making. European Social Investment Forum (Eurosif) is a pan-European network based in Brussels with affiliates at the national level that produces advocacy research and directly lobbies the EU to facilitate socially responsible investment practices. In 2011, the International Investors Group on Climate Change (IIGCC) mobilized 285 investors in support of a call on governments "to work towards a binding international treaty that includes all major emitters and that sets short-, mid-, and long-term greenhouse gas emission reduction targets" (IIGCC 2011). In 2007, the Investor Network on Climate Risk (INCR), affiliated with CERES, lobbied the SEC to issue rules that make it mandatory for companies to disclose their climate risks (Richardson 2008: 139).

With regards to assessing the impact of these practices on financial markets, an important measure is whether financial institutions are allocating a greater share of their capital toward companies that make profits from environmentally sustainable technologies and activities. In aggregate, only a small share of global capital is invested in economic activities that are environmentally sustainable in the long term (Kolk *et al.* 2008; Harmes 2011; OECD 2011). One cause is that a growing share of financial activity finds its purpose in generating returns through short-term buying and selling of stocks. This has become more widespread with the emergence of information and communication technologies that provide greater opportunities for investors – including those with mandates to generate long-term returns – to make money from financial volatility through arbitrage. As an indication, over half of the volume of share trading in US stock markets is undertaken by computer programs that buy and sell at speeds and volumes that exceed human capabilities (MacKenzie 2011). Around-the-clock financial news coverage has further encouraged investors to make decisions on the basis of single market events rather than long-term market trends. As a result, companies are increasingly under pressure to report on their financial results at shorter and shorter intervals and pursue investment strategies that maximize quarterly earnings (Vogel 2006: 67).

While there is no broader market momentum toward environmental investing (Haigh and Shapiro 2011), financing for renewable energy and clean technology is growing rapidly, including among institutional investors. In 2011, global investment reached a new record of US$260 billion, a more than fivefold increase since 2004 (Bloomberg New Energy Finance 2011). The growth was driven by investments in utility-scale renewable energy projects and rooftop photovoltaics. To some extent, pension funds and sovereign wealth funds have contributed to this rise. For example, the California Public Employees' Retirement System (CalPERS) has invested US$500 million in a new clean energy fund tracking a climate change investment index, while Danish, Dutch, and US pension funds purchased large stakes in onshore and offshore wind farms (Bloomberg New Energy Finance 2011). Meanwhile, the governments of China, Abu Dhabi, and Indonesia, amongst others, have established separate clean energy funds to invest in renewable energy projects. The Norwegian Pension Fund – Global, which invests the country's oil wealth – has allocated US$4.6 billion to various environmental investments and also excludes companies from its global equities portfolio that it has found to be complicit in major environmental damages (Bahgat 2008).

These examples notwithstanding, environmental investments account for less than 1% of pension fund portfolios (OECD 2011: 6). This can be explained by factors both internal and external to pension funds. Many pension fund trustees do not regard environmental considerations to be within the parameters of their fiduciary duty to generate financial returns for pension beneficiaries (Woods 2009). Drawing on modern portfolio theory, most large funds wishing to generate stable, long-term, average returns have adopted so-called "passive" investment strategies that are based on wide diversification within and across stocks, bonds, and other asset classes. This is implemented through index investing, in which capital may be provided to all companies listed on particular stock exchanges. By implication, returns on investment will correlate strongly with the stock market as a whole. Given that industries associated with long-term environmental harm – such as oil, gas, and mining – are also among those that have historically generated the highest and most stable returns on investment, funds may find it hard to omit them without changing their returns expectations. This explains why large funds structured to generate long-term average market returns rarely use environmental data in a systematic way for portfolio selection purposes (Haigh and Shapiro 2011).

Moreover, the structure of global institutional investment, and, relatedly, the process of investment management favored by most large funds, creates some challenges to incorporating long-term environmental concerns into investment decisions. Most institutional investors have delegated investment decisions to other financial institutions, so-called fund managers (Sullivan and MacKenzie 2006). They do this to save administrative costs and benefit from the knowledge of fund managers with specialized expertise in particular markets. While providing many benefits, this approach creates agency problems. In particular, while fund owners may wish to invest long term, they may also find it necessary to at least annually review the financial performance of fund managers in order to hold them accountable. In practice, most fund owners – or pension beneficiaries for that matter – would not accept poor short-term returns over an extended period of time (Woods 2009). In turn, most fund managers are not encouraged to make investment decisions based on long-term time horizons and follow investment approaches that conflict with market norms (Harmes 2011).

These challenges internal to the investment process are compounded by an institutional environment that does not sufficiently encourage companies and investors to support sustainable development (Rowlands 2005; Newell and Paterson 2010; OECD 2011). Notwithstanding notable developments in domestic and international environmental policy, companies are not being sufficiently discouraged by regulations, taxes, and fiscal policies to harm the environment, and, conversely, they are not being sufficiently rewarded to invest in economic activities that are comparatively environmentally benign. Symptomatically, non-hydro renewable energy has attained the greatest market share in countries where fiscal regimes and government financing have provided companies and investors with risk guarantees to develop other energy sources, such as wind, solar, geothermal, and biomass (Bloomberg New Energy Finance 2011). Markets are being given mixed signals, since governments are also pursuing other, and sometimes competing, energy policy objectives through fiscal instruments. In 2010, subsidies to fossil fuel consumption reached US$409 billion, compared to US$66 billion for renewable energy, as a result of governments wanting to boost domestic economic output, maintain employment, develop technology, and alleviate energy poverty, alongside promoting renewable energy (IEA 2011: 508). In part encouraged by these government programs, the 70 leading oil and gas companies invested more than US$500 billion in oil and gas exploration and production in 2010, which is expected to rise to an average of US$620 billion per year until 2035 (IEA 2011: 142–144).

Even if public policy encourages companies to pay for the cost of their pollution, asymmetric information may prevent capital markets from reallocating capital to those that are more environmentally friendly. Besides being largely qualitative, of lesser quality, and often unaudited, environmental information is typically only reported on an annual basis. In comparison, financial results and projections are in most cases released at least quarterly, while changes to stock prices are updated instantly. Moreover, many companies do not explain the relevance of their environmental impacts and risks to their overall financial performance. The difference in reporting intervals and the lack of integration with financial metrics further solidify the impression that environmental information is of lesser relevance and not material to the financial performance of the company. In turn, fund managers that are asked to invest on the basis of financial valuations of stocks can in most cases discount environmental information from their company analysis without being sanctioned in the form of lesser returns, at least in the short term.

Project Finance, Export Credits, and the Environment

Whereas pension funds, and to a much lesser extent sovereign wealth funds, have adopted SRI policies largely in response to the ethical concerns of their beneficiaries, the global banking industry has done so mainly in response to public pressure from environmental NGOs (van Putten 2008; Park 2012). Standards development has taken place within industry networks and associations, facilitated by international organizations with explicit environmental mandates, such as the UN Environment

Programme (UNEP) and the World Bank. More than 200 banks and insurers, by virtue of being members of the UNEP Finance Initiative "regard financial institutions to be important contributors to sustainable development, through their interaction with other economic sectors and consumers and through their own financing, investment and trading activities" (UNEP FI 2011). Its main impact has been to mobilize banking and insurers around general principles of environmental stewardship, build networks of environmental finance professionals across regional markets, and generate new knowledge about how to implement commitments in practice (Park 2012). In 2003, commercial banks that provide project finance loans created the Equator Principles, a voluntary framework to harmonize environmental and social risk management practices according to those developed by the International Finance Corporation (IFC), the private-sector financing arm of the World Bank (van Putten 2008: 178–217; Wright 2012). The framework is designed to inform the way banking institutions engaged in project finance identify, assess, and mitigate the environmental and social impacts of projects, particularly in countries with weak or poorly enforced laws protecting the environment and human rights (Equator Principles 2006). It has been voluntarily adopted by more than 70 public and commercial banking institutions.

Both UNEP FI and the Equator Principles have encouraged financial institutions to develop and disclose policies for managing environmental risk. They have also established working groups and held conferences and workshops to facilitate the sharing of knowledge, ideas, and experiences between financial institutions (Park 2012; Wright 2012). UNEP FI issues policy briefs and technical guides that identify challenges facing financial institutions and methodologies for undertaking responsible investment practices. The Equator Principles Association has published best practice guidance on how financial institutions should incorporate environmental and social considerations into loan documentation, and how they should publicly report on their implementation of the Equator Principles. Although it is difficult to ascertain the direction of causality, studies have found that financial institutions that have adopted the Equator Principles are more likely to have published environmental lending policies than those who have not (BankTrack 2010).

Finally, most governments with export-oriented economies have established national ECAs to help domestic export companies sell goods and services to importers in other countries. They are commonly structured as public or semi-public institutions mandated to meet demand among national companies for export credits and risk guarantees. Their primary purpose is to promote domestic employment and growth through export subsidization. In 2010, global export credit volumes reached a record US$514 billion (Wright 2011: 134). Most financing benefited industries of strategic importance that are also exposed to significant commercial and non-commercial risk, such as commercial aircraft, aerospace technology, armaments, industrial plants, energy infrastructure, and transportation systems. Their association with these industries has subjected many ECAs to public criticism. In 2003, OECD governments negotiated the Common Approaches on Environment and Officially Supported Export Credits, a set of non-binding, consensus-based rules for harmonizing environmental and social standards for providing medium- and long-term export credits and risk insurance (OECD 2007; Schaper 2007). By doing so, they committed to have their respective agencies publish an environmental policy, adopt the environmental screening process used by multilateral development banks, and "benchmark" projects against host country standards and the IFC Performance Standards in the case of private sector projects (OECD 2007: 5–6). These rules complement existing governance arrangements negotiated by OECD governments to self-regulate their export financing practices.

It is difficult to assess the impact of these networks and associations on actual lending and export financing decisions, given the lack of transparency around specific transactions. It is easier to assess whether reforms proposed by critics have been accepted and implemented by financial institutions (BankTrack 2010). They fall into three broad categories. First, critics have called for greater transparency around financing decisions and the environmental and social conditions attached to financial instruments. The Equator Principles require financial institutions to annually report on the number of projects they have financed across three categories of environmental risk. The OECD Common Approaches require ECAs to publish an environmental policy and report annually on its implementation to the OECD Export Credit Group. But neither framework requires the disclosure of transaction-level information on environmental management. Second, critics have demanded recourse for local communities adversely affected by projects financed by signatories to the Equator Principles or ECAs governed by the OECD Common Approaches. The Equator Principles do not hold financial institutions directly accountable to local communities adversely affected by their project financing, but they do require them to demand that companies receiving their project loans establish a grievance mechanism that allows individuals to file complaints and receive a response. Meanwhile, the financing decisions made by ECAs can be challenged only if provided for by home-country laws and regulations. And finally, critics have called on financial institutions to refrain from financing projects in sensitive ecosystems or of a certain type (large dams, coal-fired power plants). Neither framework has challenged the right of financial institutions to decide for themselves how to allocate their capital, as they are allowed to support projects that do not meet the respective standards if they feel this is justified (Equator Principles 2006: 3; OECD 2007: 6).

In summary, both the Equator Principles and OECD Common Approaches are designed to address the environmental impacts of particular forms of financing and are overwhelmingly focused on mobilizing support behind general aspirations and commitments, and gaining acceptance for certain procedures for identifying, assessing, and managing environmental risks. Given that neither framework intends to dictate investment decisions, it is problematic to use the outcome of a particular financial transaction as evidence of whether a financial institution has acted on its commitments (Wright 2012). The growth of the project finance market since the emergence of the Equator Principles demonstrates that the framework has not significantly influenced which projects banks choose to finance or, conversely, whether companies developing projects likely to have significant adverse environmental impacts are able to raise the necessary financing. Similarly, the OECD Common Approaches have done little to curtail the growth of export financing to industries associated with significant environmental harm. The main impact of both frameworks has been that the standards governing the undertaking of environmental impact assessments and consultations with project-affected communities have been raised and more widely adopted.

Conclusion

This chapter has considered the relationship between finance and the environment. At the systemic level, it finds that the structure of global finance has given rise to new forms of environmental investment, but also increased financial instability and encouraged short-term investing. This has created an uncertain and unstable environment for governments, companies, and investors to make long-term decisions and plan for the future. The discussion of environmental accounting and disclosure revealed how rules remain weak and fragmented compared to those governing financial accounting and disclosure. While not all environmental problems can be solved through better corporate accounting and disclosure, the current situation can be remedied by moving away from the parallel development of financial and environmental reporting, toward an integration of institutional rules at the national and international level. And finally, at the level of financial actors, the discussion centered on the roles and impacts of pension funds, sovereign wealth funds, banks, and export credit agencies. While the adoption of environmental commitments among them has been pervasive, this has not caused a significant shift in financial activity towards environmental investing. It reflects how the mandates, strategies, and investment practices of most financial institutions contain strong biases in favor of investing in companies that provide, or depend on, natural resources that are essential for human consumption and wealth.

Rectifying this seemingly depends on government actions that cause a shift in the risk-adjusted returns from investments that cause environmental harm to those that promote environmentally sustainable development. Finance is predominately motivated by an overarching purpose of finding financial value in physical or intangible assets within a set of institutional rules and market conditions. For government-owned financial institutions, these institutional rules often reflect political imperatives. It is difficult to foresee a growth in environmental investing at a scale that is needed unless this is aggressively promoted by national and international policies. International financial regulatory reform that reduces financial market volatility and encourages long-term investing would seemingly benefit the environment. While financial institutions have tended to oppose new financial regulations that restrict or impose costs on their own financial activities, many have issued public support for international environmental policies and regulations that aim to regulate the activities of the companies they are invested in (IIGCC 2011). This suggests that long-term investors represent a nascent environmental policy constituency that could play an increasingly influential role in shaping global environmental governance through their financing activities and engagement with policy-makers and standard-setters.

References

Ascui, F. and H. Lovell. 2011. "As Frames Collide: Making Sense of Carbon Accounting." *Accounting, Auditing & Accountability Journal*, 24(8): 978–999.

Bahgat, Gawdat. 2008. "Sovereign Wealth Funds: Dangers and Opportunities." *International Affairs*, 84(6): 1189–1204.

BankTrack. 2010. *Close the Gap: Benchmarking Credit Policies of International Banks*, April. Nijmegen, the Netherlands: BankTrack, http://www.banktrack.org/download/close_the_gap/close_the_gap.pdf (accessed October 12, 2010).

Bloomberg New Energy Finance. 2011. *New Global Trends in Renewable Energy Investment 2011: Analysis of Trends and Issues in the Financing of Renewable Energy.* n.p.: United Nations Environment Programme and Bloomberg New Energy Finance.

Bowen, Frances and Bettina Wittneben. 2011. "Carbon Accounting: Negotiating Accuracy, Consistency and Certainty across Organisational Fields." *Accounting, Auditing & Accountability Journal*, 24(8): 1022–1036.

Carbon Disclosure Project. 2011. *CDP Global 500 Report 2011: Accelerating Low Carbon Growth.* London: Carbon Disclosure Project.

CERES. 2012. "Institutional Investors' Expectations of Corporate Climate Risk Management." n.p.: Investor Group on Climate Change, Institutional Investors Group on Climate Change, and Investor Network on Climate Risk.

Cerny, Phil. 2010. *Rethinking World Politics.* Oxford: Oxford University Press.

Clark, Gordon L., James Salo, and Tessa Hebb. 2008. "Social and Environmental Shareholder Activism in the Public Spotlight: US Corporate Annual Meetings, Campaign Strategies, and Environmental Performance, 2001–04." *Environment and Planning A*, 40(6): 1370–1390.

Dauvergne, Peter. 2005. "The Environmental Challenge to Loggers in the Asia-Pacific: Corporate Practices in Informal Regimes of Governance." In *The Business of Global Environmental Governance*, ed. David L. Levy and Peter J. Newell, 169–196. Cambridge, MA: MIT Press.

Environment Agency. 2010. *Environmental Disclosures: The Third Major Review of Environmental Reporting in the Annual Report and Accounts of the FTSE All-Share Companies.* Bristol: Environment Agency.

Equator Principles. 2006. "Equator Principles II," released July 6, www.equator-principles.com (accessed October 12, 2010).

Haigh, Matthew and Matthew A. Shapiro. 2011. "Carbon Reporting: Does It Matter?" *Accounting, Auditing & Accountability Journal*, 25(1): 105–125.

Harmes, Adam. 2011. "The Limits of Carbon Disclosure: Theorizing the Business Case for Investor Environmentalism." *Global Environmental Politics*, 11(2): 98–120.

Helleiner, Eric and Jennifer Clapp. 2012. "International Political Economy and the Environment: Back to the Basics?" *International Affairs*, 88(3): 485–501.

Helleiner, Eric and Stefano Pagliari. 2010. "Crisis and the Reform of International Financial Regulation." In *Global Finance in Crisis*, ed. Eric Helleiner, Stefano Pagliari, and Hubert Zimmermann, 1–17. Abingdon: Routledge.

IEA. 2011. *World Energy Outlook 2011.* Paris: International Energy Agency.

IIGCC (International Investors Group on Climate Change). 2011. "Global Investor Statement on Climate Change." n.p.: IIGCC.

Kolk, Ans, David L. Levy, and Jonatan Pinkse. 2008. "Corporate Responses in an Emerging Climate Regime: The Institutionalization and Commensuration of Carbon Disclosure." *European Accounting Review*, 17(4): 719–745.

Lederer, Markus. 2010. "Evaluating Carbon Governance: The Clean Development Mechanism from an Emerging Economy Perspective." *Journal of Energy Markets*, 3(2): Summer 2010, http://www.risk.net/journal-of-energy-markets/technical-paper/2160782/evaluating-carbon-governance-clean-development-mechanism-emerging-economy-perspective# (accessed October 20, 2012).

MacKenzie, Donald. 2011. "How to Make Money in Microseconds." *London Review of Books*, 33(10), May 19: 16–18

MacLeod, Michael and Jason Park. 2011. "Financial Activism and Global Climate Change: The Rise of Investor-Driven Governance Networks." *Global Environmental Politics*, 11(2): 54–74.

Matz, Nele. 2005. "Financial Institutions between Effectiveness and Legitimacy: A Legal Analysis of the World Bank, Global Environment Facility and Prototype Carbon Fund." *International Environmental Agreements: Politics, Law and Economics*, 5(3): 265–302.

Mercer. 2012. "Through the Looking Glass: How Investors Are Applying the Results of the Climate Change Scenarios Study," January 2012. n.p.: Mercer LLC.

Monk, Ashby. 2009. "Recasting the Sovereign Wealth Fund Debate: Trust, Legitimacy, and Governance." *New Political Economy*, 14(4): 451–468.

Morales, Rumi and Edouard van Tichelen. 2010. "Sustainable Stock Exchanges: Real Obstacles, Real Opportunities." Discussion paper prepared for the Sustainable Stock Exchanges 2010 Global Dialogue. Geneva: Responsible Research.

Newell, Peter and Mathew Paterson. 2010. *Climate Capitalism*. Cambridge: Cambridge University Press.

Nölke, Andreas. 2010. "The Politics of Accounting Regulation: Responses to the Subprime Crisis." In *Global Finance in Crisis*, ed. Eric Helleiner, Stefano Pagliari, and Hubert Zimmermann, 37–55. Abingdon: Routledge.

OECD (Organisation for Economic Co-operation and Development). 2007. "Revised Recommendation on Common Approaches on Environment and Officially Supported Export Credits, agreed by the OECD Ministerial Council on 18 December 2003." TAD/ECG/2007/9.

OECD (Organisation for Economic Co-operation and Development). 2011. *The Role of Pension Funds in Financing Green Growth Initiatives*, by Raffaele Della Croce, Christopher Kaminker, and Fiona Stewart. OECD Working Papers on Finance, Insurance, and Private Pensions, No. 10. Paris: OECD Publishing.

Park, Susan. 2012. "Bankers Governing the Environment? Private Authority, Power Diffusion and the United Nations Environment Program Finance Initiative." In *The Diffusion of Power in Global Governance: International Political Economy Meets Foucault*, ed. Stefano Guzzini and Iver Neumann, 141–171. Basingstoke: Palgrave Macmillan.

Paterson, Matthew. 2001. "Risky Business: Insurance Companies in Global Warming Politics." *Global Environmental Politics*, 1(2): 18–41.

Pattberg, Philipp. 2007. *Private Institutions and Global Governance: The New Politics of Environmental Sustainability*. Cheltenham: Edward Elgar.

Porter, Tony. 2005. *Globalization and Finance*. Cambridge: Polity.

Prakash, Aseem. 2000. *Greening the Firm*. Cambridge: Cambridge University Press.

Richardson, Benjamin. 2008. *Socially Responsible Investment Law*. Oxford: Oxford University Press.

Rowlands, Ian. 2005. "Renewable Energy and International Politics." In *Handbook of Global Environmental Politics*, ed. Peter Dauvergne, 78–94. Cheltenham: Edward Elgar.

Schaper, Marcus. 2007. "Leveraging Green Power: Environmental Rules for Project Finance." *Business and Politics*, 9(3): 1–27.

Siddy, Dan. 2009. "Exchanges and Sustainable Investment." Report prepared for the World Federation of Exchanges (WFE), August 2009. n.p.: World Federation of Exchanges.

Solomon, Jill F., Solomon Aris, Simon Norton, and Nathan L. Joseph. 2011. "Private Climate Change Reporting: An Emerging Discourse of Risk and Opportunity?" *Accounting, Auditing & Accountability Journal*, 24(8): 1119–1148.

Sullivan, Rory and Craig MacKenzie. 2006. *Responsible Investment*. Sheffield: Greenleaf Publishing.

Thistlethwaite, Jason. 2011. "Counting the Environment: The Environmental Implications of International Accounting Standards." *Global Environmental Politics*, 11(2): 75–97.

UNEP FI (United Nations Environment Programme Finance Initiative). 2011. "*UNEP Statement of Commitment by Financial Institutions (FI) on Sustainable Development*." Geneva: UNEP FI.

UN PRI (United Nations Principles for Responsible Investment). 2011. *Five Years of PRI: Annual Report of the PRI Initiative 2011*. Geneva: UN PRI.

van Putten, Maartje. 2008. *Policing the Banks*. Montreal: McGill-Queen's University Press.

Vogel, David. 2006. *The Market for Virtue: The Potential and Limits of Corporate Social Responsibility*. Washington, DC: Brookings Institution Press.

WCED (World Commission on Environment and Development). 1987. *Our Common Future*. Oxford: Oxford University Press.

Woods, Claire. 2009. "Funding Climate Change: How Pension Fund Fiduciary Duty Masks Trustee Inertia and Short-Termism." In *Corporate Governance Failures: The Role of Institutional Investors in the Global Financial Crisis*, ed. James P. Hawley, Shyam J. Kamath, and Andrew T. Williams, 242–278. Philadelphia: University of Pennsylvania Press.

Wright, Christopher. 2011. "Export Credit Agencies and Global Energy: Promoting National Exports in a Changing World." *Global Policy* (special issue), 2(s1): 133–143.

Wright, Christopher. 2012. "Global Banks, the Environment, and Human Rights: The Impact of the Equator Principles on Lending Policies and Practices." *Global Environmental Politics*, 12(1): 56–77.

Chapter 26

Energy Policy and Climate Change

Benjamin K. Sovacool

Introduction

In some ways, the twentieth century has been all about energy. From 1900 to 2000, engineers and architects built more than 75 000 power plants, at least 3.2 million kilometers of transmission and distribution lines for electricity, 5.1 million kilometers of natural gas pipelines, 300 nuclear waste storage facilities, and more than 600 refineries. The past century saw the world profoundly shaped by the automobile, truck, aircraft, and atomic energy as millions of people shifted from non-mechanized forms of transport and agriculture to reliance on automobiles and industrial food manufacturing. Electricity, once so novel that it was prized for its "healing powers" and served as a spectacle at numerous World's Fairs, moved from its infancy into the primary fuel for heating homes, powering industrial processes, energizing air conditioners (also invented during the century), and enabling the digital-telecommunications-media-computer-information age.

For example, from 1900 to 2000 the population of the earth quadrupled from 1.6 billion to 6.1 billion, but annual average supply of energy per capita grew *even more*, from 14 GJ in 1900 to roughly 60 GJ in 2000. Over this period, energy consumption more than tripled in the USA, quadrupled in Japan, and increased by a factor of 13 in China (Brown and Sovacool 2011). Global use of hydrocarbons as a fuel by humans increased 800-fold from 1750 to 2000 and 12-fold again from 1900 to 2000 (Smil 2000).

If the twentieth century was about energy, then the twenty-first century could very well be about energy governance and climate change. Issues surrounding energy supply and use connect with many of the world's most pressing public policy problems: possible conflagrations over rapid depletion of fossil-fuel reserves, the environmental consequences of climate change, and millions of communities that must

The Handbook of Global Climate and Environment Policy, First Edition. Edited by Robert Falkner.
© 2013 John Wiley & Sons, Ltd. Published 2013 by John Wiley & Sons, Ltd.

endure "energy poverty" without access to consistent sources of lighting, heating, water, mobility, or comfort (Florini and Dubash 2011; Yergin 2011; Sovacool *et al.* 2012).

This chapter introduces readers to the energy governance and climate change nexus. It details the processes, sectors, technologies, and countries responsible for greenhouse gas emissions. It then discusses a collection of barriers which explain why progress on reducing emissions has been slow to occur. It lastly elaborates on a common set of policy mechanisms that can overcome these barriers and problems as well as offers a collection of brief case studies.

Two things make the chapter unique. First, it looks at energy supply – things like power plants, pipelines, and oil rigs – alongside energy demand – things like patterns of consumption and energy use. Second, it discusses energy technologies alongside often neglected topics such as consumer behavior, social values and attitudes, politics, and governance concerns.

The Energy–Climate Change Nexus

According to the most recent data available from the Intergovernmental Panel on Climate Change (IPCC 2008), human sources emitted 49 billion tonnes of carbon dioxide equivalent into the atmosphere in 2004. Global greenhouse gas (GHG) emissions grew by 70% from 1970 to 2004, and if trends continue could increase by 130% by 2040. Yet the climate-related impacts of these emissions could last longer than Stonehenge, time capsules, and perhaps even high-level nuclear waste. For each ton of carbon dioxide we leave in the atmosphere today, one quarter of it will still be affecting the atmosphere a thousand years from now (Archer 2009). Put another way, the climate system is like a bathtub with a very large tap and a small drain (Victor *et al.* 2009). As Figure 26.1 shows, four interrelated areas – electricity supply, transport, agriculture and forestry, and waste and water – are responsible for most of these dangerous emissions.

As the following sections demonstrate, sources of emissions come roughly from the following major categories: transportation, buildings, the industrial sector, and electricity supply.

Transportation

On a global scale, the transportation of people and goods accounts for approximately one quarter of the world's energy consumption and 28% of its energy-related CO_2 emissions (IPCC 2008). Over the next few decades the transportation sector is expected to be one of the fastest-growing sources of GHG emissions. Much of the projected increase is attributed to the rapidly growing demand for petroleum-based transportation fuels in non-OECD economies, which are forecast to increase more than 2% per year; as compared with the OECD countries, which are forecast to increase at less than 1% per year (EIA 2006).

Buildings

The built environment – consisting of residential, commercial, and institutional structures – accounts for about one third of primary global energy demand and is the

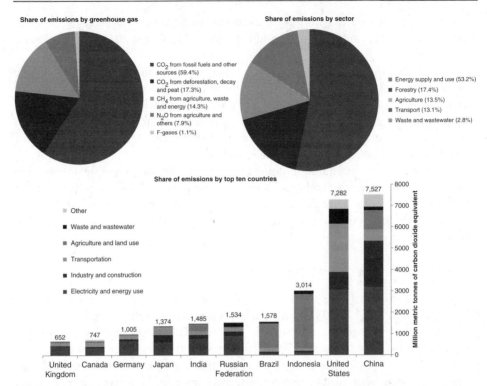

Figure 26.1 Global greenhouse gas emissions by gas, sector, and country.
Source: Brown, Marilyn A. and Benjamin K. Sovacool. 2011. *Climate Change and Global Energy Security: Technology and Policy Options.* Cambridge, MA: MIT Press. © 2011 MIT.

source of 35% of global energy-related CO_2 emissions (International Energy Agency 2010). Over the long term, buildings are expected to continue to be a significant component of energy use and emissions, driven in large part by the continuing trends of urbanization, population, GDP growth, and the longevity of building stocks. A growing body of evidence suggests that improving the energy efficiency of the existing building stock and new construction is a low-cost approach to mitigating GHG emissions (IPCC 2008).

Industry

The industrial sector is the largest consumer of energy worldwide, accounting for an estimated 36% of global primary energy in 2006 and producing a slightly larger share of CO_2 emissions, partly due to the use of fossil fuels as feedstocks in the production of chemicals and other industrial products such as the release of CO_2 in the production of cement (International Energy Agency 2010). Global energy consumption and CO_2 emissions from this sector are projected to increase rapidly through 2030, driven by the expansion of China, India, and other emerging economies such as Brazil and South Africa.

A handful of large industries are highly energy-intensive in most countries of the world where they operate. These include, for instance:

- petroleum refining and the production of chemicals and fertilizers
- the metals industries (including iron, steel, and aluminum)
- pulp and paper
- mineral products (including cement, lime, limestone, and soda ash)
- glass.

Light manufacturing, which includes the manufacture and assembly of automobiles, appliances, electronics, textiles, and food and beverages, generally requires less energy per dollar of shipped product. As a result, these plant managers pay much less attention to their energy requirements, even though light manufacturing remains a large fraction of economic output and contributes significantly to global emissions.

Electricity Supply

Globally, electricity generation is the largest contributor to climate change, producing more than 10 billion t. of carbon dioxide every year, the greatest contribution from any given industry or sector. As world population grows and standards of living rise, the global demand for electricity is projected to continue its rapid expansion in both developing and industrialized economies. Nearly 2 billion people do not have access to the electric grid, but expectations are that this share will continue to shrink, resulting in a rapid expansion of electricity demand. As a result, the electric grid will need an infusion of transmission and distribution (T&D) system investments. Worldwide, the International Energy Agency (2010) forecasts an investment of US$6.8 trillion in T&D upgrades between 2007 and 2030.

Changes in Land Use

Globally, agricultural sources of methane (CH_4) and nitrous oxide (N_2O) account for nearly 60% of non-CO_2 emissions and 48% of US non-CO_2 GHGs primarily from crop and livestock production (Brown and Sovacool 2011). Enteric fermentation is the largest anthropogenic source of methane emissions in the USA, accounting for nearly a quarter of the total. This source of methane continues to increase with the growth of livestock operations. CH_4 emissions from manure management have also been increasing over the past decade, mostly as the result of an increase in the use of liquid systems in swine and dairy cow products.

The livestock sector accounts for 18% of global GHG emissions and 80% of anthropogenic land use. One study projected that deforestation and a small amount of desertification were responsible for 35% of livestock-associated GHG emissions. Put another way, an area the size of Russia and Canada *combined* is currently used exclusively as pasture or cropland to grow animal feed. If this land was converted instead to growing vegetables for human consumption or into forests, it would soak up such large amounts of carbon dioxide that it could cut compliance costs with the Kyoto Protocol in half, rather than being a source of emissions, as it is today (Stehfest *et al.* 2009).

Much of the world's farming, livestock production, and changes in land use have taken place in former forests and tropical forests. Thus forests can be a sink of emissions but also a source, depending on how they are managed. It is helpful to view forests through the lens of stocks and flows. The total stock of carbon in all *tropical* forests equals about 300 billion t., about 1.5 billion t. is converted into 6 billion t. of CO_2 through deforestation that is emitted into the atmosphere (Boucher 2009). In other words, tropical forests alone contribute to about 20% of overall human caused CO_2 emissions per year, making them the largest emitter of carbon in the world after the energy sector. This amount is equivalent to the total emissions of China or the USA, and it is more than the emissions produced by every car, truck, plane, ship, and train on Earth.

High Global Warming Potential Gases

There are numerous highly potent anthropogenic GHGs that are entirely human-made. Three of these also deplete the stratospheric ozone layer: chlorofluorocarbons (CFCs), hydrochlorofluorocarbons (HCFCs), and bromofluorocarbons (i.e. halons). These ozone-depleting substances (ODSs) are controlled under the Montreal Protocol of 1987 on Substances that Deplete the Ozone Layer, and as a result their impact on both ozone and the greenhouse effect has been greatly reduced. Other anthropogenic fluorine-containing halogenated substances do not deplete stratospheric ozone but are potent GHGs. The most important of these are collectively called the "F-gases" and include hydrofluorocarbons (HFCs), perfluorocarbons (PFCs), and sulfur hexa-fluoride (SF_6). Numerous other minor trace gases complete the inventory of GHGs. Emissions of high global-warming potential gases are expected to increase significantly worldwide due to growing demand for refrigeration and air conditioning and the industrialization of developing economies.

Barriers to Low-Carbon Technologies

Although we have made progress in terms of understanding the forces behind climate change and available solutions to it, a mesh of obstacles and impediments towards low-carbon technologies and practices exist. The barriers facing these practices and technologies are tenacious, interconnected, and deeply embedded in our social fabric, institutional norms, and modes of production across the world. The most significant of these relate to cost-effectiveness; fiscal, regulatory, and statutory barriers; intellectual property; and "other" cultural, social, and institutional barriers summarized in Table 26.1.

Cost-effectiveness Barriers

Types of cost-effectiveness barriers that impede market introduction and penetration of less carbon-intensive technologies and practices include un-priced externalities, high costs, and technical and market risks. Externalities can make it difficult for clean energy technologies to compete in today's market, where GHG emission

Table 26.1 Typology of barriers to GHG mitigation technologies.

Cost effectiveness	Fiscal, regulatory, and statutory barriers	Intellectual property (IP) barriers	Other cultural, social, and institutional barriers
External benefits and costs	Competing fiscal policies	Anti-competitive patent practices	Incomplete and imperfect information
High costs	Fiscal uncertainty	IP transaction costs	Infrastructure limitations
Technical risks	Competing regulatory policies	Weak international patent protection	Industry structure
Market risks	Regulatory uncertainty	University, industry, government perceptions	Policy uncertainty
Lack of specialized knowledge	Competing statutory policies	Statutory uncertainty	Misplaced incentives

reductions have only limited market value. Clean energy technologies also often have inherently higher up-front costs due to the need for additional features and subsystems required to achieve GHG reductions. These can increase the capital to operating expense ratio. For example, SF_6 is a high GWP gas used in the magnesium industry as a cover gas. SO_2 is being considered as an alternative, but it is more toxic and therefore requires additional monitoring (and cost) to deal with the health and safety issues. There are no simple drop-in substitutes (Brown and Sovacool 2011: 154).

The efficient operation of energy markets is compromised by the existence of external benefits and costs. Externalities occur when important societal benefits and costs are "external" to, or un-priced in, the marketplace. As law professor Noah Sachs notes:

> Think of externalities as a second price tag on every product we consume, representing the real costs of disposing the product and the environmental impacts directly flowing from the existence of that product. The price tag may be less than a cent for some products and several dollars for others, but because this price is never actually "paid" by consumers or producers, the price becomes externalized as a social cost (Sachs 2006: 56).

Indeed, less carbon-intensive technologies may be difficult to deploy (without public intervention) if their principal benefits are entirely societal and external to the marketplace. For low-carbon technologies across most of the USA and many areas around the world, GHG mitigation is not currently governed by explicit regulatory legislation and it is not rewarded in the marketplace. When the developer of a low-carbon technology cannot capture all of the benefits that might accrue to society, the result is underinvestment in its development and a sub-optimal supply of the technology. Because polluters do not pay for their societal damages, the "free rider" makes it difficult for the higher-priced clean energy technologies to compete. In general, goods generating positive externalities are underproduced and goods generating negative externalities are overproduced (Dunn 2008).

Specific examples of externalities in the energy and climate sectors are striking. In the electricity industry, the generation costs from a coal power plant may appear low, but they do not include the costs of coal-mine dust that kills thousands of workers each year; black lung disease that has imposed at least US$35 billion in health-care costs; and coal emissions that cause acid deposition, smog, and global warming and also contribute to asthma, respiratory and cardiovascular disease, and premature mortality. These external costs would easily double the price of coal if they were incorporated into its price (Jacobson and Masters 2001). The negative externalities associated with electricity generation overall amount to about US$13.46 /kWh or more than US$2 trillion in global damages each year (Brown and Sovacool 2011). The global chemical industry would have to spend eight times its profits each year (more than US$20 billion in the 1990s) to pay to incinerate the waste from its top 50 products (Hawken 1998). In the transport sector, vehicle crashes are the leading cause of injury-related deaths in the US for people between the ages of one and 65, causing 40 000 deaths, 2 million injuries, and US$150 billion in economic losses each year that are not reflected in the price of a new vehicle (Brown and Sovacool 2011). Gasoline is cheap because its price does not incorporate the cost of smog, acid rain, and their effects on health and the environment.

Fiscal, Regulatory, and Statutory Barriers

Unfortunately, just as markets can fail, so can public policies designed to correct them. Government action has its own set of problems, too. Public policies can provide broad societal benefits that increase overall economic welfare, for example, but can also inadvertently disfavor certain segments of the economy, including, in some cases, inhibiting the commercialization and deployment of clean energy technologies. When applied to the context of this chapter, these policies are referred to as "competing policies" and considered a barrier to deployment. Many competing priorities result from policies established years ago for a public purpose that could be better addressed in other ways today.

Competing priorities also arise as a result of legal inertia. For example, regulations take a long time to adopt and modify; as a result, they can be slow to adapt to technology advances and therefore inhibit innovation. Similarly, environmental standards that propelled the large-scale reduction of acid rain in the 1980s now enable the continued operation of some of the most polluting power generators in the USA far beyond their normal life and disincentivize investing in plant upgrades. Competing policies caused by outdated fiscal rules include the IRS tax depreciation schedules put into place more than two decades ago as part of the IRS Tax Reform Act of 1986. These rules have not kept up with technology breakthroughs and inhibit the advance of some modern low-carbon technologies. Back-up generators (which provide reliability at the expense of energy efficiency and clean air) are depreciated over 3 years, while a new combined heat and power system (providing both reliability and energy efficiency) is depreciated over 20 years (Brown and Sovacool 2011: 160).

Intellectual Property Barriers

Generally, intellectual property law is intended to stimulate innovation, entrepreneurship, and technology commercialization. However, its application can also impede the innovation process. For example, patent filing and other transaction costs associated with strong patent enforcement and protection, as well as the anti-trust challenges related to technological collaboration and patent manipulation can be serious barriers to technology diffusion. Anti-competitive business practices can also play their part in impeding cleaner energy systems.

Patent warehousing, suppression, and blocking, for instance, are anti-competitive practices undertaken by incumbent firms that impose barriers to technological change. Patent warehousing is a form of patent manipulation that involves owning the patent to a novel technology but never intending to develop the technology (Sabety 2005). Patent suppression involves refusing to file for a patent so that a novel process or product never reaches the market. As one example, in 1977 Tom Ogle developed an automotive system for Ford Motor Company that used a series of hoses to feed a mixture of gas vapors and air directly into the engine. Ford built a small number of prototypes that averaged more than 100 miles per gallon at 55 miles per hour (2.35 liters/100 km), but the technology was ultimately suppressed (Saunders and Levine 2004). Patent blocking occurs when firms use patents to prevent another firm from innovating. While Ford has used Toyota technology (in the Ford Escape), Ford has resisted purchasing Toyota's technology for hybrid vehicles because of hefty licensing fees, and likewise, Honda has not been able to successfully negotiate a license to use nickel metal hydride batteries in their hybrid vehicles. General Electric has also used its patent on variable-speed wind turbines to prevent Mitsubishi (a Japanese firm) and Enercon (a German firm) from entering the US market (Sovacool 2008).

Other Cultural, Social, and Institutional Barriers

Many additional barriers inhibit the deployment of GHG mitigation technologies in ways that are not captured by the categories discussed thus far. These barriers stem in part from the cultural traits that impact the behavior and choices of individuals. The influence of lifestyle and tradition on energy use is most easily seen by cross-country comparisons. For example, cold water is traditionally used for clothes washing in China, whereas hot water washing is common in the USA and Europe. Similarly, there are international differences in how lighting is used at night; the preferred temperatures of food, drink, and homes; and the operating hours of commercial buildings.

The provision of information is subject to a classic public-goods problem. If one person generates useful information, it creates a positive externality because it provides knowledge to others. Those that have information may have strategic reasons to manipulate its value; self-interested sellers have incentives to provide misinformation about their products; and well-distributed misinformation can often overpower the distribution of unbiased and more accurate information.

Technologies that are otherwise expected to be successful may still face difficulties penetrating the market due to infrastructure limitations. These include a wide range of supply-chain shortfalls ranging from inadequate physical systems and facilities; shortage of key complementary technologies that improve the functionality of a new technology; insufficient supply and distribution channels; and inadequate operation and maintenance (O&M) support (Brown and Sovacool 2011).

Solutions and Policy Mechanisms

Thankfully, a slew of public policy mechanisms can overcome these types of barriers. Government interventions have generally fallen into two broad classifications: supply-push mechanisms focus primarily on "pushing" technologies into the market through direct subsidies, and demand-pull mechanisms focus primarily on "pulling" technologies into the market by creating demand for them. Common examples of supply-push strategies include: (a) conducting basic and applied research and development on energy technologies; (b) building large test or prototype facilities; (c) having the government procure large amounts of an experimental technology; and (d) investor tax credits that spur innovation on a given technology. Common examples of demand-pull strategies include: (a) creating markets for technologies through production tax credits; (b) establishing rate-based or purchase-based incentives such as higher rates of return or tariffs; and (c) promoting technologies through training or information and awareness campaigns (Blumstein *et al.* 1980; Loiter and Nornerg-Bohm 1999; Espey 2001; Haas *et al.* 2004; Vandenbergh 2004; Lindén *et al.* 2006). Synthesizing from this literature, and updating it to today, the most basic and elementary policy tool is putting a price on carbon; secondary measures include everything from renewable portfolio standards and feed-in tariffs to building codes and appliance standards.

Putting a Price on Carbon

One of the simplest actions countries and international institutions such as the United Nations could take to provide an equity-increasing and welfare-maximizing response to climate change is to create a market price for GHG emissions and charge emitters for the cost of climate mitigation technologies. Putting a price on GHG emissions is accomplished with various policies including energy and carbon taxes and cap-and-trade systems (see also Chapter 27 in this volume). An extensive academic literature suggests that macroeconomic efficiency favors a carbon tax with socially productive revenue recycling over other forms of regulation. The choice of policy instrument, however, is less important than having an effectively designed instrument. In a cap-and-trade program, sources of GHG emissions covered under the program receive allowances that determine the amount of emissions they can produce. Based on that amount, sources of emissions can design their own emission-control strategy using any of several emission-reduction options such as: adopting new technology, purchasing offsets, or trading in the emissions market.

This flexibility provides numerous advantages. Because emissions trading uses markets to determine how to deal with the problem of pollution, cap and trade is often touted as an example of effective free market environmentalism. Markets encourage low-cost solutions rather than mandating specific technologies. While the cap is usually set by a political process, individual companies are free to choose how or if they will reduce their emissions. In theory, firms will choose the least-cost way to comply, creating incentives that reduce the cost of achieving a pollution reduction goal. Putting a price on carbon is a critical "core" policy because it addresses the principal market failure that has prevented individuals and firms from responding effectively to the damages precipitated by GHG emissions. Some have argued that, in fact, putting a price on carbon is all that is needed. Evidence is mounting, however, that complementary policies are required as well.

Energy Supply Options

Complementary energy supply policy options include renewable energy obligations, such as renewable portfolio standards and real-time pricing for electricity, as well as reducing fossil-fuel subsidies and passing feed-in tariffs. Some of these instruments have become quite popular, with 85 countries having some type of policy target for renewable energy in 2009, a jump from only 45 in 2005 (REN21 2010). Europe's target of 20% of final energy by 2020 is predominant among countries belonging to the OECD, Brazil is targeting 75% renewable electricity by 2030, China 15% final energy by 2020, India 20 000 MW of solar by 2022, and Kenya 4000 MW of geo-thermal by 2030. In early 2010, no fewer than 50 countries and 25 states and provinces had some type of feed-in tariff, and 46 countries were home to renewable portfolio standards for electricity. A number of towns and municipalities around the world – including Güssing (Austria), Dardesheim (Germany), Moura (Portugal), Varese Ligure (Italy), Samsø (Denmark), Thisted (Denmark), Frederikshavn (Denmark), and Rock Port (United States) – have already implemented 100% renewable energy sectors or will implement them by 2015. Table 26.2 provides an overview of these policies at the national level around the world.

As Table 26.2 also demonstrates, planners have adopted a cornucopia of other types of policies to promote renewable energy, many in combination. Direct capital investment subsidies, grants, and rebates are offered in 45 countries; tax credits, import duty reductions, and other tax incentives are offered in more than 30 countries; net metering laws now exist in 10 countries and in 43 states in the USA.

As one example of an innovative program, the city of Ellensburg, Washington, USA, started promoting virtual net metering to incentivize residents to invest in a municipal-scale community solar PV system. The city built a 36 kW solar array in 2006 and asked interested residents to contribute to its capital cost; in return, participants receive a credit on their electricity bill apportioned to their level of investment (Coughlin and Cory 2009).

Transport Options

Transport policies include mandates for biofuel blending along with investments in alternative transport and carbon dioxide standards for cars or airplanes. Biofuel blending mandates exist in 41 states and provinces and 25 countries as of 2010, with most requiring a blending of 10 to 15% ethanol with gasoline or 2 to 5% biodiesel

Table 26.2   Renewable energy promotion policies as of 2010.

Country	Feed-in tariff	Renewable portfolio Standard/Quota	Capital subsidies, grants, rebates	Investment or other tax credits	Sales tax, energy tax, excise tax, or VAT reduction	Tradable RE certificates	Energy production payments or tax credits	Net metering	Public investing, loans, or financing	Public competitive bidding
EU-27										
Austria	X		X	X		X			X	
Belgium		(*)	X	X	X	X		X		
Bulgaria	X		X						X	
Cyprus	X		X							
Czech Republic	X		X	X	X	X		X		X
Denmark	X		X	X	X	X		X	X	
Estonia	X		X		X		X			
Finland	X		X		X	X	X			
France	X		X	X	X	X			X	X
Germany	X		X	X	X	X		X	X	
Greece	X		X	X	X			X	X	
Hungary	X		X	X	X				X	X
Ireland	X		X	X	X	X			X	X
Italy	X	X	X	X	X	X		X		X
Latvia	X		X		X			X	X	
Lithuania	X		X	X	X				X	X
Luxembourg	X		X	X	X				X	
Malta			X		X			X		

Country									
Netherlands		X	X	X	X	X			
Poland	X	X		X	X			X	X
Portugal	X		X	X				X	X
Romania	X	X		X	X			X	
Slovakia		X	X	X	X			X	
Slovenia		X	X	X	X	X		X	X
Spain		X	X	X	X			X	
Sweden	X	X	X	X		X		X	
United Kingdom	X	X		X	X	X		X	
Other developed/transition countries									
Australia	(*)	X		X		X		X	
Belarus					X			X	
Canada	(*)	X	X	X		X	X	X	X
Israel	X			X					X
Japan	X	X	X	X	X		X		
Macedonia	X								
New Zealand		X		X	X			X	
Norway		X	X	X	X	X		X	
Russia		X							
Serbia	X								
South Korea	X	X	X	X		X		X	
Switzerland	X	X		X					
Ukraine	X								
United States	(*)	X	X	X	(*)	X	(*)	(*)	(*)

(continued)

Table 26.2 (Continued).

Country	Feed-in tariff	Renewable portfolio Standard/Quota	Capital subsidies, grants, rebates	Investment or other tax credits	Sales tax, energy tax, excise tax, or VAT reduction	Tradable RE certificates	Energy production payments or tax credits	Net metering	Public investing, loans, or financing	Public competitive bidding
Developing Countries										
Algeria	X			X	X					
Argentina	X		X	(*)	X		X		X	X
Bolivia					X					
Brazil			X	X					X	X
Chile		X	X	X	X				X	X
China	X	X	X	X	X		X		X	X
Costa Rica							X			
Dominican Republic	X		X	X	X					
Ecuador	X		X	X						
Egypt					X					X
El Salvador				X	X				X	
Ethiopia				X	X					
Ghana			X	X	X				X	
Guatemala				X	X					
India	(*)	(*)	X	X	X	X	X		X	
Indonesia	X			X	X					
Iran				X			X			
Jordan					X			X	X	
Kenya	X			X						
Malaysia									X	

Mauritius						X				
Mexico	X				X		X		X	X
Mongolia							X			
Morocco	X		X	X					X	X
Nicaragua	X		X	X						
Pakistan	X				X	X				
Palestinian Territories			X		X					
Panama								X		
Peru	X		X	X				X	X	X
Philippines	X	X	X	X	X		X	X	X	X
Rwanda							X			
South Africa	X		X				X		X	X
Sri Lanka	X									
Tanzania	X		X							
Thailand	X		X				X		X	
Tunisia	X		X							
Turkey	X		X				X		X	
Uganda	X	X	X				X		X	
Uruguay										X
Zambia	X									

Source: Brown, Marilyn A. and Benjamin K. Sovacool. 2011. *Climate Change and Global Energy Security: Technology and Policy Options*. Cambridge, MA: MIT Press.© 2011 MIT.

Note: Entries with an asterisk (∗) mean that some states/provinces within these countries have policies but there is no national-level policy. Only enacted policies are included.

with diesel. Biofuels targets exist in more than 10 countries plus the European Union, and exemptions for fuel taxes and production subsidies are also common.

In Israel, the government has started an ambitious program to promote plug-in hybrid electric vehicles (PHEVs): Project Better Place. The government has teamed up with automobile and battery manufacturers to distribute PHEVs, construct recharging facilities, and create service stations that can quickly replace depleted batteries. Renault and Nissan provide the cars (at a discounted price comparable to gasoline vehicles due to an Israeli subsidy), and Project Better Place provides lithium-ion batteries that are capable of traveling 124 miles per charge. The government provides the infrastructure needed to keep the cars going, such as small plugging stations on city streets, much like parking meters, and at service stations and highways. When batteries no longer perform well, drivers can visit a "car-wash like" station and have them replaced in a few minutes. To get drivers interested, the government offered generous tax incentives and has also invested US$200 million in public funds on electric vehicle infrastructure. Drivers get an electric vehicle at a greatly reduced price and then pay a fixed monthly fee for mileage for the electricity they use (Brown and Sovacool 2011).

Building Options

Building policy options include regulatory approaches such as appliance standards and building codes as well as demand-side management programs operated by electric and gas utilities and incentives for energy service companies (ESCOs). Cities and local governments around the world are becoming especially involved in setting building standards that require the installation of renewable energy. For example, in 2008, Spain became the first county to mandate solar water heating nationwide. And in Jiangsu, one of the most populous provinces in China, all new residential buildings of 12 stories and below must use solar water heating.

As an instance of efforts to promote building energy efficiency at the municipal scale, the city of Minneapolis, Minnesota, USA, and CenterPoint Energy operated a series of innovative neighborhood energy workshops in the early 1990s. Staff working for the city identified and trained volunteers to serve as block captains who then invited their neighbors to energy workshops. These workshops emphasized providing information about energy use habits, the energy efficiency and consumption of domestic appliances, and techniques that could be implemented quickly to save energy such as caulking or adding insulation (Harrigan 1994).

Japan has been especially successful at promoting appliance standards, with minimum energy performance standards beginning in 1983 for refrigerators and air conditioners, and later expanded to virtually all appliances, including the underrated electric toilet-seat warmer. The appliance standards effectively reduced electricity consumption over a short period of time. Average electricity use for refrigerators, for example, declined by 15% from 1979 to 1997, while average refrigerator size increased by 90%. Japanese regulators also applied their performance standards to imported technology ranging from automobiles and televisions to air conditioners and computers, demanding that the efficiency level of new imported products had to meet the best-performing product in the Japanese market, in some cases requiring energy-efficiency improvements of more than 50% (Geller et al. 2006).

Industry Options

Industrial policy options include mandatory performance standards or audits for manufacturers along with voluntary agreements and the provision of benchmarking information.

For example, the Netherlands has taken a proactive stance on industrial energy efficiency, beginning with their Long-Term Agreements on Energy Efficiency with industry starting in 1992. These agreements were established through an understanding by the industry that the government is closely observing energy consumption and would not initiate strong regulations so long as industry met the agreed targets. A second phase of this program, launched in 2000, is benchmarking the most energy intensive industries to comparable industries worldwide. The affected industries must be best in class in energy efficiency, and in return, the government will not implement additional stringent climate change policies (Brown and Sovacool 2011).

As is true of most developing countries, India's industrial makeup is dominated by small and medium-sized companies. To achieve the ambitious goal of reducing their energy intensity by 5% each year, India has introduced an energy efficiency trading program. It is expected that this market will be worth US$15 billion and will cover nine sectors by 2015 (Brown and Sovacool 2011). Analogously, from 1980 through 2000, China cut its national energy intensity by 65%, as the result of process and technological changes, as well as structural shifts throughout Chinese industry. Its Energy Conservation Law was revised, its tax policy was modified for export products, tax credits for efficiency investments were granted, and the Top-1000 Energy Consuming Enterprises program was initiated to promote energy efficiency throughout large-scale industry. The end result of these policies has placed China on a path towards reaching its mandates and reducing energy intensity once again (Lin *et al.* 2006).

Agriculture and Forestry Options

Agricultural and forestry options include land-use regulation and harvest quotas for timber alongside financial incentives for improved land management or increased forest area and Payments for Ecosystem Services (PES). One method is to provide financial incentives for organic fertilizer. In the state of Tamil Nadu, India, tea plantation owners have utilized bio-organic fertilization to replenish degraded land, restore soil fertility, and improve productivity. After decades of excessive chemical fertilizer and pesticide application had depleted the soil fertility and crippled the productivity of tea plantations (in some cases instigating crop losses as high as 70% of ordinary yield) plantation managers coordinated with university researchers and a fertilizer company to use natural methods to restore the land. Researchers placed vermicultured earthworms in trenches between tea rows, and relied on tea prunings and high-quality horticultural waste from nearby farms to create organic fertilizer that they then distributed to six large tea estates. The combination of earthworm trenching and organic fertilization increased tea yields from 76 to 239% and saw profits rise significantly (Bennack *et al.* 2003).

Another is Costa Rica's strategy of PES, which distributes payments to the owners of forests and forest plantations in exchange for their preservation and management of the land. The Costa Rican program, passed under their 1996 Forest Law and termed the Private Forest Project, recognizes four services provided by forests – protection of biodiversity, sequestration and fixation of carbon, erosion prevention and water purification, and scenic beauty – and then pays landowners using revenue from activities that threaten those services. A 5% tax on gasoline creates about US$16 million per year used to enhance biodiversity protection, the sale of carbon credits (called Certifiable Tradable Offsets) helps pay for carbon sequestration, and donations from private hydropower companies sponsor hydrologic services. During the first two years of the program more than 1000 landowners signed contracts to receive payments averaging US$120 per hectare per year for plantations, US$60 for forests, and US$45 for forest management and reforestation. The combined taxes and donations now produce about US$16 to US$20 million per year and generate 4 million t. of carbon credits to be brokered on the international market (Brown and Sovacool 2011).

As an example of quotas and forest management in Malaysia, home to less than 0.25% of the world's forests but 10% of its total number of plants and 7% of its species, regulators passed a National Forest Act in the 1980s to classify forests and set limits on harvesting and deforestation. The rules mandated that only trees of a certain length and age could be felled (protecting both young and old trees), prohibited harvesting of timber and wood within an extensive network of reserves, set strict quotas, and relied on surveillance (now performed by satellites) to track compliance. In 2007, the maximum harvest quota was 50 000 cubic meters, and newer standards require that those forests that have been harvested undergo regeneration and restoration efforts. Collectively, such policies have seen the amount of forest area grow from 58.7% of land area in 2000 to 63.6% of land area in 2005 (Brown and Sovacool 2011).

Waste and Water Options

Policy options for waste and water include waste management regulations and volumetric water pricing as well as incentives for waste incineration or anaerobic digestion, cleaner production processes, and extended producer responsibility. One promising approach is to use "cradle to cradle" design that intends to reuse and recycle products back into the manufacturing process at the end of their useful life. Another tool is the promotion of methane capture and biomethanization. In Brazil, for example, the Bandeirantes Landfill Gas to Energy Project captures methane that would otherwise be vented into the atmosphere and converts it to electricity. The city of São Paulo produces nearly 15 000 t. of waste per day and half of it goes to the Bandeirantes landfill. As a result, Bandeirantes is also one of the world's largest landfills, with a current capacity of about 30 million t. (or a size of 175 football fields filled with up to 8 m, or 26 ft., of trash). The Bandeirantes can hold about 20 years' worth of Brazilian rubbish, but it was also responsible for emitting a staggering 808 450 t. of carbon dioxide equivalent per year. Working with the city, Biogás Energia Ambiental SA built a system of degassers, pipes, heat exchangers, and 24 Caterpillar engines to capture the methane and use it to generate about 20 MW of electricity, enough to run the homes of about 400 000 people. From its inception in 2006 the facility has worked with a flare efficiency of 99.997% (meaning it captured almost

100% of the methane) and has reduced the metropolitan region's entire carbon emissions by 11% (Brown and Sovacool 2011).

At a much larger scale, the European Union (EU) has begun to address the pollution coming from discarded products through a principle known as Extended Producer Responsibility (EPR). First enacted in Germany and then expanded into an EU directive in 2001, EPR assigns long-term responsibility for the environmental impacts of products (such as lawnmowers and household paints, computers, batteries, and cellular telephones, to name just a few) from consumers to their manufacturers. It requires that manufacturers take back their products or charge consumers a small fee to pay for collection and recycling (Sachs 2006). While member countries have implemented the EPR directive differently, four types are most prevalent within the EU:

- economic, which requires manufacturers to pay all or a portion of end management and disposal or recycling costs;
- physical, which requires manufacturers to take possession of discarded goods to ensure that materials and components are recycled;
- informative, which requires manufacturers to publish information about where consumers can recycle their product; and
- legal, which makes manufacturers liable for the environmental damage resulting from their products, including costs for remediation, cleanup, and disposal.

The central premise behind EPR is manufacturers should be made responsible for their goods at the source. As a result of EPR legislation in Europe, manufacturers have designed products to be more recyclable and/or with less environmentally damaging raw materials; improved their efforts to collect discarded goods; incorporated recycled components and materials back into their production processes; adopted modular designs that are easier to disassemble; and unified and harmonized standards for various types and grades of materials and plastics.

Global Policy Options

Due to the scale and complexity of energy and climate challenges, a final set of approaches operates at the level above nation-states – at the supranational scale. Intergovernmental organizations (IGOs), for example, are created and funded by national governments, which have secretariats that answer to a governing body, but operate within the global system. Some of these, such as the International Energy Agency or International Atomic Energy Agency, deal with exclusively with energy.

Sometimes, organizations like the United Nations will specifically adopt resolutions aiming to enhance the attention to energy issues, such as the 2012 "International Year for Sustainable Energy for All," which was adopted in December 2010. The initiative seeks to engage governments, companies, and other civil society actors to achieve three goals by 2030: universal access to modern energy services, reducing global energy intensity by 40%, and increasing renewable energy use globally to 30% of total primary energy supply.

Multilateral financial institutions give loans and financial support for infrastructure projects intended to promote economic development, often involving energy systems and technology. The best known of these banks is the World Bank Group, which consists of five Washington, DC-based institutions, the three most important being: the International Bank for Reconstruction and Development, the International Finance Corporation, and the International Development Association. The role that these banks play in shaping national energy programs through financing and technical assistance has come under intensive scrutiny in the past few years, with lending from these banks often exceeding hundreds of millions of dollars per project.

Governments sometimes form treaties dealing with energy. The Energy Charter Treaty, for instance, places an obligation on its 51 current members to facilitate safe transit of energy fuels across territories, with the aim of creating a transparent and efficient energy market. It also offers dispute settlement over energy-transit-related issues, seeks to protect European foreign investments in energy, and promotes free-flowing trade of energy commodities (Sovacool and Florini 2012).

Other supranational institutions focus on setting global technology standards. The International Partnership for the Hydrogen Economy establishes common codes and standards conducive to the global adoption of hydrogen systems through its 17 member countries. Similarly, the Collaborative Labeling and Appliance Standards Program is funded by a variety of organizations including the US government, World Bank Group, and United Nations. It assists with the implementation of various standards and labels relating to energy and energy efficiency technologies and services (Sovacool and Florini 2012).

Lastly, hybrid entities form partnerships, often between public- and private-sector organizations, to accomplish their energy-related goals. One example is the Renewable Energy and Energy Efficiency Partnership (REEEP), which is dedicated to reducing greenhouse gas emissions, improving the access to reliable and clean forms of energy in developing countries, and promoting energy efficiency. The 2008 program year saw REEEP running 145 projects worth a total cumulative investment of €65 million, most of this leveraged from REEEP partners through equity financing, and plans for 37 new projects. These new projects included the promotion of solar water heaters in Uganda, energy-efficient lighting in India, rural biomass development in China, renewable energy financing in Mexico, and assessing the regulatory framework for renewable energy in Argentina (Florini and Sovacool 2009).

Conclusions

Unfortunately, complementary policies such as undertaking R&D, adjusting subsidies, internalizing externalities, promulgating standards, and improving information will not work in isolation. Changing R&D practices without removing subsidies for carbon-intensive technologies, for instance, would have to swim against the current created by existing incentives and momentum. Removing subsidies without promoting public information and education will ensure that consumers remain uninformed about other options and the inefficiency of their current practices. Some energy services fulfill social functions independent of cost, so that people will ignore price changes for as long as possible until it becomes completely prohibitive and a threshold is passed. Consumers want to preserve their lifestyles and often do so until costs become prohibitive, and manufacturers will protect their current practices against any changes that might threaten to disrupt productivity or profitability.

Policy-makers and regulators must design policy mechanisms that match the technical-economic-political-socio-cultural dimensions of current society. Once recognized, they must consistently pursue a variety of policy mechanisms that simultaneously alter R&D practices, fine-tune subsidies, price externalities, and better inform the public if they are to affect consumer demand and promote sustainable energy practices at the speed, scope, and scale required. With this in mind, three conclusions are offered.

First, the energy and climate change issues confronting the world are neither technical nor social, but *socio-technical*. That is, they involve not only technologies including physical devices, objects, infrastructures, systems, and tools, but also people who are motivated by human values, habits and routines, cognitive limitations, and cultural beliefs. This simple conclusion has somewhat profound implications for energy and climate research. Technology research and commercialization efforts must be coupled with attempts to educate and inform consumers, overcome biases and apathy, shift cultural values and behavior, and incentivize people to use new technologies along with old ones that already work. Individuals making relatively simple changes to their lifestyles, such as consuming less energy at home, cycling instead of driving to work, eating less meat, and purchasing second-hand or used items, can in aggregate add up to significant climatic benefits. In short the socio-technical dimension of energy and climate change necessitates holistic and complementary solutions that avoid looking at only one face of the socio-technical coin. Our own individual behavior can be just as important as developing new technology.

Second, the complex socio-technical nature of climate and energy challenges offers a robust justification for government intervention. Numerous market failures and barriers exist on both social and technical planes, including externalities, high costs, infrastructural limitations, and technical risks (technical obstacles) as well as policy failures, utility monopolies, energy price volatility, and lack of knowledge, training, and information (social obstacles).

The good news is that governments can do much to overcome these impediments, from putting a price on carbon to a range of innovative and effective complementary policies, some regulatory and others voluntary. If targeted to overcome behavioral barriers – such as loss aversion, asymmetric information, habits, and heuristics to deal with overwhelming deliberation costs – these policies can transform markets. Options include increasing research expenditures for key technologies, sponsoring neighborhood workshops to personalize information about clean energy choices, reforming subsidies and designing incentives to overcome social impediments, and implementing payments for ecosystem services along with extended producer responsibility.

However, to achieve the levels of market transformation needed to match the challenges faced, a much deeper understanding of policy barriers and drivers is essential. Shifts to individual and institutional behavior are instrumental so that marketable and effective energy and climate technologies and policies become widely adopted.

Third and finally, while intervention by governments is important, it is often much more effective when implemented at a variety of scales in cooperation with a plurality of actors, and with the speed, scope, and scale required to repair the planet. Individuals, cities, corporations, and other groups must act alongside regulators and government officials. Or, as the philosopher Jürgen Habermas once wrote, "in the process of enlightenment there can only be participants." The same holds true for climate change: we must *all* participate.

Individuals, however, can alter many of their daily practices to substantially reduce emissions: they can, for instance, use less energy-intensive goods and services, drive more efficient cars, purchase better electrical appliances, eat less meat, and conserve water. They should not be viewed as passive recipients loosely connected to climate change, but as active participants whose lifestyles play a central role in contributing to energy and climate problems. The situation brings to mind the words of Rachel Carson (1962: ix), who wrote that "the human race is challenged more than ever to demonstrate our mastery – not over nature, but of ourselves."

References

Archer, David. 2009. *The Long Thaw*. Princeton, NJ: Princeton University Press.

Bennack, Dan, George Brown, Sally Bunning, and Mariangela Hungria da Cunha. 2003. "Soil Biodiversity Management for Sustainable and Productive Agriculture: Lessons from Case Studies." In *Biodiversity and the Ecosystem Approach in Agriculture, Forestry, and Fisheries*, 196–223. Rome: United Nations Food and Agricultural Organization.

Blumstein, Carl, Betsy Krieg, Lee Schipper, and Carl York. 1980. "Overcoming Social and Institutional Barriers to Energy Conservation." *Energy*, 5: 355–371.

Boucher, Doug. 2009. *Money for Nothing? Principles and Rules for REDD and Their Implications for Protected Areas*. Washington, DC: Tropical Forest and Climate Initiative of the Union of Concerned Sciences.

Brown, Marilyn A. and Benjamin K. Sovacool. 2011. *Climate Change and Global Energy Security: Technology and Policy Options*. Cambridge, MA: MIT Press.

Carson, Rachel. 1962. *Silent Spring*. New York: Houghton Mifflin.

Coughlin, Jason and Karlynn Cory. 2009. *Solar Photovoltaic Financing: Residential Sector Deployment*. NREL/TP-6A2-44853, March. Golden, CO: National Renewable Energy Laboratory.

Dunn, William N. 2008. *Public Policy Analysis: An Introduction*, 4th edn. Upper Saddle River, NJ: Pearson Prentice Hall.

EIA (Energy Information Administration). 2006. *International Energy Outlook*. DOE/EIA-0484. Washington, DC: Department of Ecology.

Espey, S. 2001. "Renewables Portfolio Standard: A Means for Trade with Electricity from Renewable Energy Sources?" *Energy Policy*, 29: 557–566.

Florini, Ann and Navroz K. Dubash. 2011. "Introduction to the Special Issue: Governing Energy in a Fragmented World." *Global Policy*, 2(s1): 1–5.

Florini, Ann and Benjamin K. Sovacool. 2009. "Who Governs Energy? The Challenges Facing Global Energy Governance." *Energy Policy* 37(12): 5239–5248.

Geller, Howard, Philip Harrington, Arthur H. Rosenfeld *et al*. 2006. "Policies for Increasing Energy Efficiency: Thirty Years of Experience in OECD Countries." *Energy Policy*, 34: 556–573.

Haas, R., W. Eichhammer, C. Huber *et al*. 2004. "How to Promote Renewable Energy Systems Successfully and Effectively." *Energy Policy*, 32: 833–839.

Harrigan, Merrilee. 1994. "Can We Transform the Market without Transforming the Consumer?" *Home Energy*, 11(1): 17–23.

Hawken, Paul. 1998. *The Ecology of Commerce: A Declaration of Sustainability*. Washington, DC: Island Press.

International Energy Agency. 2010. *World Energy Outlook 2010*. Paris: OECD.

IPCC (Intergovernmental Panel on Climate Change). 2008. *Climate Change 2007: Synthesis Report*. Geneva: IPCC.

Jacobson, Mark Z. and Gilbert M. Masters. 2001. "Exploiting Wind versus Coal." *Science*, 293: 1438–1439.

Lin, J., Nan Zhou, Mark D. Levine, and David Fridley. 2006. *Achieving China's Target for Energy Intensity Reduction in 2010: An Exploration of Recent Trends and Possible Future Scenarios*. Berkeley, CA: Lawrence Berkeley National Laboratory.

Lindén, Anna-Lisa, Annika Carlsson-Kanyama, and Björn Eriksson. 2006. "Efficient and Inefficient Aspects of Residential Energy Behavior: What Are the Policy Instruments for Change?" *Energy Policy*, 34: 1918–1927.

Loiter, J.M. and V. Nornerg-Bohm. 1999. "Technology Policy and Renewable Energy: Public Roles in the Development of New Energy Technologies." *Energy Policy*, 27: 85–97.

REN21. 2010. *Renewables 2010 Global Status Report*. Paris: REN21 Secretariat.

Sabety, Ted. 2005. "Nanotechnology Innovation and the Patent Thicket: Which IP Policies Promote Growth?" *Albany Law Journal of Science & Technology*, 15: 477–515.

Sachs, Noah. 2006. "Planning the Funeral at the Birth: Extended Producer Responsibility in the European Union and the U.S." *Harvard Environmental Law Review*, 30: 51–98.

Saunders, Kurt M. and Linda Levine. 2004. "Better, Faster, Cheaper – Later: What Happens When Technologies Are Suppressed." *Michigan Telecommunications and Technology Law Review*, 11: 23–69.

Smil, Vaclav. 2000. "Energy in the Twentieth Century: Resources, Conversions, Costs, Uses, and Consequences." *Annual Review of Energy and Environment*, 25: 21–51.

Sovacool, Benjamin K. 2008. "Placing a Glove on the Invisible Hand: How Intellectual Property Rights May Impede Innovation in Energy Research and Development (R&D)." *Albany Law Journal of Science & Technology*, 18(2): 381–440.

Sovacool, Benjamin K., Christopher Cooper, Morgan Bazilian *et al.* 2012. "What Moves and Works: Broadening the Consideration of Energy Poverty." *Energy Policy*, 42: 715–719.

Sovacool, Benjamin K. and Ann Florini. 2012. "Examining the Complications of Global Energy Governance." *Journal of Energy and Natural Resources Law*, 30(3): 235–263.

Stehfest, Elke, Lex Bouwman, Detlef P. van Vuuren *et al.* 2009. "Climate Benefits of Changing Diet." *Climatic Change*, 95(1–2): 83–102.

Vandenbergh, Michael P. 2004. "From Smokestack to SUV: The Individual as Regulated Entity in the New Era of Environmental Law." *Vanderbilt Law Review*, 57: 515–610.

Victor, David, Granger Morgan, John Steinbruner, and Kate Ricke. 2009. "The Geoengineering Option: A Last Resort against Global Warming?" *Foreign Affairs*, 88: 61–68.

Yergin, Daniel. 2011. *The Quest: Energy, Security, and the Remaking of the Modern World*. New York: Penguin Books.

Chapter 27

Economic Instruments for Climate Change

Jonas Meckling and Cameron Hepburn

Climate policy instruments have proliferated around the world – there are now several thousand climate policy interventions – but the major policy debate remains focused on the choice between different types of "economic instruments" to put a price on carbon dioxide emissions. Economic instruments include carbon taxes, which directly create an explicit price on emissions of carbon dioxide, and emissions trading schemes, which indirectly create an explicit price, through the creation of a market in licenses or permits to pollute. Economic instruments might be contrasted with so-called "command-and-control" regulatory measures, which do not create an explicit price on pollution (Hepburn 2006).[1]

The pervasiveness of greenhouse gas (GHG) emissions in modern economies and the substantial associated mitigation cost has led policy-makers to focus on cost-effective economic instruments for climate change. A broad price incentive is viewed as a necessary (but likely not sufficient) policy intervention. Since the late 1980s, climate policy debates have therefore focused in on the relative merits of taxes and trading schemes. Thirty years later, both instruments are in use, and new carbon taxes and emissions trading schemes continue to be created in different countries around the world. However, emissions trading appears to have emerged as the dominant economic instrument in climate policy mixes around the globe (Meckling 2011b), and reflects a broader trend toward market-based environmental policy (Newell 2008).

The EU Emission Trading Scheme (EU ETS) is currently the largest cap-and-trade scheme in the world. In addition, the Clean Development Mechanism (CDM) under the Kyoto Protocol allows developing countries to participate in emission-reducing, credit-generating activities. A number of other industrialized and emerging economies are designing or implementing cap-and-trade schemes, notably Australia and South Korea. In 2010, the carbon markets were worth US$142 billion (Point

The Handbook of Global Climate and Environment Policy, First Edition. Edited by Robert Falkner.
© 2013 John Wiley & Sons, Ltd. Published 2013 by John Wiley & Sons, Ltd.

Carbon 2011), although trading values have crashed, along with market prices, in 2011 and 2012 due to, among other things, weak economic activity in the Eurozone.

This chapter discusses the economics, politics, and governance of economic instruments for climate change mitigation. We first review the economic arguments for carbon taxes and emissions trading. In particular, we compare the two instruments along a number of criteria, including economic efficiency, effectiveness, flexibility and credibility, administrative cost, industrial dynamics, and international aspects. Thereafter, the chapter offers an overview of the politics of carbon tax and emissions trading proposals in the international negotiations, the EU, and the USA. We will discuss the political economy of climate policy instruments, explaining why emissions trading could mobilize a larger political constituency than carbon taxes. Next we outline the current landscape of carbon markets, including their geographic scale, financial scope, their performance and governance. Finally we offer our conclusions.

The Economics of Pricing Carbon: Emissions Trading vs. Pollution Taxes

Standard economic theory holds that the problem with pollution is that polluters do not incur the costs of their actions, so they pollute excessively. Economics suggests several solutions. Government can intervene by "command and control" – firms can be required to reduce pollution by a certain amount. However, it is very difficult for governments to determine how much each individual firm should optimally contribute to the total reduction in pollution, as this requires detailed information on individual firms' costs. This leads to potentially vast inefficiencies, particularly for a problem such as climate change where abatement costs vary widely.

There are two simple solutions that can achieve the optimum allocation of pollution reduction between firms, without requiring unmanageable amounts of information and government planning. Pigou (1920) proposed direct taxation of pollution, creating a fixed and explicit price on pollution. Alternatively, Coase (1960) noted that capping the total quantity of pollution and allowing firms to trade in a market would yield an (indirect) market price on pollution, and an efficient allocation of abatement between firms.[2] Hybrids between the two are possible, and even in their pure form these two simple economic instruments can be implemented in a wide variety of different ways.

Pollution taxes can be levied "upstream," near to the point of extraction of the polluting resource (e.g. fossil fuels), "downstream" (e.g. at the point of emission by consumers using gasoline in vehicles), or somewhere in between. Taxes are often imposed as a flat rate per unit of pollution (e.g. $/tCO_2$), but an increasing (or decreasing) schedule of tax rates could also be imposed, not unlike income tax schedules. Finally, as with any tax regime, exemptions may be granted to certain sectors or groups, often with the aim of protecting internationally exposed industries, or helping poorer or more vulnerable consumers.

Cap-and-trade systems can also be implemented in a variety of ways. Regulated entities can be upstream, downstream, or in between. In most carbon-trading schemes, such as the EU ETS, the regulated entities are direct sources of emissions. For trade-exposed industries, government might hand out some permits for free. Indeed, in most environmental trading schemes, a very high proportion of allowances are given for free to polluters.

Trading schemes can provide further flexibility in how regulated companies meet their obligations (Fankhauser and Hepburn 2010a, 2010b). Offsets from projects that reduce emissions outside the regulated sectors (or in other countries) can be permitted as a way of further reducing the cost of achieving a given environmental goal (Hepburn 2007). Regulators may allow firms to "bank" permits from one period to the next, so that they can choose to make more emission reductions early and sell or use their permits later if they believe prices will rise (Fankhauser and Hepburn 2010a). One of the concerns about trading schemes is that while they fix a specific quantity of emissions, the pollution price is uncertain and potentially sometimes quite volatile. To address these concerns, price ceilings and floors can be imposed on the market, to create a so-called "hybrid" system, which blend features of tax and trading schemes.

So, are pollution taxes "better" in theory than trading schemes? Do hybrids offer the best of both worlds? Unsurprisingly, the answer depends upon the particular pollutant and the specific domestic and international political context. The relevant criteria by which policy-makers might choose the suitable instrument include: (i) efficiency (under uncertainty and policy lags); (ii) environmental effectiveness; (iii) credibility and flexibility; (iv) market dynamics; (v) administrative costs; (vi) international considerations; (vii) political issues; and (viii) governance challenges. Political issues are considered in the section on "Emissions Trading vs. Pollution Taxes"; governance challenges in the section on "The State and Performance of Carbon Markets." Here we examine the first five considerations.

Efficiency

There is a basic symmetry between taxes and trading. Taxes directly set an explicit price, while a trading scheme indirectly creates an explicit price, revealed by the market. Under idealized conditions, if the regulators are aiming at the same objectives, the market price under the trading scheme will equal the level of the optimum tax (Weitzman 1974). A looser cap translates into setting a lower tax, and vice versa. Under idealized conditions, there is a one-to-one correspondence between taxes and trading, and their implications for economic efficiency are identical.

However, the real world is far from ideal, and there are various reasons one might expect taxes and trading schemes to have different implications for economic efficiency. For instance, climate change is an inherently international problem; to minimise costs, carbon prices would be the same in all countries. Yet if each nation imposed carbon taxes, in their own currencies, those tax rates would need to be continuously adjusted to reflect changing foreign exchange rates. In contrast, usual processes of market arbitrage would ensure that a global emissions market implied an equivalent permit price in all relevant currencies from day to day and even hour to hour.

Another reason is that policy-makers do not know what the "optimum" pollution price or cap is going to be in advance – this depends upon how much it costs companies to clean up the pollutant, and how damaging it is. Both can be estimated, but are not known with certainty. When abatement costs are uncertain, Weitzman (1974) showed that the basic symmetry between taxes and trading is lost. Neither instrument can be certain to be optimal. The aim is to minimize the *expected* efficiency

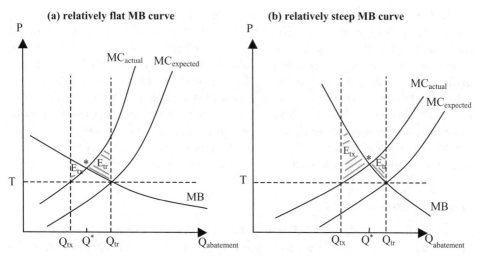

Figure 27.1 Trading has lower expected efficiency loss with a steep marginal benefit curve. Source: Adapted from Hepburn, Cameron. 2006. "Regulation by Prices, Quantities, or Both: A Review of Instrument Choice." *Oxford Review of Economic Policy*, 22(2): 226–247.

loss. Weitzman (1974) demonstrated that under certain conditions, the efficiency loss depends upon the relative slopes of the marginal costs (MC) and the marginal benefits (MB) of abating pollution. Trading is more efficient when the marginal benefit of pollution reduction increases rapidly as more pollution is emitted (or the less pollution is abated), relative to the marginal costs of abating the pollution. On the other hand, taxes are likely to be more efficient when the MB is reasonably constant, that is, when one unit of pollution does around as much damage as any other unit of pollution.

Figure 27.1 provides an illustration where the actual marginal costs of abatement are higher than originally expected. Here, the tax (T) generates too little abatement ($Q_{tx} < Q^*$) leading to efficiency loss shown by the shaded area E_{tx}. In contrast, the trading scheme with cap Q_{tr} leads to too much abatement ($Q_{tr} > Q^*$) and efficiency loss E_{tr}. As Figure 27.1 shows, the tax has a lower efficiency loss compared to trading ($E_{tx} < E_{tr}$) when the MB curve is relatively flat, and vice versa.

This analysis is limited in various ways: it doesn't consider uncertainty in the marginal benefit curve, nor does it consider the possibility that policy-makers will realize their error and adjust policy to correct for it, nor does it consider transitional efficiency losses. Critically, it also assumes that there are no known "tipping points" in the climate system. However, a broad conclusion that can be drawn from this analysis is that for a stock pollution problem like climate change, if policy is adjusted over short periods, and in the absence of known "tipping points," then taxes are the more efficient policy instrument under uncertainty.

Environmental Effectiveness

A simple but important consideration is whether the policy instrument will actually achieve the intended objective of reducing emissions. One of the major disadvantages of price instruments such as carbon taxes is that they cannot provide this guarantee, unlike cap-and-trade systems. Returning again to Figure 27.1, if the international community agrees that Q_{tr} is the appropriate level of abatement, an emissions trading scheme will achieve that, irrespective of whether costs and benefits shift. Indeed, one of the reasons the price in the EU ETS is currently so low is because economic activity in Europe has collapsed, as have emissions, and hence the cost of abatement is considerably lower than previously. As the target has not moved, the permit price has fallen. If economic activity picks up, the price rises in order to ensure the target is achieved. With a fixed tax, in contrast, changes in abatement costs (e.g. due to a recession or a boom) would lead to abatement which is above or below the target.

Credibility and Flexibility

One of the major challenges of climate policy is the long-term nature of the response. Policy incentives that are allegedly supposed to last for several decades need to be credible before the private sector will make investments in reliance upon them. This credibility problem is acute in nations when climate policy is highly politicized – each new administration finds it expedient to roll back the policies of the previous one.

Helm *et al.* (2003) define the credibility problem as being caused by conflicts between multiple objectives (e.g. energy costs, emissions, energy security), the irreversible nature of the necessary capital investments, and the scope and incentive for ex post reneging on earlier policies. Of course, there is also merit in policy flexibility to respond to new events. This does not necessarily imply discretion, however. Policy can be designed with clarity over the rules that would guide adjustments in the light of new information. If it is suspected that politics creates risks that such rules will not be followed, delegation to an independent agency, as with the delegation of monetary policy to central banks, can provide a solution (Helm *et al.* 2003).

These considerations have a bearing on the design and implementation of the relevant economic instrument. In the UK for instance, power was delegated to the CCC to advise on emissions budgets rather than on tax rates, partly because HM Treasury guards taxation powers closely. An independent agency with power to set (or advise on) taxes may have been less credible. In the EU ETS, changes to the cap, for instance, require negotiation and agreement by the relevant EU member-states. This implies that the cap is difficult to adjust, for better or worse. This enhances credibility but reduces flexibility. It is unclear whether adjustments to an EU-wide tax would be more or less straightforward. It can be argued that trading schemes such as the EU ETS might further increase their credibility by incorporating some kind of mechanism to stabilize prices, whether in a hybrid model with floors and ceilings, auction reserve prices (Hepburn *et al.* 2006), or gateways (Fankhauser and Hepburn 2010a).

Industrial Dynamics

In a broader sense, different price instruments generate different industrial dynamics, a feature not often commented upon by economists. For instance, carbon taxes provide a stable price signal that favors investment by risk-averse firms. They create incentives for greater activity by accountants. For better or worse, taxes tend to promote business-as-usual to a greater extent than trading. Taxes also promote greater market concentration in oligopolistic industries (Hepburn *et al.* forthcoming), although this effect should also be observed from the price incentives arising from trading schemes. The overarching discourse in an environmental tax regime is one of "tax minimization" and the industrial focus is on the stick rather than the carrot.

In contrast, carbon trading can lead to a more volatile, higher-risk environment, with greater potential for creative destruction, with both its good and bad aspects. Prices can move wildly, and the financial derivatives created to manage price risk also allow market participants to speculate on price movements to make leveraged gains or losses. However, a new market also creates the possibility of new business models, where entrepreneurs can grow clean energy firms that, through their activities, acquire tradable property rights.

Similarly, with a trading scheme, the discourse within financial and industrial firms is not merely about "compliance," but also focuses on "profit opportunities," either from trading or by identifying previously unknown abatement opportunities and making a profit margin when these are cheaper than market prices. Finally, the large emitters who are granted emissions permits for free find themselves with a new (and often substantial) asset on their balance sheet, which they can use to secure finance for new initiatives, clean or otherwise.

Administrative Costs

Administrative costs – which are largely a deadweight loss – can vary widely from one policy choice to another (Krutilla and Krause 2011). The application of a policy "downstream," placing compliance obligations on individuals, can create an enormous burden (Kahn and Franceschi 2006). For instance, if every individual (including the young and the elderly) had a tradable "personal carbon allowance," the IT costs would be enormous, not to mention the transaction costs and delays of ensuring that an adequate number of allowances were retired with each fuel purchase. One suspects that grandparents have better things to do than to trade their personal carbon allowances on a market. Equally, the privacy implications of a personal carbon-trading scheme may be significant.

At the other end of the spectrum, selecting the largest emitters and controlling pollution "upstream" can reduce transactions costs (Smith 2007). For instance, the Australian emissions trading scheme passed in November 2011 applies to roughly the largest 500 emitters in the country. The EU ETS applies to just over 10 000 installations, rather than the several hundred million European citizens. These policy choices considerably reduce transaction costs.

As between trading and taxes, there is little doubt that the administrative costs of taxes are lower than of setting up a trading scheme (Kahn and Franceschi 2006). Existing government taxation infrastructure can be deployed which, while non-trivial, is nowhere near as complex as the infrastructure required for the establishment of a fully functioning and well-regulated market, with the various elements of the industrial ecosystem that this entails. Once the market infrastructure has been set up, however, trading costs can be relatively low, and bid-ask spreads relatively low, allowing the system to reasonably efficiently work so that that those who value the permits most end up holding them. Additional features of trading, such as offset mechanisms, can also involve high administrative costs – although they reduce the overall costs of abatement. For instance, the administrative burden of the CDM implies that projects must reduce emissions by around 30 000–40 000 t. of carbon dioxide a year at a minimum before it is worth bothering with the costly and lengthy validation, registration, and verification processes.

International Considerations

Finally, the choice of policy instrument also depends upon international considerations. Does the instrument dovetail with obligations under international agreements? Is it consistent with the policy choices of trading partners? In the climate context, the fact that international negotiations tend to be conducted in quantities (e.g. 2020 emission reduction targets) rather than prices (e.g. carbon tax rates) can make it easier to transpose these obligations domestically by implementing a cap-and-trade regime. As noted above, it is far easier to use international markets to harmonize carbon prices and account for fluctuating exchange rates than it is to rely upon periodic intergovernmental meetings to adjust carbon taxes. It is entirely possible to achieve an international quantity target using a series of (potentially different) price instruments domestically, but it provides less certainty, and it also does not achieve global efficiency if prices differ between countries.

If trading partners also elect to regulate climate change by emissions trading, then linking markets together can further reduce costs by exploiting spatial variation in abatement costs (Fankhauser and Hepburn 2010b), and provide the benefits of deeper and more liquid markets (which also reduce transaction costs).

Overall

An economic analysis of instrument choice for climate change can be summarized as follows. First, for such a vast and challenging policy problem, cost-effectiveness is critical so economic instruments should be deployed. Second, for a problem like climate change with a relatively flat marginal damage function (unless and until a tipping point is located at a specific concentration), carbon taxes are more efficient under uncertainty. Third, carbon taxes likely involve lower administrative costs than the creation of a market. Fourth, taxes are likely to provide a more stable signal for investors. However, fifth, only emissions trading guarantees a particular environmental outcome. Sixth, trading appears to fit better with the international nature of the problem. Seventh, trading creates clearer opportunities for entrepreneurs to find new ways to reduce emissions and reduce costs of mitigation. Eighth, trading leverages the profit motive of firms who are more likely to support it.

The net result of this analysis is that the choice between trading and taxes is an important but second-order consideration to the need to get a carbon price in place through whatever mechanism is most politically feasible. As we will argue in the next section, the evidence so far suggests that trading has more appealing political features (cf. Meckling 2011b). Further, as Fankhauser and Hepburn (2010a, 2010b) argue, slight tweaks to make an ETS more "tax-like" – including long commitment periods, banking, and some kind of "price management" – can help an ETS to gain some of the advantages of taxes without the concomitant disadvantages.

The Politics of Pricing Carbon: Emissions Trading vs. Pollution Taxes

The previous section examined the economic theory of instrument choice. This section considers the politics. We provide a political history of the debates on taxes and trading and explore the political economy reasons for the current dominance of trading over taxes. In Europe, carbon trading has clearly been the most significant policy intervention on climate change, with greater impact than command-and-control regulation, voluntary initiatives, and carbon taxes. But in a historical context, the rapid initial adoption and the current, if somewhat unsteady, trend towards globalization of GHG emissions trading is puzzling given the strong initial opposition from EU governments, the majority of environmental groups, and parts of industry. Different explanatory accounts of the rise of carbon trading have focused on the role of transnational coalitions of firms, state actors, and green groups (Meckling 2011a, 2011b), point to the role of liberal norms (Bernstein 2001), the role of states or supranational institutions – especially the European Commission (Skjærseth and Wettestad 2008), the role of global capital (Matthews and Paterson 2005; Newell and Paterson 2010), and the role of financial service centers, such as New York and London (Knox-Hayes 2009), as driving forces. These narratives are not always mutually exclusive or mutually consistent, yet a comprehensive discussion is beyond the scope of this chapter. We focus here on laying out key historical steps in the political battle over pricing carbon, before we discuss the political economy of different market-based climate policy instruments.

The Politics of Carbon Tax Proposals in the EU and the USA, 1991–1993

In the early 1990s, environmental groups in both the EU and the USA were largely in favor of a carbon tax to address global climate change. After the UN Framework Convention on Climate Change was signed in 1992, the domestic battle over climate policy unfolded. After President Clinton took office in January 1993, he announced the US target of reducing GHG emissions to 1990 levels by 2000. In support of this stabilization target, the administration proposed a tax to be based on the heat content of the fuel. The tax was rejected by the Senate, which at that time had a Democratic majority. The US oil industry played an important role in killing the tax proposal (Newell 2000: 100). As a consequence of its defeat in the case of the carbon tax, the Clinton administration became increasingly inaccessible to the oil industry but instead consulted more closely with the environmental movement (Skjærseth and Skodvin 2003). The political cleavage increasingly ran between business, on the one hand, and environmental groups and the administration, on the other.

A similar battle was fought in the EU (Meckling 2011b). Preparing for the Rio conference, the European Commission proposed a package on climate policy including a carbon/energy tax in 1991. This was fiercely opposed by European industry associations spearheaded by Business Europe (formerly UNICE), the umbrella organization of 34 business associations, and by EUROPIA (Skjærseth and Skodvin 2003). The latter rejected any new tax on fossil-fuel products. The lobbying campaign proved successful, as the implementation of the tax was made conditional upon other OECD countries following suit, which was not going to happen. Furthermore, the UK rejected it decisively in 1993, and other member-states did not come out with strong support for the proposal, either. This meant the *de facto* burial of an EU-wide carbon tax. While some national carbon taxes have been implemented since the 1990s, notably in Denmark, the story has been a complicated patchwork of efforts more notable for their failures than their successes. The experience with the defeat of carbon/energy tax proposals had a lasting effect on policy-makers on both sides of the Atlantic. They acknowledged that some form of business support for the choice of instrument was crucial in order to be able to pass mandatory climate policy.

The Emergence of Emissions Trading on the International Agenda, 1994–1999

While international greenhouse gas emissions trading had been discussed among a small group of scholars and policy-makers since 1989, it emerged as a viable policy option only in the international negotiations in the mid-1990s. The First Conference of the Parties (COP) of the UNFCCC in Berlin in 1995 put international emission reduction targets and timetables firmly on the agenda. The so-called Berlin Mandate represented a watershed in the negotiations and was a major success for the environmental movement (Alcock 2008). At COP 2 of the UNFCCC in Geneva the next year, the US delegation proposed the use of "trading mechanisms" in implementing the emission reduction target. Emissions trading had been officially placed on the agenda of international climate politics. The US proposal arose from a number of mostly domestic processes. First, the US administration preferred emissions trading as a market-based policy since the successful and highly cost-effective implementation of the domestic sulfur dioxide trading scheme. Second, a new informal transnational alliance of firms and green groups had emerged that promoted market mechanisms (Meckling 2011b). European oil major BP and the green group Environmental Defense spearheaded a new political strategy among business and environmental groups that focused on the promotion of market-based climate policy. Market mechanisms appeared as the compromise solution between industry's reluctance to accept any kind of mandatory emissions targets and the environmental community's preference for command-and-control policies. The new advocacy strategy of some firms and environmental groups and the Clinton administration's foreign climate policy co-evolved, giving momentum to market-based mechanisms.

Initially, the proposal met with strong resistance from the EU and developing countries (Bodansky 2001). European governments lacked experience with the instrument and their environmental constituency perceived emissions trading to be granting a "license to pollute." Developing countries were mostly concerned that emissions trading would allow industrialized countries to escape domestic emission reductions. Yet during the 1997 Kyoto conference, the USA and its allies among business and environmental groups actively promoted the idea of market mechanisms in particular among European governments. In the end, flexibility mechanisms were included in the protocol as part of a compromise deal between the EU and the USA: while the EU accepted emissions trading, the USA agreed to an internationally binding emission reduction target. Developing countries were – with the exception of Brazil's support for the CDM – not in favor of emissions trading. This is noteworthy, as it would later change, when the CDM market channeled funds to emerging economies in particular. Once emissions trading was a constitutive part of the international climate policy framework, the political focus shifted to the ratification and implementation of the protocol at the national level, especially within the key entities such as the EU and the USA. The EU surprisingly took the lead.

The Creation of the EU Emission Trading Scheme, 2000–2005

Despite its initial opposition to tradable permits, the EU designed and implemented the first cross-border emissions trading scheme. The political momentum to go ahead with emissions trading in Europe grew from the bottom up, starting in the UK. Under the influence of BP, UK oil and power companies saw an opportunity to put emissions trading firmly on the European agenda if the UK pioneered a trading scheme. They therefore set up the UK Emissions Trading Group (UK ETG) with the support of the UK government (Nye and Owens 2008). An advocacy coalition at heart, the UK ETG developed the UK ETS, which became operational in 2002. The UK ETG and corporate leaders on emissions trading in general were arguably driven by an anti-taxation agenda, and as such, were mainly pursuing a pro-regulatory risk-management strategy.

The pioneering work in the United Kingdom subsequently spurred action at the EU level, *inter alia*, to prevent regulatory fragmentation within the EU. The European Commission became a powerful driver of an EU-wide scheme in its own right. The Commission had a number of reasons for supporting emissions trading in Europe (Zapfel 2005; Skjærseth and Wettestad 2008). First, the European Commission felt that a carbon tax was doomed to fail, as it had in the early 1990s because of business opposition. Business, in turn, was aware of the political will to implement mandatory climate regulation in the EU. Second, the development of the UK and Danish trading schemes spurred fears of regulatory fragmentation in the EU, which could have undermined integration achievements with regard to the internal market and environmental policy. Third, officials were critically aware of the fact that the acceptance of emissions trading was the price the EU had to pay to get the USA to ratify the Kyoto Protocol.

The advocates of carbon trading in Europe did not remain unchallenged. German industry and energy-intensive manufacturing industries were particularly opposed to a European trading scheme (Christiansen and Wettestad 2003). Unlike in the USA, environmental groups were not among the carbon-trading champions but were instead on the fence. They became watchdogs for the environmental effectiveness of the scheme only at a relatively late stage in the process. After the EU ETS had entered its implementation phase in early 2005, the coalition supporting it became increasingly at risk of fragmenting, as business was divided over the stringency of the system. The financial services industry started to advocate a more stringent EU ETS, which the big emitters from the energy industry and energy-intensive manufacturing industries tried to avoid.

The Proliferation of Trading Schemes: The USA and Asia-Pacific, 2006–2011

After Kyoto, the USA implemented voluntary climate policies for a number of years. During the Clinton administration, the Senate was the major hurdle for any mandatory climate policy. After the Bush administration withdrew the USA from the Kyoto Protocol in 2001, the executive and legislative branches were aligned regarding voluntary climate policy.

Yet business and state activities on the ground began to put emissions trading back on the US agenda. In 2003, the Chicago Climate Exchange, a private initiative, established the first GHG emissions trading scheme in the USA. In the same year, a number of northeastern US states set out to develop a regional emissions trading system for the power sector – what would become the RGGI, which started trading in 2009. Neither of these initiatives could be described as great successes, but they got emissions trading under way in what can be seen as an experimental period. In 2006, California passed the Global Warming Solutions Act, which kicked off a process to develop an economy-wide, state-level emissions trading scheme to be implemented by 2013. With business and US states moving forward with mandatory and market-based climate policy, pressure to enact mandatory emissions cuts at the national level increased.

The Democrats' win in the 2006 midterm elections increased the momentum for climate legislation in both the House and Senate, leading to a phase of heightened legislative activity on climate policy that further accelerated when President Obama entered office in 2009. Legislative activity in the US Congress culminated in the passage of the Waxman–Markey Bill – a comprehensive cap-and-trade bill – by the House of Representatives in 2010. It was supported by a large alliance of environmentalists and firms who had organized in the US Climate Action Partnership. Yet the bill never came to a vote in the Senate. Health-care reform ranked higher on the political agenda, the economic crisis led to concerns about the costs of climate legislation, and the financial crisis, low permit prices, and fraud in the EU ETS led to questions regarding the value of carbon trading in general. The window for federal climate legislation closed as political parties entered campaign mode ahead of the presidential elections in 2012.

Meanwhile, a similar trend towards adopting domestic emissions trading schemes can be observed in the Asia-Pacific region, notably in New Zealand, Australia, and Japan. In September 2008, New Zealand passed legislation on the New Zealand Emission Trading Scheme. New Zealand was the first country outside Europe to have a mandatory, economy-wide emissions trading scheme. In Australia, a prolonged political battle over climate policy arguably took the political scalps of two prime ministers and two leaders of the opposition, but eventually resulted in the passage of comprehensive climate legislation in November 2011 (Siegel 2011). In July 2012, the Australian government introduced a carbon tax as a first step. It is intended to transition to a cap-and-trade scheme in 2015.[3] In October 2008, Japan's government launched a voluntary trial carbon-trading scheme which is supposed to pilot a mandatory cap-and-trade scheme (Maeda 2008). In March 2010, the Japanese government proposed the Basic Act on Global Warming Countermeasures, which foresees a mandatory cap-and-trade scheme, a carbon tax, and a feed-in tariff for renewable energy sources (World Bank 2010). Other countries working on emissions trading schemes include China, Mexico, and South Korea (World Bank 2011).

The Political Economy of Instrument Choice

The central political-economic question regarding market-based climate policy is why carbon trading trumped carbon taxes, despite considerable opposition to the instrument from a range of actors. The brief answer is that the history of domestic and international climate politics shows that relatively broad coalitions can be mobilized for cap-and-trade proposals, but not for carbon taxes. The political economy of the choice of environmental policy instrument is driven by the distributional and environmental effectiveness of competing instruments. The distributional effects relate to wealth transfers between the private and public sectors, between different industry sectors and firms, and between different national economies. The environmental effects depend on different design characteristics of the policy instruments. In the following, we discuss how the distributional and environmental effects of emissions trading and carbon taxes respectively shape the policy preferences of major stakeholders, including governments, business, and environmental groups.

Governments have long been divided over how to price carbon. Throughout the first period of climate politics, the EU and developing countries opposed emissions trading, whereas the USA favored it in principle. The EU's initial preference for command-and-control policies, or a carbon tax, might be seen to be due to the preferences of the strong environmental movement in Europe, but also due to regulatory preferences in the coordinated market economies of continental Europe (Meckling 2011b). The USA's early preference for emissions trading reflected the successful introduction of the sulfur dioxide trading program in the domestic power sector. The cost-effectiveness of emissions trading has been a key driver of its appeal to the US government. Both carbon taxes and the auctioning of permits in a cap-and-trade scheme can create significant wealth transfers to government, which is unsurprisingly resisted by industry. However, industry worked on the (likely correct) assumption that taxes would generate greater revenues from them than permit trading schemes, which have usually been combined with the free allocation of permits (Hepburn 2006).

The reasons for business supporting emissions trading vary mostly by industry sector. Energy companies and energy-intensive manufacturing firms have viewed emissions trading as imposing lower burdens on them, compared with carbon taxes. While most emission-intensive firms have spent a lot of political energy in fighting caps in general, some firms – once faced with the inevitability of emission caps – supported emissions trading (Levy and Egan 2003; Meckling 2008). The early campaign for emissions trading by leaders such as BP was arguably an anti-taxation campaign, through which firms in emission-intensive industries tried to hedge their regulatory risk. The perception of the cost-effectiveness of carbon trading built mostly on the success of the sulfur dioxide trading scheme in the USA, which reached its environmental goals at 30% of the projected cost (US National Science and Technology Council 2005). Next to cost-effectiveness, the option of grandfathering emission permits, that is, handing out permits for free, has been attractive to business. Grandfathering can serve to contain compliance costs and to entrench incumbent advantage (US National Science and Technology Council 2005; Hepburn et al. forthcoming). Industry supported grandfathering in particular in cases of unilateral carbon regulation, as in the EU. It was seen to be able to mitigate potential negative competitiveness effects that would occur through carbon leakage, that is, the shift of production to unregulated territories. Yet theoretically a carbon tax that recycles revenues could achieve the same. Emissions trading has not only been more attractive than a tax to regulated entities, but also to the financial services sector due to its market-creating effect. Market service providers such as investment banks and law firms started seeing a business opportunity in carbon markets later on in the political process, as the first trading schemes went operational. Initially they were represented alongside the emitters in the International Emissions Trading Association (IETA). They later created their own trade association, the Climate Markets & Investment Association.

Environmental groups were – with the exception of the Environmental Defense Fund – not early advocates of emissions trading. They favored a carbon tax in the early phase of climate politics, as the politics of the carbon/energy tax proposals in the EU and the USA reflect. Yet in the mid-1990s a split emerged in the global climate change movement, when some green groups threw their political weight behind emissions trading (Alcock 2008). Reasons included the need to find a policy solution that could mobilize some business support and the notion of quantity certainty. Environmental groups came to like carbon trading because it set fixed emission reduction goals, which ensured the environmental integrity of the instrument. Henceforth, green groups in Europe and the USA played the role of advocates for a stringent form of cap and trade.

In sum, while pricing carbon is generally a hard political sell often facing significant opposition, emissions trading garnered more support than taxes. This is largely because the distributional and environmental effects of emissions trading are more attractive to parts of the environmental community, key state actors in Europe and the USA, and to big emitters and financial intermediaries. The cost-effectiveness, the market-creating effect, and the quantity certainty offered by emissions trading have been the predominant factors aligning interests around the instrument.

The State and Performance of Carbon Markets

What followed in the decade after Kyoto was not a top-down implementation of a global trading scheme but rather a bottom-up process of trading experiments and schemes. Academics and market actors have described it as fragmented (Tangen and Hasselknippe 2005), plurilateral (Sandor 2001), decentralized, and bottom-up (Victor *et al.* 2005). In the following, we outline the key market segments, discuss the performance of existing markets, and debate questions of the governance of carbon markets. We argue that while the carbon markets are highly fragmented and actors are still going through a learning curve, the EU ETS and the CDM have demonstrated some level of effectiveness. Yet the economic efficiency and environmental integrity of carbon markets critically hinges on the ability of actors to govern them.

The Scale and Scope of the Global Carbon Market

The global carbon market can be segmented in different ways – for one, in terms of the distinction between mandatory markets, mostly resulting from Kyoto commitments, and voluntary markets. In 2010, the mandatory markets dominated the carbon market with a share of more than 99% (World Bank 2011). The backbone of the compliance market, the EU ETS, accounted for 85% of the carbon market based on the value of EU ETS Allowances or for 97% including the secondary CDM market. Carbon markets globally were valued at US$142 billion (in terms of overall total value of transactions) in 2010 (World Bank 2011).[4]

The EU ETS is the only existing multilateral trading scheme for CO_2 and the world's largest mandatory cap-and-trade scheme. As a tributary market to the Kyoto Protocol, the scheme serves to achieve the EU's Kyoto target. Legislation for the EU ETS was adopted in 2003, and actual trading began in January 2005, with a pilot phase running until 2007. The second trading period of the EU ETS ran in parallel to the first commitment period under the Kyoto Protocol from 2008 to 2012. In spring 2007, the EU heads of state decided to ensure the long-term continuity of the EU ETS by setting an emissions reduction target for 2020. Also in 2007, the EU ETS underwent a review process, which led to an institutional overhaul of the scheme, including issues such as sectoral coverage and allocation method.

Since the EU ETS is the major mandatory trading scheme, it is also the main driver for project-based mechanisms by creating a demand for credits. In 2010, the total CDM market – including primary and secondary transactions – was worth US\$19.8 billion (World Bank 2011). Early projects produced credits by reducing industrial gases that had an especially high global warming factor such as HFC-23 and N_2O, often described as "low-hanging fruit." Since 2007, more credits have increasingly resulted from renewable energy and energy-efficiency projects. Since the inception of the CDM market, China has been the largest recipient of CDM funding.

Representing the second pillar of the carbon market, the voluntary market is a credit-based trading market in which credits are generated and sold for non-compliance purposes. In 2010, the voluntary market had a financial volume of only about US\$430 million, which is miniscule compared to the total size of carbon markets (World Bank 2011).

The Performance of Carbon Markets

A comprehensive verdict on the efficiency and effectiveness of existing carbon markets is still pending. Here, we offer a provisional assessment of the effectiveness of carbon trading. The available data on the effectiveness of carbon trading are mostly limited to the trial period of the EU ETS (cap and trade) and the CDM (baseline and credit).

The trial period of the EU ETS ran from 2005 to 2007. This first period has been criticized for mainly two flaws (Ellerman and Buchner 2007). First, emissions permits to regulated entities were overallocated (Anderson and Di Maria 2011), due to a lack of accurate emissions data. Once the data were corrected, and the excessively generous allocations revealed, the carbon price plummeted. Second, electric utilities reaped significant windfall profits by passing along the costs of freely allocated allowances. The trial period of the EU ETS thus led to relatively modest emissions reductions (Ellerman and Buchner 2007). The emissions of the sectors covered by the EU ETS flattened during the 2005–2007 period despite robust gross domestic product growth. Hence, the trial period was somewhat effective in terms of reducing CO_2 emissions.

Yet as A. Denny Ellerman and Paul Joskow (2008) argue, the trial period was not meant to lead to significant emissions reductions but rather to establish the trading scheme and provide lessons for reform. The EU ETS in fact delivered on these criteria. It established the market infrastructure and created a carbon price, which companies started to incorporate into their decision-making. In December 2008, the

EU passed a reform package, which aimed to make the system more effective. In particular, it introduced partial auctioning as the allocation method and granted the European Commission stronger authority in the allocation process. The effectiveness of the scheme hinges critically on the ability of member-states and the European Commission to manage the market. The performance of the second trading period so far provides cause for cautious optimism. Anecdotal evidence suggests that higher allowance prices in 2008 led to fuel switching in the power sector and improvements in the efficiency of power plants, which resulted in emissions reductions (Ellerman *et al.* 2010).

The experience with the CDM is similar to that with the EU ETS: it underperformed regarding its environmental outcome, mostly due to design issues. Again, the imperfect performance is not surprising and does not undermine the merit of the instrument *per se*. The criticisms of the CDM relate mostly to its limited scope and the "addititionality" of emissions reductions achieved through CDM projects (Hepburn 2007; Harvard Project on International Climate Agreements 2009).[5] The CDM's limited scope is due to a couple of reasons. First, the approval of CDM projects through the CDM executive board is a bureaucratic and expensive process. Every single project has to go through this approval process in order to receive credits. Second, the CDM initially excluded a number of mitigation activities, such as the conservation of forests. With regard to additionality, there are concerns that some CDM projects would have been conducted in the absence of the CDM – that is, they are not "additional" to what would have occurred under business-as-usual (Schneider 2007).

Linked to this is the issue of what kinds of emissions-reducing projects are funded through the CDM. The CDM was meant to encourage investment into low-carbon energy infrastructure in developing countries (Wara 2007). Renewable energy projects, however, accounted for only 35% of the emissions reductions to be achieved through the CDM until 2012.[6] The largest share of emissions reductions result from capturing and destroying industrial gases such as HFC-23, N_2O, and CH_4 emitted by landfills and confined-animal-feeding operations. Hence, in these cases the CDM credits did not spur investment in low-carbon energy infrastructure. Credits of this type were banned from use in the EU system from the end of 2012. While the shortcomings of the CDM market are significant, a number of analysts suggest that institutional reform could greatly strengthen the mechanism (Victor and Cullenward 2007). The CDM is important as it engages the fastest-growing economies such as China and India in global mitigation efforts.

In sum, both the EU ETS, the first multilateral cap-and-trade scheme, and the CDM, a global baseline-and-credit scheme, produced only modest emissions reductions in the early trial period from 2005 to 2008, when market infrastructure was established and lessons were learned. While a comprehensive analysis remains to be done (to our knowledge), there is a good chance that the carbon-trading schemes will score significantly higher on environmental effectiveness over the 2008–2012 period (Grubb *et al.* 2010).

Conclusion

This chapter has examined the politics and economics of putting an explicit price on carbon dioxide emissions using economic instruments. We observed that carbon-trading schemes have, so far, trumped carbon taxes in the quest for the predominant way to price carbon. In the early 1990s, initiatives to implement carbon/energy taxes in Europe and the USA failed largely due to significant business opposition. In the mid-1990s, international GHG emissions trading emerged on the international agenda, once the USA had insisted on its inclusion in the Kyoto Protocol. We argue that support from some business and environmental groups was important in building momentum for the instrument. Subsequently, the EU – which had previously been opposed to market-based mechanisms – went ahead in implementing the first multilateral trading scheme, the EU ETS. The European scheme was henceforth the backbone of the carbon markets. Other countries in North America and Asia-Pacific followed suit by implementing schemes or kicking off legislative processes on cap-and-trade regulation.

While a large number of economists conclude that carbon taxes are more efficient under uncertainty than emissions trading, for a problem like climate change, emissions trading dominated mostly for reasons of political economy. It is able to garner support from environmental groups (due to environmental certainty), business groups (due to lower transfers to government and to new business opportunities in emissions markets), and from government actors (due to cost-effectiveness and ability to generate some revenue). The allocation of free permits also allowed government to "buy off" various resisting groups. This led to an unusual alliance of actors that promoted or at least accepted trading schemes.

The Kyoto Protocol envisioned a global trading scheme based on an international treaty, which, however, has not yet materialized. Instead, subnational, national, and regional trading schemes emerged in Europe, North America, and Asia-Pacific. Carbon markets thus remain highly fragmented and diverse. They face significant challenges with regard to their governance and market integration. Their economic efficiency and environmental effectiveness depend in particular on the ability of governments and other actors to set emission caps right, to allocate permits efficiently, and to master the information challenges related to measuring, reporting, and verifying emissions. Given the heterogeneity of emerging trading schemes, future market integration will most likely occur in an incremental and messy fashion.

Notes

1 Economists note that although they do not create an explicit price, they do create an implicit, or "shadow," price.
2 See Hepburn (2006) for a comparison of price- and quantity-based regulation.
3 For a summary of the key provisions, see Hepburn and Jotzo (2011).
4 For data on carbon markets, please see the annual *State and Trends of the Carbon Market* report by the World Bank.
5 The term "additionality" refers to the requirement that the emissions reductions are "additional" to emissions reductions that would have happened anyway if the CDM had not been in place.
6 "CDM Pipeline Overview," http://uneprisoe.org/ (accessed January 9, 2012).

References

Alcock, Frank. 2008. "Conflicts and Coalitions within and across the ENGO Community." *Global Environmental Politics*, 8(4): 66–91.
Anderson, Barry and Corrado Di Maria. 2011. "Abatement and Allocation in the Abatement Phase of the EU ETS." *Environmental and Resource Economics*, 48: 83–103.

Bernstein, Steven. 2001. *The Compromise of Liberal Environmentalism*. New York: Columbia University Press.

Bodansky, Daniel. 2001. "The History of the Global Climate Change Regime." In *International Relations and Global Climate Change*, ed. U. Luterbacher and D.F. Sprinz, 23–40. Cambridge, MA: MIT Press.

Christiansen, Atle C. and Jorgen Wettestad. 2003. "The EU as a Frontrunner on Greenhouse Gas Emissions Trading. How Did it Happen and Will the EU Succeed?" *Climate Policy*, 3: 3–18.

Coase, R.H. 1960. "The Problem of Social Cost." *Journal of Law and Economics*, 3 (October): 1–44.

Ellerman, Denny A. and Barbara K. Buchner. 2007. "The European Union Emissions Trading Scheme: Origins, Allocation, and Early Results." *Review of Economics and Policy*, 1(A): 66–87.

Ellerman, Denny A., Frank J. Convery, and Christian de Perthuis, eds. 2010. *Pricing Carbon: The European Emissions Trading Scheme*. Cambridge: Cambridge University Press.

Ellerman, Denny A. and Paul L. Joskow. 2008. *The EU Emissions Trading System in Perspective*. Washington, DC: Pew Center on Global Climate Change.

Fankhauser, Samuel and Cameron Hepburn. 2010a. "Designing Carbon Markets, Part I: Carbon Markets in Time." *Energy Policy*, 38: 4363–4370.

Fankhauser, Samuel and Cameron Hepburn. 2010b. "Designing Carbon Markets, Part II: Carbon Markets in Space." *Energy Policy*, 38: 4381–4387.

Grubb, Michael, Tim Laing, Thomas Counsell, and Catherine Willan. 2010. "Global Carbon Mechanisms: Lessons and Implications." *Climatic Change*, 104(3): 539–573.

Harvard Project on International Climate Agreements. 2009. *Options for Reforming the Clean Development Mechanism*. Issue Brief 2009-01. Cambridge, MA: Harvard Project on International Climate Agreements.

Helm, D.R., C. Hepburn, and R. Mash. 2003. "Credible Carbon Policy." *Oxford Review of Economic Policy*, 19(3): 438–450.

Hepburn, Cameron. 2006. "Regulation by Prices, Quantities, or Both: A Review of Instrument Choice." *Oxford Review of Economic Policy*, 22(2): 226–247.

Hepburn, Cameron. 2007. "Carbon Trading: A Review of the Kyoto Mechanisms." *Annual Review of Environment and Resources*, 32: 375–393.

Hepburn, Cameron, Michael Grubb, Karsten Neuhoff *et al.* 2006. "Auctioning of EU ETS Phase II Allowances: Why and How?" *Climate Policy*, 6: 137–160.

Hepburn, Cameron and Frank Jotzo. 2011. "The Australian Government's Proposals for a Carbon Pricing Policy." Commentary, London School of Economics and Political Science, Grantham Research Institute on Climate Change and the Environment, August 2.

Hepburn, C., J. Quah, and R. Ritz. Forthcoming. "Emissions Trading with Profit-Neutral Permit Allocations." *Journal of Public Economics*. doi:10.1016/j.jpubeco.2012.10.004.

Kahn, James R. and Dina Franceschi. 2006. "Beyond Kyoto: A Tax-Based System for the Global Reduction of Greenhouse Gas Emissions." *Environmental Economics*, 58(4): 778–787.

Knox-Hayes, Janelle. 2009. "The Developing Carbon Financial Service Industry: Expertise, Adaptation and Complementarity in London and New York." *Journal of Economic Geography*, 9: 749–777.

Krutilla, Kerry with Rachel Krause. 2011. "Transaction Costs and Environmental Policy: An Assessment Framework and Literature Review." *International Review of Environmental and Resource Economics*, 4(4): 261–354.

Levy, David L. and Daniel Egan. 2003. "A Neo-Gramscian Approach to Corporate Political Strategy: Conflict and Accommodation in the Climate Change Negotiations." *Journal of Management Studies*, 40(4): 803–829.

Maeda, Risa. 2008. "Japan Launches Voluntary CO2 Market." Reuters, October 21.

Matthews, Karine and Matthew Paterson. 2005. "Boom or Dust? The Economic Engine behind the Drive for Climate Change Policy." *Global Change, Peace and Security*, 17: 59–75.

Meckling, Jonas. 2008. "Corporate Policy Preferences in the EU and the US. Emissions Trading as the Climate Compromise?" *Carbon and Climate Law Review*, 2(2): 171–180.

Meckling, Jonas. 2011a. "The Globalization of Carbon Trading: Transnational Business Coalitions in Climate Politics." *Global Environmental Politics*, 11(2): 26–50.

Meckling, Jonas. 2011b. *Carbon Coalitions: Business, Climate Politics, and the Rise of Emissions Trading*. Cambridge, MA: MIT Press.

Newell, Peter. 2000. *Non-state Actors and the Global Politics of the Greenhouse*. Cambridge: Cambridge University Press.

Newell, Peter. 2008. "The Marketization of Global Environmental Governance. Manifestations and Implications." In *The Crisis of Global Environmental Governance*, ed. J. Park, K. Conca, and M. Finger, 77–95. New York: Routledge.

Newell, Peter and Matthew Paterson. 2010. *Climate Capitalism: Global Warming and the Transformation of the Global Economy*. Cambridge: Cambridge University Press.

Nye, Michael and Susan Owens. 2008. "Creating the UK Emission Trading Scheme. Motives and Symbolic Politics." *European Environment*, 18: 1–15.

Pigou, A.C. 1920. *The Economics of Welfare*. London: Macmillan and Co.

Point Carbon. 2011. *Carbon 2011*, by E. Tvinnereim, E. Zelljadt, N. Yakymenko, and E. Mazzacurati. n.p.: Thomson Reuters Point Carbon.

Sandor, Richard. 2001. "Corporate Giants to Aid Design of US Carbon Market." *Environmental Finance*, June: 11.

Schneider, Lambert. 2007. *Is the CDM Fulfilling its Environmental and Sustainable Development Objectives? An Evaluation of the CDM and Options for Improvement*. Report for the WWF. Berlin: Oeko-Institut.

Siegel, Matt. 2011. "Australian Senate Approves Emissions Trading Plan." *New York Times*, November 7.

Skjærseth, Jon Birger and Tora Skodvin. 2003. *Climate Change and the Oil Industry. Common Problems, Varying Strategies*. Manchester: Manchester University Press.

Skjærseth, Jon Birger and Jorgen Wettestad. 2008. *EU Emissions Trading. Initiation, Decision-Making and Implementation*. Aldershot: Ashgate.

Smith, Anne E. 2007. "Climate Change: Lessons Learned from Existing Cap-and-Trade Programs." Subcommittee on Energy and Air Quality, Committee on Energy and Commerce, United States House of Representatives. Washington, DC: CRA International.

Tangen, Kristian and Henrik Hasselknippe. 2005. "Converging Markets." *International Environmental Agreements: Politics, Law and Economics*, 5: 47–64.

US National Science and Technology Council. 2005. *National Acid Precipitation Assessment Program Report to Congress. An Integrated Assessment*. Washington, DC: US National Science and Technology Council, www.esrl.noaa.gov (accessed October 20, 2012).

Victor, David G. and David Cullenward. 2007. "Making Carbon Markets Work." *Scientific American*, September 24.

Victor, David G., Joshua C. House, and Sarah Joy. 2005. "A Madisonian Approach to Climate Policy." *Science*, 309, September 16: 1820–1821.

Wara, Michael W. 2007. "Is the Global Carbon Market Working?" *Nature*, 445(8): 595–596.

Weitzman, Martin. 1974. "Prices vs. Quantities." *Review of Economic Studies*, 41: 477–491.

World Bank. 2010. *State and Trends of the Carbon Market 2010*, by A. Kossoy and P. Ambrosi. Washington, DC: World Bank.

World Bank. 2011. *State and Trends of the Carbon Market 2011*, by N. Linacre, A. Kossoy, and P. Ambrosi. Washington, DC: Carbon Finance at the World Bank.

Zapfel, Peter. 2005. "Greenhouse Gas Emissions Trading in the EU. Building the World's Largest Cap-and-Trade Scheme." In *Emissions Trading for Climate Policy. US and European Perspectives*, ed. B. Hansjürgens, 162–176. Cambridge: Cambridge University Press.

Chapter 28

International Aid and Adaptation to Climate Change

Jessica M. Ayers and Achala Chandani Abeysinghe

Introduction

Although the world is now fully engaged in the climate change debate, international efforts to limit greenhouse gas emissions are not translating into a detectable slowing down of the rate of global warming. According to the Intergovernmental Panel on Climate Change (IPCC), the impacts of climate change will be severe, particularly for the poorest people in vulnerable developing countries that have the least capacity to cope (Schneider *et al.* 2007). For these groups, adaptation to the impacts of climate change is a priority. Adaptation describes the adjustment in natural or human systems in response to actual or expected climatic stimuli or their effects, which moderates harm, or exploits beneficial opportunities (IPCC 2007). It can be any process, action, or outcome in a system (ecosystem, household, community, group, sector, or region) that helps that system to better cope with, manage, or adjust to the changing conditions, stresses, hazards, risks, or opportunities associated with climate change (Smit and Wandel 2006).

Historically, adaptation was seen as a marginal policy option in global climate governance arenas, often perceived as the "poor cousin" of mitigation, which describes efforts to limit greenhouse gas emissions (Pielke *et al.* 2007). However, as both the inevitability and implications of climate change become apparent, especially in vulnerable developing countries, adaptation has risen up both the global environmental and international development policy agendas. In 2007, adaptation was adopted as one of the four "building blocks" (along with mitigation, technology cooperation, and finance) of a comprehensive climate change response under the United Nations Framework Convention on Climate Change (UNFCCC), and has also been taken up as a key priority for international development agencies working in vulnerable developing countries.

The Handbook of Global Climate and Environment Policy, First Edition. Edited by Robert Falkner.
© 2013 John Wiley & Sons, Ltd. Published 2013 by John Wiley & Sons, Ltd.

Adaptation to climate change requires huge resources. Although estimates of the costs of adaptation vary widely, recent estimates suggest that the "global price tag" for adaptation in developing countries is US$70 billion–US$100 billion per year for 2010–2050 (Narain *et al.* 2011). This presents a challenge to the international global community committed to supporting vulnerable developing countries in adaptation: where will this money come from, who should pay, and how should it be delivered?

From a global perspective, developed countries hold the greatest responsibility for climate change, given the relative contributions of historic and current greenhouse gas emissions and their greater capacity to respond, while developing countries are most in need of adaptation. In line with this argument, the UNFCCC commits developed countries to providing finance for adaptation to developing countries. Given that the international aid architecture already has well-established mechanisms for channeling resources from high-income to low- and middle-income countries, we might assume that aid finance and institutions would play a significant role in adaptation finance.

However, the principles governing adaptation finance under the global climate change regime explicitly require that adaptation funding should be *additional* to existing international aid commitments, because climate change poses an additional burden to existing development needs. Several climate funds for adaptation have already been established on this basis. This principle has resulted in a great deal of confusion over the role of international aid in funding adaptation. On the one hand, international aid should have a strong role to play in supporting adaptation, because many of the objectives of aid such as reducing poverty and improving social welfare also contribute to reducing climate vulnerability. Further, the impacts of climate change threaten the sustainability of aid investments in vulnerable developing countries, so aid institutions need to consider the implications of climate change for their development portfolios. On the other hand, at the global level, arguments for the additionality of climate finance to existing aid commitments have been used by developing countries to negotiate for fair and equitable international funding arrangements for adaptation under the UNFCCC. This creates a paradox for adaptation finance (Ayers 2011): international aid is clearly relevant for funding adaptation, but it is important that this principle of adaptation funding as additional to aid is upheld.

So, what is the role for international aid in supporting adaptation? This chapter addresses this question by exploring the question of "additionality" in principle and practice, and the challenges for financing effective adaptation that this gives rise to. We begin by discussing the synergies and conflicts between international aid and adaptation finance, including the role of development in enabling effective adaptation, and also the challenges that this relationship gives rise to at the global policy level. Next, we explore how the international institutions of climate change and international aid are dealing with these challenges, in terms of how funds for adaptation are sourced, governed, and delivered. We show that while the global governance of adaptation finance attempts to achieve a relative distinction between international aid and adaptation finance "on paper," in reality there is a complex web of funding flows for adaptation that confuse the relationship between the two. We conclude by reflecting on the implications of these challenges for achieving effective adaptation in developing countries through the global climate and development funds.

How and Why Does International Aid Matter for Adaptation to Climate Change?

This section begins by briefly outlining the cost implications of adaptation to climate change, before considering why international aid is perceived as relevant to meeting some of these costs.

The Costs of Adaptation in Developing Countries

Estimates of the costs of adaptation vary from US$4 billion to US$109 billion depending on the assumptions and methodological approaches used in different studies (Narain *et al.* 2011). A key challenge for assessing adaptation costs is that there is no uniform and agreed definition of what constitutes an adaptation intervention. Adaptation can be a "hard" intervention specifically targeted at the anticipated impacts of climate change, such as an irrigation system, or it can be a "soft" intervention such as information awareness and capacity-building. The "target" for adaptation can be managing specific climate-change risks, for example coastal infrastructure in anticipation of increased storm surges and sea-level rise. Or adaptation can try to address a range of factors underpinning vulnerability to climate and other risks, such as poverty and social marginalization, which prevent people from coping with and responding to climate impacts, in which case adaptation overlaps significantly with development approaches. A review of a wide range of projects and programs labeled "adaptation" revealed that all of these approaches are legitimate, and most adaptation sits on a scale between "development-based" approaches and "climate-impacts-focused" approaches (McGray *et al.* 2007). How adaptation is defined gives rise to very different issues and activities that need to be included in the costing process.

The most common approach to assessing the costs of adaptation is to focus on adaptation in different sectors and compare the costs of that sector under a "business-as-usual" scenario, with the cost based on projected future climate change (Haites 2011). This approach is problematic for two reasons: first, it depends on many assumptions such as perfect foresight, when in reality there is a huge amount of uncertainty over what the future impacts of climate change will be and how they might interact with a future state of any one particular sector. There are a number of changing and interacting variables in making this calculation. Second, these estimates are almost always limited to the costs of new "hard" adaptation measures, and many observers have criticized this approach for not including the significant but harder-to-measure "soft" adaptation actions (Haites 2011).

Further, estimates often assume the baseline for adaptation costs is the current development scenario. Yet, those most vulnerable to climate change are also those in development deficit situations, so maintaining the "status quo" in light of climate impacts will not lift people out of vulnerability. They will continue to live in a development deficit situation, and to experience vulnerability to *existing* climate variability – an "adaptation deficit" (Burton 2004). Confusions exist around whether to include the costs of addressing the development deficit and/or the adaptation deficit as adaptation costs or development costs.

One of the more comprehensive efforts to cost adaptation that addresses this issue is a recent study launched by the World Bank in 2009. This study defines "adaptation costs" as those additional to development due to climate change, thereby avoiding cofounding the costs of the development deficit and the implicit adaptation "deficit" (Narain *et al.* 2011). This study estimates the cost of adaptation in the developing world at US$70 billion–US$100 billion per year from 2010 to 2050 under a 2 °C increase in global temperatures by 2050 (Narain *et al.* 2011).[1] Although this range is huge, and based on a number of uncertain assumptions, it nevertheless demonstrates the scale of the adaptation finance challenge.

The Role of International Aid in Funding Adaptation in Developing Countries

There are a number of arguments for exploring the role of international aid in meeting some of the costs of adaptation. A key argument is the significant overlap between aid and adaptation objectives (Ayers and Dodman 2010). One definition of adaptation is to enable social and economic activities and to reduce their vulnerability to climate risks, including its current variability and extreme events as well as longer-term climate change (Smit 1993). Key components of vulnerability are economic, social, and cultural factors that determine whether a person, group, or system has the capacity to cope with and adapt to climate change and other risks (Blakie *et al.* 1994: 9).

International aid comes in the form of Official Development Assistance (ODA).[2] According to the Development Assistance Committee (DAC) of the Organisation for Economic Co-operation and Development (OECD), ODA is defined as financial flows that are designed to promote economic development and welfare as their main objective (OECD DAC n.d.). Following on, economic welfare and development are key components of adaptation to climate change, because they underpin vulnerability. For example, the Millennium Development Goals of reducing poverty, providing general education and health services, improving living conditions in urban settlements, and providing access to financial markets and technologies will all improve the livelihoods of vulnerable individuals, households, and communities, enabling them to better adapt to climate and other risks. An analysis of the categories of ODA activities reported by the OECD DAC countries demonstrated that more than 60% of all ODA could be relevant to reducing climate vulnerability and facilitating adaptation (Levina 2007).

Climate change also carries implications for the effectiveness of development interventions in three ways (Klein 2001): first, climate change poses direct risks to aid investments, given that the impacts of climate change will be felt first and most severely in the poorest and most vulnerable communities that are the target of international aid; second, the climate vulnerability of the community or system that is the target of aid may impinge on how the investment is implemented; and third, aid investments and their deliverables may have effects (positive or negative) on the vulnerability of communities or ecosystems to climate change (Klein 2001).

Given these synergies between adaptation and aid objectives, supporting adaptation through development makes sense (Dodman *et al.* 2009). Development assistance can reduce vulnerability to climate change; indeed on the ground, the way climate adaptation finance is spent in helping vulnerable countries adapt to climate change is in many instances indistinguishable from aid, because often actions related to poverty reduction are the best way to reduce climate change vulnerability.

Adaptation activities are therefore often regarded as synonymous with development activities and key to good development practice. As noted by Huq and Ayers:

> Good (or sustainable) development (policies and practice) can (and often does) lead to building adaptive capacity. Doing adaptation to climate change often also means doing good (or sustainable) development (Huq and Ayers 2008: 52).

A second argument for turning to aid channels to support adaptation is a pragmatic one. Adaptation finance requires fund flows from high-income countries, which have driven the causes of climate change, towards low- and middle-income countries that bear the brunt of climate-change impacts. International aid has well-established institutions, mechanisms, and principles of governing financial flows from developed to developing countries. Some observers have proposed that there are many lessons gained from the experiences of development cooperation that could be useful as climate finance (OECD DAC 2009; Bird and Brown 2010).

In particular, donors and developing countries have developed the Aid Effectiveness Principles that are embedded in the Paris Declaration and the Accra Agenda to guide their partnership. These are: support for national ownership of the development process, promotion of donor harmonization, alignment of donor systems with national systems, management for results, and mutual accountability between donor and recipient. The aid-effectiveness agenda grew out of many years of experience and lessons on aid implementation, and has developed from a retrospective view of what has been judged to be the successes and failures of aid delivery (Bird and Glennie 2011).

Such a body of experience is missing from the adaptation finance arena, and much could be learned about "good practice" in relation to the international transfer of funds for activities related to adaptation and development. The OECD DAC suggests that the principles relating to the governance arrangements for climate-change finance at the national level and how these arrangements are established to channel external sources of public finance are especially relevant for climate finance (see Box 28.1).

Box 28.1 Lessons from Development Financing Applicable to Climate-Change Financing Ownership

For development to be sustainable over the long term, developing country governments must exercise effective ownership over the development process. Developing countries must therefore take the lead in establishing and implementing their national climate-change strategies through a broad consultative process and ensuring that these strategies are fully integrated into policies, plans, and programs in all relevant sectors.

Alignment

Climate change financing needs to be integrated into countries' own planning and budgeting mechanisms, to enable the partner country to exercise genuine ownership and control over financial resources. Recording these resources in the national budget ensures that the use of these funds is subject to scrutiny by parliaments, other domestic accountability institutions, and civil society.

Capacity Development

Capacity development will be critical to ensure that partner countries have sufficient capacity to absorb and manage climate-change financing and to integrate climate-change adaptation actions into national planning.

Harmonization

Experience with aid has shown the importance of harmonization of international financial flows. When there are scores of contributors and funding mechanisms, each with its own administrative and reporting requirements, the resulting workload may place a strain on partner countries' administrative capacity. It is important for the international community to coordinate their actions, simplify procedures, and share information to avoid proliferation and duplication of funding mechanisms.

Managing for Results

The challenges posed by climate change call for effective responses, which yield actual results on the ground. This is well recognized by the Bali Action Plan, which stresses the need for the various actions undertaken by Parties to implement the Convention to be "measurable, reportable and verifiable."

Source: OECD DAC (2009).

However, there are strong counter-arguments for turning to aid to fund adaptation. Although it makes sense to support adaptation through international aid from an operational perspective, from a global policy perspective there are important reasons for separating out international aid and adaptation finance (Klein 2008; Persson *et al.* 2009).

From a global policy perspective, negotiations around climate adaptation finance are based on a fundamental equity principle of "common but differentiated responsibilities and respective capabilities." In relation to the global negotiations around adaptation finance, this principle recognizes the relative contributions of developed and developing countries in driving greenhouse gas emissions, as well as their respective capabilities to take responsive measures. The principle implies that those with the responsibility and capacity should pay for adaptation – that is, it is the responsibility of developed countries to finance adaptation in vulnerable developing countries. This is laid out in Articles 3.1 and 4 of the UNFCCC convention text.

Upholding this principle at the global level presents three challenges for understanding the role of international aid in financing adaptation. First, under international aid paradigms, it is donor countries that have the power to define positions or institutional arrangements that govern financials flows (Bird and Glennie 2011). But

as pointed out by Bird and Glennie (2011), the narrative of "common but differentiated responsibilities" as interpreted under the UNFCCC suggests a very different
type of partnership. Financing for adaptation is not owed to poor countries as "aid"
with the accompanying implications of donor-recipient power relations, but rather
as compensation from high-emission countries for those that are most vulnerable to
the impacts, implying a very different – and more equal – partnership in determining
how the money is allocated (ActionAid 2007; Oxfam International 2007). Climate
finance should in principle offer a much more equal "seat at the table" for recipient countries to define allocation of adaptation resources. Using international aid
to finance adaptation shifts the balance of accountability back to donor countries
and institutions.

Second (and reflective of the power dynamics inherent in international aid), aid
flows have historically been voluntary transfers, defined by donor country governments and then negotiated with developing country governments (Riddell 1987).
Although a 0.7% ODA target was agreed in 1970 and has been repeatedly re-
endorsed at the highest level at international aid and development conferences,
including the most recent Rio+20 conference, this remains a target and not mandatory. Only Sweden, the Netherlands, Norway, and Denmark have managed to consistently meet this target since it was established (OECD DAC 2010). By contrast,
Bird and Glennie (2011) point out that there have been strong, early calls within
the UNFCCC negotiations to make climate finance transfers mandatory within a
legally binding global agreement. The outcome of such an agreement has yet to be
reached, but achieving it would be based on the premise that adaptation funding is an
obligation, and not a voluntary donation, from developed to developing countries.

Third, and perhaps most significantly in terms of operationalizing fund flows from
international aid for adaptation, adaptation finance under the global climate regime
should be "new and additional" finance – that is, over and above existing aid commitments. This principle is laid out in the Bali Action Plan agreed during the 13th
Conference of the Parties to the UNFCCC (COP 13) in Indonesia, which states explicitly that funding for adaptation is made available above and beyond that which is
provided as ODA. This decision was followed up two years later at COP 15 in Copenhagen, with the resulting Copenhagen Accord calling for a collective commitment by
developed countries to provide "new and additional resources . . . approaching USD
30 billion for the period 2010–2012 with balanced allocation between adaptation
and mitigation." The same holds for Cancun Decisions agreed at COP 16.

At first glance, this principle that funding for adaptation should be additional to
aid might render the role of international aid in adaptation obsolete. However, as
this chapter will show, although the role of aid in adaptation is highly contested in
principle at the level of global policy, in practice both confusion over and failure to
adhere to these principles have resulted in aid playing a significant role.

The Role of International Aid in the Global Adaptation Finance Architecture

This section begins by describing the international architecture of financing adaptation, before considering how the role of international aid within this framework
responds to the principles of adaptation funding described above.

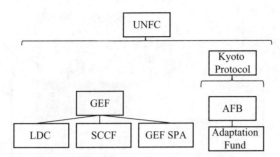

Figure 28.1   Structure of adaptation funding under the UNFCCC.

The International Architecture for Adaptation Finance

Funding for adaptation exists both under the UNFCCC and outside it. Under the UNFCCC, the 2001 Conference of the Parties to the UNFCCC meeting in Marrakesh (COP 7) established the Marrakesh Accords, which included three new funds, collectively known as the "Marrakesh Funds": the Least Developed Countries Fund (LDCF), established under the Convention, to support the 49 least developed countries to adapt to climate change, and initially used to support the design of National Adaptation Programmes of Action (NAPAs); the Special Climate Change Fund (SCCF) to support a number of climate-change activities including mitigation and technology transfer, but intended to prioritize adaptation; and the Kyoto Protocol Adaptation Fund (AF) to support concrete adaptation projects in developing countries that are party to the Protocol. This fund sits under the Kyoto Protocol, managed by the independent Adaptation Fund Board (AFB), and is financed from a levy on the Clean Development Mechanism. Decision 6 of the Marrakesh Accords further requested that the Global Environment Facility (GEF), the financial mechanism of the UNFCCC with responsibility for the transfer of funds from developed to developing countries, should fund pilot adaptation projects, leading the GEF to establish the Strategic Priority "Piloting an Operational Approach to Adaptation" (SPA) under the GEF Trust Fund (see Figure 28.1).

In addition to the UNFCCC funds, international finance for adaptation is provided through bilateral climate funds, development banks, and ODA. For example, the World Bank-established Climate Investment Funds (CIFs) have been established outside the UNFCCC process, to provide concessional loans and grants to policy reforms and investments that achieve development goals through a transition to a low-carbon development pathway and a climate-resilient economy (World Bank 2008). The Pilot Programme for Climate Resilience (PPCR) is the CIF that is most relevant to adaptation.[3] The PPCR has a target size of US$1 billion, and is aimed at increasing climate resilience in developing countries. Private-sector sources and investments also contribute, although currently these contributions are much smaller and so the remainder of this chapter focuses on public finance streams.

In principle, funds outside the UNFCCC should be aligned with the same principles, as illustrated by Article 11 of the Convention text, which states that "developed country Parties may also provide and developing country Parties avail themselves of, financial resources related to the implementation of the Convention through bilateral, regional and other multilateral channels" (UNFCCC 1992: Article 11). Thus,

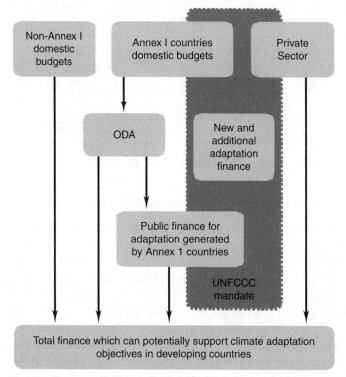

Figure 28.2 Overview of adaptation funding channels.
Source: Persson, A., R.J.T. Klein, C.K. Siebert *et al.* 2009. *Adaptation Finance under a Copenhagen Agreed Outcome*. SEI Research Report. Stockholm: Stockholm Environment Institute. With kind permissions of SEI, Stockholm.

the global climate change architecture includes funds for adaptation both within and also outside the Convention, but all funding should be aligned. However, as illustrated by Figure 28.2, the adaptation-financing landscape is highly fragmented, with a proliferation of funds and accompanying policies, rules, and procedures (Klein 2011).

What do these institutional funding arrangements mean for the role of international aid in adaptation finance? This chapter addresses this question in terms of the source, governance, and delivery of adaptation funds.

Sources of Funds for Adaptation: New and Additional?

To date,[4] a total of US$32 billion climate change finance has been pledged, of which US$2.1 billion has been disbursed. Of this disbursed finance, 21% went to adaptation, and many critics have pointed to the inadequacies of the level of funding available relative to the current and anticipated costs (Flåm and Skjærseth 2009; Pauw *et al.* 2011; Smith *et al.* 2011).

In terms of the sources of these funds, almost all funding for adaptation currently comes from public finance, drawn from international aid budgets. In 2009, developed countries pledged to provide US$30 billion "new and additional" resources for the period 2010–2012 (labeled "Fast Start" funding), with balanced allocation between mitigation and adaptation (see Box 28.2 for an example of the UK's commitment to Fast Start financing). They also committed to a goal of jointly mobilizing US$100 billion by the year 2020 to address the needs of developing counties, although they did not specify how this would be allocated. The 2011 Cancun Agreement that emerged out of COP 16 in Mexico established the Green Climate Fund, through which a significant share of this new climate finance would flow (Klein 2011).

Figure 28.2 suggests that sources of public funding for adaptation stem from Annex 1 (developed countries and economies in transition) and non-Annex 1 (developing countries) country domestic budgets. In terms of flows from Annex 1 countries, funds are seemingly divided into ODA funds (outside UNFCCC mechanisms) and "new and additional" adaptation finance, which falls under the UNFCCC. From this framework, it appears that funding under the UNFCCC *is* additional to ODA funds, whilst that outside counts as international aid. Such a framework upholds the principles of the global climate change architecture.

However, as pointed out by Brown *et al.* (2010) and Forstater and Rank (2012), interpretations of "new and additional" to ODA vary considerably. Brown *et al.* (2010) show how the current debate over what constitutes "new and additional" climate finance can be divided into four broad positions, each with different technical and political implications (see Table 28.1).

As demonstrated by Table 28.1, there is no agreement on the baseline for assessing whether adaptation finance pledges are "new and additional." Depending on the baseline chosen, between all and almost none of the amount pledged "counts" as additional (Smith *et al.* 2011). In Table 28.1, the first definition – that "new and additional" means additional to 0.7% ODA commitments – is the most common definition supported by developing countries and is also formally backed by both Norway and the Netherlands. It has the greatest cost implications for increasing financial flows to developing countries, and suggests that ODA should not be included in climate finance at all.

However, if we review the current situation of climate pledges in light of this definition, all climate finance is currently double counted as aid (Forstater and Rank 2012). For example, 96% of contributions to the GEF are recognized as ODA. The UK's £1.5 billion Fast Start commitment has been reported to be reallocated from existing aid budgets (Forstater and Rank 2012; see also Box 28.2). Thus, in reality most of the adaptation funding within the UNFCCC funds is in fact sourced from international aid.

As shown in Box 28.2, much international climate finance is channeled through funds outside the UNFCCC, which Figure 28.2 suggests do not carry the same expectations of "new and additional" contributions. For example, the PPCR has a total of US$972 million pledged by 12 countries, with the UK as the largest contributor. Funds are sourced from international aid budgets. However, one controversy over the sources of funds in the PPCR is that some of this money is provided in loan rather than grant form. The idea of "loans for adaptation" raises the same ideological challenge: if climate impacts occur mainly due to historical and current high contributions of developed countries and are additional to existing development needs, developing countries should not be expected to pay back funding for addressing them, even if this investment can lead to a return. Although many developing countries have now

Table 28.1 The four definitions of climate finance additionality.

	Definition	Technical considerations	Political considerations
1	Aid that is additional to (over and above) the 0.7% ODA target	Easy to track given that it is measuring an increase at disbursement level and technically feasible but raises same questions around the validity of the ODA tracking system and what gets counted as climate finance.	Most countries have difficulty reaching the 0.7% target in the first place, so politically challenging to raise the target. Supported by international development community.
2	Increase in climate finance on 2009 ODA levels directed at climate change activities	Easy to track given that it is measuring an increase at disbursement level and technically feasible but current issues with ODA tracking. There will be no diversion from development objectives for donors who have already met their 0.7%, but may not be the case for those who have not.	Some issues with setting 2009 as financial baseline – implies different things depending on if donor has met the 0.7% target or not. Those donors who have not given to ODA-related climate finance before 2009 will have a lower baseline compared to those who have, implying equity issues.
3	Rising Official Development Assistance (ODA) which includes climate change finance but limited (e.g. to X%)	Aid diverted to climate finance causes changing the composition of finance if overall levels of ODA are not raised sufficiently. Issues around how to know what percentage is the right level – and should ideally only apply to governments who have already met their 0.7% so that the percentage of ODA spending going to climate change is above the 0.7% for development related efforts. Still need to secure additional channels of funding over and above a percentage of ODA, especially if limited to only 10% as is the case with UK proposal.	Countries which have already met their 0.7% target will not want those who have not to sacrifice this original goal for climate change objectives. It signifies a diversion in priorities. Setting the percentage in relation to ODA spending means funding is based on a country's current contributions, even if they are insufficient. Contributions are therefore not based on ability to pay, unlike one set on percentage of GNI.
4	Complete separation between ODA and CC financing	Emphasis on separation of funds at source. Need to ensure that new sources of finance are mainstreamed with existing ODA flows – technically challenging.	Would allow concerns regarding diversion of ODA funds away from development goals to be allayed. Politically challenging to agree what a new financial mechanism would look like, who should be in charge of the tracking, and how it should be tracked.

Source: Brown, J., N. Bird, and L. Schalatek, L. 2010. "Climate Finance Additionality: Emerging Definitions and Their Implications". Climate Finance Policy Brief 2. Washington, DC: Heinrich Böll Foundation North America and ODI. Used by permission.

welcomed highly concessional loan contributions from the PPCR, this has proved to be a sticking point for negotiations around the establishment of the fund (Ayers and Huq 2009) and also the way in which it has been delivered (Ayers *et al.* 2011), as will be discussed later in this chapter.

Box 28.2   UK International Climate Fund and UK Fast Start Climate Finance

The UK Government is providing £1.5 billion in Fast Start finance for climate change from 2010 to 2012, making the UK the biggest EU donor to Fast Start. This funding makes up part of the £2.9 billion for climate finance approved by the UK government for the period 2011–2015 under the UK International Climate Fund (ICF). The ICF commits UK finance for two years beyond the Fast Start period.

The money allocated to the ICF comes from ODA sources. As such, spending from the ICF is consistent with the DAC definition of ODA, and be in line with the overall purpose of UK development assistance, which is poverty reduction.

The ICF will channel Fast Start funds through various avenues: £122 million will flow through bilateral programmes; and £934 million will flow through multilateral funds. The UK has committed £310 million to the Pilot Programme for Climate Resilience (PPCR), which includes £287 of Fast Start funding. The UK has also committed £30 million Fast Start funding to the Least Developed Countries Fund and £10 million to the Adaptation Fund.

One fund that does meet the sourcing requirements of "new and additional" is the Adaptation Fund under the Kyoto Protocol (see Figure 28.1). Although some contributions to the Adaptation Fund are counted as ODA (for example, the UK ICF contributes some funds through this source), funding is mainly sourced from a levy on CDM trading (see note 4). The revenue generated from the CDM levy is projected to be between US$160 million and US$190 million, and potentially much more depending on the volumes traded and prices as targets are set (Müller 2007).

Thus, international aid does play a significant role in the generation of funds for adaptation. Beyond the CDM levy contributions to the Adaptation Fund, and some small private-sector opportunities, the majority of money flowing through funds for adaptation both under and outside the UNFCCC is international aid. A recent DARA/CVF report suggest that only 9% of allocated Fast Start Finance can be said to be "new and additional" (DARA/CVF 2011).

Governance of Climate Adaptation Finance: What Is the Role for International Aid Institutions?

As noted, many argue that the institutions of the international aid architecture are well placed to mobilize and channel funds for adaptation, because of the well-established mechanisms for managing financial flows from high-income to lower- and middle-income countries. So what is the role of these institutions? This section will consider this question, paying particular attention to the role of the Bretton Woods Institutions (the World Bank Group and the International Monetary Fund).

The governance of the climate funds falls broadly under three models. First, the management model of the Global Environment Facility (GEF), the financial mechanism of the UNFCCC.[5] The GEF Strategic Priority Fund, the Least Developed Countries Fund (LDCF), and the Special Climate Change Fund (SCCF) are all managed under GEF guidance. Under Article 11 of the UNFCCC, the GEF is required to have "an equitable and balanced representation of all Parties within a transparent system of governance" (UNFCCC 1992). While decisions by the GEF Council are taken by consensus of all Parties to the Convention, if no consensus is available then a majority of countries, weighted by donation, is required to carry a vote. This means that GEF Council members from countries that make the largest contributions carry the most weight, essentially giving veto power to the group of five largest donor countries (Streck 2001). This lack of "one country, one vote" structure has come under criticism from civil society actors for undermining any ownership of adaptation funds by developing countries (Müller 2006; ActionAid 2007; Ayers 2009).

Second, the Adaptation Fund model. The Adaptation Fund has its own independent board with representation from the five UN regions as well as special seats for the LDCs and Small Island Developing States. The GEF provides secretariat services to the Adaptation Fund on an interim basis. Decision-making is by consensus of the board members, and if consensus fails, by a two-thirds majority vote, based on one member, one vote. Ballesteros *et al.* (2010) suggest that this balance of power in favor of developing countries on the AFB may be attributable in part to the fact that financing of the Adaptation Fund is not dependent on donor contributions.

Third, the international aid model. This model has been valued by donor agencies because of its familiarity and perceived low fiduciary risk (Tanner and Allouche 2011). The previous section suggested that the decades of experience in international aid funding, and the resulting Paris and ACCRA principles of aid effectiveness, could bring benefits to adaptation fund governance in terms of ownership and accountability that are lacking under the GEF-managed funds. But is this the case?

Although the governances systems that are channeled through ODA vary depending on the donors involved and whether funds are made available as part of a bi- or multilateral initiative (Persson *et al.* 2009), this section will address this question by examining the Climate Investment Funds (CIFs), as the largest set of funds outside the UNFCCC process.

The CIFs are managed by the World Bank. However, as pointed out by Ballesteros *et al.* (2010), the governance structure of the CIFs is a departure from the traditional donor-dominated Bretton Woods model governance structure. Although early drafts of the CIFs' governance structure were heavily criticized for not including adequate modalities for developing country decision-making (Seballos and Kreft 2011), the final agreed structure features an even division of membership and decision-making power between contributors and recipients. Each of the CIFs is governed by a Trust Fund Committee, with an equal number of contributor country representatives and recipient country representatives. Under each of the CIFs, decisions are made by consensus.

However, Seballos and Kreft suggest that the set-up of the World Bank was designed to engage the multilateral banks in adaptation finance, and as such has served to reinforce a donor-driven and top-down approach to decision-making. In their critique of the political economy of the PPCR, the authors state that there was weak inclusion of developing countries in the design of the CIFs, which

> [l]ed to a programme and structure more in tune with the donor and MDB agenda than one which seeks to respond to the needs of the most vulnerable and establish true country ownership... The overwhelming power has been retained in the hands of the World Bank and MDBs (Seballos and Kreft 2011: 39).

Of all the governance structures of adaptation funds presented, there is strong consensus among developing country Parties that the Adaptation Fund model presents the most democratic and accountable structure for meeting the principles of adaptation funding. Persson *et al.* (2009) propose that in comparison with international aid mechanisms, developing countries consider adaptation finance delivered through UNFCCC processes as promoting a greater degree of country ownership, imposing fewer conditionalities, allowing greater access, and ensuring a more equitable distribution of resources. On the other hand, the role of the Multilateral Development Banks (MDBs) and the World Bank as lending institutions means they are perceived as vehicles for developed country interests (Seballos and Kreft 2011). Indeed, critics such as the Bretton Woods Project watchdog have suggested that rather than bringing the lessons of aid effectiveness to the climate change arena, climate change finance reflects a "huge leap backwards" and has been used as a platform to reverse much of the progress around "good governance" in international aid made over the past decades (Bretton Woods Project 2008).

The Role of International Aid Mechanisms in Adaptation Finance Delivery

International aid has a significant role to play in the delivery of adaptation. In operational terms, some observers have suggested that international aid has a wider remit than specific climate adaptation funds, enabling a greater degree of flexibility when it comes to investing in the diverse range of activities that can reduce vulnerability to climate change (Ayers and Huq 2009). The climate funds that are managed by the UNFCCC have a narrow remit: to address the impacts of climate change. This is in part a repercussion of the "additionality" debate around adaptation funds at the international level. Lemos and Boyd (2009) suggest that the need to meet the "additionality" criteria of the international adaptation funding frameworks constrains the kinds of local-level adaptation options that can be developed. The result is that national and local-level decision-makers are encouraged by an international climate change discourse to segregate "adaptation" from more general "development," when in fact the most appropriate means of addressing vulnerability would be to take the two together (Lemos and Boyd 2009).

As discussed, building adaptive capacity requires actions that focus not only on the measurable and verifiable impacts of climate change but also on a wide range of factors that contribute to a broader reduction in vulnerability to climate variability and climate change (Adger and Kelly 1999; Schipper 2007; Klein 2008). It is important that funding is made available for adaptation activities that can also address these other, non-climatic "baseline" aspects of vulnerability. Such activities have traditionally been the focus of development practitioners.

In principle, then, bilateral international aid funds and funds such as the PPCR should provide a more open discourse of climate change risk that moves away from an "impacts-based" approach. Indeed, the name of the PPCR focuses on "climate resilience" and seems an explicit attempt to open up adaptation support to a broader range of activities than climate sensitivities alone. In line with this thinking, Ayers and Huq (2009) optimistically suggest that the arrival of the PPCR signified a real opportunity for development assistance to address underlying factors of vulnerability that are overlooked by a UNFCCC-based approach. The authors state:

> [The establishment of the PPCR] does point to progress in understanding the role of ODA as contributing to broader adaptive capacity – or "climate-resilient develop-ment" – rather than specific and additional climate-change adaptation... new devel-opment funds relevant to climate-change adaptation should be used to fund what the UNFCCC cannot; namely, broader resilience building, necessary for "additional" adap-tation to be successful (Ayers and Huq 2009: 682).

But has this opportunity materialized, and has it resulted in new avenues for a more inclusive approach to defining climate-change risk beyond the UNFCCC? One approach adopted both by the PPCR and other bilateral international aid mech-anisms in an attempt to deliver adaptation and development benefits together is "mainstreaming" (see Box 28.3).

Box 28.3 Mainstreaming Adaptation to Climate Change

Mainstreaming involves the integration of information, policies, and measures to address climate change into ongoing development planning and decision-making (Klein *et al.* 2003). It is seen as making more sustainable, effective, and efficient use of resources than designing and managing policies separately from ongoing activities. In theory, mainstreaming should create "no regrets" oppor-tunities for achieving development that is resilient to current and future climate impacts for the most vulnerable groups, and avoid potential trade-offs between adaptation and development strategies that could result in maladaptation in the future.

Source: Klein *et al.* (2003).

Mainstreaming adaptation into development can be approached in different ways. On the one hand, mainstreaming can be interpreted as targeting development efforts at issues that are essential for reducing vulnerability to climate and also other risks. Klein (2010) provides the example of ensuring water rights to groups exposed to water scarcity during a drought. It recognizes that adaptation involves many actors, requires creating an enabling environment by removing existing financial, legal, institutional, and knowledge barriers to adaptation, and strengthening the capacity of people and organizations to adapt (Klein 2010).

But a second and more common approach is "climate proofing" of existing development efforts, that is, ensuring that projections of climate-change impacts are considered in the decision-making about climate investments, so that the technologies are chosen or improved to withstand the future climate. For example, in an area projected to experience more intense rainfall events, water managers would fit a drainage system with bigger pipes when replacing old ones (Klein 2010).

A "climate-proofing"-only approach to mainstreaming has been widely criticized for failing to fully address the underlying drivers of vulnerability; for not addressing maladaptation; and for not realizing the potential of development interventions to achieve climate resilience (Klein 2008; Ayers *et al.* 2011; Seballos and Kreft 2011). For example, strengthening an embankment to ensure it can withstand anticipated increases in storm surges will not protect those who cannot afford to reside behind it, and may inadvertently encourage investment and settlement in a climate-vulnerable area. Yet, this approach is also more straightforward – it requires "screening" existing development efforts for climate sensitivities and then responding to those. A more holistic approach requires us to question the basis of the development intervention altogether in terms of its impact on climate vulnerability.

Unfortunately, early signs suggest that large international aid funds for adaptation such as the PPCR are favoring a "climate-proofing" approach (Ayers *et al.* 2011; Seballos and Kreft 2011). Further, Seballos and Kreft comment that:

> This climate "add-on" approach to development allows the World Bank Group and other multilateral development banks (MDBs) to claim a space in managing future climate finance flows... curtail[ing] opportunities for multi-stakeholder dialogue and thus the potential for development of broad country ownership of programmes (Seballos and Kreft 2011: 33).

There are other channels for the delivery of international aid that appear more promising for addressing adaptation and development together. Many donors are using climate change as an opportunity to review whether their bilateral portfolios are actually addressing vulnerability. For example, the Bangladesh office of the UK Department for International Development (DfID) has used climate change as an entry point for reviewing its livelihoods program, and reviewing whether it is producing sustainable development benefits over the long term, and for the climate-vulnerable poor (DfID Bangladesh, personal communication).

Further, a growing number of NGOs are channeling international aid to the grassroots through "community-based adaptation" (CBA). CBA is a growing field, which operates at the community level to identify, assist, and implement community-based development activities that strengthen the capacity of local people to adapt (Huq and Reid 2007). Proponents of a CBA approach suggest that the localized networks already in place undertaking good community-based development work are the kind of institutional design that could be used to channel international aid in ways that can identify and address the diversity and complexity of local vulnerability contexts (Jones and Rahman 2007; Ayers and Forsyth 2009).

However, some critics of this approach suggest that community-based approaches are limited in terms of spatial and temporal scale (Ribot 2002), a particular problem for managing "global" environmental risks where there is a need to connect to higher-level governance structures. As noted by Dodman and Mitlin (2011), while there has been much work on developing participatory tools and methods for enabling community-based development at the project level, relatively little attention has been paid to building up the links with political structures above the level of the settlement. Both donors and NGOs are responding to these critiques, and attempts are ongoing to "scale up" community-based efforts and link them with subnational and national climate and development planning.[6]

Thus, the institutional structures of international aid already in place provide good opportunities for delivering climate-adaptation finance in ways that address vulnerability. However, caution is needed to ensure that the principles of good development are not overlooked in the process.

Conclusions

This chapter has considered the role of international aid in adaptation. We have shown that international aid has a strong role to play in adaptation in principle, not least because of the synergies between adaptation and development, which means that tackling the two together makes sense. Focusing only on responding to climate-change impacts without also addressing the underlying factors related to development that drive vulnerability will not lift the poorest and most marginalized people out of vulnerability to climate change or other risks. Development interventions that do not consider the potential impacts of climate change risk proving maladaptive in the long term.

However, within the global climate-change arena, there are important reasons for separating out international aid from adaptation funding: climate change places an additional burden on developing countries, so additional resources should be provided. This is a fundamental equity principle underpinning many of the negotiations around international climate finance.

Yet when this principle spills over into operationalizing investments in adaptation, the role of international aid becomes confused. Indeed, there is not yet any agreement on what "new and additional" adaptation funding actually means. In terms of sourcing funds, developed countries are not meeting their international aid obligations, so almost no funds flowing through the adaptation finance architecture are additional. Further, contributions to funds outside the UNFCC process dwarf those within it.

In terms of governance, there are disputes over the role of international aid institutions. Donors favor the use of development institutions like the Multilateral Development Banks for governing and delivering climate-change finance because it affords them greater control over spending decisions (Fankhauser and Burton 2011). Many developing-country recipients of these funds dispute the role of these institutions for the same reasons. When development institutions cross over into the role of managing climate finance, it appears that many of the principles of "good governance" of international aid are left behind.

It is in the role of delivery that international aid institutions have the greatest comparative advantage. On the ground, adaptation interventions differ little from good, sustainable development. Both bilateral and NGO agencies have a wealth of experience in targeting participatory and localized development interventions that generate benefits for the poorest groups, who are also vulnerable to the impacts of climate change; there are also decades of development failures to learn from. It is critical to engage international aid institutions in delivering adaptation benefits to ensure these lessons are incorporated, and to avoid competing or duplicating efforts to reduce vulnerability on the ground.

Notes

1   A 2 °C increase above pre-industrial levels by 2050 is considered highly probable under the business-as-usual assumption of global warming (Allen *et al.* 2009; Meinshausen *et al.* 2009) and is commonly regarded in climate policy-making as the limit for avoiding "dangerous climate change." However, other prominent climate scientists have demonstrated that temperature rises of up to 4 °C by 2060 are much more likely outcomes, given the record of climate action to date and the slow foreseeable progress on future action for limiting greenhouse gas emissions, with much more severe implications for the costs of (and limits to) adaptation (Anderson and Bows 2011).

2   ODA is the largest type of international aid, consisting of aid provided by donor governments to low- and middle-income countries.

3   www.climateinvestmentfunds.org (accessed October 20, 2012).

4   For the most up-to-date figures on climate finance, see www.climatefundsupdate.org (accessed October 20, 2012).

5   The GEF was established in 1991 following the Earth Summit, to provide a mechanism to fund projects and programs that protect the "global environment." The GEF is a designated financial mechanism to the international environmental conventions of six focal areas: biodiversity; climate change; international waters; ozone; land degradation; and persistent organic pollutants, with the mandate to support the generation of "global environmental benefits" under each: www.gefweb.org.

6   See, e.g., the Local Adaptation Plans of Action (LAPA) framework development by the Government of Nepal (www.moest.gov.np); and also NGO efforts towards mainstreaming CBA into local government planning (www.arcab.org).

References

ActionAid. 2007. *Compensating for Climate Change: Principles and Lessons for Equitable Adaptation Funding*. Washington, DC: ActionAid.

Adger, N. and P.M. Kelly. 1999. "Social Vulnerability to Climate Change and the Architecture of Entitlements." *Mitigation and Adaptation Strategies to Global Change*, 4: 253–256.

Anderson, K. and A. Bows. 2011. "Beyond 'Dangerous' Climate Change: Emission Scenarios for a New World." *Philosophical Transactions of the Royal Society A: Mathematical, Physical and Engineering Sciences*, 369(1934): 20–44.

Allen, M.R., D.J. Frame, C.D. Huntington *et al.* 2009. "Warming Caused by Cumulative Carbon Emissions towards the Trillionth Tonne." *Nature*, 458: 1163–1166.

Ayers, J. 2009. "Financing Urban Adaptation." *Environment and Urbanization*, 21(1): 225–240.

Ayers, J. 2011. "Resolving the Adaptation Paradox: Exploring the Potential for Deliberative Adaptation Policy-Making in Bangladesh." *Global Environmental Politics*, 11(1), http://www.mitpressjournals.org/doi/pdf/10.1162/GLEP_a_00043 (accessed October 20, 2012).

Ayers, J. and D. Dodman. 2010. "Climate Change Adaptation and Development: The State of the Debate." *Progress in Development Studies*, 10(2): 161–168.

Ayers, J. and T. Forsyth. 2009. "Community-Based Adaptation to Climate Change: Strengthening Resilience through Development." *Environment*, 51(4): 22–31.

Ayers, J. and S. Huq. 2009. "Supporting Adaptation through Development: What Role for ODA?" *Development Policy Review*, 27(6): 675–692.

Ayers, J., N. Kaur, and S. Anderson. 2011. "Negotiating Climate Resilience in Nepal." *IDS Bulletin*, 42(3): 70–79.

Ballesteros, A., S. Nakhooda, J. Werksman, and K. Huriburt. 2010. Power, Responsibility, and Accountability: Re-thinking the Legitimacy of Institutions for Climate Finance. Final Report. Washington, DC: World Resources Institute, http://www.wri.org/publication/power-responsibility-accountability (accessed October 20, 2012).

Bird, N. and J. Brown. 2010. "International Climate Finance: Principles for European Support to Developing Countries." EDC2020 Working Paper 6. Bonn: EADI.

Bird, N. and J. Glennie. 2011. "Going beyond Aid Effectiveness to Guide the Delivery of Climate Finance." ODI Background Note, August 2011. London: Overseas Development Institute.

Blakie, P.M., T. Cannon, I. Davis, and B. Wisner. 1994. *At Risk: Natural Hazards, People's Vulnerability and Disasters*. London: Routledge.

Bretton Woods Project. 2008. "World Bank Climate Funds: 'A Huge Leap Backwards'," www.brettonwoodsproject.org/art-560997 (accessed October 20, 2012).

Brown, J., N. Bird, and L. Schalatek, L. 2010. "Climate Finance Additionality: Emerging Definitions and Their Implications." Climate Finance Policy Brief 2. Washington, DC: Heinrich Böll Foundation North America and ODI.

Burton, I. 2004. "Climate Change and the Adaptation Deficit." Adaptation and Impacts Research Group Occasional Paper 1. Quebec: Environment Canada.

DARA/CVF (Climate Vulnerable Forum). 2011. "Briefing Note to the Climate Change Vulnerable Forum, Dhaka, 2011." Madrid: DARA, http://daraint.org/wp-content/uploads/2011/10/cvf-Briefing_Notes.pdf (accessed October 20, 2012).

Dodman, D., J. Ayers, and S. Huq. 2009. "Building Resilience." In *State of the World 2009: Into a Warming World*, ed. Worldwatch Institute, 151–168. Washington, DC: Worldwatch Institute.

Dodman, D. and D. Mitlin. 2011. "Challenges to Community-Based Adaptation." *Journal of International Development*, 23(3). doi:10.1002/jid.1772.

Fankhauser, S. and I. Burton. 2011. "Spending Adaptation Money Wisely." Centre for Climate Change Economics and Policy Working Paper No. 47; Grantham Research Institute on Climate Change and the Environment Working Paper No. 37. London: Grantham Research Institute on Climate Change and the Environment.

Flåm, K.H. and J.B. Skjærseth. 2009. "Does Adequate Financing Exist for Adaptation in Developing Countries?" *Climate Policy*, 9(1): 109–114.

Forstater, M. and R. Rank. 2012. "Towards Climate Finance Transparency." n.p.: Publish What You Fund and AidInfo.

Haites, E. 2011. "Climate Change Finance." *Climate Policy*, 11: 963–969.

Huq, S. and J. Ayers. 2008. "Streamlining Adaptation to Climate Change into Development Projects at the National and Local Level." In *Financing Climate Change Policies in Developing Countries*, ed. European Parliament, 52–68. Brussels: European Parliament.

Huq, S. and H. Reid. 2007. *Community-Based Adaptation. IIED Briefing*. London: IIED.

IPCC (Intergovernmental Panel on Climate Change). 2007. "Summary for Policymakers." In *Climate Change 2007: Impacts, Adaptation and Vulnerability. Contribution of Working Group II to the Fourth Assessment Report of the Intergovernmental Panel on Climate*

Change, ed. M.L. Parry, O.F. Canziani, J.P. Palutikof *et al.*, 7–22. Cambridge: Cambridge University Press.

Jones, R. and A. Rahman. 2007. "Community-Based Adaptation." *Tiempo*, 64: 17–19.

Klein, R.T.J. 2001. *Adaptation to Climate Change in German Official Development Assistance: An Inventory of Activities and Opportunities, with a Special Focus on Africa.* Eschborn: Deutsche Gesellschaft für Technische Zusammenarbeit (GIZ).

Klein, R.T.J. 2008. "Mainstreaming Climate Adaptation into Development Policies and Programmes: A European Perspective." In *Financing Climate Change Policies in Developing Countries*, ed. European Parliament, 38–51. Brussels: European Parliament.

Klein, R.T.J. 2010. "Mainstreaming Climate Adaptation into Development: A Policy Dilemma." In *Climate Governance and Development*, ed. A. Ansohn and B. Pleskovic, 35–52. Washington, DC: World Bank.

Klein, R.T.J. 2011. *Ensuring Equity, Transparency and Accountability for Adaptation Finance. SEI Policy Brief*. Stockholm: SEI.

Klein, R.J.T., L. Schipper, and S. Dessai. 2003. "Integrating Mitigation and Adaptation into Climate and Development Policy: Three Research Questions." Tyndall Centre Working Paper 405. Norwich: Tyndall Centre for Climate Change Research.

Lemos, M.C. and E. Boyd. 2009. "The Politics of Adaptation across Scales: The Implications of Additionality to Policy Choice and Development." In *The Politics of Climate Change: A Survey*, ed. M. Boykoff, 96–110. Abingdon: Routledge.

Levina, E. 2007. *Adaptation to Climate Change: International Agreements for Local Needs.* COM/ENV/EPOC/IEA/SLT(2007)6. Paris: OECD Publishing.

McGray, H., A. Hammill, and R. Bradley. 2007. *Weathering the Storm: Options for Framing Adaptation and Development.* Washington, DC: World Resources Institute.

Meinshausen, M., N. Meinshausen, W. Hare *et al.* 2009. "Greenhouse-Gas Emissions Targets for Limiting Global Warming to 28C." *Nature*, 458: 1158–1162.

Müller, B. 2006. "Nairobi 2006: Trust and the Future of Adaptation Funding," www.oxfordenergy.org/pdfs/ EV38.pdf (accessed October 20, 2012).

Müller, B. 2007. "The Nairobi Climate Change Conference: A Breakthrough for Adaptation Funding." Oxford Energy and Environment Comment, http://www.oxfordenergy.org/wpcms/wp-content/uploads/2011/01/Jan2007-NairobiClimateChang-Conference-Benito Muller.pdf (accessed October 20, 2012).

Narain, U., S. Margulis, and T. Essam. 2011. "Estimating Costs of Adaptation to Climate Change." *Climate Policy*, 11: 1001–1019.

OECD DAC. 2009. "Key Principles to Inform Climate Change Financing," http://www.aideffectiveness.org/ClimateChangeFinance#keylessons (accessed October 20, 2012).

OECD DAC. 2010. "History of the 0.7% ODA Target." Original text from DAC Journal 2002, 3(4): III-9–III-11. Revised June 2010, http://www.oecd.org/dac/aidstatistics/45539274.pdf (accessed October 20, 2012).

OECD DAC. n.d. "Official Development Assistance." In DAC Glossary of Key Terms and Concepts, http://www.oecd.org/dac/dacglossaryofkeytermsandconcepts.htm#ODA (accessed October 20, 2012).

Oxfam International. 2007. Adapting to Climate Change: What's Needed in Poor Countries and Who Should Pay. Oxfam Briefing Paper 104. Oxford: Oxfam.

Pauw, P., C. Ifejika Speranza, I. van de Sand *et al.* 2011. "Climate Change Adaptation. Challenges in Institutionalization and Financing." SEF Policy Paper 35. Bonn: Stiftung Entwicklung und Frieden (SEF).

Persson, A., R.J.T. Klein, C.K. Siebert *et al.* 2009. *Adaptation Finance under a Copenhagen Agreed Outcome. SEI Research Report.* Stockholm: Stockholm Environment Institute.

Pielke, R., G. Prins, S. Rayner, and D. Sarewitz. 2007. "Lifting the Taboo on Adaptation." *Nature*, 445(8): 578–579.

Ribot, J.C. 2002. Democratic Decentralization of Natural Resources: Institutionalizing Popular Participation. Washington, DC: World Resources Institute.

Ribot, J.C. 2010. "Vulnerability Does Not Just Fall from the Sky: Toward Multi-scale Pro-poor Climate Policy." In *Social Dimensions of Climate Change: Equity and Vulnerability in a Warming World*, ed. R. Mearns and A. Norton, 47–74. Washington, DC: World Bank.

Riddell, R.C. 1987. *Foreign Aid Reconsidered*. Baltimore, MD: Johns Hopkins University Press.

Schipper, L. 2007. "Climate Change Adaptation and Development: Exploring the Linkages." Tyndall Centre Working Paper Series 107. Norwich: Tyndall Centre for Climate Change Research.

Schneider, S.H., S. Semenov, A. Patwardhan *et al.* 2007. "Assessing Key Vulnerabilities and the Risk from Climate Change." In *Climate Change 2007: Impacts, Adaptation and Vulnerability. Contribution of Working Group II to the Fourth Assessment Report of the Intergovernmental Panel on Climate Change*, ed. M.L. Parry, O.F. Canziani, J.P. Palutikof *et al.*, 779–810. Cambridge: Cambridge University Press.

Seballos, F. and S. Kreft. 2011. "Towards an Understanding of the Political Economy of the PPCR." *IDS Bulletin* 42(3): 33–41.

Smit, B. 1993. *Adaptation to Climatic Variability and Change: Report to the Task Force on Climatic Adaptation*. Canadian Climate Change Program Occasional Paper. Ontario: Department of Geography, University of Guelph.

Smit, B. and J. Wandel. 2006. "Adaptation, Adaptive Capacity and Vulnerability." *Global Environmental Change*, 16: 282–292.

Smith, J.B., T. Dickinson, J.D.B. Donahue *et al.* 2011. "Development and Climate Change Adaptation Funding: Coordination and Integration." *Climate Policy*, 11: 987–1000.

Streck, C. 2001. "The Global Environment Facility: A Role Model for Global Governance?" *Global Environmental Politics*, 1: 17–18.

Tanner, T. and J. Allouche. 2011. "Towards a New Political Economy and Climate Change and Development." *IDS Bulletin*, 42(3): 1–14.

UNFCCC (United Nations Framework Convention on Climate Change). 1992. "Article 11," United Nations Framework Convention on Climate Change, New York, May.

World Bank. 2008. "Q&A: Climate Investment Funds," July 1. Washington DC: World Bank.

Index

Note: page numbers in *italics* refer to figures; those in **bold** to tables.

The Handbook of Global Climate and Environment Policy, First Edition. Edited by Robert Falkner.
© 2013 John Wiley & Sons, Ltd. Published 2013 by John Wiley & Sons, Ltd.